ENCYCLOPEDIA OF PHYSICS

EDITED BY

S. FLÜGGE

VOLUME X

STRUCTURE OF LIQUIDS

WITH 41 FIGURES

Springer-Verlag Berlin Heidelberg GmbH
1960

HANDBUCH DER PHYSIK

HERAUSGEGEBEN VON

S. FLÜGGE

BAND X

STRUKTUR DER FLÜSSIGKEITEN

MIT 41 FIGUREN

Springer-Verlag Berlin Heidelberg GmbH
1960

ISBN 978-3-662-23039-8 ISBN 978-3-662-25003-7 (eBook)
DOI 10.1007/978-3-662-25003-7

Springer-Verlag Berlin Heidelberg 1960
Originally published by Springer-Verlag OHG Berlin Göttingen Heidelberg 1960.
Softcover reprint of the hardcover 1st edition 1960

Contents.

Page

The Structure of Liquids. By Dr. Herbert S. Green, Professor of Mathematical Physics, University of Adelaide (South Australia). (With 7 Figures) 1

I. General nature of liquid structure 1

a) Molecular constitution of liquids 1
b) Liquid models 7
c) Actual structure of liquids 15

II. The quantitative description of liquid structure 25

a) Distribution functions and their properties 25
b) Molecular mechanics and liquid structure 33

III. Structure of uniform liquids 47

a) The radial distribution function 47
b) Structure-dependent properties 61
c) Structure in the surface zone 77

IV. Structure of non-uniform liquids 81

a) Effect of irreversible processes 81
b) Deformation of structure by viscosity 89
c) Theory of non-uniform liquids 96
d) Kirkwood's theory of dissipative processes 112

V. Structure of quantum liquids 119

a) Liquid helium 119
b) Quantum theory of structure 126

Molecular Theory of Surface Tension in Liquids. By Dr. Syu Ono, Associate Professor of Physics, University of Tokyo and Dr. Sohei Kondo, Chief on Radiation Laboratory, National Institute of Genetics, Misima, Sizuoka-ken. (Japan). (With 32 Figures) 134

A. Thermodynamics and quasithermodynamics 136

I. Thermodynamics 136

a) Thermodynamic quantities of interface layer 136
b) Plane interface 138
c) Spherical interface 145

II. Hydrostatic approach 157
III. Quasithermodynamics 163
IV. Application of thermodynamics of irreversible processes 168
V. Empirical equations for temperature dependence of surface tension 172

B. Statistical mechanics 177

I. Statistical thermodynamic method 177

a) Canonical ensemble 177
b) Grand canonical ensemble 190

Page

II. Mechanical definition of surface tension 208

III. Numerical calculations 216

a) Pure liquid 216

b) Electrolyte solution 223

c) Gas adsorption on a solid surface 231

IV. Quantum statistical mechanics 237

C. Lattice theory approaches 240

I. Pure liquids 240

a) Free volume theory 240

b) Hole theory 249

II. Solutions of non-electrolytes 262

a) Regular solutions 262

b) Polymer solutions 272

References. 277

The Theory of Capillarity. By Dr. FRANK P. BUFF. Associate Professor of Chemistry, University of Rochester, New York (USA). (With 2 Figures). 281

Introduction 281

A. General theory 281

B. Applications 295

General references. 304

Sachverzeichnis (Deutsch-Englisch) 305

Subject Index (English-German) 313

The Structure of Liquids.*

By

H. S. Green.

With 8 Figures.

I. General nature of liquid structure.

a) Molecular constitution of liquids.

1. Introduction. A large part of modern physics is concerned with the relations between the physical behaviour of a composite system and that of its constituent parts. It is desired, for example, to explain the properties of an atomic nucleus in terms of the elementary neutrons and protons of which it is believed to consist; to explain the properties of an atom in terms of the nucleus and electrons from which it is constructed; to explain the properties of a molecule in terms of its constituent atoms; and finally to explain the properties of macroscopic matter in terms of its molecular structure. The kinetic theory of matter, and of liquids in particular, is designed to provide the last link in this chain.

The object of relating the physical behaviour of a liquid with the properties of the molecules which it contains is two-fold. In the first instance, it may happen that the structure of the individual molecules of the liquid and the nature of their mutual interactions, can be determined from purely theoretical considerations; this is so in the case of hydrogen, helium, neon, argon, etc. Then, by study of molecular structure, one can hope to predict macroscopic properties of the corresponding liquid, such as its equation of state, its coefficients of viscosity and thermal conduction, etc. In the second instance, the molecular constitution may be too complicated to be determined by theoretical considerations alone. Then the unknown features can often be inferred from the observed macroscopic behaviour of the substance, with the help of the general molecular theory which has been developed. In addition, one may hope to reach a more profound understanding of the fundamental mechanism governing the behaviour of liquids.

The molecular theory of the liquid state is very much more difficult than that of the gaseous or solid state. This could be expected from a purely superficial comparison of the physical behaviour of the phases. There is no parallel for liquids of the gas laws, or Debye's law concerning the specific heat of solids. Instead one is confronted with anomalies such as the negative coefficient of expansion of water near its melting point. Some liquids obey Andrade's law of viscosity precisely, while others deviate from it very widely. There is no simple way of representing the coefficient of thermal conduction of any wide class of liquids. Obviously there is little hope of describing in the same way the behaviour of the glasses at low temperatures, and other liquids near their critical point. It is the thesis of this and the following chapters that such complexities of physical behaviour correspond very closely to complexities of molecular

---

* The original of this contribution was communicated in April, 1956. Important work published after this date has been mentioned in additions to Sects. 21 α, 23 α and 23, made in proof.

structure. Regularities of behaviour will be found only within groups of liquids with similar molecular constitution.

The group of liquids which offer the most favourable conditions for theoretical study is that consisting of the noble gases in their condensed state, particularly liquid neon, liquid argon and liquid krypton. The spherical symmetry and lack of reactivity of the atoms of these substances reduces the intricacies of molecular structure to a minimum, but even so the quantitative treatment of this structure remains one of the most difficult problems of statistical mechanics.

Next in complexity come substances like liquid nitrogen and methane, with molecules which are chemically inert, though they do not possess perfect spherical symmetry. Some of the properties of liquids—the specific heat and thermal conductivity are the main exceptions—do not depend sensitively on the structure of the individual molecules, which can therefore be disregarded when such properties are considered. This is unfortunately no longer true if the departure from spherical symmetry is great, as in the long chain-like molecules of the higher homologues of the paraffin series. Nor is it true of molecules which have a dipole or quadrupole moment, still less of molecules with localized centres of attraction for neighbouring molecules. The molecular theory of liquids of these types must for some time be either rather formal and qualitative, or else based on models so crude that their permanent value is very doubtful.

A quite tractable generalization of the theory of simple liquids with inert, spherically symmetrical molecules can be applied to mixtures of liquids of the same type. Here one is interested in the dependence of the physical properties of the mixtures on the concentrations of its components, and also in the possibility of additional types of irreversible processes: diffusion and thermal diffusion, the last of which is called the Soret effect in liquids. Just as in simple fluids, such macroscopic phenomena must be related to the underlying molecular structure.

The molecular structure of a liquid obviously depends to some extent on externally imposed conditions, such as temperature and pressure. Equally obviously, it must depend on the nature of the molecules themselves: not simply on their structure, but on their mutual interactions. In this work, it is necessary to consider not so much how the intermolecular forces arise as what their effect is on the disposition of the molecules relative to one another. However, these questions are not quite separate from one another, and it will be found useful to pay some attention to the first before proceeding to a detailed study of liquid structure.

2. Intermolecular forces. On a sub-microscopic scale, a liquid may be regarded as a giant assembly of atomic nuclei and electrons. These do not form isolated molecular groups, as in gases; on account of the greater density, the chances of finding such a group which is not in continual interaction with other neighbouring groups is negligible. From a certain point of view, therefore, the liquid constitutes one giant molecule. However, the existence of certain atomic groups, which remain essentially unchanged, in spite of the kinetic motion of the individual particles, allows one to regard the liquid as truly composite. The molecules are those irreducible collections of atoms which are very rarely, if ever, mutually separated, through a wide range of temperature. One does not, accordingly, regard as a single molecule a structure which, though sufficiently permanent at low temperatures, is broken up by the kinetic motion when the temperature is somewhat raised.

However, the concept of separate molecular groups of nuclei and electrons is sometimes of restricted value when one wishes to consider the interaction of two or more molecules, since the electrons, at least those which do not form closed shells, have to be regarded as common property of the interacting molecules. The interaction energy has to be computed on a quantum mechanical basis, which takes into account not only the classical Coulomb forces between charged particles, but also the "exchange" forces required by quantum statistics. The difficult problems which consequently arise are considered elsewhere[1]; the following, somewhat simplified picture emerges.

The interaction between inert molecules, which are not ionized or polarized and do not possess unsaturated chemical bonds, is of the simplest type: it is, on the average, attractive at sufficiently great intermolecular distances, owing to the mutual polarization of different molecules. The attraction increases as the molecules approach one another, the negative potential energy between two molecules increasing approximately as the inverse sixth power of their distance apart. At shorter distances, however, a repulsive force arises, which finally becomes the dominant feature of the intermolecular field. This is due to the repulsion between the outer closed electron shells of the molecules concerned, and is very great at short distances, but decreases exponentially with their distance apart. Probably the best simple approximation to the interaction $\phi(r)$ between two spherically symmetrical molecules whose mass-centres are at distance r apart is

$$\phi(r) = -\mu r^{-6} + N e^{-(r/\varrho)}, \tag{2.1}$$

with constants μ, N and ϱ depending on the nature of the molecules. From a theoretical calculation for a pair of helium molecules, SLATER and KIRKWOOD[2] deduced the values $\mu = 0.68\varrho^6 \times 10^{-10}$ erg, $N = 7.7 \times 10^{-10}$ erg, and $\varrho = 0.527 \times 10^{-8}$ c. These values can be verified by calculating physical quantities like the second virial coefficient and comparing with the experimentally determined quantities; the agreement is generally excellent. Conversely, where the quantum mechanical calculations are too difficult to allow the direct calculation of the interaction energy, the constants μ, N and ϱ can be inferred from experimental data, such as the second virial coefficient and its temperature dependence.

Even with the simple radial dependence of the interaction energy indicated by (2.1), calculations for liquids are apt to be unmanageable, and many authors have preferred to assume

$$\phi(r) = -\mu r^{-6} + \nu r^{-12}. \tag{2.2}$$

If the values of μ and ν are suitably chosen, there is not much difference between (2.1) and (2.2), except for very small values of r, which are not realized physically, and where (2.1) does not apply anyway. It is important to keep in mind, however, that some properties of liquids are very sensitive to the exact interaction energy, especially the repulsive term which is represented only approximately by either (2.1) or (2.2). For this reason it is desirable in theoretical calculations to keep the form of interaction potential as general as possible. Values of the constants μ and ν in (2.2) have been determined for many types of molecules, from data on the second virial coefficient, following the method of LENNARD-JONES[3]; but, though these are very useful, more accurate determinations of the intermolecular forces will soon be required.

[1] e.g., see Chap. 8 to 9 of J. C. SLATER: Quantum Theory of Matter. New York, Toronto and London 1951. — Also J. C. SLATER in Vol. XIX of this Encyclopedia.

[2] J. C. SLATER and J. G. KIRKWOOD: Phys. Rev. **37**, 682 (1931).

[3] J. E. LENNARD-JONES: Proc. Roy. Soc. Lond., Ser. A **106**, 463 (1924). — Proc. Phys. Soc. Lond. **43**, 461 (1931); also R. A. BUCKINGHAM: Proc. Roy. Soc. Lond., Ser. A **168**, 264 (1938)

Another assumption which is commonly made, and will be adopted for most practical purposes in what follows, is that the force experienced by any one of a group of molecules is the resultant of the forces exerted by each of the others separately. On this assumption, the interaction energy of a group of molecules can be expressed in terms of the mutual potential energy of a pair of molecules; specifically for a group of q molecules whose mass centres are at the points $\boldsymbol{x}^{(1)}$, $\boldsymbol{x}^{(2)}, \ldots \boldsymbol{x}^{(q)}$, the interaction energy will be

$$\Phi_q = \sum_{j>i}^{q} \phi(|\boldsymbol{x}^{(j)} - \boldsymbol{x}^{(i)}|), \tag{2.3}$$

where $|\boldsymbol{x}^{(j)} - \boldsymbol{x}^{(i)}|$ represents the distance between $\boldsymbol{x}^{(i)}$ and $\boldsymbol{x}^{(j)}$. Quantum mechanical considerations indicate that this assumption is not rigorously justified, but that it should involve no serious error.

The restriction to molecules which possess spherical symmetry is one which often has to be made in the interests of mathematical simplicity, but one which is found only occasionally in nature. The configuration of a molecule must usually be described by reference to a set of internal coordinates ϑ_1, ϑ_2 etc., as well as the position $\boldsymbol{x}$ of its mass centre, and the interaction energy of a pair of molecules is then a function $\phi(|\boldsymbol{x} - \boldsymbol{x}'|, \boldsymbol{\vartheta}, \boldsymbol{\vartheta}')$ of both sets of internal coordinates as well as the distance $|\boldsymbol{x} - \boldsymbol{x}'|$ between the mass-centres. It will still often be possible to express the interaction energy of a group of q molecules approximately in the form

$$\Phi_q = \sum_{j>i}^{q} \phi^{(ij)}, \tag{2.4}$$

where

$$\phi^{(ij)} = \phi(|\boldsymbol{x}^{(i)} - \boldsymbol{x}^{(j)}|, \boldsymbol{\vartheta}^{(i)}, \boldsymbol{\vartheta}^{(j)}), \tag{2.5}$$

i.e., as a sum of the mutual potential energies of pairs of molecules. A possible exception arises where chemical bonds are formed between different molecules. This does not necessarily exclude liquids whose molecules have a complex structure, or polar liquids.

Water should most likely be regarded as an example of a polar liquid. The H_2O molecule has a pair of electrons in the outer shell which are available for bond formation. Probably the bonds are not of the usual chemical type; according to LENNARD-JONES and POPLE[1], who describe them as "lone-pair" bonds, they can be understood in terms of classical electrostatics. As a first approximation, the water molecule has a tetrahedral structure, with lone-pair electrons at two vertices, and the hydrogen bonds at the other two vertices. According to this picture, the edge of the tetrahedron connecting the lone-pair electrons is a region of predominantly negative charge, and the edge connecting the imperfectly screened protons is a region of predominantly positive charge. Water has, therefore, the essential feature of a polar liquid, that its molecules should possess a dipole or quadrupole moment which is largely independent of the presence of neighbouring molecules. In any liquid the structure of a molecule is deformed to some extent by its neighbours, and its electric moments are therefore subject to fluctuations; but, in a polar liquid, these should be secondary to the moments of the unperturbed molecule.

Some polar liquids are referred to also as associated liquids, especially those, like water, where the molecule has localized regions with an excess of positive or negative charge. They have properties very similar to those of a general type of associated liquids, whose molecules possess unsaturated chemical bonds.

[1] J. LENNARD-JONES and J. A. POPLE: Proc. Roy. Soc. Lond., Ser. A **205**, 155 (1951).

Such bonds arise through the presence of unpaired electrons on the molecules; association is caused by the pairing of electrons on different molecules. When the pairing is effected, the bond is saturated: that is to say, no further molecule can be held. Consequently, if there is just one unpaired electron per molecule, the molecules will associate in pairs. If, on the other hand, there are two unpaired electrons per molecule, the molecules can associate in chains. If there are more, complicated branching associations can be formed. The interaction energy of such an association of molecules cannot be exactly represented by a formula like (2.4); but in view of the close resemblance between chemical and certain electrostatic bonds, probably to assume the correctness of (2.4) would not lead to serious error in a wide range of circumstances.

It is an essential feature of association that somewhere in the liquid state the bonds formed between molecules can be broken sufficiently frequently by the thermal motion; otherwise one would regard the liquid as made up of complex molecules. The glasses[1] illustrate what happens when branching associations are formed between the molecules. As the temperature is reduced, the difficulty in breaking the intermolecular bonds gradually increases and it is hard to say at what point the liquid first becomes an amorphous solid. Rather than make a purely arbitrary distinction, a glass will here be regarded as a liquid so long as it does not crystallize.

3. Structure of liquids, solids and gases. The essential features of the liquid state can be described best by comparison with those of the solid and gaseous states. A liquid is usually conceived as a substance with no rigidity but possessing a free surface. However, the example just considered of the glasses, which are perfectly rigid at low temperatures, but cannot without arbitrariness be distinguished from other liquids, shows that rigidity is not a universal criterion. Similarly the state of a highly compressed substance just above the critical temperature has no free surface but is indistinguishable from corresponding states below the critical temperature which may or may not possess a free surface; so that the second criterion is also inadequate. Indeed, there seems to be no satisfactory way of defining a liquid except by reference to its molecular structure.

If attention is restricted to substances with a well defined molecular constitution, a solid will be taken to mean a crystalline solid. The molecular structure of a crystal is easy to visualize: it consists of an orderly arrangement of similarly spaced, and often similarly oriented molecules extending over a macroscopic region. Apart from occasional irregularities, the atoms are distributed so that just one of a given kind will be found in the neighbourhood of a lattice site. There is a high degree of correlation between the positions of even widely separated atoms in the crystal. The solid state is, accordingly, an ordered state. The liquid state, by contrast, is a disordered state, and the phenomenon of melting consists of an order-disorder transition.

This need not imply that the liquid state is completely disordered; even the molecules of a gas possess a degree of short-range order. The implication is simply that in a liquid there is no system of occupied lattice positions extending for more than a few molecular diameters. There is little or no correlation between the positions of widely separated atoms or molecules; in melting, long-range order is replaced by a special type of short-range order. Together with the transition, there is a discontinuous change in the internal energy, specific heat and density of the substance. The discontinuities are much smaller than those

[1] Cf. the contributions on glasses by J. STEVELS in Vols. XIII and XX of this Encyclopedia.

associated with condensation, except near the critical point, and it has sometimes been argued[1] that on this account there cannot be much difference between the solid state and the liquid state near the melting point. Actually, of course, there is a very fundamental qualitative difference between structures with and without long-range order, and any reasoning which ignores this difference is in need of separate and special justification. Presently the merits of the lattice model of a liquid will be considered in detail, but first the distinction between the gaseous and liquid states must be observed.

A fundamental concept of the kinetic theory of gases is the molecular free path. Since the intermolecular forces are electrostatic in origin, their range is in principle infinite. But in practice the deflection suffered by the path of a gas molecule, on account of its neighbours, is negligible when the nearest of these is beyond a certain distance r_0. This distance may be called the range of the intermolecular forces; it is somewhat arbitrary, depending on what deflection is considered negligible, but on the other hand its exact value is not important. In a rare gas, the mean distance between neighbouring molecules is very much greater than r_0, and there is little ambiguity in defining the free path of a molecule as that part of its trajectory which lies outside the range of interaction with neighbouring molecules. The average length of free path between successive interactions (the mean free path) is inversely proportional to the density at low densities. The probability of finding a molecule within the range of interaction of q other molecules is proportional to the q-th power of the density. So, as the density increases, the mean free path contracts and the incidence of clusters of two or more molecules increases.

At the point of condensation, clusters are of frequent occurrence but the mean free path is still finite. The phenomenon of condensation is due, on the molecular scale, to a tendency for clusters exceeding a critical size to grow at the expense of the smaller clusters, when the fluid is compressed and the temperature fixed. A liquid drop is, in fact, a giant cluster, and is therefore characterized by the fact that any one of its molecules is within the range of interaction of one or more of the neighbouring molecules. Within the liquid, the concept of the free path ceases to apply: the mean free path, if one insists, is rigorously zero.

The distinction which has just been made between the molecular structures of the vapour and the liquid is equally applicable in the region above the critical point where there is no discontinuous change in the macroscopic properties of the fluid, up to the point where it solidifies. The transition from gas to liquid may be supposed to occur at the point where the mean free path vanishes. Owing to the arbitrariness already noticed in what one regards as the range of the intermolecular forces, the division so effected is not completely unambiguous; but it suffices for practical purposes.

Comparison with solids and gases leads one to conclude that liquids have a molecular structure devoid of long-range order but sufficiently closely packed to ensure that any molecule is in continual interaction with its neighbours. This conclusion, however, is hardly sufficient to provide a detailed picture of molecular structure. The true picture is very complex, and it is very natural that attempts should have been made to simplify it in various ways. In this way the model theories have been created, which attempt to account for the properties of liquids by idealizing certain features of their molecular constitution. These theories will now receive some attention.

---

[1] e.g. in Chap. III, § 1—2, J. FRENKEL: Kinetic Theory of Liquids. Oxford 1946.

b) Liquid models.

4. The lattice-cell model. The models which have been proposed to describe the molecular structure of liquids range by small gradations from the lattice model, in which the analogy is drawn between a liquid and a crystalline solid, to the kinetic model, which was suggested originally by the study of condensing gases. The first of these models, the lattice model, is indistinguishable in its premises and consequences from a simple kind of cell model, and the two will therefore be considered here together.

The lattice model attempts to represent molecular structure as a modification of the structure of a perfect crystal. As a first approximation the molecules are imagined to be distributed with regular orientations over a set of evenly spaced lattice sites, some of which may, however, be left vacant. The number of vacancies, or holes as they are often called, can be adjusted, having regard to the distance between the lattice sites, so that the mean molecular density is the same as for the liquid state which one wishes to describe. To take account of the molecular disorder of the actual liquid, each molecule is then displaced from its lattice site, and—except for liquid crystals or liquids with spherically symmetrical molecules—allowed to rotate.

If the rotations and the displacements of the molecules from their lattice configurations could be arbitrarily large, and were suitably correlated, it would obviously be possible to represent the structure of liquids exactly in this way. But the model requires that the dissolution of the original lattice should not proceed too far, and in particular that a molecule should rarely be displaced, say, more than half a lattice distance in any direction; one supposes that each molecule for most of the time executes small oscillations about its lattice position. Also the correlations, which might exist between the displacements of neighbouring molecules, are entirely neglected.

The evidence which suggests such a model might be a fairly accurate representation of a liquid has been well presented by FRENKEL[1]. He regards melting not so much a transition from an ordered to a disordered state, as a thermodynamic phenomenon, a necessary consequence of the existence of two different homogeneous states at the same temperature and pressure. On this view, the disappearance of order is a gradual process, beginning while the substance is still crystalline and still not complete in the liquid. It is admitted that the crystal possesses long-range order, while the liquid does not; but it is suggested that only the local molecular order is important in determining most thermodynamic properties of the substance. The fact that the density and specific heat of most substances change only slightly on melting is advanced to uphold the idea that the local order is not much changed by fusion. Early experiments on the scattering of X-rays by liquids, and the observation that the diffraction patterns obtained show a marked resemblance to those produced by crystals, also appeared to support the idea. STEWART[2] was one of the first to draw attention to this evidence, and proposed the term "cybotaxis" to describe the association of groups of molecules of a liquid in an ordered structure similar to that of the crystal.

Today one is inclined to be more cautious in adopting a quasi-crystalline model for a liquid. It is now known that the structure of liquids has much more complexity and variety than such a simple model could convey. In spite of the superficial resemblance between the X-ray diffraction patterns of liquids[3] and

[1] J. FRENKEL: Kinetic Theory of Liquids. Oxford 1946.
[2] G. W. STEWART: Rev. Mod. Phys. **2**, 116 (1930).
[3] Cf. G. FOURNET's contribution on X-ray diffraction in liquids, Vol. XXXII of this Encyclopedia.

crystals, the quantitative analysis of these data reveals important differences of a fundamental kind. One is therefore disposed to interpret results obtained by means of the lattice model in a somewhat different way. Instead of regarding the structure of a liquid as obtained by modifying that of a crystal, one imagines the volume of the liquid to be broken up into small regions called cells. These cells are usually regarded as having the same size and shape, so that their centres form a lattice. If the volume of each cell is equal to or less than the mean volume per molecule in the liquid, the probability of finding more than one molecule in a cell is rather small, and, for the purposes of the model is neglected. The correlation between the positions of molecules in different cells is also ignored. Under these conditions, the cell model is obviously precisely equivalent to the lattice model. The difference in the point of view is, however, important, for the restrictive conditions under which the equivalence of the lattice and cell models is apparent can be relaxed, and more realistic models obtained. First, however, it is appropriate to consider what can be achieved within the limitations of the lattice model, or the lattice-cell model, as it might be described.

There are two important variants of the lattice-cell model. In the first of these, which was developed originally by EYRING and his collaborators[1], the cell size is variable and somewhat less than the volume per molecule, so empty cells, or holes, can occur. One is then confronted with the initial problem of finding the mean number of vacant cells adjoining an occupied cell—or, in the terminology of the lattice model, the mean number of vacant lattice sites next to an occupied site. The best solution to this problem, on the basis of statistical mechanics, is obtained by means of the "quasi-chemical approximation" of GUGGENHEIM[2]. The next problem is to determine the relative probability that a molecule will be found with a displacement $\boldsymbol{r}$, say, from the centre of its cell (or from its lattice site). According to statistical mechanics, this will be proportional to $\exp\{-\beta\psi(\boldsymbol{r})\}$, where $\beta kT=1$, k is BOLTZMANN's constant, T the absolute temperature, and $\psi(\boldsymbol{r})$ the mean potential energy of the molecule with the displacement $\boldsymbol{r}$. The function $\psi(\boldsymbol{r})$ has usually been calculated by assuming that in the occupied neighbouring cells the molecules are at the centres of the cells.

The thermodynamics of the lattice-cell model has been investigated quantitatively mainly in the neighbourhood of the critical point and the condensation region. If the model should be found to be applicable there, it would also presumably be applicable to the whole of the liquid state, in which conditions approximate more nearly to those presupposed by the model. Following CERNUSCHI and EYRING[1], ONO[3], PEEK and HILL[4], and ROWLINSON and CURTISS[5] have successively refined the model, with improving agreement with experiment. The best values predicted for the critical temperature and density were correct within 10%, and ONO obtained a good value for the critical pressure, but the calculated critical pressures were usually incorrect by a factor of 2 or 3. ROWLINSON and CURTISS found that their version of the model was not self-consistent, in the sense that it required a cell size too great to accommodate not more than one molecule

[1] H. EYRING: J. Chem. Phys. **4**, 283 (1936). — H. EYRING and J. HIRSCHFELDER: J. Phys. Chem. **41**, 249 (1937). — J. HIRSCHFELDER, D. STEVENSON and H. EYRING: J. Chem. Phys. **5**, 896 (1937). — J. F. KINCAID and H. EYRING: J. Chem. Phys. **6**, 620 (1938). — F. CERNUSCHI and H. EYRING: J. Chem. Phys. **7**, 547 (1939). — J. WALTER and H. EYRING: J. Chem. Phys. **9**, 393 (1941).

[2] E. A. GUGGENHEIM: Proc. Roy. Soc. Lond., Ser. A **169**, 134 (1938).

[3] S. ONO: Mem. Fac. Eng., Kyushu Univ., Japan **10**, No. 4, 196 (1947).

[4] H. M. PEEK and T. L. HILL: J. Chem. Phys. **18**, 1252 (1950).

[5] J. S. ROWLINSON and C. F. CURTISS: J. Chem. Phys. **19**, 1519 (1951).

per cell. MAYER and CARERI[1] later drew attention to another inconsistency of the model by pointing out that the pressure calculated from the dependence on density of the free energy was not in agreement with that obtained with the help of the virial theorem, except when the cell volume is determined in a particular way.

A second variant of the model was initiated by LENNARD-JONES and DEVONSHIRE[2]. They assumed a cell volume equal to the volume per molecule in the liquid; this assumption, though more restrictive than required by EYRING'S version of the model, eliminates the difficulties associated with the distribution of holes. The quantitative predictions of this variant were again in fairly good agreement with experiment, except for the critical pressure, which came out about four times too large. The results were, in fact, better than warranted by the inaccuracy associated with some of the approximations involved. More recently, KIRKWOOD[3] has shown how to improve systematically on these approximations. He demonstrated that, if one assumes exactly one molecule per cell and neglects the correlation between the positions of molecules in neighbouring cells, the best possible thermodynamic results are obtained by making the free energy of the model liquid approach its minimum value. MAYER and CARERI[1] used this principle in their work on EYRING'S variant of the model; however, the value which they found for the critical pressure was still more than three times too large, and the calculated vapour pressures were too large by almost an order of magnitude. The variant of LENNARD-JONES and DEVONSHIRE was similarly studied by WENTORF, BUEHLER, HIRSCHFELDER and CURTISS[4], who made extensive calculations for a wide range of liquid densities, and found that the resulting equation of state was in poor agreement with experiment except near the freezing point. The conclusion is that the lattice-cell model for a fluid is in some respects very inaccurate, at least for low densities.

The most serious weakness of the model is undoubtedly its failure to reproduce correctly the fluctuations in density in a liquid, which are a consequence of the molecular disorder. Owing to the arbitrary limitation imposed on the displacements of the molecules from the cell centres (or lattice sites), large fluctuations are suppressed—whereas in liquids, particularly near the critical point and in the condensation region, fluctuations in density are known to be large. These fluctuations are also intimately associated with the thermodynamic properties of the liquid. It will, in fact, be shown later [cf. Eq. (14.9)] that the pressure is completely determined by them, so that if the fluctuations are not properly represented by a model, the equation of state will turn out well only by accident. Independent evidence from X-ray scattering shows that the cell-lattice model also fails to give correctly the mean number of "nearest neighbours" of a molecule in the liquid. The model requires that the number of empty cells should be small and slowly varying with temperature and density, whereas the experimental evidence indicates that the "shell" containing the nearest neighbours of a molecule in a liquid varies very widely in its occupation.

In spite of these defects, the lattice model has a domain of application near the melting point of many liquids, particularly in the metastable, supercooled state. Although liquid structure cannot have long-range order, it seems likely

[1] J. E. MAYER and G. CARERI: J. Chem. Phys. **20**, 1001 (1952).

[2] J. E. LENNARD-JONES and F. DEVONSHIRE: Proc. Roy. Soc. Lond., Ser. A **163**, 53; **165**, 1 (1938).

[3] J. G. KIRKWOOD: J. Chem. Phys. **18**, 380 (1950).

[4] R. H. WENTORF, R. J. BUEHLER, J. O. HIRSCHFELDER and C. F. CURTISS: J. Chem. Phys. **18**, 1484 (1950).

that, when freezing is incipient, small groups of molecules with an ordered structure similar to that of the crystal are formed, and STEWART's concept of cybotaxis is realized. Cybotactic groups of molecules in a liquid near its melting point probably develop in much the same way as "clusters" of molecules in condensing gases. The X-ray scattering evidence is not now, in the main, regarded as affording proof of the phenomenon in ordinary liquids. But other evidence relating to supercooled liquids is quite convincing. This evidence has been summarized by GREENWOOD and MARTIN[1], with special reference to supercooled complexes of boron trifluoride. When any of a large group of liquids is cooled without freezing below its melting point, discontinuities in the temperature dependence of structural properties such as the viscosity, electrical conductivity and dielectric constant are observed. Other properties, such as the density, specific heat and vapour pressure, are not palpably affected, but these are just the physical characteristics which are not very different in the liquid and crystalline states. The fact that those discontinuities which do occur take place at the normal freezing point, suggests inevitably the widespread formation of cybotactic groups on the molecular scale. These groups cannot, however, fill the whole liquid; otherwise it would behave like a polycrystal, an aggregate of ultra-microscopic crystals with different orientations, or like an amorphous solid. The presence of cybotaxis, on the other hand, does not affect the fluid properties of a liquid to a marked degree.

Even above the melting point, the lattice-cell model has the advantage of greater mathematical tractability over other models. Its quantitative predictions are considerably worse than those attained by other methods at lower densities, but not near the melting point. Also the model can be used for quantitative purposes outside the present scope of other models and methods: for example, it has been applied with some success not only to simple monatomic liquids, but to mixtures and solutions, and to polar and associated liquids. On the theoretical side, it has led to the development of the more accurate cell-cluster model, which will next be considered.

5. The cell-cluster model. The sources of error in the lattice-cell model of a liquid are twofold. Firstly, the number of molecules per cell is not allowed to exceed one; secondly, the correlation between the positions of molecules in neighbouring cells is neglected. Within these limitations, one cannot hope for an exact evaluation of the partition function, which determines the thermodynamic properties of the liquid. The cell-cluster model has been developed in the attempt to remove the limitations, and so the inaccuracy of the cell model.

KIRKWOOD[2] was the first to show that the results of the cell model appeared by applying a well-defined method of approximation to the evaluation of the partition function. While his work did not show how to take account of correlations between the positions of molecules, it opened the way for admitting more than one molecule per cell. Subsequently JANSSENS and PRIGOGINE[3] and POPLE[4] independently showed how to generalize the cell model of LENNARD-JONES and DEVONSHIRE to take account of multiply occupied cells. But it soon appeared that the influence of doubly occupied cells on this model was insufficient to effect any great improvement, and cells more than doubly occupied would be still less important at liquid densities. A cell not larger than the molecular volume will

[1] N. N. GREENWOOD and R. L. MARTIN: Proc. Roy. Soc. Lond., Ser. A **215**, 46 (1952).
[2] J. G. KIRKWOOD: J. Chem. Phys. **18**, 380 (1950).
[3] P. JANSSENS and I. PRIGOGINE: Physica, Haag **16**, 895 (1950).
[4] J. A. POPLE: Phil. Mag. **42**, 459 (1951).

not readily accommodate two or more molecules when there is an effective barrier near the walls of the cell, created by the repulsive interactions of molecules in other cells. There is no doubt that the neglect of the correlation between configurations in neighbouring cells creates such a barrier where in actual liquids it may not exist. If in a real liquid one cell is doubly occupied, it is quite likely that one of the neighbouring cells will be empty, and then there is no barrier between the two cells. So the cell model will not be improved very much by taking account of multiply occupied cells, if the cell size remains small and the correlation between configurations in neighbouring cells is still neglected.

Taking for granted the possibility of multiply occupied cells, there are, therefore, two ways of improving on the cell model. One way is simply to increase the size of the cell. Probably the neglect of correlations between cells would not have any important effect if the volume of a cell were, say, ten times the molecular volume. It is quite easy to modify the theory of the cell model in this direction. But the computational difficulties associated with the model are thereby increased enormously, and the principal merit of the cell model—the ease of the calculations to which it leads—is lost. Nevertheless, as computational techniques improve, the giant cell model, as this variant of the model might be called, may increase in importance.

The second way of improving on the cell model is to take account of correlations between configurations in neighbouring cells. If this could be done, the inaccuracy of the model would be eliminated. In principle, it can be done, but again at the cost of computational difficulties in proportion to the degree of accuracy required. However, it is quite practicable to calculate corrections for correlations between configurations between molecules in small groups, or clusters of adjoining cells. This was first pointed out by DE BOER[1], who introduced what he described as a "cell-cluster" model, as an improvement on the model of LENNARD-JONES and DEVONSHIRE. DE BOER's version of this model did not envisage multiple occupation of cells and no attempt was made to secure a minimum value of the free energy. A similar way of accounting for correlations between the positions of the molecules was devised by BARKER[2], who also showed how to allow for multiple occupation of the cells. The author[3] has independently shown how to remove the restriction to singly occupied cells from DE BOER's cell cluster model, and also how to minimize the free energy to obtain an improved first approximation. The model can at last be used as the basis for a precise method of evaluating the partition function. The applications of the variant of the cell-cluster model refined in this way will be described in Sect. 20 below.

6. The hole model. It has been seen that one of the chief difficulties with the cell model is that, by its nature, the model cannot accurately represent the fluctuations in density which occur in real liquids. One variant of the cell model admits the existence of vacant cells, or holes, and MAYER and CARERI have shown how best to treat the holes, within the framework of the cell model; but still the model is not satisfactory judged by quantitative standards. The hole model of a liquid is designed to give a more satisfactory account of density fluctuations. The underlying idea is to fix attention on microscopic regions of the liquid which are either empty or have a density appreciably less than the average value. The statistical frequency of such holes depends on the thermodynamic properties of the liquid, and conversely, if one can calculate the probability of the

---

[1] J. DE BOER: Physica, Haag **20**, 655 (1954).
[2] J. A. BARKER: Proc. Roy. Soc. Lond., Ser. A **230**, 390 (1955).
[3] H. S. GREEN: J. Chem. Phys. **24**, 732 (1956).

occurrence of holes of a prescribed shape and size, the thermodynamical properties of the liquid can be deduced.

The hole model is partly based on well established results of statistical thermodynamics. The probability of finding a region of volume v and surface s in a liquid completely empty is $\exp\{-\beta(pv+\gamma s)\}$ where p is the pressure and γ the proper surface tension of the liquid. The probability of finding the region in the same state as the saturated vapour is approximately $\exp\{-\beta(p'v+\gamma' s)\}$ where p' is the excess pressure of the liquid over that of the vapour, and γ' is the surface tension relative to the vapour. To make use of these facts the hole model developed by ALTAR[1] and FÜRTH[2] introduces approximations based on the analogy between the microscopic holes and macroscopic bubbles in a liquid. Unfortunately it has not yet been possible to justify these approximations by showing how they arise in the rigorous evaluation of the partition function. However, they are to some extent justified by the numerical predictions which follow from them. ALTAR found that the model would provide a fairly good account of the equation of state for liquid argon, and FÜRTH showed that the compressibilities and coefficients of thermal expansion of many liquids could also be predicted fairly accurately.

The important question from the present point of view is whether it is correct to imagine a liquid as containing myriads of tiny vapour bubbles. If such a picture is valid at all, it might be expected to be so in the neighbourhood of the boiling point. The question is really very similar to whether microscopic crystals are formed in a liquid near its freezing point. The second question has already been answered in the sense that, though configurations rather similar to the ordered state of the crystal are common in liquids above the freezing point, no truly crystalline structures are formed except in supercooled liquids. The answer to the first question seems to be very similar. In liquids below the boiling point there are certainly regions of much lower density than the average; these regions however, bear only a superficial similarity to the vapour bubbles formed in boiling or superheated liquids. FRENKEL[3] has presented arguments which might be construed rather differently, but it is clear from his later work[4] that he regarded these arguments as somewhat over-simplified. The empty or nearly empty regions in a liquid must be regarded as ill-defined and unstable fissures created by the kinetic fluctuations, rather than vapour bubbles of the type familiar to macroscopic experience.

The density fluctuations in liquids can, of course, be investigated quantitatively with the help of statistical mechanics. MAYER[5], using the powerful methods of the grand partition function, has developed apparatus suitable for calculating the probabilities of special types of configurations within finite regions of a liquid. Though computational difficulties have so far prevented the development of this work to the point where numerical predictions can be made, a qualitative picture can be constructed which is in agreement with that already drawn. The probability of finding precisely q molecules in a region of volume v is $z^q \exp\{-\beta(pv+E_q)\}$, where z is the thermodynamic activity per molecule, p is the pressure, and E_q the energy of the molecules in their particular configuration. In a sufficiently small region numbers of molecules may quite probably

[1] W. ALTAR: J. Chem. Phys. 5, 577 (1937).

[2] R. FÜRTH: Proc. Cambridge Phil. Soc. 37, 252, 276, 281 (1941); see also R. FÜRTH, L. S. ORNSTEIN and J. M. W. MILATZ: Proc. Amsterdam Acad. Sci. 42, 107 (1939).

[3] J. FRENKEL: J. Chem. Phys. 7, 538 (1939).

[4] J. FRENKEL: Kinetic Theory of Liquids, pp. 174—177. Oxford 1946.

[5] J. E. MAYER: J. Chem. Phys. 10, 629 (1942); 15, 187 (1947).

be found deviating appreciably from the mean, but there is no special preference for densities in the neighbourhood of the vapour density or any other density except the mean. Only in the condensation region the activity z is independent of density over a wide range and the vapour density may be found as well as any other in the range.

7. The kinetic model. The final model which will be considered is one which has developed through attempts to extend the kinetic theory of gases into the liquid region and is therefore appropriately called the kinetic model. It is not a model today in the sense that it is based on a simplified picture of the infra-microscopic structure, though it has been in the past. There was at one time, for example, a tendency to assume that the molecular mechanism responsible for viscosity and thermal conduction is the same in liquids and gases; but this assumption is now recognized as erroneous. Similarly the initial success of VAN DER WAALS' equation of state led to a mistaken impression that the physical concepts on which it was based were really applicable to the liquid state. Yet there is a continuity of liquid and gaseous states in the neighbourhood of the critical point which suggests that they are not so essentially distinct as the liquid and crystalline states, for example, so that a theory of dense gases might without serious modification apply also to liquids. ENSKOG[1] was the author of a theory of dense gases which cannot in fact be applied to liquids, but this is because it retained the concept of the molecular free path. The essential difference between a liquid and a gas is that in a liquid the free path does not exist: a molecule is in continual interaction with its neighbours. Consequently it is this concept which must be eliminated from the kinetic theory if it is to apply to the liquid state.

The kinetic theory, freed of the hypothesis of binary or multiple encounters, regards a fluid as a system of molecules in continual motion under the influence of their mutual attractions and repulsions. The fundamental problem is the statistical one of finding the most probable state of the system, subject to externally imposed conditions; this is the state to which actual fluids are supposed to approximate. In thermodynamic equilibrium, statistical mechanics provides a solution in principle to the problem, but one which is almost useless since it refers to the system as a whole, and not directly to individual molecules or groups of molecules. YVON[2] was the first to formulate the problem of finding the relative probability of configurations of a small group of molecules in a fluid; he derived equations which in the hands of later workers furnished the most accurate calculations of thermodynamic functions yet achieved. Unfortunately YVON's equations are not sufficient to solve the problem for liquids, but have to be supplemented by approximations of a statistical character. The only approximation which is simple enough to be immediately useful but, at the same time, sufficiently exact to enable satisfactory predictions to be made, is the superposition approximation introduced by KIRKWOOD and BOGGS[3]. The meaning of this approximation cannot be made clear independently of the statistical analysis of liquid structure, but will be discussed in Sect. 21. KIRKWOOD and BOGGS used the superposition approximation to investigate the structure of a hypothetical liquid of spherical molecules. It was later used by BORN and GREEN[4] to determine the

[1] D. ENSKOG: Kgl. svenska Vetensk. Akad. Handl. **63**, No. 4 (1921).

[2] J. YVON: La Théorie Statistique des Fluides et l'Equation d'Etat. (Actualités Scientifiques et Industrielles No. 203. Paris). Hermann & Cie. 1935.

[3] J. G. KIRKWOOD and E. M. BOGGS: J. Chem. Phys. **10**, 394 (1942).

[4] M. BORN and H. S. GREEN: Proc. Roy. Soc. Lond., Ser. A **188**, 10 (1946). — H. S. GREEN: Proc. Roy. Soc. Lond., Ser. A **189**, 103 (1947).

structure of actual monatomic liquids and an approximation to their equation of state. This work was taken up by RODRIGUEZ[1], McLELLAN[2] and CHENG[3] who showed that quantitatively significant results could be obtained for the thermodynamic functions of a fluid in and about the condensation region. The computations for the liquid were, however, very difficult, and were finally performed accurately with the help of an electronic computer by KIRKWOOD, LEWINSON and ALDER[4]. Results are in fairly good agreement with experiment except near the freezing point of the liquid, where the cell model is superior.

The future progress of the kinetic model of liquids in thermodynamic equilibrium will evidently depend on improving the superposition approximation. It is easy to see how this could be done in principle, but only at the cost of complicating the numerical calculations. In the meantime, the thermodynamics of monatomic liquids appears to be well accounted for by one or other of the models which have been discussed.

The kinetic model has the advantage that it is not restricted to thermodynamic equilibrium. Attempts have been made to account for the viscous flow and conduction of heat with the help of other models, but they have mostly been founded on incorrect conceptions of irreversible processes in liquids. Theories of viscosity and thermal conduction were developed independently, on the basis of the kinetic model, by KIRKWOOD and his collaborators[5], and by BORN and GREEN[6]. These theories are in complete agreement concerning the mechanism of viscosity and thermal conduction. Viscosity in liquids is due only to a very small extent to the transfer of momentum by the molecules themselves, though this is the principal mechanism in gases. Instead, liquid viscosity is due mainly to a deformation of the molecular structure by the flow, which produces a dragging action, on account of the intermolecular forces, between adjacent layers moving with different velocities. Similarly the energy transfer associated with thermal conduction is not, for the most part, accomplished by the motion of the molecules, as in gases; instead it is produced mainly by the intermolecular forces, acting *via* the correlations between the velocities of neighbouring molecules, which are set up by a temperature gradient.

The theories of KIRKWOOD and BORN and GREEN differ mainly in the way they are developed for practical purposes. The main difficulty lies in the way the irreversibility of natural processes should enter into the model. Real irreversible processes are distinguished from other possible processes only by an overwhelming statistical preponderance, the implications of which on the molecular scale are not easy to assess. In face of this difficulty, KIRKWOOD introduced approximations, suggested by the theory of the Brownian motion, which were time-irreversible in character. BORN and GREEN, on the other hand, followed more closely the procedures of the kinetic theory of gases; these do not, by themselves, solve the problem of the influence of irregularities of flow and temperature on the structure of the fluid. Clearly none of these devices could be entirely satisfactory, since the basic statistical problem was thereby evaded.

[1] A. E. RODRIGUEZ: Proc. Roy. Soc. Lond., Ser. A **196**, 73 (1948).
[2] A. G. MC LELLAN: Proc. Roy. Soc. Lond., Ser. A **210**, 509 (1952).
[3] K. C. CHENG: Proc. Phys. Soc. Lond. A **63**, 1028 (1950).
[4] J. G. KIRKWOOD, V. A. LEWINSON and B. J. ALDER: J. Chem. Phys. **20**, 929 (1952); see also J. G. KIRKWOOD, E. K. MAUN and B. J. ALDER: J. Chem. Phys. **18**, 1040 (1950).
[5] J. G. KIRKWOOD: J. Chem. Phys. **14**, 180 (1946). — J. G. KIRKWOOD, F. P. BUFF and M. S. GREEN: J. Chem. Phys. **17**, 988 (1949). — R. W. ZWANZIG, J. G. KIRKWOOD, I. OPPENHEIM and B. J. ALDER: J. Chem. Phys. **22**, 783 (1954).
[6] M. BORN and H. S. GREEN: Proc. Roy. Soc. Lond., Ser. A **188**, 10 (1946); **190**, 455 (1947).

The kinetic theory of gases, as formulated by BOLTZMANN and MAXWELL[1], did not encounter any similar difficulty, as the features which guarantee the irreversibility of the transport phenomena are implicit in its formulation. The isolation of the statistical postulate which is time-irreversible in the kinetic theory of gases, and its generalization for liquids, were required in order to supply the missing element in the kinetic model of liquids. The author[2] has recently shown how these requirements can be fulfilled, and the kinetic theory of liquids can now be regarded as reasonably complete in its essentials; but of course formidable computational difficulties remain to be overcome.

c) Actual structure of liquids.

8. Evidence of the diffraction experiments[3]. The study of liquid models is not the most direct way to the understanding of the structure of liquids. Admittedly, the more successful the predictions based on a particular model, the more likely it is that the model should be a faithful representation. But it is obviously more profitable to take the experimental facts as a starting point and discover their implications concerning the molecular structure, than to make more or less plausible assumptions concerning the structure, and look for experimental confirmation where it can be found.

To appreciate the full significance of the X-ray and neutron diffraction experiments, it will be necessary to resort to a more precise description of molecular structure than the model theories could provide. By the structure of a liquid one means the typical disposition of other molecules relative to a particular molecule in the liquid. When the molecules possess radial symmetry and are all of the same kind, the structure is implicitly specified by the mean number of molecules situated at various displacements from a given molecule. A quantitative specification is provided by what is known as the radial distribution function, and denoted by $g(r)$. This may be defined as the ratio of the mean molecular density at distance r from the particular molecule, to the average molecular density. So, if n is the average molecular density, the average number of molecules at a distance between r and $r+dr$ from the central molecule will be $n g(r) \cdot 4\pi r^2\, dr$. The definition of $g(r)$ assumes that radial symmetry exists in the liquid, which means that the liquid may not be in a state of non-uniform motion, nor may the influence of anisotropic forces, such as gravity or surface tension, be great enough to cause asymmetry. Actually the influence of gravity is usually negligible, since appreciable changes of density due to its action occur only through distances very much greater than the mean distance between neighbouring molecules.

The radial distribution function naturally depends on the temperature and density of the liquid. It provides a fairly complete description of the standard environment of a molecule, and the way this varies with the macroscopic state of the liquid. The description is, of course, statistical in character; in actual liquids the molecular structure varies from point to point and from time to time, and there may be quite large fluctuations from the mean. But such fluctuations are of secondary interest, serving only to fill in the finer detail of the sub-microscopic picture; they do not have much independent influence on the macroscopic behaviour of the liquid.

---

[1] See S. CHAPMAN and T. G. COWLING: Mathematical Theory of Non-Uniform Gases. Cambridge 1939.

[2] H. S. GREEN: Proc. Phys. Soc. Lond. A **69**, 269 (1956).

[3] For more details cf. Vol. XXXII of this Encyclopedia.

The radial distribution function can be determined experimentally by measuring the intensity of X-radiation or neutrons scattered by a liquid, at various angles with the incident beam. It is important that the beam should be as nearly as possible monochromatic, and that the wavelength should be of the same order of magnitude as the intermolecular distances. When these conditions are fulfilled, the radiation scattered by different but neighbouring molecules interferes to form a diffraction pattern, which depends on the radial distribution. The method will be explained fully in Sect. 19; at present it need only be stated that the radial distribution function can be calculated from the scattered intensity and its angular variation.

For obvious reasons, suitable neutron diffraction experiments are a comparatively recent development; but it was recognized as long ago as 1923, by KEESOM and DE SMEDT[1] that the X-ray diffraction techniques could be a source of quantitative information on molecular structure for liquids, just as they are for solids. The detailed analysis which was required for the interpretation of the diffraction patterns was developed by DEBYE[2] and MENKE[3], and ZERNIKE and PRINS[4]. The radial distribution function has been determined in this way for liquid helium, lithium, nitrogen, oxygen, sodium, aluminium, phosphorus, sulphur, chlorine, argon, potassium, zinc, gallium, germanium, selenium, rubidium, cadmium, indium, tin, caesium, gold, mercury, thallium, lead and bismuth. Excellent reviews of this work were made in 1943 by GINGRICH[5] and in 1949 by GLOCKER[6]. For liquid helium, REEKIE[7] has considerably improved on earlier results; his work will be discussed in Sect. 44. An important advance has also been made by PRINS[8] in the elucidation of the structure of liquid sulphur; he has shown that the rather abrupt change of structure, reported previously at about 160° C, does not exist; instead, the structure changes more gradually and may be more complex than at one time believed. A more accurate study of the structure of liquid mercury, and its variation with temperature, has been made by CAMPBELL and HILDEBRAND[9], who also furnished original data for liquid xenon[10]. As an example of the results obtained, the radial distribution function in liquid argon, as determined by EISENSTEIN and GINGRICH[11], will be discussed in some detail.

All the curves shown in Fig. 1 have certain features in common; these are, indeed, features of the radial distribution function in any monatomic liquid. To begin with small values of r, it will be noticed that $g(r)$ is effectively zero, meaning that the centres of two molecules cannot approach within a certain distance of one another. This is a consequence of the strong repulsive forces between the outer electron shells of the two molecules, which prohibit more than a small degree of interpenetration.

The next feature of the radial distribution function is a rapid increase, culminating in a peak which is often the region of greatest density in the whole distribution.

---

[1] W. H. KEESON and J. DE SMEDT: Proc. Acad. Amsterd. **27**, 112 (1923).
[2] P. DEBYE: Phys. Z. **28**, 135 (1927).
[3] H. MENKE: Phys. Z. **33**, 593 (1932).
[4] P. ZERNIKE and J. A. PRINS: Z. Physik **41**, 184 (1927).
[5] N. S. GINGRICH: Rev. Mod. Phys. **15**, 90 (1943).
[6] R. GLOCKER: Ergebn. exakt. Naturw. **22**, 186 (1949).
[7] J. REEKIE: Proc. Roy. Soc. Lond., Ser. A **228**, 363 (1954).
[8] J. A. PRINS: Physica, Haag **20**, 124 (1954); also J. A. PRINS and N. J. POULIS: Physica, Haag **15**, 694 (1949).
[9] J. A. CAMPBELL and J. H. HILDEBRAND: J. Chem. Phys. **11**, 330 (1943).
[10] J. A. CAMBPELL and J. H. HILDEBRAND: J. Chem. Phys. **11**, 334 (1943).
[11] A. EISENSTEIN and N. S. GINGRICH: Phys. Rev. **58**, 307 (1940); **62**, 261 (1942)

The molecules in the vicinity of this peak are to be regarded as the nearest neighbours of the central molecule, and form what is known as the first coordination shell. The peak is located in the attractive part of the intermolecular field of force, and it is natural to regard the first coordination shell as made up of molecules captured by the attraction of the central molecule. This, however, is only part of the truth, as a similar shell is found in the radial distribution of a collection of impenetrable spheres with no attractive field. Experiments have been carried out[1] with macroscopic spheres, which are found to exhibit a radial distribution function not very different in appearance from those shown in Fig. 1. The first coordination shell must therefore be regarded as due to the pushing of molecules away from the central region as well as the attraction from outside.

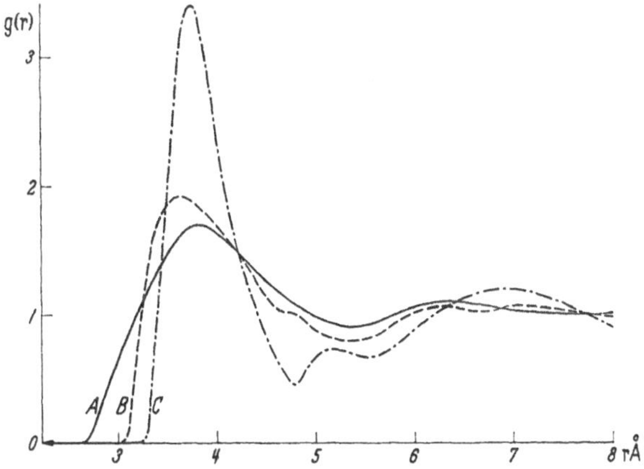

Fig. 1. The radial distribution function in liquid argon. Curve *A*: 149.3° K, 43.8 Atm. Curve *B*: 126.7° K, 18.3 Atm. Curve *C*: 84.4° K, 0.8 Atm. The pressure is that of the saturated vapour. The curves are normalized to unity at large values of *r*.

Beyond the first peak, the function $g(r)$ falls sharply to a minimum, which is sometimes, but not always, deeper than any subsequent minimum. This minimum represents a region partially excluded to other molecules on account of the proximity of the molecules in the first coordination shell. Sometimes the gap is partially filled by molecules of the second coordination shell, consisting of second nearest neighbours; this will occur especially if the first coordination shell is not more than half complete compared with the crystalline state, and then the minimum beyond the second peak is the deepest. Under these circumstances it is also difficult to distinguish clearly between the first and second coordination shells, or to say precisely how many molecules each contains. The following table, showing estimates of the distances and occupation numbers of the first and second co-ordination shells, does, however, provide a reliable guide to the nature of these shells.

Table 1. *r_1 and r_2 are the distances of the first and second coordination shells and n_1 and n_2 their occupation numbers.* All observations taken at the saturated vapour pressure, and the temperature in the first column, in argon.

	r_1	n_1	r_2	n_2
crystal	3.82 Å	12	5.4 Å	3
84.4° K	3.79 Å	10.5	5.3 Å	
91.8° K	3.79 Å	7.0	4.7 Å	4
126.7° K	3.8 Å	6.0	4.8 Å	
144.1° K	3.8 Å	4.2	5.4 Å	
149.3° K	4.5 Å	6		
vapour	4.1 Å	2		

It will be seen that the first coordination shell is not complete, even at 84.4° K, near the melting point, and at most other temperatures along the saturated vapour curve it is only about half complete. The distance of this peak does not change much in the liquid, except for an apparent increase at 149.3° K, near the critical point. The second coordination shell has a tendency to approach the first as soon as the first is depleted, though as the density decreases it moves away again.

It is interesting to observe that the inner coordination shell appears to be nearer to the central molecule in the liquid at the melting point than in the

[1] e.g. by W. E. Morrell and J. H. Hildebrand: J. Chem. Phys. 4, 224 (1936).

neighbouring crystalline state, indicating a significant change in structure in the process of melting. A similar collapse of the inner shell appears to take place in the process of boiling, as one can see by comparing the values of r_1 at 149.3° K and at the same temperature in the vapour.

The interpretation of the X-ray scattering data in the case of argon is particularly simple because the liquid is monatomic. Other elements, such as nitrogen and chlorine, form diatomic liquids; liquid phosphorus has four atoms per molecule, and the molecular constitution of liquid sulphur is even more complicated. For these liquids, the usual procedure has been to determine directly not the molecular radial distribution function, but the atomic radial distribution function, which describes the mean distribution of atoms about one of the atoms of the molecule. One can usually identify without difficulty peaks due to atoms in the same molecule, and the remainder of the distribution can then be attributed to other molecules. In this way one obtains information about the atomic structure of the molecules in the liquid, as well as the molecular structure of the liquid itself. When the molecule is almost radially symmetric, however, it is possible to determine the molecular radial distribution function, in the same way as for monatomic liquids. This is true not only for liquid elements like phosphorus, but even for liquids like carbon tetrachloride whose molecules contain more than one kind of atom. Difficulties of interpretation arise only for liquids with asymmetric molecules or molecules of different kinds, when the different orientations of the molecules or interference between radiation scattered by dissimilar molecules complicate the diffraction pattern.

Since the development of nuclear reactors with a large neutron flux, an alternative diffraction method for the investigation of molecular structure has become available[1]. A monochromatic component is separated from the neutron beam by reflection from a crystal plane and allowed to irradiate the liquid whose structure is to be determined. By measuring the intensity of the scattered neutron beam at various angles of scattering, data are obtained from which the radial distribution function can be reconstructed. This method may be expected to be particularly useful for the investigation of deuterium and deuterated compounds of hydrogen, and also of heavy elements which absorb X-radiation strongly but do not readily capture neutrons. One distinct advantage in the use of neutrons is that they are scattered at only one point in each atom, so that one does not need to compute atomic structure factors. Also there is little incoherent scattering as compared with the use of X-radiation.

The first application of the method was made by CHAMBERLAIN[2] to determine the structures of liquid sulphur, lead and bismuth. Somewhat more accurate measurements for lead and bismuth were made later by SHARRAH and SMITH[3], who reported a first coordination shell containing an average of 9.4 atoms, at a distance of 3.40 Å, in lead at 350° C, which was hardly changed at 550° C; and a first coordination shell containing an average of 7.7 atoms, at a distance of 3.35 Å, in bismuth at 300° C, which also was hardly changed at 550° C. Other liquids which have been investigated are liquid helium, by HENSHAW and HURST[4]; nitrogen, oxygen and argon, by HENSHAW, HURST and POPE[5]; and mercury by VINEYARD[6].

[1] See G. R. RINGO in Vol. XXXII of this Encyclopedia.
[2] O. CHAMBERLAIN: Phys. Rev. **77**, 305 (1950).
[3] P. C. SHARRAH and G. P. SMITH: J. Chem. Phys. **21**, 228 (1953).
[4] D. G. HENSHAW and D. G. HURST: Phys. Rev. **91**, 1222 (1953).
[5] D. G. HENSHAW, D. G. HURST and N. K. POPE: Phys. Rev. **92**, 1229 (1953).
[6] G. H. VINEYARD: J. Chem. Phys. **22**, 1665 (1954).

9. Relation between structure and properties. Although the diffraction experiments enable one to form an excellent picture of the general structure of liquids, they are not always sufficiently accurate for many quantitative purposes. One cannot at present, for example, compute the pressure in liquid argon corresponding to a given temperature and density from the experimentally determined function $g(r)$, though this ought to be possible inprinciple. The difficulty is that many macroscopic properties of liquids depend very sensitively on the radial distribution function in the region where it rises from zero to its first maximum, so that even small errors here may lead to serious discrepancies. For this reason the detailed structure of the liquid is often inferred with the help of macroscopic data other than the diffraction data. Thus, to obtain the form of the radial distribution function accurately enough for their calculation of the viscosity of liquid argon, KIRKWOOD, BUFF and GREEN[1] used data on the equation of state as well as diffraction data. The procedure is to assume a functional dependence for $g(r)$ of a kind similar to that determined by the X-ray diffraction technique, but arbitrary to the extent of a number of parameters α_1, α_2 etc., i.e., to assume

$$g(r) = g[r, \alpha_1, \alpha_2, \ldots] \tag{9.1}$$

where the right-hand side is determinate apart from the values of the parameters. In this section a review will be made of various results which enable the parameters to be determined.

To begin with, suppose the pressure is expressed as a function $p(n, \beta)$ of the mean molecular density n, and $\beta = (kT)^{-1}$, where k is BOLTZMANN's constant and T the absolute temperature. Then $(\partial p/\partial n)^{-1}$ is a measure of the isothermal compressibility of the liquid, and, according to a subsequent equation, (23.27), is given by

$$\beta \frac{\partial p}{\partial n}\left[1 + 4\pi n \int_0^\infty \{g(r) - 1\} r^2 \, dr\right] = 1. \tag{9.2}$$

This particular relation is fulfilled quite well by the experimentally determined radial distribution function and compressibility. Near the critical point and the condensation region, the compressibility is exceptionally large, so the integral in (9.2) should be large as well. In fact (cf. curve C of Fig. 1) one does find an unusual preponderance of positive values of $g(r) - 1$ at moderately large values of r from diffraction data in regions of high compressibility. This is associated with unusually large fluctuations in density, which cause a marked increase in the scattering of radiation through small angles.

It has been assumed that the molecules have radial symmetry, so that the mutual potential energy of two molecules can be expressed as a function $\phi(r)$ of the distance between their centres. According to a subsequent equation, (15.14), the pressure in a liquid at rest is given by

$$p = \frac{n}{\beta} - \frac{2}{3}\pi n^2 \int_0^\infty g(r)\, \phi'(r)\, r^3 \, dr, \tag{9.3}$$

where $\phi'(r)$ means the derivative of $\phi(r)$. The integral in this formula depends much more sensitively on the values of $g(r)$ in the repulsive part of the intermolecular field, where $\phi'(r)$ is large. So (9.3) is useful to determine a parameter which fixes the behaviour of $g(r)$ where it first begins to increase from zero.

[1] J. G. KIRKWOOD, F. P. BUFF and M. S. GREEN: J. Chem. Phys. **18**, 901 (1950).

For monatomic liquids the internal energy per unit volume is

$$U = \frac{3}{2}\frac{n}{\beta} + 2\pi n^2 \int_0^\infty g(r)\,\phi(r)\,r^2\,dr, \tag{9.4}$$

according to (16.6); this formula is also sensitive to the behaviour of $g(r)$ in the repulsive part of the intermolecular field. It is well known that the thermodynamic properties of a liquid are completely determined by the way the pressure and internal energy depend on temperature and density. According to (9.3) and (9.4) these are determined as well by the radial distribution function and its dependence on temperature and density; it is apparent, therefore, that all the thermodynamic properties of a liquid are directly related to its structure.

It should, perhaps, be remarked that the formulae (9.3) and (9.4) rest on the validity of classical mechanics, and apply at very low temperatures, where quantum effects are important, only in a modified form. The required modification is discussed in Sect. 46; it amounts to the replacement of β by a function of both temperature and density, which approaches a finite value at absolute zero. At ordinary temperatures this function is practically equal to β.

The elastic constants of liquids might also be used to provide information on the radial distribution function. In quasi-static processes, of course, there is no shearing elasticity, and the only elastic constant is the ordinary compressibility. But in sufficiently rapid processes, there is an elastic response, which, according to (25.6), is characterized by a shear modulus of elasticity

$$\mu = \frac{4\pi n^2}{30} \int_0^\infty g(r)\,\frac{d}{dr}\{r^4\,\phi'(r)\}\,dr. \tag{9.5}$$

The technique of measuring the shear modulus of mobile liquids is, however, in its infancy, and this formula is of mainly theoretical interest at present.

Information concerning the molecular structure in the surface zone, which is difficult to study by diffraction methods, can be obtained indirectly from the observed behaviour of the surface tension and surface energy. The theoretical results which connect these macroscopic quantities with the molecular structure will be given in Sect. 27.

When a liquid is in non-uniform motion, so that viscosity is brought into play, the molecular structure of the liquid is deformed by the motion. In fact, the mean distribution of molecules about a given molecule, which is described in a liquid at rest by the function $g(r)$, is no longer radial, but rather ellipsoidal, to an extent depending on the gradient of the macroscopic velocity of flow. The anisotropy of the molecular structure is associated with an anisotropy of the macroscopic stress distribution. The observed relation between the stress and velocity gradient, expressed by the law of viscosity and measured, in Newtonian liquids, by the coefficient of viscosity, is therefore governed by the molecular response to the macroscopic conditions. Further, the coefficient of viscosity is in a true sense a measure of the deformation of the molecular structure by the flow, and can be used to gain information about the structure of fluids in non-uniform motion. The detailed analysis of the relationship between the deformed structure and the flow will be presented in Sect. 33 below.

Other irreversible processes in liquids are also accompanied by changes in the molecular structure, in a broad sense. Thermal conduction does not alter the radial distribution of molecules relative to a given molecule, but it does require a progressive migration of the molecules in consequence of which the molecules in the neighbourhood of a given point are constantly changing. Pairs of neighbouring molecules aligned in the direction of the temperature gradient drift towards the higher temperature and, to compensate, pairs of molecules aligned in a direction normal to the temperature gradient drift towards the lower temperature. It will appear in Sect. 32 that the coefficient of thermal conduction is a measure of the efficacy of this molecular process.

The connections which have been mentioned between the molecular structure and macroscopic properties of liquids assist not only to determine the structure of liquids for which experimental data on the equation of state, etc. are available, but also to predict the properties of liquids whose molecular structure can be determined theoretically. The possibility of obtaining experimental verification has indeed supplied much impetus to theoretical investigations of molecular structure, based only on a knowledge of the intermolecular forces. Unfortunately such theoretical investigations are possible at present only for liquids with a particularly simple molecular composition. Where the molecular composition is at all complicated, it is necessary to make maximum use of the experimental data to determine the molecular structure.

10. Liquids with complex molecular composition. The progress which has been made towards the elucidation of molecular structure in liquids whose composition is not simple will now be reviewed. Complexity may arise for two entirely different reasons. Firstly, it may be due to a complicated structure of the individual molecules, so that the idea of central forces between pairs of molecules is inapplicable. This happens, for example, in polar and associated liquids. Secondly, though the molecules may be simple, more than one kind of molecule may be present. Both complications may of course co-exist; but for the sake of clarity they will first be discussed separately.

α) Polar liquids. Many kinds of molecules possess electric (or magnetic) moments, which may be classified as dipole, quadrupole, or multipole in character. The force between two molecules then depends on their mutual orientation as well as the distance between them. Much depends on whether the polarity is localized on certain regions near the surface of the molecule, or whether it is approximately concentric with the mass centre of the molecule. Water is usually regarded as a typical example of a polar liquid, consisting of molecules which are approximately tetrahedral in shape and have electric polarity of opposite signs localized near a pair of opposite edges. Ammonia is supposed to have a somewhat similar molecular constitution. Molecular phenomena in such liquids have been described in a semi-qualitative but convincing way by LENNARD-JONES and POPLE[1]. What happens is that bonds are formed between regions of different molecules with opposite polarity. These bonds are, in the liquid, constantly broken and reformed, as a result of the thermal agitation; but, at any particular instant, probably quite a high percentage of them is intact. This applies particularly near the melting point; at very much higher temperatures there may be difficulty in saying whether a bond exists, on account of its short duration. It is important to notice that the formation of polar bonds of this type does not completely fix the mutual orientation of the molecules, though the energy of any bond naturally depends on the degree of "bending". In water one can understand, in such terms, why the density of the liquid near the freezing point is greater than that of ice. In the solid state, no appreciable bending of the bonds is possible, but when the lattice structure is destroyed, the bending of the bonds

[1] J. LENNARD-JONES and J. A. POPLE: Proc. Roy. Soc. Lond., Ser. A **205**, 155 (1951). — J. A. POPLE: Proc. Roy. Soc. Lond., Ser. A **205**, 163 (1951).

allows the second and third neighbours of a molecule to approach it more closely under pressure, so that the volume of the configuration in the liquid is less.

Probably the most significant macroscopic characteristic of a polar liquid is its large dielectric constant, and considerable theoretical work has been devoted to the study of the dielectric properties of liquids. One of the things which has to be taken into account is the tendency of polar molecules to further polarize one another, so that the degree of polarization depends very sensitively on the structure and therefore on the temperature and density of the liquid. For a general account of results obtained up to the year 1949, the reader is referred to FRÖHLICH's monograph[1]. If the molecules of a liquid have a dipole moment which can be localized at the mean centre of the molecules, KIRKWOOD[2] has shown that the dielectric constant depends on the mean cosine of the angle between two molecules at a given distance, as well as the mean number of molecules at this distance from a given molecule. SABRY[3] has described a method for calculating these mean values approximately, and POPLE[4] later published a somewhat different method. Both authors were able to account satisfactorily for the abnormally large dielectric constant of water, residual discrepancies being attributed to the fact that the water molecules are not truly point-dipoles, as assumed by KIRKWOOD's theory. JACOBS and LAWSON[5] have examined the pressure dependence of the dielectric constant on the basis of several rather crude models, with results which rather support POPLE's theory of the bending of the intermolecular bonds, under the influence of pressure, in polar liquids.

Another important characteristic of a polar liquid is the abnormal depth of its surface zone, in which the structure of the liquid is modified to conform with the physical boundary conditions. There is a tendency for solid boundaries, especially those formed by polar or other associated substances, to orient the molecules of the liquid near the boundary, in a direction bearing a special relation to the orientation of the boundary itself. These oriented molecules in turn orient those somewhat further from the boundary, and under favourable conditions the orientation may extend for very many intermolecular distances from the surface. Similar effects can be induced by the local anisotropy at the surface between a liquid and its vapour. The extreme of these phenomena is found in liquid crystals, where the special orientation of the molecules extends throughout the liquid, which therefore exhibits anisotropic properties in bulk, though it remains fluid and does not possess the long range spatial order found in real crystals. Thermal agitation tends to disturb the co-operative orientation of the molecules, so that such effects are modified on increasing the temperature of the liquid.

A comprehensive review of the experimental evidence pertaining to deep surface orientation of the molecules in polar liquids has been given by HENNIKER[6]. This will be discussed in some detail in Sect. 28, but it should be mentioned here that, as well as in liquid crystals, optical anisotropy, abnormally high viscosity, great tensile strength, and various other abnormal properties are found at distances up to thousands of angstrom from the surface of a variety of polar liquids. The evidence amounts to a striking demonstration of the dependence of various physical properties on molecular structure, as well as the extensive effect of a local source of anisotropy in orienting the molecules of polar liquids.

---

[1] H. FRÖHLICH: Theory of Dielectrics. Clarendon Press: Oxford 1949.
[2] J. G. KIRKWOOD: J. Chem. Phys. **7**, 911 (1939).
[3] A. A. SABRY: Proc. Phys. Soc. Lond. A **63**, 716 (1950).
[4] See footnote 1, p. 21.
[5] I. S. JACOBS and A. W. LAWSON: J. Chem. Phys. **20**, 1161 (1952).
[6] J. C. HENNIKER: Rev. Mod. Phys. **21**, 322 (1949).

The orientation of the individual molecules may be of significance on account of their shape as well as on account of polar forces between different molecules. If the molecules are not approximately spherical, it is impossible to represent the interaction of two molecules by central forces depending only on their separation; and one would therefore expect liquids consisting of such molecules to have some of the characteristics of polar liquids. It has to be remembered, however, that dipolar forces have a much larger range than the van der Waals forces; so that cooperative effects are much more in evidence in polar liquids than liquids in which other types of molecular asymmetry are present.

β) Ionic and associated liquids. It has been mentioned in Sect. 2 that chemical linkage may be responsible for a somewhat different type of intermolecular force from that found in polar liquids. The distinction arises from the fact that the chemical bond is capable of saturation, and once formed between two molecules is no longer effective in binding other molecules. The attraction between two ions of opposite charge has a similar property, and will therefore be appropriately regarded as a special type of chemical bond. The forces between dipolar molecules, where the polarity is localized in certain regions near the "surface" of each molecule, are also similar, and it is therefore quite legitimate to regard at least certain polar liquids as associated liquids: the adjectives "polar" and "associated", though not synonymous are also not mutually exclusive. The term "associated liquid" will, however, here be reserved for a liquid in which the intermolecular forces possess the property of saturation which is especially associated with chemical linkage. It is essential that the energy of the chemical bond in the liquid environment should not be greater than the fluctuations in the thermal energy; otherwise the association would be permanent and the liquid should be regarded as consisting of complex molecules formed by the association. Sometimes the association is almost permanent at sufficiently low temperatures but not at higher temperatures; in the transition zone the liquid can be regarded as a mixture of different components consisting of the dissociated and associated molecules, the chemical equilibrium of which depends on the temperature.

The chemical bond is, according to the theory of valency, a consequence of the fact that certain molecules possess unpaired electrons, meaning that, of the two spin states available to the electrons in the outermost shells surrounding the molecule, only one is occupied. The nature of the resulting association depends on the number of such unpaired electrons. When there is only one, the molecules merely associate in pairs; when there are two, they may associate in chains and rings; when there are more than two, very complicated linkages are possible, such as are found in the glasses. Irregular structures formed by association tend to persist when the temperature is lowered and inhibit the possibility of freezing. Associated liquids can, in fact, often be obtained in a supercooled state, which, though technically metastable, is remarkably persistent; acetic acid is a well known example.

As might be expected, the special structure of associated liquids is reflected in their physical properties. The viscosity of a liquid is a particularly good index of the degree of association on the molecular scale; so is the surface tension, and the vapour pressure. It is not hard to understand in a qualitative way the connection between these macroscopic quantities and the molecular configuration; but quantitatively little has been achieved to estimate or predict them.

γ) Mixtures. Two liquids are miscible, provided the energy of the mixture is less under the same conditions of temperature and pressure or volume than that of the separate liquids. Since the energy of a liquid or liquid mixture is due primarily to the potential energy arising from the intermolecular forces, and this in turn depends on the actual structure of the liquid, the very existence of the mixture is closely connected with its structural properties. On mixing, most liquids suffer a change of volume and entropy, depending on their relative concentrations, which it is the endeavour of the statistical molecular theory to predict. Although the statistical mechanics of liquid mixtures has received much attention, the exact connection between their molecular structure and thermodynamical properties was first investigated in detail by MCMILLAN and MAYER[1], and somewhat more simply by KIRKWOOD and BUFF[2]. For the purpose of making numerical predictions, however, it is necessary to resort to the model theories. Early calculations were mostly based on the crude lattice model, and in addition to the unsatisfactory features already mentioned in Sect. 4, led to the conclusion that the entropy of mixing should always be negative, in contradiction with experiment. LONGUET-HIGGINS[3] showed that this incorrect result was a feature of the model, and that more refined considerations could satisfactorily correlate much of the experimental data on the heats, entropies and volume changes of mixing of a wide variety of liquids. Improved calculations have been made with cell models of the type of LENNARD-JONES and DEVONSHIRE, by PRIGOGINE and his collaborators[4], ROWLINSON[5], and POPLE[6]. The results so obtained are qualitatively satisfactory, but the model employed will have to be still further refined to be quantitatively reliable. On the other hand, where one or both of the liquids contained in a mixture is a polar or associated liquid, a refined model is hardly useful at present. POPLE[7] and MÜNSTER[8] have both made useful contributions to the theory of mixtures in which the orientations of the molecules are important, as they are for example in polar liquids. Mixtures of associated liquids have been treated on the basis of a simple lattice model by BARKER and SMITH[9].

Outside the thermodynamical properties of mixtures, the irreversible processes of diffusion, including thermal diffusion (the Soret effect) and diffusion under external forces have to be considered. These effects are directly related to the distribution of molecular velocities rather than the actual molecular structure, but the velocity and spatial distributions of the molecules are so intimately related that it is impossible to ignore one in relation to the other. It is, in fact, the closely packed structure of liquids which impedes the translatory motion of the molecules and makes diffusion a slow process in comparison to gaseous diffusion. Most of the work devoted to the understanding of diffusion in liquids has been based on semi-phenomenological theories, especially the "thermodynamics of irreversible processes"[10]. For an account of this work,

---

[1] W. G. MCMILLAN and J. E. MAYER: J. Chem. Phys. **13**, 276 (1945).

[2] J. G. KIRKWOOD and F. P. BUFF: J. Chem. Phys. **19**, 774 (1951).

[3] H. C. LONGUET-HIGGINS: Proc. Roy. Soc. Lond., Ser. A **205**, 247 (1951).

[4] I. PRIGOGINE and G. GARIKAN: Physica, Haag **16**, 239 (1950); I. PRIGOGINE and V. MATHOT: J. Chem. Phys. **20**, 49 (1952); I. PRIGOGINE and A. BELLEMANS: J. Chem. Phys. **21**, 561 (1953).

[5] J. S. ROWLINSON: Proc. Roy. Soc. Lond., Ser. A **214**, 192 (1952).

[6] J. A. POPLE: Trans. Faraday Soc. **49**, 591 (1953).

[7] J. A. POPLE: Disc. Faraday. Soc. **15**, 35 (1953).

[8] A. MÜNSTER: Trans. Faraday Soc. **46**, 165 (1950).

[9] J. A. BARKER: J. Chem. Phys. **20**, 1526 (1952). — J. A. BARKER and F. SMITH: J. Chem. Phys. **22**, 375 (1954).

[10] S. R. DE GROOT: Thermodynamics of Irreversible Processes. Amsterdam: North-Holland Publishing Co. 1951. — In this Encyclopedia, J. MEIXNER in Vol. III, part 2.

the reader is referred to the papers of DENBIGH[1], RUTHERFORD and DRICKAMER[2], WIRTZ and HIBY[3] and PRIGOGINE, DE BROUKERE and AMAND[4]. The first attempt to construct a rigorous molecular theory was due to YANG[5], who compared the motion of molecule through the liquid to that of a Brownian particle. The thermal motion of each molecule is continually changed by the forces exerted by the surrounding molecules, and the statistical effect of these forces can only be determined by very complicated mathematical analysis. A concise account of the progress achieved will be given in Sect. 30 below.

II. The quantitative description of liquid structure.

a) Distribution functions and their properties.

11. Molecular distribution functions. The evidence reviewed in the previous chapter has allowed one general conclusion: that the molecular structure of liquids is not invariable, but changes from place to place and from time to time in a disorderly way. It can therefore only be described statistically, and its quantitative description is entirely in terms of probabilities. A probability is a calculated quantity, which is intended to predict the behaviour of a physical system. The prediction is as follows. Suppose an experiment performed on the system can lead to the mutually exclusive results R_1, R_2, R_3 etc. Then to say that the probability of obtaining the result R_i is p_i is to *predict* that, if the experiment is performed a sufficiently large number of times, the fraction of this number for which the result R_i is obtained will be very nearly p_i. It is essential that the material conditions under which the experiment is performed should always be the same, for these are the conditions on which the calculation of p_i is based.

The obvious but essential properties of probabilities can be briefly summarized as follows. The sum of a complete set of probabilities for mutually exclusive results is unity: $p_1+p_2+p_3+\cdots=1$. The probability of obtaining a result R'_i from one experiment and a result R'_j from a second experiment on the same system will be the product of the probabilities p'_i and p'_j for obtaining these results separately, if and, in general, only if the result of the first experiment cannot affect the result of the other. If the results of two experiments are causally related, so that each determines the other, then the probabilities for obtaining these results will be equal. If the different possible results R_1, R_2, R_3, etc. of an experiment are expressed numerically as values of a variable R, the expectation value of R is *defined* as $p_1R_1+p_2R_2+p_3R_3+\cdots$, in terms of the probabilities. This is the predicted average value of the experimental results.

The chief application of these principles for the present will be in the definition of the molecular distribution functions. Consider a very small space element within a fluid: so small that its volume d^3x may be regarded as infinitesmal. This element will be regarded as *occupied* if it contains the mass centre of a molecule. For simplicity it will be supposed for the present that the fluid is *simple*, in that all the molecules are alike. The necessary generalization for fluid mixtures will be considered in Sect. 18. Owing to the strong repulsion which exists between two molecules at close approach, it is practically impossible that an element should be occupied by more than one molecule, in the sense described above.

[1] K. G. DENBIGH: Trans. Faraday Soc. **48**, 1 (1952).
[2] W. M. RUTHERFORD and H. G. DRICKAMER: J. Chem. Phys. **22**, 1157 (1954).
[3] K. WIRTZ and J. W. HIBY: Phys. Z. **44**, 369 (1943). — K. WIRTZ: Naturwiss. **31**, 349 (1943).
[4] I. PRIGOGINE, L. DE BROUKERE and R. AMAND: Physica, Haag **16**, 577, 851 (1950).
[5] L. M. YANG: Proc. Roy. Soc. Lond., Ser. A **198**, 94, 471 (1949).

Further, the probability that the element should be occupied by one molecule is proportional to the volume d^3x, and will be written $n\, d^3x$. The coefficient n is then the local molecular number density, defined as the expectation value of the number of molecules per unit volume. For, if one conducts a series of experiments under similar conditions to determine whether the space element is occupied, one will find that it does contain one molecule in a fraction $n\, d^3x$ of the total number of experiments. The average number of molecules in a finite region is $\int n\, d^3x$, where the integration extends over the three-dimensional region concerned.

Within a fluid at rest and in thermodynamical equilibrium, the molecular number density is everywhere the same, provided no external force, such as gravity, acts on the fluid. In the presence of a conservative system of forces, the number density varies in such a way (see Sect. 24) that the sum of the thermodynamic potential (GIBBS' free energy) and the potential energy of the external force system is everywhere the same. However, this applies only if there are no differences of temperature within the fluid and the fluid is at rest. Where there are temperature gradients, or the fluid is accelerated, the number density n is a function of the position $\boldsymbol{x}$ in the fluid, and the time t. Even there, however, the variation is slow by molecular standards. Clearly the number density gives little information about the detailed molecular structure. But a distribution function, denoted by n_2, can be defined in an analogous way, which is much more important in this respect.

Consider now *two* space elements, of volumes $d^3x^{(1)}$ and $d^3x^{(2)}$, situated at the points $\boldsymbol{x}^{(1)}$ and $\boldsymbol{x}^{(2)}$ respectively in the fluid. The probability that these elements are both occupied is proportional to the product $d^3x^{(1)}\, d^3x^{(2)}$, and will be written $n_2\, d^3x^{(1)}\, d^3x^{(2)}$. It is important that the probability of finding a molecule near $\boldsymbol{x}^{(2)}$ is very much affected by the presence of a molecule at $\boldsymbol{x}^{(1)}$, when these points are near together, on account of the intermolecular forces. So only when the distance $r = |\boldsymbol{x}^{(2)} - \boldsymbol{x}^{(1)}|$ is much greater than the range of the molecular interaction will n_2 approximate very nearly to the product of the molecular number densities at $\boldsymbol{x}^{(1)}$ and $\boldsymbol{x}^{(2)}$. For a fluid at a given temperature and density, in thermal and mechanical equilibrium, and free from the action of external forces, n_2 depends only on the distance r between the two molecules and will be called the radial distribution function; n_2/n^2 is the correlation function $g(r)$ introduced in Sect. 8 of the previous chapter. It has been seen that experimentally, values of the radial distribution function are easily obtained from the scattered intensities associated with X-ray or neutron diffraction.

As the temperature and density vary, the radial distribution function is found to vary in a complicated way which mirrors the changing structure of the fluid. The principal effect of an external force system is to cause a variation of the density and hence of the radial distribution function from point to point in the fluid; this variation is, however, usually very small in distances comparable with the range of the molecular interaction. Only at the boundary of the fluid may the forces exerted by the containing wall cause a serious distortion of the radial character of the function n_2. When a fluid is subjected to shearing stresses, there are also small but important departures from radial symmetry in the molecular structure, as a result of the non-uniform motion which follows. Under such conditions it is, in fact inappropriate to refer to n_2 as a radial distribution function, and it will be called simply a molecular distribution function. Changes in molecular structure are almost invariably associated with changes in the macroscopic properties of the fluid. The deformation of the molecular structure by non-uniform fluid motion, for example, is associated with momentum transfer which can be observed macroscopically as viscosity (Sect. 33). Clearly the molecular distribution function n_2 provides an important quantitative measure of effects of this kind.

The number density n and the molecular distribution function n_2 are only the first two of a hierarchy of molecular distribution functions $n_1 = n$, n_2, n_3, n_4, etc. The properties of these functions were first studied in detail by YVON[1]. The function n_q is defined by allowing $n_q\, d^3x^{(1)}\, d^3x^{(2)} \ldots d^3x^{(q)}$ to denote the probability that each of the volume elements $d^3x^{(1)}, d^3x^{(2)}, \ldots d^3x^{(q)}$, situated at the points $\boldsymbol{x}^{(1)}, \boldsymbol{x}^{(2)}, \ldots \boldsymbol{x}^{(q)}$, should be occupied in the sense already explained. If the fluid is in thermal and mechanical equilibrium, and there are no external forces, n_q is a function of the distances $|\boldsymbol{x}^{(i)} - \boldsymbol{x}^{(j)}|$ between the volume elements. It is worth noticing too that it is always a symmetrical function of the positions $\boldsymbol{x}^{(1)} \ldots \boldsymbol{x}^{(q)}$, because the interchange of $\boldsymbol{x}^{(i)}$ and $\boldsymbol{x}^{(j)}$ corresponds merely to a change in the order in which the space elements are enumerated, in defining n_q.

It is interesting to consider the behaviour of n_q when q becomes very large. It is sometimes supposed that one knows exactly the number of molecules in the fluid, and if q exceeded this number n_q would then obviously have to vanish. In most of the subsequent considerations however, it will be supposed that one knows the molecular number density throughout the fluid, and therefore only the average total number of molecules. When q considerably exceeds this average total number, n_q will become extremely small but will never completely vanish, as there will exist a very small chance of a fluctuation, however large, from the average. The advantage of this point of view is that it permits the use of powerful methods, closely analogous to those of the grand partition function in statistical mechanics.

12. Velocity distribution functions. So far attention has been confined to the description in statistical terms of the instantaneous structure of a fluid. This is sufficient for the problems of thermal and mechanical equilibrium, since although the local molecular structure of an actual fluid is always changing, such changes disappear on averaging and do not affect the macroscopic behaviour of the fluid. But when one wishes to consider the molecular structural changes which accompany departures from equilibrium, and are intimately associated with irreversible processes, it will be essential to take into account the statistics of the molecular velocities. For this reason, it is necessary to introduce a second set of distribution functions, called velocity distribution functions. The simplest of these, denoted by f, is defined in relation to a space element d^3x, situated at the point $\boldsymbol{x}$, and an elementary range of velocities $d^3\xi$ in the neighbourhood of the velocity $\boldsymbol{\xi}$. It has the property that $f\, d^3x\, d^3\xi$ is the probability of finding the space element d^3x occupied by a molecule with its velocity in the elementary range $d^3\xi$. The function so defined is completely equivalent to that introduced by BOLTZMANN and MAXWELL in formulating the kinetic theory of gases[2]. One has obviously the relation

$$\int f\, d^3\xi = n\,. \tag{12.1}$$

Also the average velocity $\boldsymbol{u}$ of a molecule situated at the point $\boldsymbol{x}$ is given by

$$\int f\, \boldsymbol{\xi}\, d^3\xi = n\boldsymbol{u}; \tag{12.2}$$

and, if m is the molecular mass, its average kinetic energy is

$$\tfrac{1}{2}\, m\, n^{-1} \int f\, \xi^2\, d^3\xi\,.$$

---

[1] J. YVON: La Théorie Statistique des Fluides et l'Equation d'Etat. (Actualités Scientifiques et Industrielles No. 203.) Paris: Hermann & Cie. 1935.

[2] See S. CHAPMAN and T. G. COWLING: The Mathematical Theory of Non Uniform Gases. Cambridge 1939. Especially the Historical Summary (p. 380.)

From classical statistical mechanics, this is known to be $\frac{3}{2}kT$, where k is BOLTZMANN's constant and T is the local thermodynamic temperature on KELVIN's scale. So, for a liquid at rest,

$$m\int f\,\xi^2\,d^3\xi = 3\,n\,kT. \tag{12.3}$$

The quantum theory (see Sect. 47) predicts deviations from this last formula, which, however, only become appreciable at abnormally low temperatures. To anyone familiar with quantum mechanics it might appear unlikely that the function f could be defined without violating the uncertainty principle. However, it will be shown in Sect. 47 that a function f always exists such that the average value $\langle\varphi\rangle$ of any function $\varphi(\xi)$ of the velocity ξ of a molecule at the point $\boldsymbol{x}$ is given by

$$\int f\,\varphi(\xi)\,d^3\xi = n\langle\varphi\rangle. \tag{12.4}$$

In the quantum theory one merely has to relinquish the interpretation of f as a probability density in phase space; this is necessary, because f can become negative in some parts of phase space.

The velocity distribution function f, like the molecular number density n, is the lowest member of a hierarchy of functions $f_1 = f, f_2, f_3, \ldots$ extending to infinity. The definition of f_q is fairly obvious; in the classical theory it has the property that $f_q\,d^3x^{(1)}\,d^3\xi^{(1)}\ldots d^3x^{(q)}\,d^3\xi^{(q)}$ is probability of finding the space elements $d^3x^{(1)}\ldots d^3x^{(q)}$ occupied by molecules with velocities in the elementary ranges $d^3\xi^{(1)}\ldots d^3\xi^{(q)}$ respectively. The function f_q obviously satisfies

$$\overset{(q)}{\int\cdots\int} f_q\,d^3\xi^{(1)}\ldots d^3\xi^{(q)} = n_q, \tag{12.5}$$

and the mean velocity $\boldsymbol{u}_q^{(i)}$ of a molecule at $\boldsymbol{x}^{(i)}$, in an identified cluster whose positions are $\boldsymbol{x}^{(1)}\ldots\boldsymbol{x}^{(q)}$, is given by

$$\overset{(q)}{\int\cdots\int} f_q\,\xi^{(i)}\,d^3\xi^{(1)}\ldots d^3\xi^{(q)} = n_q\,\boldsymbol{u}_q^{(i)}. \tag{12.6}$$

For a fluid in thermal and mechanical equilibrium, the form of the function f_q can be written down immediately with the help of statistical mechanics. Since the total kinetic energy of a set of q molecules with velocities $\xi^{(1)}\ldots\xi^{(q)}$ is $\frac{1}{2}m(\xi^{(1)2}+\cdots+\xi^{(q)2})$, and their mean potential energy depends only on their positions, the velocity dependence of f_q will be described by a factor

$$\exp\{-\tfrac{1}{2}\beta m(\xi^{(1)2}+\cdots+\xi^{(q)2})\}$$

where $\beta = (kT)^{-1}$, and the remaining factor can be inferred from (12.5) above. So

$$f_q = \left(\frac{\beta m}{2\pi}\right)^{\frac{3}{2}} n_q \exp\{-\tfrac{1}{2}\beta m\,(\xi^{(1)2}+\cdots+\xi^{(q)2})\}. \tag{12.7}$$

The mean velocity $\boldsymbol{u}_q^{(i)}$ calculated from (12.6) and (12.7) vanishes, irrespective of the positions of the molecules. However, if the fluid is not at rest in mechanical equilibrium, (12.7) no longer holds; and even if the fluid is at rest it will fail if thermal equilibrium is not attained. In fact, it will appear later (Sect. 32) that thermal conduction in the fluid results in non-vanishing values of $u_2^{(1)}$ and $u_2^{(2)}$, as defined by (12.6).

13. Molecular statistics in a finite region. In this section interest will be centred on a finite region of a simple fluid, which will generally be regarded, though this is not essential, as part of a volume of fluid of much greater extent. One wishes to know the probability of finding exactly q molecules within the volume v, and further the probability that these molecules will occupy specified space elements and perhaps also that their velocities should fall within specified elementary ranges. Some of the work of MAYER[1] has been devoted to answering questions of this type, within the framework of statistical mechanics. Here a treatment will be given which avoids as far as possible the assumption of equilibrium conditions in the fluid.

It will be found convenient to define a new set of molecular distribution functions N_q in such a way that $N_q\, d^3x^{(1)} \ldots d^3x^{(q)}$ is the probability of finding the volume elements $d^3x^{(1)} \ldots d^3x^{(q)}$ within the finite volume v all occupied, and no other molecules within v. The previously defined functions n_q, which were quite independent of any special region of the fluid, can evidently be expressed in terms of the N_q; one has

$$n_q = N_q + \int N_{q+1}\, d^3x^{(q+1)} + \frac{1}{2!} \iint N_{q+2}\, d^3x^{(q+1)}\, d^3x^{(q+2)} + \cdots \tag{13.1}$$

where the spatial integrations are limited to the volume v. If one writes

$$N_{q,j} = \overset{(j)}{\int \cdots \int} N_{q+j}\, d^3x^{(q+1)} \ldots d^3x^{(q+j)}, \tag{13.2}$$

the above relation may be written concisely:

$$n_q = \sum_j N_{q,j}/j! \tag{13.3}$$

where the summation is over all values of j, and $N_{q,0}$ means the same as N_q. This relation holds also for $q=0$, provided $N_{0,j}/j!$ is the probability that the volume v should be occupied by precisely j molecules, independent of their position. It is to be expected that the terms of the summation on the right-hand side of (13.3) will increase to a very sharp maximum in the neighbourhood of $j = nv$, if v is of macroscopic magnitude. This conjecture will, in fact, shortly be confirmed.

If one defines, by analogy with (13.2),

$$n_{q,k} = \overset{(j)}{\int \cdots \int} n_{q+k}\, d^3x^{(q+1)} \ldots d^3x^{(q+j)}, \tag{13.4}$$

it follows from (13.3) that

$$n_{q,k} = \sum_j N_{q,j+k}/j!, \tag{13.5}$$

and, further, that, whatever the value of x,

$$\sum_k n_{q,k}\, x^k/k! = \sum_j N_{q,j}\,(x+1)^j/j!. \tag{13.6}$$

Setting $x = -1$ in this equation, one obtains

$$N_q = \sum_k (-1)^k\, n_{q,k}/k!. \tag{13.7}$$

This is the solution of the set of simultaneous equations (13.3) and expresses every N_q in terms of the distribution functions of the previous section. It follows further that

$$N_{q,j} = \sum_k (-1)^k\, n_{q,j+k}/k! \tag{13.8}$$

and also that the probability a_j of finding precisely j molecules in the volume v is

$$a_j = N_{0,j}/j! = \sum_k (-1)^k\, n_{0,j+k}/(j!\,k!). \tag{13.9}$$

[1] J. E. MAYER: J. Chem. Phys. 10, 629 (1942).

The last formula enables one to calculate the average density directly, and also the magnitude of fluctuations in density. The average number of molecules in the region considered is

$$\sum_j j\, a_j = \sum_{j,k} \frac{(-1)^k n_{0,j+k}}{(j-1)!\, k!} = n_{0,1}, \tag{13.10}$$

as it should be, and the mean square number of molecules is

$$\sum_j j^2 a_j = \sum_{j,k} \left\{ \frac{(-1)^k}{k!\,(j-1)!} + \frac{(-1)^k}{k!\,(j-2)!} \right\} n_{0,j+k} = n_{0,1} + n_{0,2}. \tag{13.11}$$

The mean square fluctuation in the number of molecules from the average is thus $n_{0,1} + n_{0,2} - n_{0,1}^2$ and the mean square fluctuation in density is

$$v^{-2} \int \{n^{(1)} + \int (n_2 - n^{(1)} n^{(2)})\, d^3x^{(2)}\}\, d^3x^{(1)} \tag{13.12}$$

where $n^{(1)}$ and $n^{(2)}$ are the molecular number densities at $\boldsymbol{x}^{(1)}$ and $\boldsymbol{x}^{(2)}$ respectively. The integral within the brackets is independent of the volume v, provided this is sufficiently large, because n_2 approaches the value $n^{(1)} n^{(2)}$ when $\boldsymbol{x}^{(1)}$ and $\boldsymbol{x}^{(2)}$ are far apart. The principal integrand of (13.12) is therefore a function of the local density and temperature, which normally varies only slowly with $\boldsymbol{x}^{(1)}$; and the whole expression is inversely proportional to v. The formula was due in principle to ORNSTEIN and ZERNIKE[1] who discussed also the relation of the density fluctuations to the scattering of light; however, the systematic treatment in terms of the molecular distribution function can be attributed to YVON[2].

The remaining considerations of this section apply only under equilibrium conditions, since they require the use of statistical mechanics. This allows one to write down explicit formulae for the N_q. For instance, the energy required to remove all the molecules from a region of volume v is $p v$, where p is the pressure; so the probability of finding the volume empty is

$$N_0 = \exp(-\beta p v). \tag{13.13}$$

The relation

$$N_1 = z N_0 \tag{13.14}$$

may be regarded as defining the thermodynamical activity z. If q is any small number one will have

$$N_q = z^q N_0 \exp(-\beta \Phi_q) \tag{13.15}$$

where Φ_q is the potential energy of the group of q molecules. Unfortunately the convergence of the series (13.3) is so poor at liquid densities, particularly at low temperatures, that substitution from (13.13) and (13.15) does not provide an effective method of calculating the molecular distribution functions. Even for condensed gases, and the region above the critical point, some transformations are desirable which will be described in the following section.

14. The theory of clusters. Finally a set of functions will be introduced, depending, like the distribution functions n_q and N_q, on a number of points in the fluid, but having the property that (except for $q = 1$) they vanish if any two

[1] L. S. ORNSTEIN and F. ZERNIKE: Amst. Proc. **17**, 793 (1914). — F. ZERNIKE: Amst. Proc. **18**, 1520 (1916). — L. S. ORNSTEIN and F. ZERNIKE: Phys. Z. **19**, 134 (1918); **27**, 761 (1926).

[2] J. YVON: Actualités Scientifiques et Industrielles No. 542. Paris: Hermann & Cie. 1937.

of the points are far apart. This set of functions is defined inductively by

$$\left.\begin{aligned} n_1 &= l_1 \\ n_2 &= l_2 + l_1^{(1)}\, l_1^{(2)} \\ n_3 &= l_3 + l_2^{(23)}\, l_1^{(1)} + l_2^{(31)}\, l_1^{(2)} + l_2^{(12)}\, l_1^{(3)} + l_1^{(1)}\, l_1^{(2)}\, l_1^{(3)} \\ &\dots\dots\dots\dots \\ n_q &= \sum_{\text{part}} \prod l_s \end{aligned}\right\} \tag{14.1}$$

where the superfixes indicate where required the positions on which the functions depend, and $\sum\limits_{\text{part}} \prod$ indicates a sum over all partitions of the points between the factors of the product. The important property of l_q so defined is that the integral

$$l_{1,q-1} = \overset{(q-1)}{\int \dots \int} l_q\, d^3x^{(2)} \dots d^3x^{(q)} \tag{14.2}$$

is practically independent of the volume v, provided this is sufficiently large, and $l_{0,q}$ is approximately proportional to the volume. [The integrand of (13.12) could be written $l_1 + l_{1,1}$, and the integral $l_{0,1} + l_{0,2}$.]

If the expression (14.1) for n_q is integrated to form $n_{0,q}$, there results

$$n_{0,q} = q!\, \sum{}' \prod_s \frac{1}{k_s!} \left(\frac{l_{0,s}}{s!}\right)^{k_s}, \tag{14.3}$$

where $\sum'$ denotes summation over all products for which $\sum\limits_s s k_s = q$, and $l_{0,0}$ is zero if it appears at all. So, whatever the value of x,

$$\sum_q n_{0,q}\, x^q/q! = \exp\Bigl(\sum_s l_{0,s}\, x^s/s!\Bigr). \tag{14.4}$$

It follows in a similar way that

$$\sum_q n_{1,q}\, x^q/q! = \Bigl(\sum_q l_{1,q}\, x^q/q!\Bigr) \exp\Bigl(\sum_s l_{0,s}\, x^s/s!\Bigr) \tag{14.5}$$

and

$$\sum_q n_{2,q}\, x^q/q! = \Bigl(\sum_q l'_{2,q}\, x^q/q!\Bigr) \exp\Bigl(\sum_s l_{0,s}\, x^s/s!\Bigr) \tag{14.6}$$

in which $l'_{2,q}$ is defined by

$$l'_{2,q} = l_{2,q} + \sum_s \binom{q}{s} l_{1,s}\, l_{1,q-s}. \tag{14.7}$$

Setting $x = -1$ in (14.4) and using (13.7) with $q = 0$, one has

$$N_0 = \exp\Bigl\{\sum_s (-1)^s\, l_{0,s}/s!\Bigr\}. \tag{14.8}$$

Comparing with (13.13), the formula

$$\beta p v = -\sum_s (-1)^s\, l_{0,s}/s! \tag{14.9}$$

follows for the pressure in thermodynamical equilibrium. From (14.5) there follows in a similar way

$$N_1 = N_0 \sum_q (-1)^q\, l_{1,q}/q!, \tag{14.10}$$

or, by comparison with (13.14),

$$z = \sum_q (-1)^q\, l_{1,q}/q!. \tag{14.11}$$

The value of $l_{1,1}$ can be inferred, though not yet with great accuracy, from experiments on the scattering of light, X-rays or neutrons by the fluid (see Sect. 19). Some information on the value of $l_{1,2}$ can be obtained from accurate measurements of the dielectric constant (Sect. 26); $l_{1,0}$ is, of course, just the molecular number density, and $l_{0,s}/v$ is the average value of $l_{1,s-1}$ within the volume considered. So the first few terms in the series (14.9) and (14.11) may be regarded as known. The indications are that these series converge fairly quickly, even in a liquid, but the experiments mentioned do not provide enough information to afford a good test of these formulae.

Considerably more use has been made of a set of formulae involving the functions L_s, defined by

$$N_q = N_0 \sum_{\text{part}} \prod L_s, \tag{14.12}$$

which is obviously analogous to (14.1). Analogues of (14.4), (14.5) and (14.6) are easily written down, as follows:

$$\sum_q N_{0,q}\, y^q/q! = N_0 \exp\Big(\sum_s L_{0,s}\, y^s/s!\Big); \tag{14.13}$$

$$\sum_q N_{1,q}\, y^q/q! = N_0 \Big(\sum_q L_{1,q}\, y^q/q!\Big) \exp\Big(\sum_s L_{0,s}\, y^s/s!\Big), \tag{14.14}$$

and

$$\sum_q N_{2,q}\, y^q/q! = N_0 \Big(\sum_q L'_{2,q}\, y^q/q!\Big) \exp\Big(\sum_s L_{0,s}\, y^s/s!\Big) \tag{14.15}$$

where

$$L'_{2,q} = L_{2,q} + \sum_s \binom{q}{s} L_{1,s} L_{1,q-s}. \tag{14.16}$$

Setting $y=1$ in (14.13) and using (13.5), one obtains

$$N_0 \exp\Big(\sum_s L_{0,s}/s!\Big) = 1, \tag{14.17}$$

or, by virtue of (13.13),

$$\beta p v = \sum_s L_{0,s}/s!. \tag{14.18}$$

Similarly one obtains from (14.14), with the use of (13.3) and (14.18),

$$n_1 = \sum_q L_{1,q}/q! \tag{14.19}$$

while from (14.15) it follows that

$$\left.\begin{aligned} n_2 &= \sum_q L'_{2,q}/q! \\ &= \sum_q L_{2,q}/q! + n_1^{(1)} n_1^{(2)}. \end{aligned}\right\} \tag{14.20}$$

The series obtained in this way are quickly convergent at gaseous densities, but below the critical temperature the number of terms required to obtain accurate results for the liquid is of the same order as the number of molecules in the region considered; they are therefore of theoretical interest only for the liquid phase. Practical methods of computing the equation of state and the radial distribution function for the liquid will be considered in the next chapter (Sects. 20 to 23). It is worth, in the meantime, noticing the physical significance of the terms of the series (14.18) to (14.20). The term $L_{0,q}/q!$ of (4.18), for example, represents a contribution to the pressure of clusters of q molecules in the region considered, which are sufficiently isolated from their neighbours not to interact with them appreciably. The predominating size of such clusters will obviously increase enormously in the process of condensation, a fact which is the basis of MAYER's theory of condensation[1]. The terms of the series (14.9) and (14.10), on the other hand, represent contributions of molecular clusters not necessarily isolated from their neighbours, and there is no reason to believe that the character and rate of convergence of the series are drastically different in the liquid and the gas.

b) Molecular mechanics and liquid structure.

15. Momentum density and flux. In classical hydrodynamics, the conservation of momentum within a liquid is expressed by the equation of motion of the liquid. The velocity of flow $\boldsymbol{u}$ is regarded as a function of position and time, and its rate of change following the motion is then

$$\frac{d}{dt}\boldsymbol{u} = \frac{\partial}{\partial t}\boldsymbol{u} + \boldsymbol{u}\cdot\frac{\partial}{\partial \boldsymbol{x}}\boldsymbol{u}. \tag{15.1}$$

There may be external forces such as gravity acting on the body of the liquid; the external force per unit mass will be denoted by $\boldsymbol{F}$. In addition there is the internal force arising from the stress distribution, which is described by means of the pressure tensor p with the nine components $p_{11}, p_{12}, p_{13}, p_{21}, \ldots, p_{33}$. This tensor is symmetrical, so that $p_{ij} = p_{ji}$, and the internal force per unit volume is the vector with components

$$\left(\frac{\partial}{\partial \boldsymbol{x}}\cdot\mathsf{p}\right)_i = \frac{\partial p_{1i}}{\partial x_1} + \frac{\partial p_{2i}}{\partial x_2} + \frac{\partial p_{3i}}{\partial x_3}. \tag{15.2}$$

The equation of motion of the liquid may therefore be written

$$\varrho\frac{d}{dt}\boldsymbol{u} + \frac{\partial}{\partial \boldsymbol{x}}\cdot\mathsf{p} = \varrho\boldsymbol{F}, \tag{15.3}$$

where ϱ is the mass density. It is the purpose of this section to analyse the implications of the above equation on the molecular scale. The procedure which will be adopted is to formulate the law of conservation of momentum as it applies directly to a group of molecules, and by comparison with (15.3) to express the pressure tensor in terms of the intermolecular forces and the structure of the liquid. It will be assumed for the present that the molecules are all alike and that the mean mutual potential energy of a pair of molecules is a function $\phi(r)$ of the distance between them.

The momentum density at any particular time is $\varrho\boldsymbol{u}$, which according to Sect. 11, can also be written $m n \boldsymbol{u}$, where m is the molecular mass and n the mean number density. The mean momentum instantaneously contained in a region R of the liquid is therefore $\int_{\mathrm{R}} \varrho\,\boldsymbol{u}\,d^3x$, where $\int_{\mathrm{R}} \ldots d^3x$ denotes spatial integration restricted to R. If it is stipulated that the boundary of the region moves everywhere with the local velocity of flow, the rate of change of momentum within it is

$$\int_{\mathrm{R}} \varrho\frac{d}{dt}\boldsymbol{u}\,d^3x.$$

This change of momentum can be attributed to the external and internal forces, of which the external forces contribute $\int_{\mathrm{R}} \varrho\boldsymbol{F}\,d^3x$ per unit time. The flux of

[1] J. E. MAYER: J. Chem. Phys. 5, 67 (1937); see also M. BORN and K. FUCHS: Proc. Roy. Soc. Lond., Ser. A **166**, 391 (1938) and G. E. UHLENBECK and B. KAHN: Physica, Haag **5**, 399 (1938).

momentum through the boundary, attributed macroscopically to internal forces, is

$$\boldsymbol{M} = \int\limits_{R} \frac{\partial}{\partial \boldsymbol{x}} \cdot \mathsf{p}\, d^3 x = \int\limits_{S} \mathsf{p} \cdot \boldsymbol{dS}, \tag{15.4}$$

where $\boldsymbol{dS}$ is a vector element of the surface S enclosing R.

On the microscopic scale, it can be seen that the momentum flux (15.4) arises from two separate causes: the motion of molecules with velocities different from $\boldsymbol{u}$ across S, and the action of the molecules outside S on those within. Consider first the momentum flux $\boldsymbol{M}_{\mathrm{k}}$ due to the kinetic motion of the molecules. According to the definition of Sect. 12, the mean density of molecules with velocities in the range ξ, $d^3\xi$ is $f\, d^3\xi$, so the contribution of such molecules to the flux of momentum across the surface element $\boldsymbol{dS}$ is $m\,\xi(\xi - \boldsymbol{u}) \cdot \boldsymbol{dS}\, f\, d^3\xi$. Therefore

$$\boldsymbol{M}_{\mathrm{k}} = \iint\limits_{S} f\, m\, \xi\, (\xi - \boldsymbol{u}) \cdot \boldsymbol{dS}\, d^3\xi. \tag{15.5}$$

In addition, there is the contribution $\boldsymbol{M}_{\mathrm{p}}$ to $\boldsymbol{M}$ from the action of the intermolecular forces. The probability that two volume elements d^3x and d^3y at the points $\boldsymbol{x}$ and $\boldsymbol{y}$ respectively are both occupied is defined in Sect. 11 as $n_2(\boldsymbol{x}, \boldsymbol{y}) \times d^3x\, d^3y$, so the mean force acting on the volume element d^3x, due to the intermolecular forces, is

$$\left\{\int \frac{\partial}{\partial \boldsymbol{y}} \phi(|\boldsymbol{y} - \boldsymbol{x}|)\, n_2(\boldsymbol{x}, \boldsymbol{y})\, d^3y\right\} d^3x. \tag{15.6}$$

There is, consequently, a loss of momentum in R at the rate

$$\boldsymbol{M}_{\mathrm{p}} = -\int\limits_{R} \left\{\int \frac{\partial \phi(r)}{\partial \boldsymbol{r}}\, n_2(\boldsymbol{x}, \boldsymbol{x} + \boldsymbol{r})\, d^3r\right\} d^3x. \tag{15.7}$$

This integral can be transformed to an integral over the surface S, by expressing n_2 as a function of

$$\left.\begin{aligned} \boldsymbol{r} = \boldsymbol{y} - \boldsymbol{x} \quad \text{and} \quad \boldsymbol{X} &= \tfrac{1}{2}(\boldsymbol{x} + \boldsymbol{y}): \\ n_2(\boldsymbol{x}, \boldsymbol{y}) &= \bar{n}_2(\boldsymbol{r}, \boldsymbol{X}). \end{aligned}\right\} \tag{15.8}$$

As the variation of $\bar{n}_2$ with the position of the mean centre $\boldsymbol{X}$ is under most conditions very slow by macroscopic standards, one can perform the expansion

$$\bar{n}_2(\boldsymbol{r}, \boldsymbol{X}) = \bar{n}_2(\boldsymbol{r}, \boldsymbol{x}) + \frac{1}{2}\boldsymbol{r} \cdot \frac{\partial}{\partial \boldsymbol{x}} \bar{n}_2(\boldsymbol{r}, \boldsymbol{x}) + \cdots \tag{15.9}$$

and neglect the terms not shown. It is worth noticing that since $n_2(\boldsymbol{x}, \boldsymbol{y}) = n_2(\boldsymbol{y}, \boldsymbol{x})$ by definition, $\bar{n}_2(\boldsymbol{r}, \boldsymbol{x}) = \bar{n}_2(-\boldsymbol{r}, \boldsymbol{x})$, and $\bar{n}_2(\boldsymbol{r}, \boldsymbol{x})$ is an even function of $\boldsymbol{r}$. Thus, when (15.9) is substituted into (15.7), the first term gets multiplied by an odd function of $\boldsymbol{r}$ and gives a vanishing contribution to the integral:

$$\left.\begin{aligned} \boldsymbol{M}_{\mathrm{p}} &= -\int\limits_{R} \left\{\frac{1}{2} \frac{\partial}{\partial \boldsymbol{x}} \cdot \int \boldsymbol{r} \frac{\partial \phi}{\partial \boldsymbol{r}}\, \bar{n}_2(\boldsymbol{r}, \boldsymbol{x})\, d^3r\right\} d^3x \\ &= -\int\limits_{S} \left\{\frac{1}{2} \int \frac{\partial \phi}{\partial \boldsymbol{r}}\, \boldsymbol{r}\, \bar{n}_2(\boldsymbol{r}, \boldsymbol{x})\, d^3r\right\} \cdot \boldsymbol{dS}. \end{aligned}\right\} \tag{15.10}$$

By comparison of (15.4) with (15.5) and (15.10), it is seen that

$$\mathsf{p} = m \int \xi(\xi - \boldsymbol{u})\, f\, d^3\xi - \frac{1}{2} \int \frac{\partial \phi}{\partial \boldsymbol{r}}\, \boldsymbol{r}\, \bar{n}_2\, d^3r. \tag{15.11}$$

It is worth emphasing that this result depends on the legitimacy of cutting off the series (15.9) at the second term. In the treatment of normal steady-state phenomena no doubts concerning this procedure need be entertained, but if shock waves or ultrasonics are being considered, it is possible that (15.11) will lead to error. Also at the surface of a liquid, the variation of $\bar{n}_2(\boldsymbol{r}, \boldsymbol{x})$ with $\boldsymbol{x}$ is likely to be so rapid that the expansion effected in (15.9) converges slowly; the use of (15.11) in connection with surface phenomena therefore requires careful justification. The formula can, however, be made exact by substituting the integral expression

$$\text{“}\bar{n}_2(\boldsymbol{r}, \boldsymbol{x})\text{”} = \frac{1}{(2\pi)^3} \iint \frac{\sin(\frac{1}{2}\boldsymbol{r}\cdot\boldsymbol{s})}{(\frac{1}{2}\boldsymbol{r}\cdot\boldsymbol{s})} \exp\{i(\boldsymbol{X}-\boldsymbol{x})\cdot\boldsymbol{s}\}\bar{n}_2(\boldsymbol{r}, \boldsymbol{X})\, d^3X\, d^3s \tag{15.12}$$

for $\bar{n}_2(\boldsymbol{r}, \boldsymbol{x})$.

The pressure in the liquid is defined as $\frac{1}{3}(p_{11}+p_{22}+p_{33})$, *i.e.*, as the mean of the diagonal components of the pressure tensor, and is therefore

$$p = \tfrac{1}{3} m \int (\xi - u)^2 f\, d^3\xi - \tfrac{1}{6}\int r\,\phi'(r)\, n_2\, d^3r \tag{15.13}$$

where $\phi'(r)$ is the scalar derivative of the function $\phi(r)$, and in forming the first term use has been made of (12.2). On account of (12.3), this formula can be reduced to

$$p = \frac{n}{\beta} - \frac{1}{6}\int r\,\phi'(r)\, n_2\, d^3r. \tag{15.14}$$

In a uniform liquid at rest n_2 is a function $n^2 g(r)$ of the distance r, and the 3-dimensional integration $\int \ldots d^3r$ reduces to $4\pi\int_0^\infty \ldots r^2\,dr$, giving the result stated in (9.3). The two terms in the right-hand side of (15.14) are usually of the same order of magnitude in a liquid, the first, of course, being somewhat greater than the second. The fact that the pressure is expressed as the difference of two large terms provides one reason why it is difficult to calculate accurately by molecular theory. Another reason is that the integral contained in the second term is also the difference between two large contributions of opposite sign, the negative contribution arising from the repulsive part of the intermolecular force field, where n_2 is small and ill-determined but $\phi'(r)$ is large and negative. An accurate computation of the equation of state of a liquid therefore requires a very precise knowledge of the radial distribution function $n_2(r)$ at such distances that the electron shells of the adjacent molecules are in contact. The diffraction experiments which provide information on $n_2(r)$ are not, in fact, sufficiently accurate at present to allow a very satisfactory test of (15.14), and it has been left to theoretical calculations (Sects. 20 to 22) to confirm the formula.

16. Energy density and flux. Just as the macroscopic equation of motion of the liquid expresses conservation of momentum, the equation of energy transfer expresses conservation of energy. The energy per unit mass of liquid depends in general on time and position, and will be denoted by E; it includes the energy of the macroscopic motion ($\frac{1}{2}\boldsymbol{u}^2$ per unit mass) and the internal energy:

$$E = \tfrac{1}{2}\boldsymbol{u}^2 + U. \tag{16.1}$$

The rate of change of E following the motion is

$$\frac{dE}{dt} = \frac{\partial E}{\partial t} + \boldsymbol{u}\cdot\frac{\partial E}{\partial \boldsymbol{x}} \tag{16.2}$$

and is partly due to the action of the external force field, partly due to the action of the internal stress system, and partly due to the flow of thermal energy by conduction etc. One has, in fact

$$\varrho \frac{dE}{dt} + \frac{\partial}{\partial \boldsymbol{x}} \cdot (\mathsf{p} \cdot \boldsymbol{u} + \boldsymbol{q}) = \varrho \boldsymbol{u} \cdot \boldsymbol{F}, \tag{16.3}$$

where the pressure tensor p and the external force per unit mass $\boldsymbol{F}$ are the same as in (15.1), and $\boldsymbol{q}$ is the thermal flux vector. The internal energy per unit mass U satisfies the somewhat simpler equation

$$\varrho \frac{dU}{dt} + \left(\mathsf{p} \cdot \frac{\partial}{\partial \boldsymbol{x}}\right) \cdot \boldsymbol{u} + \frac{\partial}{\partial \boldsymbol{x}} \cdot \boldsymbol{q} = 0, \tag{16.4}$$

which can be obtained from (16.3) by subtracting (15.3) after multiplying the latter by $\boldsymbol{u}$.

From (16.3) it is clear that the energy flux across the surface S of any region R moving with the liquid is

$$K = \int_{\mathrm{R}} \frac{\partial}{\partial \boldsymbol{x}} \cdot (\mathsf{p} \cdot \boldsymbol{u} + \boldsymbol{q})\, d^3 x = \int_{\mathrm{S}} (\boldsymbol{u} \cdot \mathsf{p} + \boldsymbol{q}) \cdot d\boldsymbol{S}. \tag{16.5}$$

Like the momentum flux $\boldsymbol{M}$, this can be analysed at the molecular level as arising from two processes: the motion of the individual molecules across the surface, and the interaction of molecules inside and outside the surface. The separate contributions of these processes will be denoted by L_{k} and L_{p} respectively.

The energy of the liquid can be regarded as the sum of the energies, kinetic and potential, of the individual molecules. However, it is necessary for this purpose to establish the convention that the mutual potential energy of a pair of similar molecules is shared equally between the two; otherwise it would be counted twice in summing the energies of the molecules. On this convention, the energy of a molecule is the sum of its kinetic energy $\frac{1}{2} m \xi^2$ and one half of its potential energy due to all other molecules. Since the mean kinetic energy of a molecule is $\frac{3}{2} kT$ and the probability of finding a molecule in a volume element $d^3 y$ at the displacement $\boldsymbol{r} = \boldsymbol{y} - \boldsymbol{x}$ from a molecule at the point $\boldsymbol{x}$ is $n_2(\boldsymbol{x}, \boldsymbol{y}) \times d^3 y/n$, this leads to the formula

$$m E = \left(\tfrac{3}{2} kT + \tfrac{1}{2} m \boldsymbol{u}^2\right) + \tfrac{1}{2} \int n_2(\boldsymbol{x}, \boldsymbol{y})\, \phi(r)\, d^3 y/n \tag{16.6}$$

for the mean energy per molecule. The rate of escape of energy from R due to the motion of molecules across the surface is

$$\left.\begin{aligned} K_{\mathrm{k}} &= \iint_{\mathrm{S}} f \cdot \tfrac{1}{2} m \xi^2 (\boldsymbol{\xi} - \boldsymbol{u}) \cdot d\boldsymbol{S}\, d^3 \xi \\ &\quad + \iint_{\mathrm{S}} n_2 \tfrac{1}{2} \phi(r) (\boldsymbol{u}_2^{(1)} - \boldsymbol{u}) \cdot d\boldsymbol{S}\, d^3 y \end{aligned}\right\} \tag{16.7}$$

where $\boldsymbol{u}_2^{(1)}(\boldsymbol{x}, \boldsymbol{y})$, according to the definition of (12.6), is the mean velocity of a molecule at $\boldsymbol{x}$ when a second molecule is at the point $\boldsymbol{y}$.

The rate of loss of energy due to the interaction of molecules inside and outside the surface must now be assessed. The mean velocities of two molecules at the points $\boldsymbol{x}$ and $\boldsymbol{y}$ are $\boldsymbol{u}_2^{(1)}(\boldsymbol{x}, \boldsymbol{y})$ and $\boldsymbol{u}_2^{(2)}(\boldsymbol{x}, \boldsymbol{y})$ respectively. The rate of working of the interaction energy $\phi(r)$ is therefore

$$\boldsymbol{u}_2^{(1)} \cdot \frac{\partial \phi}{\partial \boldsymbol{x}} - \boldsymbol{u}_2^{(2)} \cdot \frac{\partial \phi}{\partial \boldsymbol{y}}$$

and this increase of energy is shared by the two molecules. The rate of loss of energy by the molecules in R is therefore

$$K_{\mathrm{p}} = \iint\limits_{\mathrm{R}} n_2 \cdot \frac{1}{2} \left(\boldsymbol{u}_2^{(1)} + \boldsymbol{u}_2^{(2)}\right) \cdot \frac{\partial \phi}{\partial \boldsymbol{x}}\, d^3 x\, d^3 y. \tag{16.8}$$

This expression must now be transformed to a surface integral, in the same way as the right-hand side of (15.7). The variables are changed from $\boldsymbol{x}$ and $\boldsymbol{y}$ to $\boldsymbol{r} = \boldsymbol{y} - \boldsymbol{x}$ and $\boldsymbol{X} = \frac{1}{2}(\boldsymbol{x} + \boldsymbol{y})$, and $\overline{\boldsymbol{u}}_2^{(1)}$ and $\overline{\boldsymbol{u}}_2^{(2)}$ defined by

$$\boldsymbol{u}_2^{(1)}(\boldsymbol{x}, \boldsymbol{y}) = \overline{\boldsymbol{u}}_2^{(1)}(\boldsymbol{r}, \boldsymbol{X}); \qquad \boldsymbol{u}_2^{(2)}(\boldsymbol{x}, \boldsymbol{y}) = \overline{\boldsymbol{u}}_2^{(2)}(\boldsymbol{r}, \boldsymbol{X}). \tag{16.9}$$

One can then introduce the expansion

$$\left.\begin{aligned} \{\bar{n}_2(\overline{\boldsymbol{u}}_2^{(1)} + \overline{\boldsymbol{u}}_2^{(2)})\}(\boldsymbol{r}, \boldsymbol{X}) &= \{\bar{n}_2(\overline{\boldsymbol{u}}_2^{(1)} + \overline{\boldsymbol{u}}_2^{(2)})\}(\boldsymbol{r}, \boldsymbol{x}) \\ &+ \frac{1}{2}\boldsymbol{r} \cdot \frac{\partial}{\partial \boldsymbol{x}} \{\bar{n}_2(\overline{\boldsymbol{u}}_2^{(1)} + \overline{\boldsymbol{u}}_2^{(2)})\}(\boldsymbol{r}, \boldsymbol{x}) + \cdots, \end{aligned}\right\} \tag{16.10}$$

which, cut off at the second term and substituted into (16.8), gives

$$\left.\begin{aligned} K_{\mathrm{p}} &= -\int\limits_{\mathrm{R}} \left\{\frac{1}{4} \frac{\partial}{\partial \boldsymbol{x}} \cdot \int \boldsymbol{r} \frac{\partial \phi}{\partial \boldsymbol{r}} \cdot (\overline{\boldsymbol{u}}_2^{(1)} + \overline{\boldsymbol{u}}_2^{(2)})\, \bar{n}_2\, d\boldsymbol{r}\right\} d^3 x \\ &= -\int\limits_{\mathrm{S}} \left\{\frac{1}{4} \int \frac{\partial \phi}{\partial \boldsymbol{r}} \cdot (\overline{\boldsymbol{u}}_2^{(1)} + \overline{\boldsymbol{u}}_2^{(2)})\, \boldsymbol{r}\, \bar{n}_2\, d\boldsymbol{r}\right\} \cdot d\boldsymbol{S}. \end{aligned}\right\} \tag{16.11}$$

The comparison of (16.5), (16.7) and (16.8) shows that

$$\left.\begin{aligned} \boldsymbol{q} + \mathsf{p} \cdot \boldsymbol{u} &= \int \frac{1}{2} m\, \xi^2 (\xi - \boldsymbol{u})\, f\, d^3\xi + \\ &+ \frac{1}{2} \int \phi(r)\, (\boldsymbol{u}_2^{(1)} - \boldsymbol{u})\, n_2\, d^3 r - \frac{1}{4} \int \boldsymbol{r} \frac{\partial \phi}{\partial \boldsymbol{r}} \cdot (\overline{\boldsymbol{u}}_2^{(1)} + \overline{\boldsymbol{u}}_2^{(2)})\, \bar{n}_2\, d^3 r. \end{aligned}\right\} \tag{16.12}$$

The pressure tensor can be eliminated by using (15.11) and the result expressed in the following form

$$\left.\begin{aligned} \boldsymbol{q} &= \int \frac{1}{2} m (\xi - \boldsymbol{u})^2 (\xi - \boldsymbol{u})\, f\, d^3\xi + \\ &+ \frac{1}{2} \int \phi(r)\, (\boldsymbol{u}_2^{(1)} - \boldsymbol{u})\, n_2\, d^3 r - \\ &+ \frac{1}{4} \int \boldsymbol{r} \frac{\partial \phi}{\partial \boldsymbol{r}} \cdot (\overline{\boldsymbol{u}}_2^{(1)} + \overline{\boldsymbol{u}}_2^{(2)} - 2\boldsymbol{u})\, \bar{n}_2\, d^3 r. \end{aligned}\right\} \tag{16.13}$$

Again certain limitations to the validity of this formula require attention. The truncation of the series (16.10) is likely to lead to error if temperature variations are appreciable within a few intermolecular distances. The assumption that the potential energy of two molecules depends only on the distance between them and the suppression of internal degrees of freedom appear to limit the domain of application of (16.13) rather drastically to the monatomic liquids formed by the noble gases (neon, argon and krypton; a quantum mechanical treatment is required for helium). For the energy of the internal vibrations and the rotation of the molecules may be an important part of the internal energy of a liquid. In this respect energy and momentum transfer appear to be somewhat different; the internal structure of the molecules is clearly ignorable in so far as it affects the translatory motion of the molecules, but not obviously so where their energy is concerned. The influence of the internal degrees of freedom will be considered in the next section, where it will be shown that the results of this and the preceding section continue to hold in a slightly modified form. The thermal flux vector, as it is expressed in (16.13), consists of three terms. The first, arising from the kinetic motion of the molecules, is important only at low densities and is practically negligible in monatomic and most other liquids though it is the only contribution considered in the kinetic theory of gases. The second and third terms on the right-hand side of (16.13), therefore, are those chiefly contributing to thermal conduction in liquids. The mechanism of the energy transfer is seen to depend on the motion of the mean centre of a pair of molecules relative to the velocity of flow. The average value of $\bar{\boldsymbol{u}}_2^{(1)} + \bar{\boldsymbol{u}}_2^{(2)} - 2\boldsymbol{u}$ must of course be zero:

$$\int \bar{n}_2 (\bar{\boldsymbol{u}}_2^{(1)} + \bar{\boldsymbol{u}}_2^{(2)} - 2\boldsymbol{u})\, d^3r = 0, \tag{16.14}$$

but it does not vanish at all distances, except in thermodynamic equilibrium. If a pair of molecules tends to drift, relative to the general flow, in one direction when their mutual potential energy is high, and in the opposite direction when it is low, it is clear that a very efficient transport of energy will result. This is, in fact, what happens in liquids.

The results of this and the previous section are fundamental in establishing the relations between phenomena on the molecular scale and the measurable properties of liquids. The thermodynamics of a simple liquid is completely determined if the pressure and internal energy are given as functions of density and temperature. Consequently, if one knows how the function $n_2(r)$ depends on temperature and density, the thermodynamical properties are determined by (15.14) and (16.6), taken together. Further, the only steady-state irreversible processes in simple liquids are viscosity and thermal conduction; and the coefficients of viscosity and thermal conduction can be inferred from the relations (15.11) and (16.13), by means of which the dependence of the pressure tensor p and the thermal flux vector $\boldsymbol{q}$ on assigned macroscopic velocity and temperature gradients in the liquid are determined. Before any attempt is made to exploit these relations, however, their generalization to more complex types of liquids will be examined.

17. Polar and associated liquids. The preceding considerations require some generalization before they can be applied to polar and associated liquids, where the internal structure of the molecules is involved as well as the molecular structure of the liquid. The configuration of a molecule requires for its complete specification more than the three coordinates of its mass-centre. Following the author's thesis[1], suppose the internal coordinates are $x_4, \ldots, x_a$, so that the state of a molecule depends on the generalized position vector $\mathfrak{x}$ with a components, of which the first three refer to the mass-centre. Classically, and even quantum-mechanically (see Sect. 46), one can define a generalized velocity vector with a components $\mathfrak{\xi}$ representing the rate of change of $\mathfrak{x}$. The mechanical properties of a single molecule are then described by a Lagrangian function $\mathfrak{L}(\mathfrak{x}, \mathfrak{\xi})$, from which the canonical momentum vector $\mathfrak{p}$ and the energy $\mathfrak{H}$ can be derived:

$$\mathfrak{p} = \frac{\partial \mathfrak{L}}{\partial \mathfrak{\xi}}; \qquad \mathfrak{H} = \mathfrak{p} \cdot \mathfrak{\xi} - \mathfrak{L}. \tag{17.1}$$

The interaction energy of a pair of molecules is a function $\Phi(\mathfrak{x}, \mathfrak{y})$ of both sets of co-ordinates. The Lagrangian function of a set of q molecules is then

$$\mathfrak{L}_q = \sum_i \left(\mathfrak{L}^{(i)} - \tfrac{1}{2} \sum_j \Phi^{(ij)}\right) \tag{17.2}$$

[1] H. S. Green: Edinburgh Dissertation (1947, unpublished).

where $\mathfrak{L}^{(i)} = \mathfrak{L}(\mathfrak{x}^{(i)}, \boldsymbol{\xi}^{(i)})$ is the Lagrangian of the i-th molecule, and $\Phi^{(ij)} = \Phi(\mathfrak{x}^{(i)}, \mathfrak{x}^{(j)})$, the mutual potential energy of the i-th and j-th molecules. Also, the energy of the set of q molecules is

$$\mathfrak{H}_q = \sum_i \left(\mathfrak{H}^{(i)} + \tfrac{1}{2}\sum_j \Phi^{(ij)}\right). \tag{17.3}$$

A generalized type of molecular distribution function, depending on all the a components of the generalized position vectors can be defined in such a way that $\mathfrak{n}_q\, d\mathfrak{x}_q$ is the probability of finding each of q volume elements occupied by the mass-centres of molecules with assigned internal configurations. The distribution functions defined previously (in Sect. 11) can be obtained from the $\mathfrak{n}_q$ by integration over the internal coordinates:

$$n_q = \int \cdots \int \cdots \int \cdots \int \mathfrak{n}_q\, d x_4^{(1)} \ldots d x_a^{(1)} \ldots d x_4^{(q)} \ldots d x_a^{(q)}. \tag{17.4}$$

Also, an average interaction energy $\phi(\boldsymbol{x}, \boldsymbol{y})$ between two molecules whose mass-centres are at $\boldsymbol{x}$ and $\boldsymbol{y}$ can be defined by

$$n_2 \phi = \int \cdots \int\int \cdots \int \mathfrak{n}_2 \Phi\, d x_4 \ldots d x_a\, d y_4 \ldots d y_a. \tag{17.5}$$

The function ϕ so defined may depend weakly on the position of the mean centre $\boldsymbol{X}$ of $\boldsymbol{x}$ and $\boldsymbol{y}$ as well as on their relative displacement, and its gradient $\partial\phi/\partial\boldsymbol{x}$ may differ slightly from the mean of $\partial\Phi/\partial\boldsymbol{x}$. But these effects are clearly of minor importance, and if they are neglected, no change whatever is required in the considerations of Sect. 15, or in the formula for the pressure tensor which was derived there. The argument and conclusions of Sect. 16 are, however, also in need of re-examination for liquids whose molecules cannot be regarded as point centres of force.

There is a generalized velocity distribution function $\mathfrak{f}_q$, which bears a relation to f_q similar to that which is borne by $\mathfrak{n}_q$ to n_q, and depends on all q components of the generalized position vectors and velocities. Since classical mechanics often cannot be applied to the internal degrees of freedom of molecules, it is important to observe that there is a well defined analogue of $\mathfrak{f}_q$ in quantum mechanics, which can be used for computing statistical averages (see Sect. 46). The mean energy of a molecule at the point $\boldsymbol{x}$, not including its energy of interaction with other molecules, is given by

$$H = \int \ldots \int\int \mathfrak{f}\, \mathfrak{H}\, d^a \xi\, d x_4 \ldots d x_a / n, \tag{17.6}$$

and this must replace the term $\frac{3}{2}\mathrm{k}T$ on the right-hand side of (16.6). Thus

$$E = H/m + \tfrac{1}{2}\int n_2 \phi\, d^3 y/\varrho \tag{17.7}$$

is the energy per unit mass of the liquid at the point $\boldsymbol{x}$. If classical mechanics were applicable, one could equate H to $\frac{1}{2} m \boldsymbol{u}^2 + \frac{1}{2} a \mathrm{k} T + \phi_0$, where ϕ_0 is the mean potential energy of a molecule in isolation. Quantum-mechanical effects modify the explicit result, but do not affect the obvious conclusion that an appreciable fraction of the internal energy of liquids whose molecules have many internal degrees of freedom may be associated with these internal modes. An important question which now arises is how easily this energy can be transferred in the liquid, either by the thermal motion of the molecules, or by communication from molecule to molecule. Qualitatively the answer is clear. In liquids, especially where the molecules themselves have a complicated structure, the translatory motion of the molecules relative to the mass flow is very much impeded by the closeness with which they are packed. Moreover, only the energy of the mass centre really contributes to the transfer, so, even when the energy carried by the molecules is large, the thermal motion does not provide a very efficient means of transport. Whether the internal modes are effective carriers of energy from one molecule to another depends largely on their natural frequencies in relation to the temperature. If the excitation energy of any mode is greater than the energy of the thermal fluctuations, that mode is automatically suppressed; otherwise it is effective, and by its coupling to similar modes in other molecules can contribute to the transport of energy in the liquid. The conclusions of Sect. 16 are therefore, in the main, littele altered, and can be reformulated as follows.

In place of (16.7) one has

$$\left.\begin{aligned} K_{\mathrm{k}} &= \iint\cdots\int\int_S \mathfrak{f}\,\mathfrak{H}\,(\boldsymbol{\xi}-\boldsymbol{u})\cdot d\boldsymbol{S}\,dx_4\ldots dx_a\,d^a\xi\,+\\ &\quad+\iint\ \iint\cdots\int\int_S \mathfrak{f}_2\,\tfrac{1}{2}\,\Phi\,(\boldsymbol{\xi}-\boldsymbol{u})\cdot d\boldsymbol{S}\,dx_4\ldots dx_a\,d^a\mathfrak{y}\,d^a\xi\,d^a\eta\,.\end{aligned}\right\}\qquad(17.8)$$

On account of an analogue of (12.2), however, the first term reduces to a form identical with the first term of (16.7), and if one makes the plausible approximation of replacing Φ by its mean value ϕ defined by (17.5), there is no difference in the second term either. Similarly, if ϕ is regarded as a mean, the form of (16.11) is also unchanged. Hence the formula (16.13) for the thermal flux vector is substantially the same, whether the molecules possess a complicated internal structure or not.

It deserves to be emphasised, however, that in spite of the formal analogies which have been established, the internal structure of the molecules does affect the pressure tensor and thermal flux vector very radically in a quantitative way, through the behaviour of the distribution functions n_2, f, and f_2. No detailed calculations have been made for liquids, but the behaviour of f has been studied for gases with asymmetric molecules and molecules with rotational degrees of freedom[1], and the qualitative behaviour of these functions is easily visualized once their physical implications are understood. In particular the properties of the function n_2 are well known for liquids of the type contemplated, from the analysis of the X-ray diffraction data. The sensitivity of the distribution functions to the details of the molecular structure is a guarantee that even minor details will manifest themselves in the macroscopic behaviour of the liquid.

18. General properties of mixtures. If molecules of different kinds are present in a liquid, chemical reactions may take place which alter the proportions of the various primary constituents. From the standpoint of molecular theory, the change in the amount present of any constituent is an undesirable complication, which will be avoided here by regarding the liquid as fundamentally a mixture of the various radicals and other atomic groups which are unaffected by any reaction. There is then no need to discuss mixtures of reagents separately, provided one understands that the term "radical" may replace the term "molecule" anywhere in this section. Attention will accordingly be restricted to mixtures in which the total mass of any constituent remains the same, though the proportion of the various constituents may change from place to place.

The macroscopic properties of mixtures have been discussed by many writers[2]. If $\varrho_1, \varrho_2, \ldots, \varrho_b$ are the mass densities of the various constituents, so that the

[1] See Chap. 11 of S. CHAPMAN and T. G. COWLING: Mathematical Theory of Non-Uniform Gases. Cambridge 1939.

[2] e.g. by C. ECKART: Phys. Rev. **58**, 269 (1940), who states that his conclusions are practically identical with those of E. LOHR, Wien. Denkschr. **93**, 339 (1917); **99**, 11, 59 (1924).

actual mass density of the liquid is

$$\varrho = \varrho_1 + \varrho_2 + \cdots + \varrho_b, \tag{18.1}$$

the conservation of mass of the j-th constituent is expressed by the equation

$$\frac{\partial \varrho_j}{\partial t} + \frac{\partial}{\partial \boldsymbol{x}} \cdot (\varrho_j \boldsymbol{u}_j) = 0, \tag{18.2}$$

where $\boldsymbol{u}_j$ is the velocity of diffusion of this constituent. The fluid velocity $\boldsymbol{u}$ is defined as the velocity of mass transfer in the liquid:

$$\varrho \boldsymbol{u} = \varrho_1 \boldsymbol{u}_1 + \varrho_2 \boldsymbol{u}_2 + \cdots + \varrho_b \boldsymbol{u}_b, \tag{18.3}$$

and it is relative to this velocity that the time derivative following the motion is defined:

$$\frac{d}{dt} = \frac{\partial}{\partial t} + \boldsymbol{u} \cdot \frac{\partial}{\partial \boldsymbol{x}}. \tag{18.4}$$

If α_j is the local mass fraction of the j-th constituent of the mixture, defined by

$$\alpha_j = \frac{\varrho_j}{\varrho}, \tag{18.5}$$

and the velocity of this component relative to the mass motion is

$$\boldsymbol{w}_j = \boldsymbol{u}_j - \boldsymbol{u},$$

then one can re-write (18.2) in the form

$$\varrho \frac{d\alpha_j}{dt} + \frac{\partial}{\partial \boldsymbol{x}} \cdot (\varrho_j \boldsymbol{w}_j) = 0. \tag{18.6}$$

The equation of motion of the liquid as a whole must of course be the same whether it is a simple liquid or a mixture. It is possible, however, to write down separate equations of motion for each constituent, of the type

$$\varrho \frac{d}{dt} (\alpha_j \boldsymbol{u}_j) + \frac{\partial}{\partial \boldsymbol{x}} \cdot \mathsf{p}_j = \varrho_j \boldsymbol{F}_j, \tag{18.7}$$

where p_j is the partial pressure tensor of the j-th constituent, and $\boldsymbol{F}_j$ the local external force per unit mass acting on this constituent alone. The resultant pressure tensor and $\boldsymbol{F}$, the resultant external force per unit mass acting on the liquid, are defined by

$$\mathsf{p} = \mathsf{p}_1 + \mathsf{p}_2 + \cdots + \mathsf{p}_b \tag{18.8}$$

and

$$\varrho \boldsymbol{F} = \varrho_1 \boldsymbol{F}_1 + \varrho_2 \boldsymbol{F}_2 + \cdots + \varrho_b \boldsymbol{F}_b. \tag{18.9}$$

By using these relations in summing (18.7) with respect to j, one obtains the equation of motion of the liquid in the form of (15.3).

The equation of energy transfer can be split up in a similar way. If E_j is the energy per unit mass, and $\boldsymbol{Q}_j$ the partial energy flux vector, of the j-th constituent of the liquid mixture,

$$\varrho \frac{d}{dt} (\alpha_j E_j) + \frac{\partial}{\partial \boldsymbol{x}} \cdot (\mathsf{p}_j \cdot \boldsymbol{u} + \boldsymbol{Q}_j) = \varrho_j \boldsymbol{u}_j \cdot \boldsymbol{F}_j. \tag{18.10}$$

The total energy per unit mass E and the resultant energy flux vector $\boldsymbol{Q}$ are naturally given by

$$\varrho E = \varrho_1 E_1 + \varrho_2 E_2 + \cdots + \varrho_b E_b,$$

$$\boldsymbol{Q} = \boldsymbol{Q}_1 + \boldsymbol{Q}_2 + \cdots + \boldsymbol{Q}_b,$$

so that, by summation of (18.10) with respect to j one has

$$\left.\begin{aligned}\varrho\frac{dE}{dt}+\frac{\partial}{\partial\boldsymbol{x}}\cdot(\mathsf{p}\cdot\boldsymbol{u}+\boldsymbol{Q})&=\varrho\boldsymbol{u}\cdot\boldsymbol{F}+R,\\ R&=\varrho_1\boldsymbol{w}_1\cdot\boldsymbol{F}_1+\cdots+\varrho_b\boldsymbol{w}_b\cdot\boldsymbol{F}_b.\end{aligned}\right\}\tag{18.11}$$

If $E=U+\frac{1}{2}\boldsymbol{u}^2$, so that U is the internal energy per unit mass, one has finally

$$\varrho\frac{dU}{dt}+\left(\mathsf{p}\cdot\frac{\partial}{\partial\boldsymbol{x}}\right)\cdot\boldsymbol{u}+\frac{\partial}{\partial\boldsymbol{x}}\cdot\boldsymbol{Q}=R.\tag{18.12}$$

The energy flux vector $\boldsymbol{Q}$ in mixtures is usually regarded as the resultant of two terms, one of which, $\sum_j\frac{\partial U}{\partial\alpha_j}\varrho_j\boldsymbol{w}_j$, is the flux of internal energy due to the relative diffusion of different components, and the second is more closely analogous to the vector represented by the symbol $\boldsymbol{q}$ in Sect. 16. The thermal flux will be identified with this second component of the resultant energy flux.

Attention will now be directed to the interpretation of the above macroscopic equations on the molecular scale. It will be necessary at the outset to introduce distribution functions relating to q_1 molecules of the first species, q_2 molecules of the second species, ..., and q_b molecules of the b-th species. The numbers $q_1, q_2, \ldots q_b$ may be regarded as components of a b-fold vector $\mathfrak{q}$. If $\mathfrak{e}_1, \mathfrak{e}_2, \ldots, \mathfrak{e}_b$ are the independent unit vectors,

$$\mathfrak{q}=q_1\mathfrak{e}_1+q_2\mathfrak{e}_2+\cdots+q_b\mathfrak{e}_b.\tag{18.13}$$

Then one can define $n_{\mathfrak{q}}\,d^3x_1^{(1)}\ldots d^3x_b^{(q_1+q_2+\cdots+q_b)}$ as the probability that the volume elements $d^3x_1^{(1)}\ldots d^3x_1^{(q_1)}$ are occupied by molecules of the first species, that $d^3x_2^{(q_1+1)}\ldots d^3x_2^{(q_1+q_2)}$ are occupied by molecules of the second species, etc. A corresponding velocity distribution function $f_{\mathfrak{q}}$, which determines the relative probability that such molecules have assigned velocities, can be defined in a similar way. In this notation, the velocity distribution function for a molecule of the j-th species is $f_{\mathfrak{e}_j}$ and the distribution function for a pair of molecules of j-th and k-th species respectively is $n_{\mathfrak{e}_j+\mathfrak{e}_k}$.

If m_j is the mass of a molecule of the j-th species, the local mass density of this species is

$$\varrho_j=m_j n_{\mathfrak{e}_j}\tag{18.14}$$

and the mean velocity $\boldsymbol{u}_j$ is given by

$$\varrho_j\boldsymbol{u}_j=m_j\int f_{\mathfrak{e}_j}\boldsymbol{\xi}_j\,d^3\xi_j.\tag{18.15}$$

Also, if $\phi_{jk}(r)$ is the interaction energy between two molecules of the j-th and k-th species at distance r, the energy per unit mass E_j proper to molecules of the j-th species is given by

$$\varrho_j E_j=\tfrac{1}{2}m_j\int f_{\mathfrak{e}_j}\xi_j^2\,d^3\xi_j+\tfrac{1}{2}\sum_k\int n_{\mathfrak{e}_j+\mathfrak{e}_k}\phi_{jk}\,d^3x_k.\tag{18.16}$$

It is now easy to retrace the arguments of Sects. 15 to 16 and show that the partial pressure tensor p_j and the partial thermal flux vector $\boldsymbol{q}_j$ are given by

$$\mathsf{p}_j=m_j\int f_{\mathfrak{e}_j}(\boldsymbol{\xi}_j-\boldsymbol{u})\,\boldsymbol{\xi}_j\,d^3\xi_j-\frac{1}{2}\sum_k\int\bar{n}_{\mathfrak{e}_j+\mathfrak{e}_k}\boldsymbol{r}\frac{\partial\phi_{jk}}{\partial\boldsymbol{r}}\,d^3r\tag{18.17}$$

and

$$\left.\begin{aligned} \boldsymbol{q}_j &= \tfrac{1}{2} m_j \int f_{e_j}(\xi_j - \boldsymbol{u})(\xi_j^2 - 2\xi_j\cdot\boldsymbol{u})\, d^3\xi_j + \\ &+ \tfrac{1}{2}\sum_k \int n_{e_j+e_k}(\boldsymbol{u}^{(1)}_{e_j+e_k} - \boldsymbol{u})\,\phi_{jk}\, d^3x_k - \\ &- \tfrac{1}{4}\sum_k \int \bar{n}_{e_j+e_k}(\bar{\boldsymbol{u}}^{(1)}_{e_j+e_k} + \bar{\boldsymbol{u}}^{(2)}_{e_j+e_k})\cdot\frac{\partial\phi_{jk}}{\partial \boldsymbol{r}}\,\boldsymbol{r}\, d^3r \end{aligned}\right\} \tag{18.18}$$

where

$$n_{e_j+e_k}\boldsymbol{u}^{(i)}_{e_j+e_k} = \iint f_{e_j+e_k}\,\xi^{(i)}\, d^3\xi_j^{(1)}\, d^3\xi_k^{(2)}. \tag{18.19}$$

These are obvious generalizations, to a mixture of b constituents, of formulae first derived for a binary mixture by YANG[1].

While considering the distribution functions for mixtures, it may be noticed that there is no difficulty in defining the functions $N_{\mathfrak{q}}$ and $F_{\mathfrak{q}}$ which relate to the possibility that a finite region should be occupied by precisely q_1 molecules of the first species, etc., with assigned positions (and, in the instance of $F_{\mathfrak{q}}$, with assigned velocities as well). The probability that precisely q_j molecules of the j-th species, for each value of j, will be found in a finite volume V is

$$P_{\mathfrak{q}} = \frac{\int N_{\mathfrak{q}}\, d\boldsymbol{x}_{\mathfrak{q}}}{q_1!\, q_2! \ldots q_b!}, \tag{18.20}$$

where $\int \ldots d\boldsymbol{x}_{\mathfrak{q}}$ means $\int \ldots \int \ldots d^3x_1^{(1)} \ldots d^3x_b^{(q_1+\cdots+q_b)}$ and the integration is restricted to V. The mean number of molecules of the j-th kind in this volume is

$$\langle q_j\rangle = \int n_{e_j}\, d^3x_j = \sum_{\mathfrak{q}} q_j P_{\mathfrak{q}} \tag{18.21}$$

and the mean product of the fluctuations $q_j - \langle q_j\rangle$ and $q_k - \langle q_k\rangle$ is

$$\langle q_j q_k\rangle - \langle q_j\rangle\langle q_k\rangle = \iint (n_{e_j+e_k} - n_{e_j} n_{e_k})\, d^3x_j\, d^3x_k = \sum_{\mathfrak{q}} (q_j - \langle q_j\rangle)(q_k - \langle q_k\rangle) P_{\mathfrak{q}}. \tag{18.22}$$

These averages will later (Sect. 23) be seen to have important physical significance for the thermodynamics of the liquid. In the next section it will be seen that they also arise in the theory of the diffraction of particles by a liquid.

19. Theory of diffraction in liquids. The value of the information concerning molecular structure which has been obtained by X-ray and neutron diffraction techniques has already been stressed (Sect. 8), and it has been mentioned that a close relation exists between the structure and the diffraction patterns obtained. The purpose of this section is to examine this relation from a quantitative point of view. Some important differences between X-ray and neutron diffraction should first be mentioned. X-rays are scattered almost exclusively by the electrons associated with the molecules, and as only electromagnetic interactions are responsible, the use of BORN's approximation in scattering theory is rigorously justified. The wave-length of the radiation has to be comparable with the intermolecular distances and the scattering is then practically elastic, i.e., the change of frequency on scattering is negligible. Neutrons, on the other hand, are scattered by the nuclei of the molecules, and especially since only slow neutrons have a wavelength of the right order, BORN's approximation is apparently inapplicable. However, the scattering from a single nucleus is isotropic, and one can construct an interaction with zero range which is practically equivalent to the real interaction between neutron and nucleus, and to which BORN's approximation can be applied. Unfortunately the scattering of slow neutrons is not elastic, in the

[1] L. M. YANG: Proc. Roy. Soc. Lond., Ser. A **198**, 471 (1949).

sense that neutrons scattered through a particular angle can have different energies even if the incident beam were monochromatic. The usual theory of X-ray diffraction[1] is therefore only approximately applicable to neutron diffraction, and the theory given here will be a modified version due in principle to VAN HOVE[2], which is equally valid for the scattering of X-rays and neutrons.

It is usual to separate the intensity of the radiation scattered in any direction into its coherent and incoherent components. The coherent part exhibits the typical interference phenomena, while the incoherent part is what one would expect if there were no interference between the radiation from different scatterers. The incoherent radiation gives no information concerning the structure of the liquid, and its intensity has to be subtracted from the resultant scattered intensity before analysis. An important contribution to the incoherent radiation arises if the spin of a scattered particle in a particular direction is reversed by scattering; such spin-flip effects can and will, however, be ignored in the theoretical discussion since they are also rejected in the analysis of the experimental data.

The essential result of scattering theory which will be required is expressed by the formula

$$M_{AB}(\boldsymbol{k}_0, \boldsymbol{k}) = \iint F(\boldsymbol{k}_0, \boldsymbol{k})\, \varrho_{AB}(\boldsymbol{x}, t) \exp\{i(\boldsymbol{\varkappa}\cdot\boldsymbol{x} - \omega t)\}\, d^3x\, dt. \tag{19.1}$$

This can be understood in the following terms. The probability amplitude that the liquid in a pure quantum-mechanical state A will scatter an incident particle (photon or neutron) with momentum $\hbar\boldsymbol{k}_0$, in such a way that the liquid is left in the pure state B and the final momentum of the scattered particle is $\hbar\boldsymbol{k}$, is represented by $M_{AB}(\boldsymbol{k}_0, \boldsymbol{k})$. The probability amplitude that, on the assumption that the particle is scattered at the point $\boldsymbol{x}$ and time t, its final momentum will be $\hbar\boldsymbol{k}$, is $F(\boldsymbol{k}_0, \boldsymbol{k}) \exp\{i(\boldsymbol{\varkappa}\cdot\boldsymbol{x} - \omega t)\}$, where $\hbar\boldsymbol{\varkappa}$ and $\hbar\omega\,[\boldsymbol{\varkappa} = \boldsymbol{k} - \boldsymbol{k}_0$ and $\omega = \varepsilon(k) - \varepsilon(k_0)]$ are the change of momentum and energy of the scattered particle. Finally, the probability amplitude that the scattering will occur in the volume element d^3x and the time interval dt, effecting a transition of the liquid from the state A to the state B, is $\varrho_{AB}(\boldsymbol{x}, t)\, d^3x\, dt$. The probability that the particle will be found eventually in the momentum state $\boldsymbol{k}$ is obtained by squaring the modulus of the probability amplitude $M_{AB}(\boldsymbol{k}_0, \boldsymbol{k})$, then multiplying by the probability p_A of finding the liquid in the state A, and lastly summing over all states A and B. The result, per unit time and per unit volume of liquid, is

$$S(\boldsymbol{k}_0, \boldsymbol{k}) = |F(\boldsymbol{k}_0, \boldsymbol{k})|^2 \iint \Gamma(\varrho, \tau) \exp\{i(\boldsymbol{\varkappa}\cdot\boldsymbol{\varrho} - \omega\tau)\}\, d^3\varrho\, d\tau \tag{19.2}$$

where

$$\Gamma(\varrho, \tau) = \sum_{A,B} \frac{p_A}{VT} \iint \varrho_{AB}\left(\boldsymbol{x} - \tfrac{1}{2}\varrho, t - \tfrac{1}{2}\tau\right) \varrho^{*}_{AB}\left(\boldsymbol{x} + \tfrac{1}{2}\varrho, t + \tfrac{1}{2}\tau\right) d^3x\, dt \tag{19.3}$$

and, in (19.3), V is the volume of liquid and T the time allowed for scattering, corresponding to the implicit ranges of integration with respect to $\boldsymbol{x}$ and t.

Now $S(\boldsymbol{k}_0, \boldsymbol{k})$ is proportional to the number of particles scattered between the momentum states $\boldsymbol{k}_0$ and $\boldsymbol{k}$ in the diffraction experiments, and $F(\boldsymbol{k}_0, \boldsymbol{k})$ is a simple function given by the relevant scattering theory; the ratio

$$R(\boldsymbol{\varkappa}, \omega) = \iint \Gamma(\varrho, \tau) \exp\{i(\boldsymbol{\varkappa}\cdot\boldsymbol{\varrho} - \omega\tau)\}\, d^3\varrho\, d\tau \tag{19.4}$$

can therefore be obtained directly from experimental data; it depends only on $\boldsymbol{\varkappa}$ and ω, which may be regarded as independent if $\boldsymbol{k}_0$ is allowed to vary. It

[1] See, e.g. H. S. GREEN: The Molecular Theory of Fluids, pp. 57–62. Amsterdam: North-Holland Publishing Co. 1952.

[2] L. VAN HOVE: Phys. Rev. 95, 249 (1954).

should be noticed, however, that X-ray scattering is practically elastic, so that it is $R(\varkappa, 0)$ which is effectively determined. Also, even though the dependence of $R(\varkappa, \omega)$ on ω could, in principle, be determined in neutron diffraction experiments, the measurement of neutron energies has not been made with sufficient accuracy at the time of writing to allow this to be done. There is little doubt that neutron diffraction techniques will, in future, provide very precise information on molecular structure; but these will require the measurement of the energies of the neutrons as well as their angular distributions.

The precise relation of the function $\Gamma(\varrho, \tau)$ to the molecular structure must now be examined. From its definition in terms of $Q_{AB}(\boldsymbol{x}, t)$ [Eq. (19.3)] it is obviously directly connected with the distribution of scatterers. If the scatterers are electrons, they are all equivalent, but if they are nuclei, they may be of different species and differ in their scattering power. For this reason the amplitude $Q_{AB}(\boldsymbol{x}, t)$ must be decomposed into a number of terms

$$Q_{AB}(\boldsymbol{x}, t) = \sum_{\mu} a_{\mu}\, q_{AB,\mu}(\boldsymbol{x}, t) \qquad (19.5)$$

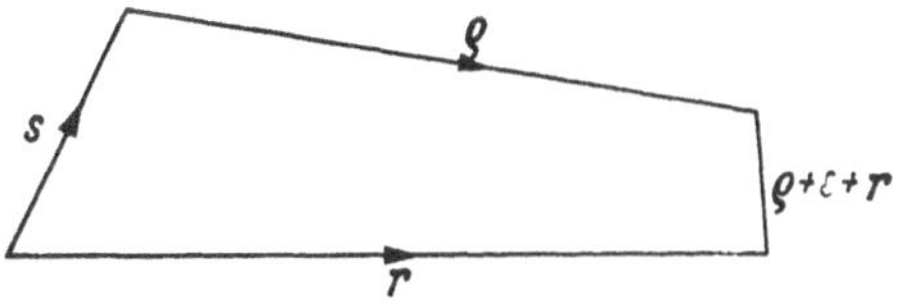

Fig. 2.

where the μ-th term corresponds to the possibility that a scatterer of the μ-th species was responsible for the scattering, and a_{μ} is proportional to the "scattering length" of this scatterer. The coefficients a_{μ} can be so chosen that if

$$\Gamma(\varrho, \tau) = \sum_{\mu,\nu} a_{\mu}\, a_{\nu}\, \gamma_{\mu\nu}(\varrho, \tau), \qquad (19.6)$$

then $\gamma_{\mu\nu}(\varrho, \tau)\, d^3\varrho$ represents the probability per unit volume of finding a scatterer of the μ-th species at any point P at time t, and a scattered of the ν-th species in the volume element $d^3\varrho$ at the displacement ϱ from P at time $t+\tau$.

To proceed further, it will be necessary to assume that the internal configurations of two molecules are uncorrelated with their relative displacement and orientation. In certain liquids—for example, polar liquids in which the molecules polarize one another—this assumption may be unwarranted, but, without it, it is very difficult to disentangle the effects of the molecular structure of the liquid and the internal structure of the molecules on the diffraction pattern; indeed, it would be pointless to try to do so. Suppose, then, in a single molecule the probability of finding a scatterer of the μ-th species in the volume element $d^3 s$ at the displacement $\mathbf{s}$ from the mass-centre of the molecule is $h_{\mu}(\mathbf{s})\, d^3 s$, independently of the environment of the molecule.

Then (see Fig. 2)

$$\gamma_{\mu\nu}(\varrho, 0) = \iint \{n\,\delta(\boldsymbol{r}) + n_2(\boldsymbol{r})\}\, h_{\mu}(\mathbf{s})\, h_{\nu}(\varrho + \mathbf{s} - \boldsymbol{r})\, d^3 r\, d^3 s \qquad (19.7)$$

where the term $n\,\delta(\boldsymbol{r})$ corresponds to the possibility that the two scatterers referred to in the definition of $\gamma_{\mu\nu}(\varrho, 0)$ are associated with the same molecule, and the term $n_2(\boldsymbol{r})$ to the possibility that they are associated with different molecules. When $\mathbf{s}$ becomes large, $h_{\mu}(\mathbf{s})$ must clearly approach zero; on the other hand, when $\boldsymbol{r}$ is large, $n_2(\boldsymbol{r})$ approaches the value n^2. Therefore, when ϱ is large, $\gamma_{\mu\nu}(\varrho, 0)$ approaches the value $\gamma_{\mu}\gamma_{\nu}$, where

$$\gamma_{\mu} = n \int h_{\mu}(\mathbf{s})\, d^3 s. \qquad (19.8)$$

Defining now the Fourier transforms

$$\left.\begin{aligned} \gamma_{\mu\nu}^F(\boldsymbol{k}) &= \int \{\gamma_{\mu\nu}(\varrho, 0) - \gamma_\mu \gamma_\nu\} \exp(i\,\boldsymbol{k}\cdot\varrho)\, d^3\varrho\,, \\ h_\mu^F(\boldsymbol{k}) &= \int h(\boldsymbol{s}) \exp(i\,\boldsymbol{k}\cdot\boldsymbol{s})\, d^3 s\,, \\ n_2^F(\boldsymbol{k}) &= \int \{n_2(\boldsymbol{r}) - n^2\} \exp(i\,\boldsymbol{k}\cdot\boldsymbol{r})\, d^3 r\,, \end{aligned}\right\} \tag{19.9}$$

it follows from (19.7) and (19.8) that

$$\gamma_{\mu\nu}^F(\boldsymbol{k}) = \{n + n_2^F(\boldsymbol{k})\}\, h_\mu^F(\boldsymbol{k})\, h_\mu^F(-\boldsymbol{k})\,. \tag{19.10}$$

Now, from (19.4) one has

$$\left.\begin{aligned} \int R(\boldsymbol{k}, \omega)\, d\omega &= 2\pi \int \Gamma(\varrho, 0) \exp(i\,\boldsymbol{k}\cdot\varrho)\, d^3\varrho \\ &= 2\pi \int \Big\{\Gamma(\varrho, 0) - \Big(\sum_\mu a_\mu \gamma_\mu\Big)^2\Big\} \exp(i\,\boldsymbol{k}\cdot\varrho)\, d^3\varrho + \\ &\quad + (2\pi)^4 \Big(\sum_\mu a_\mu \gamma_\mu\Big)^2 \delta(\boldsymbol{k})\,. \end{aligned}\right\} \tag{19.11}$$

The first term in this last expression is finite for all values of $\boldsymbol{k}$, and the second is zero except near $\boldsymbol{k}=0$, when it is very large; however, in the diffraction experiments, the contribution of the second term would be indistinguishable from that due to the incident particles which have not been scattered, and it may therefore be dropped. Using (19.6), one has, therefore

$$\int R(\boldsymbol{k}, \omega)\, d\omega = 2\pi \sum_{\mu,\nu} a_\mu a_\nu \gamma_{\mu\nu}^F(\boldsymbol{k})\,. \tag{19.12}$$

From (19.10) and (19.12), $n_2(\boldsymbol{k})$, and therefore its Fourier inverse $n_2(\boldsymbol{r}) - n^2$ can be determined:

$$n_2(\boldsymbol{r}) - n^2 = \frac{1}{(2\pi)^4} \int \Big\{ \int \frac{R(\boldsymbol{k}, \omega)\, d\omega}{\big|\sum_\mu a_\mu h_\mu^F(\boldsymbol{k})\big|^2} - n \Big\} \exp(-i\,\boldsymbol{k}\cdot\boldsymbol{r})\, d^3 k\,. \tag{19.13}$$

The present practice in the analysis of the diffraction data amounts to substituting for $\{\int R(\boldsymbol{k}, \omega)\, d\omega\}\, d^3 k$ the total intensity (less the incoherent fraction already mentioned) scattered into the momentum range $d^3 k$, when the incident particle beam is as nearly monochromatic as possible. This procedure is perfectly correct when the scattering is elastic, as it certainly is with X-rays, but may introduce errors if neutrons are used. To correct these errors, it will be necessary to repeat the scattering experiments with neutrons of various energies, measuring the energy spectrum of the scattered beam, and thus determine the variation of $R(\boldsymbol{k}, \omega)$ with ω. The data so obtained will provide information not merely on the functions $\Gamma(\varrho, 0)$ and $n_2(\boldsymbol{r})$, but also on the time dependence of $\Gamma(\varrho, \tau)$, from which additional structural properties of the liquid can be deduced.

The preceding discussion has assumed that the molecules of the liquid are all alike, but can be extended without difficulty to liquid mixtures. If $h_{\mu,j}(\boldsymbol{s})$ is the distribution function for scatterers of the μ-th species in molecules of the j-th species, and $h_{\mu,j}^F(\boldsymbol{k})$ is its Fourier transform, the formula (19.10) is replaced by

$$\left.\begin{aligned} \gamma_{\mu\nu}^F(\boldsymbol{k}) &= \sum_j n_{e_j} h_{\mu,j}^F(\boldsymbol{k})\, h_{\nu,j}^F(-\boldsymbol{k}) + \\ &\quad + \sum_{j,k} n_{e_j+e_k}(\boldsymbol{k})\, h_{\mu,j}^F(\boldsymbol{k})\, h_{\nu,k}^F(-\boldsymbol{k})\,, \end{aligned}\right\} \tag{19.14}$$

where

$$n_{e_j+e_k}(\boldsymbol{k}) = \int (n_{e_j+e_k}(\boldsymbol{r}) - n_{e_j} n_{e_k}) \exp(i\,\boldsymbol{k}\cdot\boldsymbol{r})\, d^3 r\,. \tag{19.15}$$

It can be seen from this that the diffraction experiments alone cannot determine more than one of the functions $n_{e_j+e_k}$ when the others are known: although the molecular structure of liquid mixtures determines the diffraction pattern, the converse is not true.

It may be noticed here that the above theory needs little or no modification if the incident radiation is not X-radiation or a neutron beam, but some other type of radiation which is scattered by the molecules of the liquid. The wavelength of light is long compared with that of X-radiation, and is therefore unsuitable for the determination of molecular structure. However, light is scattered by liquids, and the scattered intensity is relatively large at small scattering angles. Setting $\boldsymbol{k}=0$ in (19.14) and (19.15), it can be seen that the intensity of the coherent scattering depends on the integrals

$$\int (n_{e_j+e_k}(\boldsymbol{r}) - n_{e_j} n_{e_k})\, d^3 r$$

which determine the fluctuations in density of the liquid, or of the components of a liquid mixture. It is known that such fluctuations are particularly large in the neighbourhood of the critical point and in the region of condensation from the gas. It is therefore to be expected that the scattering of light by the fluid will be particularly marked in such regions, where in fact the phenomenon of opalescence is observed.

III. Structure of uniform liquids.

a) The radial distribution function.

20. Calculations based on the cell model. It will now be necessary to review the methods available for the theoretical calculation of the radial distribution function in liquids. An account will be given of the most accurate and highly developed methods in their present state, but the more primitive methods which foreshadowed them will also be described for the sake of their simplicity, even though their range of validity may be considerably restricted.

The cell-lattice model has inspired several such primitive calculations. The special feature of the model consists in the division of the liquid into cells of similar size and shape, usually with a volume equal to or somewhat less than the molecular volume $1/n$. If $g(r)$, i.e. $n_2(r)/n^2$ is the correlation function, the mean molecular density at distance r from a molecule at a certain point in, say, the a-th cell, is $n g(r)$—independent of a, and can be expressed as a sum of contributions from all cells:

$$n\, g(r) = \sum_b \varrho(r, r_{ab}). \tag{20.1}$$

Here $\varrho(r, r_{ab})$ is the contribution from molecules in the b-th cell, determined by the distance r_{ab} between the centres of the two cells as well as the distance r between the molecules.

Prins[1], who was one of the first to assail the problem, assumed that $\varrho(r, r_{ab})$ should be proportional to $\exp\{-c(r-r_{ab})^2/r_{ab}\}$, where c is some constant. This assumption was not, however, supported by the more detailed analysis of Wall[2] and Coulson and Rushbrooke[3], who, neglecting the correlation between the positions of molecules in different cells, introduced the distribution function

[1] J. A. Prins: Naturwiss. 19, 435 (1931).

[2] C. N. Wall: Phys. Rev. 54, 1062 (1938). Wall's theory has been applied to liquid argon by L. H. Lund: J. Chem. Phys. 13, 317 (1945).

[3] C. A. Coulson and G. S. Rushbrooke: Phys. Rev. 56, 1216 (1939).

$\sigma(s)$ for the displacements of the molecules from the centres of their cells. To be precise, $\sigma(s)\, d^3s$ is the probability of finding a molecule in a volume element d^3s at a displacement $\mathbf{s}$ from the centre of its cell. Then it is clear from the figure that, if a and b are not the same,

$$\varrho(r, r_{ab})\, d^3r = \int \sigma(s_a)\, \sigma(s_b)\, d^3s_a\, d^3s_b$$

or, setting $d^3s_a = 2\pi x\, dx\, s_a\, ds_a/r_{ab}$ and $d^3s_b = 2\pi r\, dr\, s_b\, ds_b/u$,

$$4\pi r^2 \varrho(r, r_{ab})\, dr = (2\pi)^2 r\, dr\, r_{ab}^{-1} \int_0^\infty p(r, u)\, p(u, r_{ab})\, du \tag{20.2}$$

where

$$p(r, u) = \int_{|r-u|}^{r+u} \sigma(s)\, s\, ds. \tag{20.3}$$

WALL used a form of $\sigma(s)$ appropriate to rigid spherical molecules, and obtained quite good agreement with the experimentally determined curve for $g(r)$ in liquid sodium just above its melting point, but only by making somewhat arbitrary assumptions about the number of molecules in the various co-ordination shells, or, equivalently, the normalization of the functions $\varrho(r, r_{ab})$. The most refined calculation yet made on the basis of the cell model is due to RUSHBROOKE[1], who observed correctly that $\sigma(s)$ must be proportional to $\exp\{-\beta\Psi(s)\}$, where $\Psi(s)$ is the mean potential energy of a molecule at the distance s from the centre of its cell. The essential difficulty is to calculate $\Psi(s)$; RUSHBROOKE assumed there would be precisely one molecule per cell near the melting point and adopted an estimate of $\Psi(s)$ due to LENNARD-JONES and DEVONSHIRE. The curve he obtained for $g(r)$ in liquid argon at 90 °K resembles that deduced from X-ray diffraction data (Sect. 8); the method does in fact appear to be quite a good one for liquids just above their melting point. It is not, however, able to account for the wide variation in the number of molecules in the various co-ordination shells which occurs at higher temperatures. Apparently an adequate calculation of the radial distribution function will have to take account of doubly and perhaps even trebly occupied cells, and also correlations between the positions of molecules in different cells. A way to achieve this will be elaborated, which is adapted from a procedure described by the author[2], primarily intended for the calculation of the thermodynamic functions of liquids. RUSHBROOKE's calculation will be made to appear as the first step in the more accurate determination which is now possible.

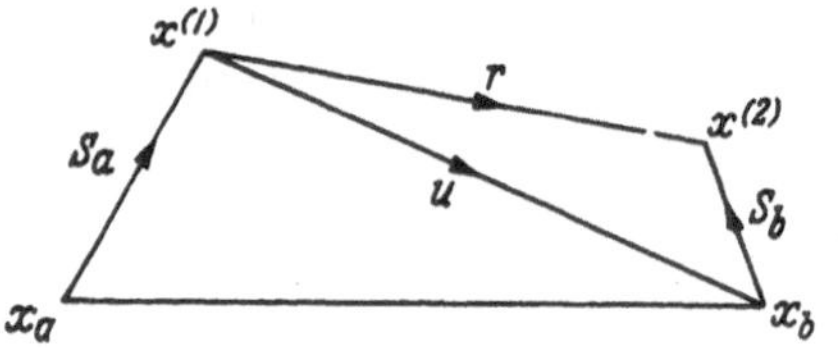

Fig. 3.

The starting point for the method is the formula (13.1), which, with $q = 2$, gives

$$n_2 = \sum_{j=0}^{\infty} \int \cdots \int (N_{j+2}/j!)\, d^3x^{(3)} \ldots d^3x^{(j+2)}, \tag{20.4}$$

where, according to (13.15) and (13.13),

$$N_q = \exp(-\beta p v)\, z^q \exp(-\beta \Phi_q). \tag{20.5}$$

[1] G. S. RUSHBROOKE: Proc. Roy. Soc. Edinburgh, Ser. A **60**, 182 (1940).
[2] H. S. GREEN: J. Chem. Phys. **24**, 732 (1956).

It will be recalled that z is the thermodynamic activity per molecule, p is the pressure, v the volume over which the integrations are taken in (20.4), and Φ_q is the total potential energy of q molecules in the assigned positions $\boldsymbol{x}^{(1)} \ldots \boldsymbol{x}^{(q)}$. It may be assumed that

$$\Phi_q = \sum_{j>i}^{q} \phi^{(ij)} + \sum_{i=1}^{q} \Psi^{(i)} \tag{20.6}$$

where $\phi^{(ij)}$ is, as before, the energy of interaction between molecules at $\boldsymbol{x}^{(i)}$ and $\boldsymbol{x}^{(j)}$, and $\Psi^{(i)}$ the mean energy of interaction between a molecule at $\boldsymbol{x}^{(i)}$ and molecules outside the volume v.

The usefulness of the cell model lies in the method it suggests for the evaluation of the integral in (20.4), which is otherwise quite intractable, even for fairly small values of j. Suppose that $q(a)$ of the q co-ordinates $\boldsymbol{x}_q$ (i.e. $\boldsymbol{x}^{(1)}, \boldsymbol{x}^{(2)}, \ldots, \boldsymbol{x}^{(q)}$) fall in the a-th cell; then Φ_q can be decomposed as follows:

$$\Phi_q = \sum_a \Phi_{q(a)} + \sum_{b>a} \Phi_{q(a)\,q(b)} \tag{20.7}$$

where $\Phi_{q(a)}$ is the total potential energy of the $q(a)$ molecules in the a-th cell, and $\Phi_{q(a)q(b)}$ is the energy of interaction between the $q(a)$ molecules in the a-th cell and the $q(b)$ molecules in the b-th cell. Obviously $\sum_a q(a) = q$.

Before proceeding further, it will be necessary to define a mean energy of interaction between the $q(a)$ molecules in the a-th cell and the molecules in the b-th cell, irrespective of their number or position. If the correlation between the positions of molecules in different cells is ignored, this is

$$\Psi_{q(a),b} = \sum_{q(b)} \frac{1}{q(b)!} \int N_{q(b)}\, \Phi_{q(a)q(b)}\, dx_{q(b)}, \tag{20.8}$$

where $N_{q(b)}\, dx_{q(b)}$ is the probability of finding $q(b)$ molecules in the b-th cell, in the assigned volume elements. Clearly, by analogy with (20.5),

$$N_{q(b)} = \exp(-\beta\, \pi\, v_b)\, \zeta^{q(b)} \exp\Big\{-\beta\Big(\Phi_{q(b)} + \sum_{a(\neq b)} \Psi_{q(b),a}\Big)\Big\} \tag{20.9}$$

where v_b is the volume of the cell and π and ζ are parameters, not identical with p and z because the cell volume is small and surface effects are not negligible. However, π and ζ can be determined by using the relations

$$\left.\begin{aligned} \sum_{q(b)} \frac{N_{0,q(b)}}{q(b)!} &= 1 \\ \sum_{q(b)} \frac{q(b)\, N_{0,q(b)}}{q(b)!} &= n\, v_b \end{aligned}\right\} \tag{20.10}$$

where, in conformity with (13.2), $N_{0,q(b)} = \int N_{q(b)}\, dx_{q(b)}$.

It should be noticed that (20.8) is really an integral equation to determine $\Psi_{q(a),b}$, since, according to (20.9), $N_{q(b)}$ depends explicitly on $\Psi_{q(b),a}$. The integral equation is solved by an iteration procedure, as follows. One first makes a reasonable guess at the form of $N_{q(b)}$, which might be called the *zero*th approximation and denoted by $N_{q(b)}^{|0|}$. This is used in (20.8) to determine a *first* approximation $\Psi_{q(a),b}^{|1|}$ to $\Psi_{q(a),b}$. One then proceeds to determine first approximations $\pi^{|1|}$ and $\zeta^{|1|}$ to π and ζ from (20.10), substituting for $N_{q(b)}$ from (20.9), with $\Psi_{q(b),a}^{|1|}$ instead of $\Psi_{q(b),a}$ on the right-hand side. The *first* approximation $N_{q(b)}^{|1|}$ to $N_{q(b)}$ is then given by (20.9), with $\pi^{|1|}$, $\zeta^{|1|}$ and $\Psi_{q(b),a}^{|1|}$ instead of π, ζ and $\Psi_{q(b),a}$ on the right-hand side. The whole procedure could be repeated indefinitely, to obtain $\Psi_{q(a),b}$ as accurately as desired. There is, of course, no difficulty in calculating the sums of the series in (20.8) and (20.10), or in evaluating the integrals, because only the first two or perhaps three terms of the series contribute appreciably to the sums.

In practice, no calculation has yet been carried beyond the first approximation to $\Psi_{q(a),b}$, and for this reason the initial choice of $N_{q(b)}^{|0|}$ has been of considerable importance. The procedure of LENNARD-JONES and DEVONSHIRE[1], which was adopted by RUSHBROOKE, amounted to taking $N_{q(b)}^{|0|}=0$ for $q>1$, $N_{1(b)}^{|0|}=n v_b \times \delta(\boldsymbol{x}-\boldsymbol{x}_b)$, where $\boldsymbol{x}_b$ is the centre of the b-th cell, and $N_{0(b)}^{|0|}=1-n v_b$. (Actually LENNARD-JONES and DEVONSHIRE took $v_b=n^{-1}$, so $N_{0(b)}^{|0|}=0$.) The choice of the singular delta-function for $N_{1(b)}^{|0|}$ is not, however, the best, except possibly at low temperatures and using classical mechanics; the procedure of MAYER and CARERI[2] is in this respect to be preferred. They assumed $N_{1(b)}^{|0|}$ is proportional to $\exp\{-c \times (\boldsymbol{x}-\boldsymbol{x}_b)^2\}$, with a constant c chosen so that $N_{1(b)}^{|0|}$ and $N_{1(b)}^{|1|}$ are in agreement as nearly as possible.

The above analysis is already sufficient to determine the displacement distribution function $\sigma(s)$ contained in (20.3), neglecting only the correlations between the positions of molecules in neighbouring cells. If $\boldsymbol{x}_b$ is the centre of the b-th cell, one has, in fact

$$\sigma(|\boldsymbol{x}^{(1)}-\boldsymbol{x}_b|)=\sum_{j(b)}\frac{N_{1,j(b)}}{j(b)!} \tag{20.11}$$

where $N_{1,j(b)}$, in conformity with (13.2), represents the integral of $N_{(j+1)b}$ over all co-ordinates except $\boldsymbol{x}^{(1)}$. In principle, therefore, one can use (20.2) to calculate all contributions $\varrho(r, r_{ab})$ to $n g(r)$, except the one arising when a and b are the same. This is given by

$$n\,\varrho(r,0)=\sum_{j(b)}\frac{\langle N_{2,j(b)}\rangle}{j(b)!} \tag{20.12}$$

where $\langle N_{2,j(b)}\rangle$ means an average value, obtained by first writing $\boldsymbol{x}^{(2)}=\boldsymbol{x}^{(1)}+\boldsymbol{r}$ in $N_{2,j(b)}$ and then performing the integration $\frac{1}{v_b}\int\ldots d^3x^{(1)}$ to calculate the mean value for the cell.

If it is required to make accurate calculations based on the cell model, however, account must be taken of correlations between the positions of molecules in different cells. This can be done by making use of the concept of cell clusters, i.e., small groups of neighbouring cells containing molecules whose positions are correlated. Cell clusters of each type will contribute a characteristic correction to the radial distribution function, calculated in the way just described. To determine these corrections, one writes

$$N_q=C\prod_a N_{q(a)}, \tag{20.13}$$

when C is a factor depending on the co-ordinates, though presumably not very different from 1 in most configurations of the molecules. By substituting (20.5) and (20.9) into (20.13), one finds

$$C=C_0\Big\{\prod_a (z/\zeta)^{q(a)}\Big\}\prod_{a>b}\left(1+\Theta_{q(a)q(b)}\right), \tag{20.14}$$

[1] J. E. LENNARD-JONES and F. DEVONSHIRE: Proc. Roy. Soc. Lond., Ser. A 163, 53 (1938).
[2] J. E. MAYER and G. CARERI: J. Chem. Phys. 20, 1001 (1952).

where

$$C_0 = \exp\left\{\beta(\pi - p)\,v + \beta \sum_{a>b} \mathrm{X}_{ab}\right\} \tag{20.15}$$

and

$$1 + \Theta_{q(a)q(b)} = \exp\{-\beta(\Phi_{q(a)q(b)} - \Psi_{q(a),b} - \Psi_{q(b),a} + \mathrm{X}_{ab})\}. \tag{20.16}$$

The disposable numbers X_{ab} are chosen so as to make the mean value of $\Theta_{q(a)q(b)}$ as small as possible, and will be defined here by

$$\left.\begin{aligned} \exp(\beta\,\mathrm{X}_{ab}) = \sum_{q(a)q(b)} \iint \frac{N_{q(a)}N_{q(b)}}{q(a)!\,q(b)!} \times \\ \times \exp\{\beta(\Psi_{q(a),b} + \Psi_{q(b),a} - \Phi_{q(a)q(b)})\}\, dx_{q(a)}\, dx_{q(b)}. \end{aligned}\right\} \tag{20.17}$$

At sufficiently high temperatures, however, the exponentials may be expanded, and terms in β^2, β^3, etc. neglected on each side; then, with the help of (20.8) and the first of Eqs. (20.10), one obtains the approximation

$$\mathrm{X}_{ab} \approx \sum_{q(a)} \frac{1}{q(a)!} \int N_{q(a)} \Psi_{q(a),b}\, dx_{q(a)}. \tag{20.18}$$

Obviously X_{ab} represents the mean energy of interaction between molecules in the a-th and b-th cells, irrespective of their number or positions.

According to (20.14), one can write

$$C = C_0 \left\{\prod_a (z/\zeta)^{q(a)}\right\} (1 + \Theta^{\{2\}} + \Theta^{\{3\}} + \cdots), \tag{20.19}$$

where

$$\left.\begin{aligned} \Theta^{\{2\}} &= \sum_{a>b} \Theta_{q(a)q(b)}, \\ \Theta^{\{3\}} &= \sum_{a>b>c} (\Theta_{q(a)q(b)}\Theta_{q(a)q(c)} + \Theta_{q(b)q(c)}\Theta_{q(b)q(a)} + \Theta_{q(c)q(a)}\Theta_{q(c)q(b)}), \end{aligned}\right\} \tag{20.20}$$

etc. When (20.13) is used for substitution in (20.4), and use is made of the development of C given by (20.19), the distribution function n_2 is obtained in the form

$$n_2 = n_2^{\{1\}} + n_2^{\{2\}} + n_2^{\{3\}} + \cdots, \tag{20.21}$$

the m-th term corresponding to the m-th term in (20.19). There are similar developments for z and p, which will be discussed in Sect. 23; here it will be said only that, for the first approximation, $z^{\{1\}} = \zeta$ and $p^{\{1\}} = \pi + \frac{1}{2}\sum_b \mathrm{X}_{ab}/v_a$. The first term of (20.21), with $z^{\{1\}}$ and $p^{\{1\}}$ substituted for z and p, and averaged with respect to the position of $\boldsymbol{x}^{(1)}$ in the cell while $\boldsymbol{r} = \boldsymbol{x}^{(2)} - \boldsymbol{x}^{(1)}$ is kept fixed, yields precisely the same result as obtained with the simple cell model. The second term represents a correction due to correlations between the positions of molecules in pairs of cells. Higher terms represent corrections for more complicated cell clusters. These corrections have not yet been investigated numerically, but their effect on the thermodynamic functions is known and will be described later (Sect. 23).

Up to the present the cell model has not provided as accurate a method for the determination of $g(r)$ as other types of approximation which will be discussed in the next two sections. However, it has not yet been thoroughly tested, and may prove to be the most effective available in the region adjacent to the melting curve.

21. Calculations based on kinetic theory. The remaining methods for the calculation of $g(r)$ in liquids are based on another type of approximation, which will be discussed before proceeding further. It can be formulated as follows. If the distribution function for a group of q molecules in the liquid is expressed in the form

$$n_q = n^q \exp(-\beta\,\omega_q), \tag{21.1}$$

then ω_q approaches the value zero when the positions $\boldsymbol{x}^{(1)}, \boldsymbol{x}^{(2)}, \ldots, \boldsymbol{x}^{(q)}$ are all far apart. According to the general principles of statistical mechanics, ω_q may be regarded as the mean energy of interaction of the molecules in their fluid environment: it includes their indirect interaction *via* other molecules as well as their direct interaction energy. Thus $-\frac{\partial \omega_q}{\partial \boldsymbol{x}^{(i)}}$ is the mean force exerted on the molecule at $\boldsymbol{x}^{(i)}$, when other molecules are situated at the other points $\boldsymbol{x}^{(1)} \ldots \boldsymbol{x}^{(q)}$. Consider now the energies

$$\left.\begin{aligned} \varepsilon_3 &= \omega_3 - \omega_2^{(23)} - \omega_2^{(31)} - \omega_2^{(12)}, \\ \varepsilon_4 &= \omega_4 - \omega_3^{(234)} - \omega_3^{(341)} - \omega_3^{(412)} - \omega_3^{(123)} + \\ &\quad + \omega_2^{(31)} + \omega_2^{(12)} + \omega_2^{(23)} + \omega_2^{(41)} + \omega_2^{(42)} + \omega_2^{(43)}, \end{aligned}\right\} \tag{21.2}$$

etc., where $\omega_2^{(23)}$ represents the value of the function ω_2 with the arguments $\boldsymbol{x}^{(2)}$ and $\boldsymbol{x}^{(3)}$ instead of $\boldsymbol{x}^{(1)}$ and $\boldsymbol{x}^{(2)}$. If the mean energy of a group of molecules in the fluid environment were, as it is outside the liquid, the sum of their mean energies in pairs, ε_3, ε_4, etc. would vanish. It should be stated at once that the ε_q do not vanish, though they must clearly become very small for sufficiently large values of q. The real question is how large q must be before it is a good approximation to set ε_q equal to zero. The neglect of ε_3 is called the superposition approximation. The same approximation can be formulated by writing

$$n^3 n_3 = n_2^{(12)}\, n_2^{(23)}\, n_2^{(31)}. \tag{21.3}$$

KIRKWOOD and BOGGS[1] seem to have been the first to make practical use of the superposition approximation, though it was suggested by KIRKWOOD[2] previously. It was later adopted by BORN and GREEN[3] in an attempt to calculate $g(r)$ for any simple liquid with central forces between the molecules. Attempts to check the approximation directly have been made by several authors[4], by calculating the virial coefficients of gases exactly and with the help of the approximation. It was found that though the second and third virial coefficients are not affected by the approximation, it entails an error of 20 to 25% in the fourth virial coefficient, and probably even more in the fifth virial coefficient. Moreover, the higher virial coefficients inferred from the formulae (9.3) and (9.2) do not agree when the superposition approximation is used to determine $g(r)$, the former giving a value larger, and the latter a value smaller than the correct value. To be set against this is the fact that the higher virial coefficients do not enter very sensitively into the equation of state, even near the point of

[1] J. G. KIRKWOOD and E. M. BOGGS: J. Chem. Phys. **10**, 394 (1942).
[2] J. G. KIRKWOOD: J. Chem. Phys. **3**, 300 (1935).
[3] M. BORN and H. S. GREEN: Proc. Roy. Soc. Lond., Ser. A **188**, 10 (1946). — H. S. GREEN: Proc. Roy. Soc. Lond., Ser. A **189**, 103 (1947).
[4] B. R. A. NIJBOER and L. VAN HOVE: Phys. Rev. **85**, 777 (1952). — G. S. RUSHBROOKE and H. I. SCOINS: Proc. Roy. Soc. Lond., Ser. A **216**, 203 (1953). — B. R. A. NIJBOER and R. FIESCHI: Physica, Haag **19**, 549 (1953).

condensation and the critical point. McLELLAN[1] found excellent agreement between isotherms which he calculated, using the superposition approximation, for liquid argon in the neighbourhood of the critical point, and those experimentally determined. On the validity of the approximation for the liquid state, little can be said *a priori*, since there is no known relation between any finite number of virial coefficients and the thermodynamical properties of the liquid. The only course open is to calculate the radial distribution function using the approximation, and to make direct comparisons with experimental data, or with the results of other and perhaps better approximations, such as that obtained by the neglect of ε_4. It is also possible to apply the approximation in different ways—for example, as described in the following Sects. 21 to 23 and thereby test the consistency of the results. The evidence which has been obtained in this way indicates that, though the error introduced by the superposition approximation is not completely negligible, it is less than that involved in any other approximation so far applied to the liquid at high temperatures though, in the region adjacent to the melting curve, the cell-lattice model will prove to be superior. The methods about to be described are therefore complementary to those outlined in the previous section.

The mean force acting on a molecule at $\boldsymbol{x}^{(1)}$, when a second molecule is situated at $\boldsymbol{x}^{(2)}$, is compounded of the force $-\frac{\partial\phi^{(12)}}{\partial\boldsymbol{x}^{(1)}}$ due to this second molecule, and the force due to all the other molecules. Since the probability of finding a third molecule in the volume elements $d^3x^{(3)}$ is n_3/n_2, the force due to all the other molecules is $-\int\frac{n_3}{n_2}\frac{\partial\phi^{(13)}}{\partial\boldsymbol{x}^{(1)}}d^3x^{(3)}$. The resultant force must also be equal to $-\frac{\partial\omega_2}{\partial\boldsymbol{x}^{(1)}}$, i.e., $\frac{1}{\beta n_2}\frac{\partial n_2}{\partial\boldsymbol{x}^{(1)}}$, according to (21.1). So

$$\frac{1}{\beta n_2}\frac{\partial n_2}{\partial\boldsymbol{x}^{(1)}}+\frac{\partial\phi^{(12)}}{\partial\boldsymbol{x}^{(1)}}+\int\frac{n_3}{n_2}\frac{\partial\phi^{(13)}}{\partial\boldsymbol{x}^{(1)}}d^3x^{(3)}=0. \tag{21.4}$$

This equation is exact, in thermodynamic equilibrium provided there are no external forces; it was originally obtained by YVON[2]. There are several other ways of deriving the equation; for example, it can be verified by using the results (13.3), with $q=2$ and 3, and (13.15). To use it for the calculation of $n_2(r)$, however, it is necessary to make some substitution for n_3. BORN and GREEN[3,4] proposed to employ (21.3), which reduces it to

$$\frac{1}{\beta}\frac{\partial}{\partial\boldsymbol{r}}\{\ln n_2(r)+\beta\phi(r)\}=\frac{1}{n^3}\int n_2(s)\,n_2(|\boldsymbol{r}+\mathbf{s}|)\,\frac{\partial\phi(s)}{\partial\boldsymbol{s}}\,d^3s, \tag{21.5}$$

where $\boldsymbol{r}=\boldsymbol{x}^{(2)}-\boldsymbol{x}^{(1)}$ and $\mathbf{s}=\boldsymbol{x}^{(1)}-\boldsymbol{x}^{(3)}$; they integrated this equation and obtained, as a result, the scalar equation

$$\ln\left\{\frac{n_2(r)}{n^2}\right\}+\beta\phi(r)=\left(\frac{\pi\beta}{n^3}\right)\int\limits_0^\infty\int\limits_{-s}^{s}(s^2-t^2)\left(1+\frac{t}{r}\right)\{n_2(|r+t|)-n^2\}\,dt\,n_2(s)\,\phi'(s)\,ds. \tag{21.6}$$

[1] A. G. McLELLAN: Proc. Roy. Soc. Lond., Ser. A **210**, 509 (1951) and communication to author (H. S. GREEN: Molecular Theory of Fluids, pp. 107—108). Amsterdam 1952.

[2] J. YVON: La Théorie Statistique des Fluides et l'Équation d'Etat. (Actualités Scientifiques et Industrielles No. 203.) Paris 1935.

[3] M. BORN and H. S. GREEN: Proc. Roy. Soc. Lond., Ser. A **188**, 10 (1946). Contrary to statements which have been published, nothing like equation (21.6) appears in the work of YVON.

[4] H. S. GREEN: Proc. Roy. Soc. Lond., Ser. A **189**, 103 (1947).

α) Approximate methods. A non-linear integral equation of the type (21.6) cannot be solved analytically; the author therefore introduced a further approximation. Writing

$$f(r) = \ln\{n_2(r)/n^2\} + \beta\phi(r), \tag{21.7}$$

and

$$\alpha(r) = \exp\{-\beta\phi(r)\} - 1, \tag{21.8}$$

products of three or more factors $f(s)$ or $f(|r+t|)$ on the right-hand side of (21.6) are neglected, so that one obtains

$$\left.\begin{aligned} r f(r) = -\pi n \int_0^\infty \int_{-s}^{s} (s^2 - t^2)(r+t)\left[a(|r+t|)\{1+f(|r+t|)\} + f(|r+t|)\right] \times \\ \times\, dt\, \alpha'(s)\{1+f(s)\}\, ds. \end{aligned}\right\} \tag{21.9}$$

RODRIGUEZ[1], who examined the application of the method in detail, estimated the error introduced in passing from (21.6) to (21.9) at about 1%. A somewhat larger error is probably involved in the final steps required to make the equation tractable. Bearing in mind that the functions $\alpha(r)$ and $\alpha'(r)$ tend rapidly to zero beyond the first co-ordination peak, $1+f(r)$ is replaced by an average value ε in the neighbourhood of this peak, where it occurs in the integrand multiplied by these functions. Also $\alpha(|r+t|)$ and $f(|r+t|)$ are replaced by $\alpha(r+t)$ and $f(r+t)$, on the understanding that $\alpha(r)$ and $f(r)$ will be defined for negative values by $\alpha(-r) = \alpha(r)$ and $f(-r) = f(r)$. Then, after an integration by parts, (21.9) becomes

$$r f(r) = 2\pi n \int_0^\infty \int_{-s}^{s} (r+t)\{f(r+t) + \varepsilon\alpha(r+t)\}\, dt\, \varepsilon\,\alpha(s)\, s\, ds. \tag{21.10}$$

This integral equation can be solved by Fourier transformation. Defining

$$\left.\begin{aligned} s\, g(s) &= i\int_{-\infty}^{\infty} r f(r)\, e^{-irs}\, dr, \\ s\beta(s) &= i\int_{-\infty}^{\infty} r\,\alpha(r)\, e^{-irs}\, dr, \end{aligned}\right\} \tag{21.11}$$

one finds that (21.10) reduces to

$$g(s) = 2\pi n\,\varepsilon\,\beta(s)\{g(s) + \varepsilon\beta(s)\}, \tag{21.12}$$

and so

$$r\{f(r) + \varepsilon\alpha(r)\} = \frac{1}{2\pi i}\int_{-\infty}^{\infty} \frac{\varepsilon\beta(s)\, e^{irs}\, s\, ds}{1 - 2\pi n\,\varepsilon\,\beta(s)}. \tag{21.13}$$

As $\beta(s)$ can be calculated from (21.11) and (21.8) when the intermolecular interaction energy is known the last equation will serve to determine $f(r)$ and therefore $n_2(r)$. It is necessary, however, to specify the parameter ε; since this was defined as an average value of $1+f(r)$, weighted with function $\alpha(r)$, one has

$$(\varepsilon - 1)\int_{-\infty}^{\infty} \alpha(r)\, r^2\, dr = \int_{-\infty}^{\infty} f(r)\,\alpha(r)\, r^2\, dr, \tag{21.14}$$

or

$$(\varepsilon - 1)\beta(0) = \int_{-\infty}^{\infty} \frac{n\{\varepsilon s\beta(s)\}^2 \beta(s)\, ds}{1 - 2\pi n\,\varepsilon\,\beta(s)} = \frac{f(0)}{2\pi n} - \frac{1}{2\pi}\int_{-\infty}^{\infty} \varepsilon\{s\beta(s)\}^2\, ds. \tag{21.15}$$

This may be regarded as a transcendental equation to determine ε; it was solved numerically for liquid argon by RODRIGUEZ, who found that in the liquid ε was appreciably different from 1 but almost independent of density.

---

[1] A. E. RODRIGUEZ: Proc. Roy. Soc. Lond., Ser. A **196**, 73 (1948).

In the vapour, n is small and there are no real roots s_k of the equation

$$2\pi\, n\, \varepsilon\, \beta(s_k) = 1. \tag{21.16}$$

When the density is increased, however, a pair of real roots $\pm s_0$ appear, first with $s_0=0$, and afterwards with increasing values of s_0. A study of the thermodynamical properties of the fluid shows that the appearance of real roots corresponds to a phase transition, and that there are therefore always a pair of real roots of (21.16) at liquid densities. It is therefore necessary to supplement the formula (21.13) with a statement of how the integral is to be evaluated in the neighbourhood of the singularities of the integrand. To ascertain the correct method of integration, it should be remarked [see (9.2)] that the integral $\int \{n_2(r) - n^2\}\, d^3 r$ determines the isothermal compressibility as well as the scattering of light and X-radiation at small angles and must therefore converge. The integral $\int_r^\infty f(r)\, r^2\, dr$ must therefore also converge; and this can only be ensured by allowing the path of integration with respect to s, in (21.13), to pass above the singularities at $\pm s_0$, into the upper half of the complex plane: The function

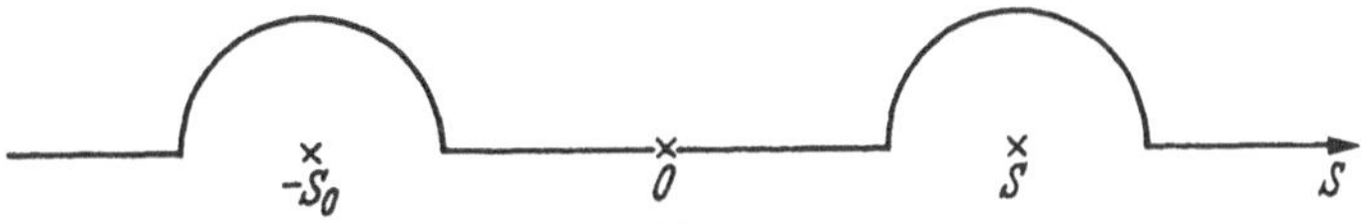

Fig. 4.

$f(r)$ obtained from (21.13) with this particular path of integration is of course not the only solution of (21.10); any expression of the form $f(r) + \{A \exp(i r s_0) + B \exp(-i r s_0)\}/r$ is also a solution. SCOINS[1] obtained a particular solution of this type. But, in order that the solution should be physically acceptable, it is necessary that A and B should vanish.

If $\beta(s)$ is not too singular on the infinite semi-circle in the upper half of the complex plane, the integral in (21.13) can be evaluated in terms of the complex roots of the equation (21.16). By the theory of residues, one has

$$r\{f(r) + \varepsilon\, \alpha(r)\} = -\sum{}' \frac{s_k \exp(i r s_k)}{(2\pi n)^2\, \varepsilon\, \beta'(s_k)}, \tag{21.17}$$

where the summation is over only the roots in the upper half of the complex plane. This formula was used by RODRIGUEZ[1] to calculate $f(r)$ for various temperatures and densities in liquid argon; he tested his results by computing the equation of state and obtained qualitative, if not quantitative agreement with experiment (see Sect. 23). In view of the extreme sensitivity of the equation of state to changes in $n_2(r)$, his results may be regarded as fairly accurate. FOURNET[2] obtained a more direct confirmation of (21.17) by using it to compute the expected intensity of X-radiation scattered from liquid argon as a function of the scattering angle. He found a respectable agreement between predictions based on the above theory and the direct experimental measurements of EISENSTEIN and GINGRICH[3]. Indeed, the evaluation of this evidence suggests that $n_2(r)$ or $g(r)$ can be computed fairly accurately by the method just described. However, for certain purposes even greater accuracy is required, and has been obtained by the direct solution of (21.6).

[1] H. SCOINS: Phil. Mag. **43**, 806 (1952).
[2] G. FOURNET: C. R. Acad. Sci., Paris **228**, 1421, 1801 (1949).
[3] A. EISENSTEIN and N. S. GINGRICH: Phys. Rev. **62**, 261 (1942).

(Added 1958.) Further evidence that the method just described affords an accurate calculation of the radial distribution function has been obtained by LING[1].

Using the method of FOURNET, he has compared the calculated curves for liquid sodium and potassium with those derived from X-ray diffraction data. The agreement between theory and observation is again remarkably good. LING has also extended this work to liquid mixtures, in particular to molten alloys of sodium and potassium. An excellent agreement between calculated and experimental values is maintained.

β) Exact methods. The work which will now be described does not require any approximation, beyond the superposition approximation used in the derivation of (21.6). First, however, mention should be made of the formal method of solution devised by CHENG[2], who transformed the integral equation (21.6) to a differential equation of infinite order by expanding part of the kernel in a power series:

$$(t+r)\{n_2(|t+r|)-n^2\}=\sum_{k=0}^{\infty}\frac{t^k}{k!}\left(\frac{d}{dr}\right)^k[r\{n_2(r)-n^2\}]. \tag{21.18}$$

The integral equation then assumes the form

$$r[\ln\{n_2(r)/n^2\}+\beta\phi(r)]=\sum_{k=0}^{\infty}A_k\left(\frac{d}{dr}\right)^k[r\{n_2(r)-n^2\}] \tag{21.19}$$

where

$$A_k=\frac{\beta\pi}{n^3 k!}\int_0^{\infty}\int_{-s}^{s}(s^2-t^2)\,t^k\,dt\,n_2(s)\,\phi'(s)\,ds. \tag{21.20}$$

The coefficients A_k vanish when k is odd, and are suitably small when k is even but large, especially at fairly high temperatures. It would therefore appear legitimate to neglect contributions to the right- hand side of (21.19) for $k=4$ or greater; the equation could then be solved numerically, or, in a linear approximation, in terms of known functions.

MCLELLAN[3] and KIRKWOOD and his associates[4] have independently obtained numerical solutions of (21.6). MCLELLAN's method has the merit of simplicity, which allows the computations to be performed by hand. On the other hand, KIRKWOOD's results, obtained with the help of specialized computing equipment, are more accurate and cover a wider region of the liquid state. Accordingly, MCLELLAN's method will be described, but the numerical results discussed will be due exclusively to KIRKWOOD and his collaborators. It may be mentioned, however, that where MCLELLAN's results are comparable with KIRKWOOD's, they are in substantial agreement.

MCLELLAN adopted a solution of (21.6) with the form

$$n_2(r)=n^2\exp\{-\beta\phi(r)\}\sum_{k=0}^{\infty}c_k r^k, \qquad (r<r_1) \tag{21.21}$$

[1] R. C. LING: J. Chem. Phys. **25**, 609, 614 (1956).

[2] K. C. CHENG: Proc. Phys. Soc. Lond. A **63**, 1028 (1950).

[3] A. G. MCLELLAN: Communication to author, 1950 (published pp. 80—81 and 107—108, H. S. GREEN: Molecular Theory of Fluids. Amsterdam 1952). — Proc. Roy. Soc. Lond., Ser. A **210**, 509 (1951).

[4] J. G. KIRKWOOD, E. K. MAUN and B. J. ALDER: J. Chem. Phys. **18**, 1040 (1950). — J. G. KIRKWOOD, V. A. LEWINSON and B. J. ALDER: J. Chem. Phys. **20**, 929 (1952). — R. W. ZWANZIG, J. G. KIRKWOOD, K. F. STRIPP and I. OPPENHEIM: J. Chem. Phys. **21**, 1268 (1953).

for values of r less than r_1, chosen about 50% greater than the distance of the minimum of the intermolecular interaction energy $\phi(r)$. For values of r greater than r_1, one could assume

$$n_2(r) = n^2 \exp\{-\beta\phi(r)\} \sum_{k=0}^{\infty} c'_k r^{-k}, \qquad (r > r_1) \tag{21.22}$$

Table of $\Psi_0(x)$.

x	$\lambda(B)$					
	1	5	10	20	27.4	33
1.00	0.095	0.368	0.587	0.859	0.980	1.040
1.02			0.570	0.834	0.957	
1.04	0.092	0.349	0.552	0.808	0.933	1.016
1.06			0.534	0.780	0.907	
1.08	0.088	0.329	0.515	0.751	0.879	0.980
1.12	0.083	0.305	0.476	0.691	0.817	0.928
1.16	0.079	0.283	0.436	0.627	0.745	0.861
1.20	0.074	0.260	0.392	0.556	0.665	0.778
1.24	0.069	0.236	0.350	0.483	0.575	0.673
1.32	0.059	0.188	0.261	0.329	0.371	0.414
1.40	0.048	0.141	0.174	0.170	0.148	0.108
1.48	0.037	0.091	0.089	0.014	−0.077	−0.204
1.56	0.027	0.048	0.013	−0.128	−0.281	−0.491
1.64	0.017	0.011	−0.052	−0.240	−0.441	−0.709
1.72	0.009	−0.020	−0.101	−0.313	−0.536	−0.828
1.80	0.002	−0.041	−0.131	−0.338	−0.548	−0.816
1.88	−0.002	−0.051	−0.133	−0.301	−0.460	−0.657
1.96	−0.004	−0.051	−0.106	−0.194	−0.264	−0.341
2.00	−0.004	−0.040	−0.077	−0.110	−0.122	−0.125
2.04	−0.003	−0.032	−0.051	−0.030	0.021	0.094
2.12	−0.002	−0.016	−0.007	0.089	0.227	0.412
2.20	−0.002	−0.003	0.021	0.155	0.333	0.580
2.28	−0.001	0.003	0.037	0.175	0.354	0.600
2.36	−0.001	0.007	0.041	0.161	0.305	0.493
2.00	0.000	0.008	0.039	0.124	0.207	0.302
2.52		0.008	0.030	0.069	0.083	0.061
2.60		0.006	0.019	0.013	−0.042	−0.173
2.68		0.004	0.008	−0.037	−0.150	−0.369
2.76		0.003	−0.003	−0.074	−0.221	−0.495
2.84		0.001	−0.010	−0.093	−0.247	−0.552
2.92		0.000	−0.012	−0.090	−0.222	−0.448
3.00		−0.001	−0.014	−0.071	−0.155	−0.283

but as the precise form of the function at large distances is unimportant, the series can be truncated near the beginning without appreciable error, and McLellan found he could replace the entire series by a constant. When the trial solutions (21.21) and (21.22) are substituted in the integral equation, the latter reduces to

$$r \ln\Big(\sum_k c_k r^k\Big) = n \sum_{k,l} c_k c_l I_{kl}(r) \tag{21.23}$$

where the integrals

$$\left.\begin{aligned} I_{kl}(r) = \beta\pi \int_0^\infty \int_{-s}^{s} (s^2 - t^2)(t + r)\,[|t + r|^k \exp\{-\beta\phi(|t + r|)\} - 1]\,dt\,\times \\ \times\, s^l \exp\{-\beta\phi(s)\}\phi'(s)\,ds \end{aligned}\right\} \tag{21.24}$$

depend implicitly on β as well as r. One first has to compute these integrals, for the values $k = 0, 1, 2, \ldots$ and $l = 0, 1, 2, \ldots$; the constants c_k in (21.23) can then be obtained by an iterative procedure.

KIRKWOOD's method, on the other hand, involves an expansion of the type

$$x \ln \{n_2(r)/n^2\} = \Psi_0(x) + \beta\varepsilon\Psi_1(x) + (\beta\varepsilon)^2\Psi_2(x) + \cdots, \quad (x > 1) \tag{21.25}$$

where $x = r/a$, and the parameters ε and a are those which appear in the interaction potential

$$\phi(r) = 4\varepsilon\left\{\left(\frac{a}{r}\right)^{12} - \left(\frac{a}{r}\right)^{6}\right\}, \tag{21.26}$$

Table of $\Psi_1(x)$

x	$\lambda^{(B)}$			
	5	10	20	27.4
1.00	−0.586	−0.789	−0.920	−0.956
1.02	−0.183	−0.382	−0.508	−0.542
1.04	0.099	−0.095	−0.218	−0.247
1.06	0.290	0.100	-0.016	−0.043
1.08	0.414	0.230	0.119	0.095
1.12	0.529	0.353	0.253	0.236
1.16	0.534	0.367	0.280	0.268
1.20	0.486	0.330	0.255	0.245
1.24	0.411	0.266	0.203	0.196
1.32	0.250	0.127	0.081	0.079
1.40	0.109	0.008	−0.019	−0.019
1.48	0.006	−0.074	−0.085	−0.082
1.56	−0.064	−0.123	−0.115	−0.109
1.64	−0.105	−0.142	−0.120	−0.108
1.72	−0.123	−0.139	−0.105	−0.086
1.80	−0.120	−0.117	−0.077	−0.055
1.88	−0.101	−0.084	−0.045	−0.027
1.96	−0.070	−0.046	−0.019	−0.011
2.00	−0.053	−0.026	−0.010	−0.009
2.04	−0.030	0.003	0.022	0.019
2.12	0.014	0.063	0.106	0.127
2.20	0.041	0.094	0.137	0.159
2.28	0.051	0.092	0.116	0.118
2.36	0.051	0.073	0.066	0.043
2.44	0.040	0.044	0.008	−0.033
2.52	0.029	0.017	−0.036	−0.091
2.60	0.017	−0.008	−0.067	−0.123
2.68	0.006	−0.026	−0.082	−0.125
2.76	−0.003	−0.035	−0.080	−0.105
2.84	−0.007	−0.037	−0.064	−0.066
2.92	−0.009	−0.033	−0.041	−0.019
3.00	−0.009	−0.022	−0.009	0.032

Table of $\Psi_2(x)$

x	$\lambda^{(B)}$			
	5	10	20	27.4
1.00	0.078	0.025	−0.038	−0.048
1.02	0.077	0.025		−0.046
1.04	0.075	0.024	−0.034	−0.044
1.06	0.074	0.024		−0.042
1.08	0.073	0.024	−0.031	−0.041
1.12	0.069	0.023	−0.028	−0.038
1.16	0.067	0.021	−0.025	−0.036
1.20	0.065	0.019	−0.023	−0.034
1.24	0.063	0.018	−0.021	−0.033
1.32	0.058	0.015	−0.017	−0.028
1.40	0.053	0.013	−0.013	−0.022
1.48	0.048	0.011	−0.008	−0.013
1.56	0.045	−0.010	−0.002	−0.001
1.64	0.044	0.010	0.003	0.010
1.72	0.044	0.012	0.009	0.024
1.80	0.046	0.015	0.018	0.040
1.88	0.049	0.020	0.029	0.054
1.96	0.055	0.027	0.039	0.069
2.00	0.059	0.030	0.043	0.074
2.04	0.062	0.031	0.034	0.056
2.12	0.069	0.033	0.020	0.025
2.20	0.067	0.033	0.017	0.014
2.28	0.057	0.023	0.010	−0.002
2.36	0.042	0.009	−0.005	−0.024
2.44	0.026	−0.005	−0.017	−0.039
2.52	0.011	−0.015	−0.022	−0.046
2.60	0.001	−0.021	−0.023	−0.038
2.68	−0.006	−0.002	−0.017	−0.022
2.76	−0.008	0.007	−0.008	0.001
2.84	−0.007	0.009	0.003	0.022
2.92	−0.005	0.005	0.013	0.039
3.00	−0.001	0.001	0.020	0.048

assumed to be of LENNARD-JONES' type. The functions $\Psi_k(x)$ are independent of temperature, and can be obtained successively by an iteration procedure, requiring the numerical solution of a linear integral equation at each step. The results are expressed in terms of a parameter $\lambda^{(B)}$, defined by

$$\lambda^{(B)} = 4\pi a^3 n_2(r = a)/n, \tag{21.27}$$

and it can be assumed that $n_2(r)$ effectively vanishes for $r < a$. Since the calculations of KIRKWOOD, LEWINSON and ALDER[1] are the most extensive and accurate completed at the time of writing their tables of $\Psi_0(x)$, $\Psi_1(x)$ and $\Psi_2(x)$ are presented, in a convenient form, below.

[1] See footnote 4, p. 56.

22. Alternative methods. In this section additional methods will be described for the computation of the radial distribution function, similar in some respects to the methods of the last section, but depending more on statistical mechanics than on an intuitively simple consideration of the intermolecular forces.

KIRKWOOD and BOGGS[1] were the authors of the first accurate method for calculating $g(r)$ with the help of the superposition approximation. It was originally applicable only to a "liquid" of rigid spherical molecules, and even now is more difficult to apply with general types of molecular interaction than the method already described; however, it is important not only for its influence on later developments, but also because it has provided a direct check on the validity of the superposition approximation. The method was based on the concept of a liquid in which the molecules are all alike, except one molecule which interacts with other molecules less strongly than any other, at the same distance. To be precise, the interaction energy of a group of q molecules, of which the exceptional molecule is at the point $\boldsymbol{x}^{(1)}$, is

$$\Phi_q(\xi) = \sum_{i=2}^{q}\Big(\sum_{j=i+1}^{q} \phi^{(ij)} + \xi\phi^{(1i)}\Big) \tag{22.1}$$

where ξ is a parameter whose value lies between 0 and 1. The radial distribution of molecules about the point $\boldsymbol{x}^{(1)}$ is then described by a function $n_2(r,\xi)$, which assumes the values $n_2(r)$ for $\xi=1$ and n^2 for $\xi=0$. From the formulae

$$n_q(\xi) = \sum_{j=0}^{\infty} \int\cdots\int \{N_{q+j}(\xi)/j!\}\, d^3x^{(q+1)}\ldots d^3x^{(q+j)} \tag{22.2}$$

and

$$N_{q+j}(\xi) = \exp(-\beta p v)\, z^q \exp\{-\beta\,\Phi_q(\xi)\}, \tag{22.3}$$

which are the analogues of (20.4) and (20.5), it follows that

$$\frac{\partial n_2(\xi)}{\partial\xi} + \beta\phi^{(12)} n_2(\xi) + \int \beta\phi^{(13)} n_3(\xi)\, d^3x^{(3)} = 0. \tag{22.4}$$

Assuming now the validity of the superposition approximation in the form

$$n^3 n_3(\xi) = n_2^{(12)}(\xi)\, n_2^{(23)}\, n_2^{(31)}(\xi), \tag{22.5}$$

one obtains the equation

$$\frac{1}{\beta}\frac{\partial}{\partial\xi}\{\log n_2(r,\xi) + \beta\xi\phi(r)\} = -\frac{1}{n^3}\int n_2(s,\xi)\, n_2(|\boldsymbol{r}+\boldsymbol{s}|)\,\phi(s)\, d^3s, \tag{22.6}$$

which can be compared with (21.5). The additional difficulty and disadvantage associated with (22.6) is that one cannot explicitly integrate it to eliminate the parameter, except when the molecules are rigid spheres, so that $\phi(r)$ vanishes when r exceeds the molecular diameter a, and is effectively infinite for $r<a$. However, KIRKWOOD and his associates have succeeded in solving the equation numerically, not only for rigid spherical molecules[2], but, at least partially, for a potential function of LENNARD-JONES' type[3]. For $\lambda^{(B)}=20$ they have computed the function $\Psi_1(x)$ as well as $\Psi_0(x)$ defined by (21.25), using the integro-differential equation (22.6) instead of (21.6). Since the superposition approximation has been used quite differently in formulating these two equations, it is not to be expected that the results obtained from them will be in perfect agreement; indeed, the discrepancy between the results of the two calculations is a measure of the

[1] J. G. KIRKWOOD and E. M. BOGGS: J. Chem. Phys. **10**, 394 (1942).
[2] J. G. KIRKWOOD, E. K. MAUN and B. J. ALDER: J. Chem. Phys. **18**, 1040 (1950).
[3] J. G. KIRKWOOD, V. A. LEWINSON and B. J. ALDER: J. Chem. Phys. **20**, 929 (1952).

error introduced by the superposition approximation. In fact, the agreement between the calculated functions is very fair, indicating that the approximation is quite trustworthy provided extreme accuracy is not essential.

To complete this survey of methods for the computation of the radial distribution function, a method will be described due to MAYER[1] and developed by KIRKWOOD and SALSBURG[2]. Although this method has not yet been put into practice, it appears to offer certain advantages, notably the possibility of dispensing with the superposition approximation and other approximations of the same type. Another feature of the method is that it provides simultaneously expressions for n_2, n_3, n_4, etc.

Again the starting point is the set of equations

$$n_q = \exp(-\beta p v) \sum_{j=0}^{\infty} (z^{q+j}/j!) \int \cdots \int \exp(-\beta \Phi_{q+j})\, d^3x^{(q+1)} \ldots d^3x^{(q+j)}. \tag{22.7}$$

The interaction energy Φ_{q+j} is split into three parts:

$$\Phi_{q+j} = \Phi_{q+j-1}^{(2\ldots q+j)} + \sum_{i=2}^{q} \phi^{(1i)} + \sum_{i=q+1}^{q+j} \phi^{(1i)}, \tag{22.8}$$

where the first part excludes contributions from the molecule at $x^{(1)}$, the second part takes account of interactions between molecules at $x^{(2)} \ldots x^{(q)}$ with the one at $x^{(1)}$, and the third part is the remainder. This third part contributes a factor

$$\prod_{i=q+1}^{q+j} (1 + u^{(1i)})$$

to the integrand of (22.7), if

$$u^{(ij)} = \exp(-\beta \phi^{(ij)}) - 1. \tag{22.9}$$

One may therefore re-write (22.7) in the form

$$\left. \begin{aligned} n_q = \exp\Big\{-\beta\big(p v + \sum_{i=2}^{q} \phi^{(1i)}\big)\Big\} \sum_{j=0}^{\infty} (z^{q+j}/j) \int \cdots \int \times \\ \times \exp(-\beta \Phi_{q+j-1}^{(2\ldots q+j)}) \prod_{i=q+1}^{q+j} (1 + u^{(1i)})\, d^3x^{(q+1)} \ldots d^3x^{(q+j)}. \end{aligned} \right\} \tag{22.10}$$

The product can then be expanded, and the co-ordinates $x^{(q+1)} \ldots x^{(q+j)}$ of integration interchanges as required to bring it into the form

$$1 + j\, u^{(1q+1)} + \tfrac{1}{2} j (j-1)\, u^{(1q+1)} u^{(1q+2)} + \cdots.$$

The formula (22.7) is finally used again to simplify the righthand side of (22.10), which reduces to

$$\left. \begin{aligned} n_q = z \exp\Big\{-\beta\big(p v + \sum_{i=1}^{q} \phi^{(1i)}\big)\Big\} \Big[n_{q-1}^{(2\ldots q)} + \int n_q^{(2\ldots q+1)} u^{(1q+1)}\, d^3x^{(q+1)} \\ + \frac{1}{2!} \int n_{q+1}^{(2\ldots q+2)} u^{(1q+1)} u^{(1q+2)}\, d^3x^{(q+1)}\, d^3x^{(q+2)} + \cdots \Big]. \end{aligned} \right\} \tag{22.11}$$

---

[1] J. E. MAYER: J. Chem. Phys. 15, 187 (1947); see also J. E. MAYER and E. MONTROLL: J. Chem. Phys. 9, 2 (1941).

[2] J. G. KIRKWOOD and Z. W. SALSBURG: Disc. Faraday Soc. No. 15, 28 (1953).

For $q=1, 2, 3 \ldots$ one obtains

$$\left.\begin{aligned}
n &= z \exp(-\beta p v)\left[1 + n \int u^{(12)}\, d^3x^{(2)} + \right.\\
&\quad \left. + \tfrac{1}{3} \iint n_2^{(23)}\, u^{(12)}\, u^{(13)}\, d^3x^{(2)}\, d^3x^{(3)} + \cdots\right],\\
n_2 &= z \exp\{-\beta(p v + \phi^{(12)})\}\left[n + \int n_2^{(23)}\, u^{(13)}\, d^3\, x^{(3)} + \right.\\
&\quad \left. + \tfrac{1}{2} \int n_3^{(234)}\, u^{(13)}\, u^{(14)}\, d^3x^{(3)}\, d^3x^{(4)} + \cdots\right],\\
n_3 &= z \exp\{-\beta(p v + \phi^{(12)} + \phi^{(23)} + \phi^{(31)})\}\left[n_2^{(23)} + \right.\\
&\quad \left. + \int n_3^{(234)}\, u^{(14)}\, d^3x^{(4)} + \cdots\right],
\end{aligned}\right\} \tag{22.12}$$

etc. It is not difficult to see that the number of terms required for the computation of the series in these equations is no greater than the number of molecules which can lie within range of interaction of the molecule at $\boldsymbol{x}^{(1)}$. This is unfortunately sufficiently large to make the exact solution of this set of integral equations a formidable proposition. However, the task is not superhuman and with the further advance of computing methods may be achieved. In the meantime, approximate methods of solution may yield interesting results.

b) Structure-dependent properties.

23. Thermodynamics. It has already been noticed in Sect. 9 that when the radial distribution function is known, the thermodynamic properties of a liquid can be determined. The theoretical methods of the last three sections therefore offer indirect methods of computing the equation of state and internal energy of a simple liquid, and hence the other thermodynamic functions. Indeed, thermodynamic variables depend so sensitively on the radial distribution function, especially in the repulsive part of the intermolecular field, that an error of 0.1% in the latter can easily lead to an error of 20% in the former. Such computations therefore provide a much more exacting test of accuracy than even direct comparison with the distribution function derived from diffraction experiments.

α) The cell model. The cell model, which will be discussed first, does not require one to determine the radial distribution function before the thermodynamic functions can be calculated. If, therefore, one is not so much interested in molecular structure as in the relation between the intermolecular forces and the macroscopic behaviour of a liquid, this model offers a slight advantage, offset by the fact that one needs an accurate value of the mean potential energy of a molecule at any point in its cell, which is closely related to the structure of the liquid.

The first step in any calculation is therefore to obtain a good approximation to the function $\Psi_{q(a),b}$, as explained in Sect. 20. It is then a straightforward matter to compute the integrals

$$I_{q(a)} = \int \exp\Big\{-\beta\Big(\Phi_{q(a)} + \sum_{b(\neq a)} \Psi_{q(a),b}\Big)\Big\}\, d x_{q(a)}, \tag{23.1}$$

and determine the parameters π and ζ using (20.9) and (20.10) in the form

$$\exp(\beta \pi v_a) = \sum_{q(a)} I_{q(a)} \zeta^{q(a)}/q(a)! \tag{23.2}$$

where

$$\sum_{q(a)} \{q(a) - n v_a\}\, I_{q(a)} \zeta^{q(a)}/q(a)! = 0. \tag{23.3}$$

If v_a does not exceed the molecular volume, it is obviously permissible to neglect terms with $q(a) > 2$ and one has

$$\beta \pi v_a \approx \ln(1 + I_1 \zeta + \tfrac{1}{2} I_2 \zeta^2) \tag{23.4}$$

with ζ equal to the greater root of the quadratic

$$(1 - \tfrac{1}{2} n v_a) I_2 \zeta^2 + (1 - n v_a) I_1 \zeta - n v_a \approx 0. \tag{23.5}$$

In the lowest approximation, the factor C defined in (20.14) is equated to unity by the neglect of the $\Theta_{q(a)q(b)}$ and setting

$$\left.\begin{aligned} z &\approx z^{\{1\}} = \zeta, \\ p &\approx p^{\{1\}} = \pi + \tfrac{1}{2} \sum_b \mathrm{X}_{ab}/v_a. \end{aligned}\right\} \tag{23.6}$$

To calculate the pressure directly from (23.6), it would be necessary to have available accurate values of I_2; in their absence, it has been usual to determine the free energy

$$F = \left\{\frac{n}{\beta} \log z - p\right\} v \approx \left\{\frac{n}{\beta} \log \zeta - \pi - \frac{1}{2} \sum_b \mathrm{X}_{ab}/v_a\right\} v \tag{23.7}$$

and then derive the pressure from the relation

$$p = -\frac{\partial F}{\partial v}. \tag{23.8}$$

In view of the approximation made, it is by no means certain that the pressure calculated in this way will agree with that obtained from (23.6); the formula (23.8) has been preferred in practice because it does not depend so sensitively on I_2. MAYER and CARERI[1] have shown, however, that when I_2 is neglected, (23.6) and (23.8) can be brought into agreement by choosing a suitable value of v_a, less than the molecular volume but depending on the temperature and density.

Before reviewing the actual calculations made on the basis of the cell model, some description will be given of the method for obtaining higher approximations, which take account of correlations between configurations in different cells. The fundamental equation is derived from (13.1) with $q = 0$:

$$1 = \sum_{q=0}^{\infty} \int N_q / q! \, dx_q, \tag{23.9}$$

using the expression (20.5) for N_q, (20.7) for Φ_q, and eliminating $\Phi_{q(a)q(b)}$ with the help of (20.16). The result is

$$\left.\begin{aligned} 1 = {} & \exp\Big\{-\beta\Big(p v + \sum_{a>b} \mathrm{X}_{ab}\Big)\Big\} \sum_{q(1),\ldots q(M)} \frac{z^{q(1)}}{q(1)!} \cdots \frac{z^{q(M)}}{q(M)!} \times \\ & \times \int \cdots \int \exp\Big\{-\beta \sum_a \Big(\Phi_{q(a)} + \sum_{b(\neq a)} \Psi_{q(a),b}\Big)\Big\} \Big[1 + \sum_{a>b} \Theta_{q(a)q(b)} + \\ & + \sum_{a>b>c} \big(\Theta_{q(a)q(b)} \Theta_{q(a)q(c)} + \Theta_{q(b)q(c)} \Theta_{q(b)q(a)} + \\ & + \Theta_{q(c)q(a)} \Theta_{q(c)q(b)}\big) + \cdots\Big] dx_{q(1)} \ldots dx_{q(M)}. \end{aligned}\right\} \tag{23.10}$$

[1] J. E. MAYER and G. CARERI: J. Chem. Phys. **20**, 1001 (1952).

Defining, by analogy with (23.1),

$$\left.\begin{aligned} I_{q(a)\,q(b)} = \iint \exp\Big\{-\beta\Big(\Phi_{q(a)} + \Phi_{q(b)} + \sum_{c(\neq a)} \Psi_{q(a),c} + \sum_{c(\neq b)} \Psi_{q(b),c}\Big)\Big\} \times \\ \times\, \Theta_{q(a)\,q(b)}\, dx_{q(a)}\, dx_{q(b)}, \end{aligned}\right\} \tag{23.11}$$

etc., and

$$\left.\begin{aligned} K_a &= \sum_{q(a)} I_{q(a)}\, z^{q(a)}/q(a)!\,, \\ K_{ab} &= \sum_{q(a),\,q(b)} I_{q(a)\,q(b)}\, z^{q(a)+q(b)}/\{q(a)!\, q(b)!\}, \end{aligned}\right\} \tag{23.12}$$

etc., (23.10) can be written

$$\exp\Big\{\beta\Big(pv + \sum_{a>b} \mathrm{X}_{ab}\Big)\Big\} = \Big(\prod_a K_a\Big)\Big\{1 + \sum_{a<b} \frac{K_{ab}}{K_a K_b} + \sum_{a<b<c} \frac{K_{abc}}{K_a K_b K_c} + \cdots\Big\}. \tag{23.13}$$

The right-hand side of this equation is the coefficient of the product $\varphi_1, \varphi_2, \ldots \varphi_M$ in the power series expansion of

$$\exp\Big(\sum_a K_a \varphi_a + \sum_{a<b} K_{ab} \varphi_a \varphi_b + \sum_{a<b<c} K_{abc} \varphi_a \varphi_b \varphi_c + \cdots\Big).$$

Also, if the cells are all alike and are enumerated in a suitable order, K_a will be independent of a, K_{ab} for $a > b$ will depend only on $a-b$, K_{abc} for $a > b > c$ will depend only on $a-b$ and $b-c$, etc. Therefore one may write $K_a = K$ and

$$\left.\begin{aligned} k_{a-b} &= K_{ab}/(K_a K_b), && (a > b) \\ k_{a-b,\,b-c} &= K_{abc}/(K_a K_b K_c), && (a > b > c) \end{aligned}\right\} \tag{23.14}$$

etc., so that (23.13) becomes

$$\exp\Big\{\beta\Big(p\,v + \sum_{a>b} \mathrm{X}_{ab}\Big)\Big\} = \{K\,\Omega(k_a, k_{ab}, \ldots)\}^M, \tag{23.15}$$

where $\{\Omega(k_a, k_{ab}, \ldots)\}^M$ is the coefficient of $\varphi_1 \varphi_2 \ldots \varphi_M$ in the expansion of $\exp\Big(\sum_a \varphi_a + \sum_{a,b} k_b \varphi_a \varphi_{a+b} + \sum_{a,b,c} k_{bc} \varphi_a \varphi_{a+b} \varphi_{a+b+c} + \cdots\Big)$. DE BOER and his associates[1] have obtained approximate expressions for $\Omega(k_a, k_{ab}, \ldots)$ which would presumably be useful if the k_a, k_{ab} were large. In the present context, they are expected to be small and an exact power series expansion of $(\log \Omega)/M$ is more accurate:

$$\log \Omega = \sum_a k_a - \Big(\sum_a k_a\Big)^2 - \tfrac{1}{2} \sum_a k_a^2 + \sum_{a,b} k_{ab} + \cdots. \tag{23.16}$$

(The author has obtained further terms in this expansion.)

To determine both p and z, an additional equation is required beyond (23.15). This can be derived from (13.1) with $q = 1$ in the same way, or more simply by using the thermodynamical relation, valid at constant temperature,

$$\beta \frac{dp}{dn} = \frac{n}{z} \frac{dz}{dn} \tag{23.17}$$

—which, in conjunction with (23.15), leads to

$$\frac{1}{K} \frac{\partial K}{\partial z} + \frac{1}{\Omega} \frac{\partial \Omega}{\partial z} = \frac{n v_a}{z}\,. \tag{23.18}$$

The above analysis enables one in principle to make calculations as exact as required. In practice a relatively simple computation results from neglecting

[1] E. G. D. COHEN, J. DE BOER and Z. W. SALSBURG: Physica, Haag 21, 137 (1955).

the k_{ab}, k_{abc}, etc., since then, by choosing the X_{ab} as suggested in (20.17), the k_a vanish identically, and the cell model as previously described can be used. Alternatively, the approximation (20.18) can be employed, and the k_a will then be small but non-vanishing quantities which can be computed, using (23.14), (23.12), (23.11) and (23.1) above. Most calculations so far carried through have, however, used expressions for the X_{ab} differing numerically from either (20.17) or (20.18) to a degree which would make corrections arising from cell clusters quite large; and it is not surprising that the calculated pressure, for example, was usually incorrect by a factor of 3 or 4. Before the true limitations of the cell model can be said to have been established, it will be necessary to revise these calculations. For that purpose some existing numerical work may, however, be useful, and this will be mentioned here.

Various approximations are available for

$$\Psi_{q(a)} = \sum_{b(\neq a)} \Psi_{q(a),b}, \tag{23.19}$$

which is the mean interaction energy between the $q(a)$ molecules with assigned positions in the a-th cell and those outside, neglecting correlations between configurations in different cells. Setting first $q(a)=1$, one has obviously

$$\left.\begin{aligned} \Psi_{1,b} &= \psi_b(\boldsymbol{x}^{(1)} - \boldsymbol{x}_a), \\ &= \int \phi(|\boldsymbol{x}^{(2)} - \boldsymbol{x}^{(1)}|)\, \sigma(|\boldsymbol{x}^{(2)} - \boldsymbol{x}_b|)\, d^3 x^{(2)} \end{aligned}\right\} \tag{23.20}$$

where $\boldsymbol{x}_a$ is the centre of the a-th cell, and $\sigma(|\boldsymbol{x}^{(2)} - \boldsymbol{x}_b|)$ is the function, given by (20.11), which measures the probability density of finding a molecule at the point $\boldsymbol{x}^{(2)}$ in the cell whose centre is $\boldsymbol{x}_b$.

On summing over b, one can average with respect to the mutual orientation of $\boldsymbol{r}_{ab} = \boldsymbol{x}_b - \boldsymbol{x}_a$ and $\boldsymbol{s} = \boldsymbol{x}^{(1)} - \boldsymbol{x}_a$, obtaining

$$\left.\begin{aligned} \Psi_1 &= \sum_b \Psi_{1,b} = \psi(s), \\ \psi(s) &= \sum_b \int \phi_{av}(s, |\boldsymbol{x}^{(2)} - \boldsymbol{x}_a|)\, \sigma(|\boldsymbol{x}^{(2)} - \boldsymbol{x}_b|)\, d^3 x^{(2)}, \\ \phi_{av}(s,t) &= \frac{1}{2st} \int_{t-s}^{t+s} \phi(r)\, r\, dr, \end{aligned}\right\} \tag{23.21}$$

or

$$\psi(s) = \sum_b \int_0^c \frac{1}{2 r_{ab} s_b} \int_{r_{ab}-s_b}^{r_{ab}+s_b} \phi_{av}(s,t)\, t\, dt\, \sigma(s_b)\, 4\pi s_b^2\, d s_b. \tag{23.22}$$

Extensive calculations have been made, using the rather poor approximation $4\pi s^2 \sigma(s) = \delta(s)$ to determine $\psi(s)$, by WENTORF, BUEHLER, HIRSCHFELDER and CURTISS[1]. They employed a slightly modified version of the model of LENNARD-JONES and DEVONSHIRE, in which each cell contains only one molecule, and presented comprehensive tables sufficient to determine I_1 and its derivatives with respect to density and temperature.

Similar but slightly more restricted tables, sufficient to determine also the second derivatives of I_1 with respect to temperature and density, have been given by SALSBURG and KIRKWOOD[2] in an application of the method of LENNARD-JONES and DEVONSHIRE to liquid mixtures. A very restricted table of I_2/I_1^2

[1] R. H. WENTORF, R. J. BUEHLER, J. O. HIRSCHFELDER and C. F. CURTISS: J. Chem. Phys. 18, 1484 (1950).

[2] Z. W. SALSBURG and J. G. KIRKWOOD: J. Chem. Phys. 21, 2169 (1953).

can be found in a paper of POPLE[1], and analytical expressions suitable for computational purposes were published by JANSSENS and PRIGOGINE[2]. These will not be reproduced here, as improved calculations will surely be available in the near future.

(Added 1958.) Considerable progress has, indeed, been made in this field. Among a large number of contributions, reference may be made to those of TAYLOR, BARKER, and LEVINE, MAYER and AROESTE[3].

A cell-cluster theory has recently been published by W. J. TAYLOR. Though of restricted value, as multiply occupied cells are excluded from consideration, it is of historical interest since it appears to have originated in 1951 and would at that time have been the first theory of its type.

BARKER has made some very useful calculations designed to test the influence of binary and multiple correlations on the thermodynamics of liquids. By comparing the calculated free energy with experimental values, he concludes that most of the discrepancy can be removed by taking account of binary correlations (i.e., those between molecules in two neighbouring cells) and cells containing not more than three molecules. It is less certain, on the basis of his later work, that multiple correlations are negligible in a liquid near its freezing point, and very doubtful that they are in a crystal.

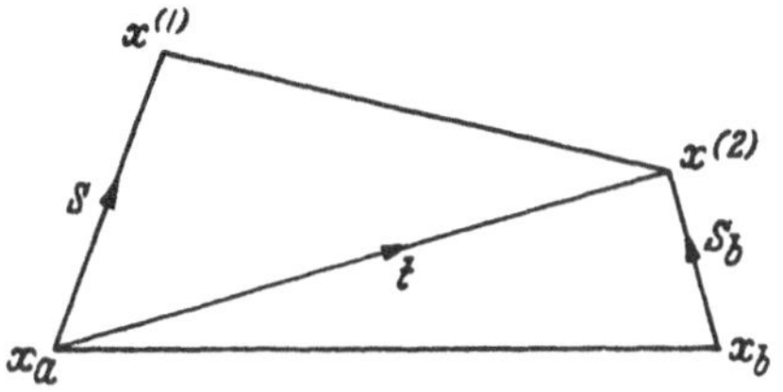

Fig. 5.

LEVINE, MAYER and AROESTE have also performed important calculations which serve to show the influence of interactions of neighbours, other than nearest neighbours, on the thermodynamic functions. It is clear that taking these into account will remove an important source of error.

β) Kinetic theory. The most accurate calculations of the thermodynamic variables so far achieved have been made using the radial distribution function, as determined by the methods of Sect. 21. They have therefore been based on the superposition approximation. The interaction energy $\phi(r)$ between two molecules has generally been assumed to be of the form $4\varepsilon\{(a/r)^{12}-(a/r)^6\}$; the results can then be expressed in terms of the dimensionless variables $n a^3$ and $\beta\varepsilon$, and applied to any liquid for which the assumed form of interaction is applicable. The monatomic liquids, with the exception of liquid helium, do appear to possess a common equation of state of the type $p a^3/\varepsilon = p^*(n a^3, \beta\varepsilon)$, where p is the pressure and p^* a function of the variables indicated, though of course the values of a and ε vary from liquid to liquid. Even inert diatomic and polyatomic liquids conform to the same equation of state fairly well.

The approximate theory of Sect. 21α was tested by RODRIGUEZ[4], who, using the formula (15.14), calculated two isotherms for argon below the critical temperature. His results were qualitatively satisfactory, though quantitatively unreliable, as was to be expected in view of the approximations, beside the superposition approximation, on which they were based. More exact numerical calculations, based on the methods of Sect. 21β, have since been completed, with

---

[1] J. A. POPLE: Phil. Mag. **41**, 459 (1951).

[2] P. JANSSENS and I. PRIGOGINE: Physica, Haag **16**, 895 (1950).

[3] W. J. TAYLOR: J. Chem. Phys. **24**, 454 (1956). — J. A. BARKER: Proc. Roy. Soc. Lond., Ser. A **237**, 63 (1956); **241**, 547 (1957). — H. B. LEVINE, J. E. MAYER and H. AROESTE: J. Chem. Phys. **26**, 207 (1957).

[4] A. E. RODRIGUEZ: Proc. Roy. Soc. Lond., Ser. A **196**, 73 (1948).

results in better quantitative agreement with experiment. To indicate the closeness of this agreement, some figures will be quoted from the work of ZWANZIG, KIRKWOOD, STRIPP and OPPENHEIM[1]. These authors employed a radial distribution function computed from the tables of $\Psi_k(x)$, reproduced on pp. 57 to 58 for $x = r/a \geq 1$, and for $x < 1$ assumed $n_2(r) = \nu^2 \exp\{-\beta\,\phi(r)\}$, with ν chosen so as to ensure the continuity of the function at $x = 1$. The values $\varepsilon = 1.653 \times 10^{-14}$ erg and $a = 3.405 \times 10^{-8}$ cm assumed for argon were those determined by MICHELS, WIJKER and WIJKER[2] by comparison of theoretical and experimental values of the second virial coefficient of the gas. The calculated critical pressure was $0.147\,\varepsilon/a^3$, against the experimental value $0.116\,\varepsilon/a^3$; the calculated critical volume was $2.89a^3$ per molecule, against the experimental value $3.17a^3$ per molecule; and the calculated critical temperature was $1.48\,\varepsilon/k$ °K, against the experimental value $1.25\,\varepsilon/k$ °K. These are all quantities which could be in error due to an incorrect choice of ε or a; the value of $\Phi = \beta p/n$ at the critical point, on the other hand, could not be in error for this reason, and in fact the calculated value is in error by less than $2\frac{1}{2}$%.

At temperatures not too far below the critical temperature, discrepancies between the calculated and experimentally determined equation of state could be removed by decreasing the value of a by a factor 1.006. At lower temperatures, i.e., below 140 °K in liquid argon, where the critical temperature is 154 °K, KIRKWOOD and his collaborators found there was a tendency for the calculated pressure to fall away and become negative, so that to obtain agreement with experiment it was necessary to allow the parameter a to vary slowly with temperature. However, even near the melting point, a decrease of only $2\frac{1}{2}$% in the value of a was needed to remove the discrepancy between theory and experiment. In view of the extreme sensitivity of the pressure to the molecular structure and therefore to the law of intermolecular interaction, it is quite possible that some of these discrepancies may be attributed to small inaccuracies in the computations or in the assumed form of $\phi(r)$. It is probable, however, that the use of the superposition approximation also leads to small errors in the calculated function $g(r)$, which are magnified in the computation of the equation of state.

To conclude this section, a direct relation will be established between the radial distribution function and the isothermal compressibility of a liquid. The reader is reminded of the formula

$$N_q = e^{-\beta p v} z^q \exp(-\beta\,\Phi_q) \tag{23.23}$$

which follows from (13.13) and (13.15). By differentiation with respect to the molecular number density, keeping the temperature fixed, one obtains

$$\frac{\partial N_q}{\partial n} = \left(-\beta\,v\,\frac{\partial p}{\partial n} + \frac{q}{z}\,\frac{\partial z}{\partial n}\right) N_q = \frac{\beta}{n}\,\frac{\partial p}{\partial n}\,(q - n\,v)\,N_q, \tag{23.24}$$

since, at constant temperature, $d(\log z) = \beta\,dp/n$. Hence, by differentiating the relation

$$\sum_q \int N_q\,dx_q/q! = 1$$

twice in succession with respect to n, one finds

$$\sum_q \frac{1}{q!}\left\{\frac{\beta}{n}\,\frac{\partial p}{\partial n}\,(q - n\,v)^2 - v\right\} \int N_q\,dx_q = 0 \tag{23.25}$$

[1] R. W. ZWANZIG, J. G. KIRKWOOD, K. F. STRIPP and I. OPPENHEIM: J. Chem. Phys. **21**, 1268 (1953).
[2] A. MICHELS, H. WIJKER and H. WIJKER: Physica, Haag **15**, 627 (1949).

or, making use of (13.3),

$$\frac{\beta}{n}\frac{\partial p}{\partial n}\left\{\iint (n_2 - n^2)\, d^3x^{(1)}\, d^3x^{(2)} + n\, v\right\} = v. \tag{23.26}$$

The integration with respect to $x^{(2)}$ merely produces a factor v, so that one has finally

$$\beta \frac{\partial p}{\partial n}\left\{n \int \{g(r) - 1\}\, d^3r + 1\right\} = 1. \tag{23.27}$$

This formula is very sensitive to errors in values of $g(r)$ calculated for large values of r, and was also used by KIRKWOOD and his collaborators to test the function computed from the tables of Sect. 21 β. ZWANZIG, KIRKWOOD, STRIPP and OPPENHEIM[1] found their calculated value for liquid argon at the normal boiling point was about 45% too high. Such a discrepancy would be difficult to attribute to the choice of an incorrect form of $\phi(r)$, and indicates that the computed values of $g(r)$ were systematically somewhat too high for large values of r. Possibly this error would be reduced if account were taken of Ψ_3, as well as Ψ_0, Ψ_1 and Ψ_2, in (21.25), but it is also possibly an essential consequence of the superposition approximation.

(Added 1958.) Meanwhile, further calculations have been made with the approximate theory of Sect. 21 α.

SUNDHEIM and RUBIN[2] have used it, in conjunction with a simple approximation of their own, to investigate the thermodynamics of the critical region. Their results are very close to those of KIRKWOOD and his associates, indicating that the approximations of Sect. 21 α may introduce a smaller error than hitherto believed. RUBIN[3] has also generalized the theory of Sect. 21 α to fluid mixtures, and investigated the phase transitions in mixtures of two components. His results appear to be in fair agreement with experimental expectations.

24. Deformation by fields of force. In the previous sections of this chapter it was assumed implicitly that the liquid was free of external forces, except possibly a pressure applied at the boundary, due either to the saturated vapour or a solid wall. Under laboratory conditions such an assumption represents an idealization in the sense that gravitational forces are present always, and possibly electromagnetic forces as well. In this section consideration will be given to the effect on the structure and thermodynamic properties of the liquid of a conservative field of force. If the field of force varies only slowly with position on the molecular scale, the principal effect is a variation in density from point to point in the liquid, which is of interest since it reflects the structure of the liquid in the presence of the force field.

The potential energy of a molecule at the point x due to the external field of force will be denoted by $\psi(x)$; the molecular number density, pressure, and other quantities in the presence of this field will be represented by n, p, etc. as usual; their hypothetical values in the absence of the field will be denoted by n^0, p^0, etc. From the equation of motion (15.3) it can be seen that under conditions of mechanical equilibrium

$$\frac{\partial p}{\partial x} = -n\frac{\partial \psi}{\partial x}, \tag{24.1}$$

[1] See footnote 1, p. 66.
[2] B. R. SUNDHEIM and E. L. RUBIN: J. Chem. Phys. 25, 785 (1956).
[3] E. L. RUBIN: Pre-publication MS (1958).

so that $dp + n\,d\psi = 0$. If U_1 and S_1 are the internal energy and entropy per molecule, the free energy and thermodynamic potential per molecule are defined by

$$\left.\begin{aligned} F_1 &= U_1 - T S_1 + \psi, \\ G_1 &= U_1 - T S_1 + p/n \end{aligned}\right\} \tag{24.2}$$

and since $d(G_1 + \psi) = d(F_1 + p/n) = -S_1\,dT$, it is clear that under isothermal conditions $G_1 + \psi$ remains the same. The activity z, defined here by

$$\log z = \beta(G_1 + \psi) = \beta(F_1 + p/n) \tag{24.3}$$

also remains unchanged.

The total potential energy of a group of q molecules is

$$\Phi_q = \Phi_q^0 + \sum_{i=1}^{q} \psi(\boldsymbol{x}^{(i)}), \tag{24.4}$$

so if

$$\vartheta^{(i)} = \exp\{-\beta\,\psi(\boldsymbol{x}^{(i)})\} - 1, \tag{24.5}$$

the partition integral for q molecules is

$$I_q = \int \exp(-\beta\,\Phi_q)\,dx_q = \int \exp(-\beta\,\Phi_q^0) \prod_{i=1}^{q} (1 + \vartheta^{(i)})\,dx_q$$

or

$$I_q = I_q^0 + q \int I_q^{(1)0}\,\vartheta^{(1)}\,d^3x^{(1)} + \tfrac{1}{2} q(q-1) \iint I_q^{(12)0}\,\vartheta^{(1)}\,\vartheta^{(2)}\,d^3x^{(1)}\,d^3x^{(2)} + \cdots, \tag{24.6}$$

where $I_q^{(1\ldots j)0}$ is the integral of $\exp(-\beta\,\Phi_q^0)$ with respect to all coordinates except $\boldsymbol{x}^{(1)}, \ldots, \boldsymbol{x}^{(j)}$. Now Eqs. (13.3) can be written in the form

$$\left.\begin{aligned} \exp(\beta\,p^0 v) &= \sum_q I_q^0\,z^q/q!, \\ n^0 \exp(\beta\,p^0 v) &= \sum_q I_q^{(1)0}\,z^q/q!, \\ n_2^0 \exp(\beta\,p^0 v) &= \sum_q I_q^{(12)0}\,z^q/q!, \end{aligned}\right\} \tag{24.7}$$

etc., so

$$\left.\begin{aligned} \exp(\beta\,p\,v) &= \sum_q I_q\,z^q/q! \\ &= \exp(\beta\,p^0 v)\left\{1 + \int n^0\,\vartheta^{(1)}\,d^3x^{(1)} + \tfrac{1}{2}\iint n_2^0\,\vartheta^{(1)}\,\vartheta^{(2)}\,d^3x^{(1)}\,d^3x^{(2)} + \cdots\right). \end{aligned}\right\} \tag{24.8}$$

On expressing the functions n_q in terms of the l_q by means of (14.1), this equation reduces to

$$\beta\,p\,v = \beta\,p^0 v + \sum_q \frac{1}{q!} \int l_q^0\,\vartheta^{(1)}\,\vartheta^{(2)} \ldots \vartheta^{(q)}\,dx_q, \tag{24.9}$$

where the l_q^0 are the field-free values of $l_1 = n$, $l_2 = n_2 - n^2$, $l_3 = n_3 - n(n_2^{(23)} + n_2^{(31)} + n_2^{(12)}) + 2n^3$, etc. If the volume v is a small macroscopic region in which the variation of ψ can be neglected, (24.9) reduces to

$$\beta\,p\,v = \beta\,p^0 v + \sum_q l_{0,q}^0\,\vartheta^q/q!, \tag{24.10}$$

showing the dependence of the pressure on the local potential of the force field. The variation of the density can be deduced from (24.1) and (24.10):

$$n = \Big(\sum_q l_{1,q}^0\,\vartheta^q/q!\Big)\,e^{-\beta\psi}. \tag{24.11}$$

It will be noticed from these results that the variation in pressure and density with the external field is very directly related to the equilibrium structure of the liquid, which determines the coefficients $l^0_{0,q}$ and $l^0_{1,q} = l^0_{0,q+1}/v$. Of course these coefficients are also connected with the equation of state; for example, $l_{1,1}$ is related to the isothermal compressibility by the formula (23.27) of the previous section. Indeed, the results (24.10) and (24.11) can be given a different interpretation, for force-free liquids: if p^0 is the pressure and n^0 the density when the activity is z, p will be the pressure and n the density when the activity is $z e^{-\beta\psi}$, at the same temperature. This conclusion was obtained by MAYER[1] in a somewhat different way.

The results of this section do not apply if the force field is rapidly changing, if the intermolecular forces are affected by the external forces—as when the molecules are polarized—, or if the forces change in a distance comparable with the intermolecular distances, as they do, for instance, near a solid boundary. Methods for treating the special problems which arise under such circumstances will occupy the remainder of this chapter.

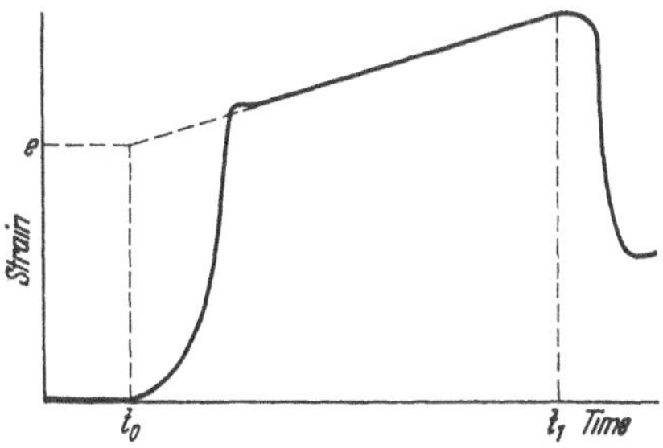

Fig. 6. The strain due to a stress suddenly applied at time t_0 and removed at time t_1.

25. Elasticity. Though it is not quite obvious that liquids possess elastic properties in the same sense as solids, a consideration of the behaviour of the glasses at various temperatures shows that there cannot be any fundamental difference. At low temperatures a glass shows the same elastic properties as other types of amorphous solids; near the surface there may even be some anisotropy. As the temperature is raised, the viscosity decreases and, although the bulk elasticity continues to be easily observed, the shearing elasticity is obscured by the increasing fluidity of the glass. The only difficulty is in the proper definition and measurement of the elastic constants. The same is true of any liquid and even of a gas, as was first pointed out by POISSON and MAXWELL[2].

Any process involving the macroscopic flow of a liquid may be separated theoretically into an elastic process and a viscous process, of which the first is ideally reversible, and the second is irreversible in the thermodynamic sense that macroscopic kinetic energy is converted into internal energy. Obviously experiments with liquids under constant stress are unsuitable for separating the two phenomena. However, one can imagine an experiment in which a shearing stress is suddenly applied to a liquid at rest, and suddenly removed after a very short time. A graph of the strain as a function of time would have the appearance of Fig. 6. The times t_0 and t_1 represent the instants at which the stress is applied and removed. The broken curve shows the contribution to the strain from the action of viscosity alone. The ordinate e, at the intersection of the tangent to the straight part of the curve with the vertical through t_0, measures the elastic strain. But in practice such an experiment would meet with insuperable difficulties unless the viscosity were very great.

The rigidity of liquids has been experimentally demonstrated by RAMAN and VENKATESWARAN[3] and FERRY[4]. The first method for the separate measurement

[1] J. E. MAYER: J. Chem. Phys. **10**, 629 (1942).
[2] See the historical remarks of C. KITTEL: Trans. Amer. Soc. Mech. Engrs. **69**, 368 (1947).
[3] C. V. RAMAN and C. S. VENKATESWARAN: Nature, Lond. **143**, 789 (1939).
[4] J. D. FERRY: J. Amer. Chem. Soc. **64**, 1323 (1942).

of the elastic and viscous constants of liquids appears, however, to be due to MASON and his collaborators[1]. They made use of the piezoelectric effect to induce torsional vibrations of high frequency in a cylindrical crystal suspended in the liquid which was under investigation. Pure shearing waves were thus generated in the liquid, and the magnitude and phase of their reaction on the crystal could be determined from the resonance frequency and damping of its vibrations. In this way both the wave length λ and the coefficient of absorption α of the shearing waves could be measured, in spite of their rapid attenuation. If viscosity alone were operative, the relation $\lambda\alpha = 1$ should be verified, as it was at low frequencies and even at quite high frequencies in ordinary mobile liquids. But in liquids such as polymerized castor oil and the more complex polyisobutylene liquids an increasing deviation from this relation was observed above a certain frequency, showing that elasticity was becoming appreciable. Probably this elasticity was associated with the internal structure of the molecules rather than the molecular structure of the liquid, but at still higher frequencies it is expected that elastic deformation of the molecular structure would also be displayed.

An elementary molecular theory of elasticity in liquids has been described by the author[2]. An attempt will be made here to exhibit the foundations on which this theory is based, and its connection with the experimental approach. It will be supposed that a liquid is subject to a variable external force which is responsible for the stress system, and that on this account a group of q molecules which in the absence of the force would have the positions $\boldsymbol{x}^{(1)}, \ldots, \boldsymbol{x}^{(q)}$ at time t have in fact the positions $\boldsymbol{x}^{(1)} + \mathfrak{s}_q^{(1)}, \ldots, \boldsymbol{x}^{(q)} + \mathfrak{s}_q^{(q)}$. The $\mathfrak{s}_q^{(i)}$, of course, depend on $\boldsymbol{x}^{(1)}, \ldots, \boldsymbol{x}^{(q)}$ as well as on the time; they may, however, if the perturbing forces are small, be treated as small quantities, so that squares and products of them may be neglected. It is worth noticing that the average value of any function $\varphi_q(\boldsymbol{x}^{(1)}, \ldots, \boldsymbol{x}^{(q)})$ of q coordinates in the perturbed state of the liquid is the same as the average value of $\varphi_q(\boldsymbol{x}^{(1)} + \mathfrak{s}_q^{(1)}, \ldots, \boldsymbol{x}^{(q)} + \mathfrak{s}_q^{(q)}$ in the unperturbed state of the liquid; and that if $n_q^0(\boldsymbol{x}^{(1)}, \ldots, \boldsymbol{x}^{q})$ is the distribution function for the liquid in its unperturbed state, the corresponding distribution function for the perturbed state will be

$$n_q = n_q^0(\boldsymbol{x}^{(1)} - \mathfrak{s}_q^{(1)}, \ldots, \boldsymbol{x}^{(q)} - \mathfrak{s}_q^{(q)}) \left(1 - \sum_{i=1}^{q} \frac{\partial}{\partial \boldsymbol{x}^{(i)}} \cdot \mathfrak{s}_q^{(i)}\right). \tag{25.1}$$

For $q = 1$, $\mathfrak{s}_q^{(1)}$ may be interpreted as the strain $\mathfrak{s}$ at the point $\boldsymbol{x}^{(1)}$.

The velocity distribution function is also changed by the external force system and the stress which it creates in the liquid. A molecule which would otherwise have the position $\boldsymbol{x}$ and the velocity $\boldsymbol{\xi}$ at time t will in fact have the position $\boldsymbol{x} + \boldsymbol{s}$ and the velocity $\boldsymbol{\xi} + \sigma$. The velocity distribution function is then

$$f = n\left(\frac{\beta m}{2\pi}\right)^{\frac{3}{2}} \exp\left\{-\frac{1}{2}\beta m(\boldsymbol{\xi} - \sigma)^2\right\}\left(1 - \frac{\partial}{\partial \boldsymbol{x}} \cdot \boldsymbol{s}\right)\left(1 - \frac{\partial}{\partial \boldsymbol{\xi}} \cdot \sigma\right). \tag{25.2}$$

With the help of (25.1) and (25.2), the stress tensor can be calculated from the general formula (15.11). One finds that

$$\left.\begin{aligned} \mathsf{p} &= p^0 \mathbf{\delta}\left(1 - \frac{\partial}{\partial \boldsymbol{x}} \cdot \mathfrak{s}\right) + m n \left(\frac{\beta m}{2\pi}\right)^{\frac{3}{2}} \int \exp\left(-\frac{1}{2}\beta m \xi^2\right)(\boldsymbol{\xi}\sigma + \sigma\boldsymbol{\xi})\, d\boldsymbol{\xi} - \\ &\quad - \frac{1}{2}\int n_2^0(r)\,(\mathfrak{s}_2^{(2)} - \mathfrak{s}_2^{(1)}) \cdot \frac{\partial}{\partial \boldsymbol{r}}\left(\boldsymbol{r}\,\frac{\partial \phi}{\partial \boldsymbol{r}}\right) d\boldsymbol{r}. \end{aligned}\right\} \tag{25.3}$$

---

[1] W. P. MASON: Trans. Amer. Soc. Mech. Engrs. **69**, 359. — W. P. MASON, W. O. BAKER, H. J. MCSKIMIN and J. H. HEISS: Phys. Rev. **75**, 936 (1949).

[2] H. S. GREEN: Proc. International Rheological Congress, 1948, 1—12. Amsterdam: North-Holland Publishing Ca 1949.

This result is quite general. To deduce the elastic constants, however, it is necessary to suppose that the stress is applied more rapidly than the liquid is able to "relax", or experimentally has a frequency much greater than the reciprocal of the "relaxation time". Suppose, then that

$$\mathfrak{s} = \mathrm{Re}\, \boldsymbol{s}_0\, e^{i\omega t} \tag{25.4}$$

where the angular frequency ω is very large, and $\boldsymbol{s}_0$ does not depend on time. It can be shown (see Sect. 36) that, under these circumstances, one can substitute

$$\left.\begin{aligned} \sigma &= \boldsymbol{\xi} \cdot \frac{\partial}{\partial \boldsymbol{x}} \mathfrak{s} + 0\left(\frac{1}{\omega}\right); \\ \mathfrak{s}_2^{(2)} - \mathfrak{s}_2^{(1)} &= \boldsymbol{r} \cdot \frac{\partial}{\partial \boldsymbol{x}} \mathfrak{s} + 0\left(\frac{1}{\omega}\right), \end{aligned}\right\} \tag{25.5}$$

so that (25.3) reduces to

$$\left.\begin{aligned} \mathsf{p} &= p^0 \boldsymbol{\delta} - 2\mu \mathbf{e}, \\ \mu &= -\frac{1}{30} \int n_2'(r)\, \phi'(r)\, r^2\, d^3 r, \\ e_{11} &= \frac{\partial \mathfrak{s}_1}{\partial x_1}, \qquad e_{12} = \frac{1}{2}\left(\frac{\partial \mathfrak{s}_1}{\partial x_2} + \frac{\partial \mathfrak{s}_2}{\partial x_1}\right), \qquad \text{etc.} \end{aligned}\right\} \tag{25.6}$$

The coefficient μ may be called the coefficient of elasticity of the liquid. It is fully analogous to the coefficient of shearing elasticity in solids, and even the formula for μ in (25.6) is analogous to the formula obtained in the dynamical theory of crystals[1]. The inference is that, within times shorter than the predominant relaxation time, the liquids which are normally mobile behave as though they had rigidity, just like the glasses at low temperatures.

It should perhaps be observed that, though liquids possess compressional elasticity even at the lowest sonic frequencies, this elasticity is almost certainly modified at frequencies of the order of the lowest relaxation frequency, so that there is no reason to assume the compressibility predicted by (25.6) to bear any simple relation to the adiabatic compressibility at low frequencies. The experiments do not yet provide accurate information on the relaxation spectra of any except liquids with rather complex molecules. Improvements in ultrasonic technique will, however, undoubtedly extend our knowledge of these interesting phenomena in the near future.

26. Dielectric behaviour. The dielectric constant of a liquid is an easily measured quantity which is also closely linked to the molecular structure. The mere order of magnitude of the dielectric constant is often quite a reliable indication of whether the liquid is polar, i.e., whether the molecules have a permanent electric moment or not. Only in non-polar liquids is it possible that the orientations of the molecules should be of little account, and because of the relative simplicity conferred by the assumption of spherical symmetry of the intermolecular forces, the theory for non-polar liquids will be considered first.

α) Non-polar liquids. The rigorous molecular theory of the dielectric constant in non-polar liquids is due in essence to YVON[2] and KIRKWOOD[3]. It has been

---

[1] M. BORN: Atomtheorie des festen Zustandes. Leipzig: J. B. Teubner 1923.

[2] J. YVON: La Constante Dielectrique. (Actualités Scientifiques et Industrielles, No. 543.) Paris: Hermann & Cie. 1937. — C. R. Acad. Sci., Paris **202**, 35 (1936).

[3] J. G. KIRKWOOD: J. Chem. Phys. **4**, 592 (1936).

presented in slightly varying forms by BROWN[1] and the author[2]; though the author's result differed somewhat from YVON'S, BROWN[3] has shown that the two are equivalent, apart from terms involving higher powers of the molecular polarizability.

When a molecule is subjected to an electric field, the mean positions of the electrons and nuclei are shifted relative to the mass centre. The basic assumption of the theory is that this displacement of the charge distribution can be adequately represented as an electric dipole moment, localized at the mass centre of the molecule, and proportional to the component of electric field intensity there due to sources outside the molecule. The constant of proportionality, which is the molecular polarizability α, will clearly depend only on the internal structure of the molecule, provided the field is not too strong.

It should be obvious that the average electric intensity $\boldsymbol{E}_1$ at the mass centre of a molecule in a liquid, due to sources outside the molecule, is different from the electric intensity $\boldsymbol{E}_m$ defined as in macroscopic electrostatics. The latter is compounded partly of the intensity $\boldsymbol{E}_0$ due to sources outside the liquid, and partly of the intensity associated with mean charge distribution induced at the surface of, and perhaps also inside the liquid, by $\boldsymbol{E}_0$. The mean dipole moment of a molecule in the liquid is $\boldsymbol{P}_1 = \alpha \boldsymbol{E}_1$; and since, the dipole moment of the liquid per unit volume is $n\boldsymbol{P}_1$, the mean charge density inside the liquid will be $-\frac{\partial}{\partial \boldsymbol{x}} \cdot (n\boldsymbol{P}_1)$, and the mean surface charge density will be $n\boldsymbol{P}_1 \cdot \boldsymbol{a}$, where $\boldsymbol{a}$ is the outward normal to the surface. One has, therefore, the relation

$$\boldsymbol{E}_m^{(1)} = \boldsymbol{E}_0^{(1)} + \int \frac{\boldsymbol{r}}{r^3} \frac{\partial}{\partial \boldsymbol{x}^{(2)}} \cdot (n^{(2)} \boldsymbol{P}_1^{(2)})\, d^3x^{(2)} - \int \frac{\boldsymbol{r}}{r^3} n^{(2)} \boldsymbol{P}_1^{(2)} \cdot d\boldsymbol{S}^{(2)}, \tag{26.1}$$

where the superfixes $^{(1)}$ and $^{(2)}$ refer to positions $\boldsymbol{x}^{(1)}$ and $\boldsymbol{x}^{(2)}$ in the liquid, $\boldsymbol{r} = \boldsymbol{x}^{(2)} - \boldsymbol{x}^{(1)}$, and $d\boldsymbol{S}^{(2)}$ is a vector surface element which must be drawn just inside the liquid if $n^{(2)}$ is the density of the liquid proper. By using GAUSS' theorem, this relation can be rewritten in the form

$$\boldsymbol{E}_m^{(1)} = \boldsymbol{E}_0^{(1)} - \frac{4\pi}{3} n^{(1)} \boldsymbol{P}_1^{(1)} + \int T^{(12)} \cdot \boldsymbol{P}_1^{(2)} n^{(2)} d^3x^{(2)} \tag{26.2}$$

where the second term is the value of a surface integral on an infinitesimally small sphere surrounding $\boldsymbol{x}^{(1)}$, and

$$T^{(12)} = \frac{3\boldsymbol{r}\boldsymbol{r} - r^2 \boldsymbol{\delta}}{r^5}. \tag{26.3}$$

It is worth noticing, before any further step is taken, that the electric displacement in the liquid is

$$\varkappa \boldsymbol{E}_m = \boldsymbol{E}_m + 4\pi n \boldsymbol{P}_1, \tag{26.4}$$

where $\varkappa$ is the dielectric constant; if, therefore, $\boldsymbol{P}_1$ can be expressed in terms of either $\boldsymbol{E}_m$ or $\boldsymbol{E}_0$, the value of $\varkappa$ will be obtained.

Now the intensity $\boldsymbol{E}_1$ is the resultant of $\boldsymbol{E}_0$ and the mean intensity arising from the polarization of all other molecules. If the mean dipole moment of a molecule at $\boldsymbol{x}^{(2)}$ is $\boldsymbol{P}_2^{(2)}$, its contribution to the field intensity experienced by a second molecule at $\boldsymbol{x}^{(1)}$ is $T^{(12)} \cdot \boldsymbol{P}_2^{(2)}$, so

$$\left.\begin{aligned} \boldsymbol{P}_1^{(1)} &= \alpha \boldsymbol{E}_1^{(1)} \\ &= \alpha \left(\boldsymbol{E}_0^{(1)} + \int T^{(12)} \cdot \boldsymbol{P}_2^{(2)} n_2\, d^3x^{(2)}/n\right). \end{aligned}\right\} \tag{26.5}$$

[1] W. F. BROWN: J. Chem. Phys. 18, 1193 (1950).

[2] Chap. 7, § 6 of H. S. GREEN: Molecular Theory of Fluids. Amsterdam 1952. (This book was sent to press in 1950.)

[3] W. F. BROWN: J. Chem. Phys. 21, 1121 (1953).

It should be noticed that the mean dipole moment $\boldsymbol{P}_2^{(2)}$ of the molecule at $\boldsymbol{x}^{(2)}$, computed in the knowledge that a second molecule is at $\boldsymbol{x}^{(1)}$, is different from $\boldsymbol{P}_1^{(2)}$, which is the unconditional mean. Indeed, $\boldsymbol{P}_2^{(2)}$ is a function of both $\boldsymbol{x}^{(1)}$ and $\boldsymbol{x}^{(2)}$, given by

$$\left.\begin{aligned} \boldsymbol{P}_2^{(2)} &= \alpha \boldsymbol{E}_2^{(2)} \\ &= \alpha \left(\boldsymbol{E}_0^{(2)} + \mathsf{T}^{(21)} \cdot \boldsymbol{P}_2^{(1)} + \int \mathsf{T}^{(23)} \cdot \boldsymbol{P}_3^{(3)} n_3 \, d^3 x^{(3)}/n_2\right), \end{aligned}\right\} \tag{26.6}$$

where $\boldsymbol{P}_3^{(3)}$ is the mean electric moment of a molecule at $\boldsymbol{x}^{(3)}$, computed in the knowledge that other molecules are present at $\boldsymbol{x}^{(1)}$ and $\boldsymbol{x}^{(2)}$. The general relation is

$$\boldsymbol{P}_q^{(q)} = \alpha \left(\boldsymbol{E}_0^{(q)} + \sum_{i=1}^{q-1} \mathsf{T}^{(q,i)} \cdot \boldsymbol{P}_q^{(i)} + \int \mathsf{T}^{(q,q+1)} \cdot \boldsymbol{P}_{q+1}^{(q+1)} n_{q+1} \, d^3 x^{(q+1)}/n_p\right). \tag{26.7}$$

The convergence of the integrals in (26.5), (26.6) and (26.7) is poor, since each receives contributions from the liquid surface, and also from the interior of the liquid if $n\boldsymbol{P}_1$ is not a constant. However, on using the relation (26.2) to eliminate $\boldsymbol{E}_0$, these contributions are cancelled, and one obtains

$$\left.\begin{aligned} \boldsymbol{P}_q^{(q)} = \alpha \Big\{ & \boldsymbol{E}_m^{(q)} + \frac{4\pi}{3} n^{(q)} \boldsymbol{P}_1^{(q)} + \sum_{i=1}^{q-1} \mathsf{T}^{(q,i)} \boldsymbol{P}_q^{(i)} + \\ & + \int \mathsf{T}^{(q,q+1)} \cdot \left(\boldsymbol{P}_{(q+1)}^{(q+1)} n_{q+1}/n_q - \boldsymbol{P}_1^{(q+1)} n^{(q+1)}\right) d^3 x^{(q+1)} \Big\} \end{aligned}\right\} \tag{26.8}$$

where now the convergence is satisfactory, since when $\boldsymbol{x}^{(q+1)}$ recedes from the points $\boldsymbol{x}^{(1)}, \ldots \boldsymbol{x}^{(q)}$, $\boldsymbol{P}_{q+1}^{(q+1)} n_{q+1}$ approaches the value $\boldsymbol{P}_1^{(q+1)} n_q n^{(q+1)}$. So, on iterating the formula (26.8), for $q = 1, 2, 3$ etc., a power series expansion is obtained for

$$\left.\begin{aligned} \left(1 - \frac{4\pi}{3} n\alpha\right) \boldsymbol{P}_1^{(1)} = \alpha \boldsymbol{E}_m^{(1)} + \alpha^2 \int \Big\{ & \mathsf{T}^{(12)} \cdot \boldsymbol{E}_m^{(2)} \left(\frac{n_2}{n} - n^{(2)}\right) \times \\ & \times \left(1 + \frac{4\pi}{3} n^{(2)} \alpha\right) + \alpha \mathsf{T}^{(12)} \cdot \mathsf{T}^{(21)} \cdot \boldsymbol{E}_m^{(1)} \frac{n_2}{n} \Big\} d^3 x^{(2)} + \\ & + \alpha^3 \iint \mathsf{T}^{(12)} \cdot \mathsf{T}^{(23)} \cdot \boldsymbol{E}_m^{(3)} \left(n_3 - n^{(3)} n_2 - n n_2^{(23)} + \right. \\ & \left. + n n^{(2)} n^{(3)}\right) d^3 x^{(3)}/n + 0(\alpha^4). \end{aligned}\right\} \tag{26.9}$$

If $\boldsymbol{E}$ and n are uniform throughout the liquid, the integrals can be simplified, using

$$\int \mathsf{T}^{(12)} (n_2 - n n^{(2)}) \, d^3 x^{(2)} = 0,$$

$$\int \mathsf{T}^{(12)} \cdot \mathsf{T}^{(21)} n_2 \, d^3 x^{(2)} = 2 I_2 \boldsymbol{\delta},$$

$$\int \mathsf{T}^{(12)} \cdot \mathsf{T}^{(23)} \left(n_3 - n^{(3)} n_2 - n n_2^{(23)} + n n^{(2)} n^{(3)}\right) d^3 x^{(2)} \, d^3 x^{(3)} = I_3 \boldsymbol{\delta}$$

where

$$\left.\begin{aligned} I_2 &= \int n_2(r)/r^6 \, d^3 r, \\ I_3 &= \iint \frac{\{n_3(\boldsymbol{r}, \boldsymbol{s}) - n n_2(r) - n n_2(s) + n^3\}\{3(\boldsymbol{r} \cdot \boldsymbol{s})^2 - r^2 s^2\} \, d^3 r \, d^3 s}{r^5 s^5}. \end{aligned}\right\} \tag{26.10}$$

The result is then

$$\left(1 - \frac{4\pi}{3} n\alpha\right) \boldsymbol{P}_1 = \{\alpha + (2 I_2 + I_3) \alpha^3/n\} \boldsymbol{E}_m + 0(\alpha^4). \tag{26.11}$$

By comparison with (26.4), the dielectric constant is found to be given by

$$(\varkappa - 1)\left(1 - \frac{4\pi}{3} n\alpha\right) = 4\pi n\alpha \{1 + (2I_2 + I_3)\alpha^2/n\} + 0(\alpha^4). \qquad (26.12)$$

It is worth remarking, for comparison with YVON's result, that the integral

$$\iint \frac{\{n_2(r) - n^2\}\{n_2(s) - n^2\}\{3(\boldsymbol{r}\cdot\boldsymbol{s})^2 - r^2 s^2\}}{r^5 s^5} d^3r\, d^3s$$

vanishes, as one can see by performing the integration over the angle between $\boldsymbol{r}$ and $\boldsymbol{s}$; so, I_3 can be expressed alternatively in the form

$$I_3 = \iint \frac{\{n_3(\boldsymbol{r}, \boldsymbol{s}) - n_2(r)\, n_2(s)/n\}\{3(\boldsymbol{r}\cdot\boldsymbol{s})^2 - r^2 s^2\}}{r^5 s^5} d^3r\, d^3s. \qquad (26.13)$$

This integral could be evaluated with the help of the superposition approximation, if $n_2(r)$ were known. BROWN[1] has suggested the use of an additional approximation which can be expressed in the form

$$\boldsymbol{P}_3^{(3)} \approx \boldsymbol{P}_2^{(3,1)} + \boldsymbol{P}_2^{(3,2)} - \boldsymbol{P}_1^{(3)}; \qquad (26.14)$$

this has the effect of reducing (26.5) and (26.6) to a pair of simultaneous integral equations, of which BROWN obtained an approximate numerical solution. The calculations which have been made are quite consistent with the rather limited experimental data at present available, but no attempt has yet been made to test the result (26.12) by the use of an accurately determined $n_2(r)$. The limitations of the theory are therefore difficult to assess, but it seems possible that it may apply quite well to many non-polar liquids, irrespective of the strict spherical symmetry of their molecules. Should such a possibility be realized, the experimental measurement of the dielectric constant will be a powerful method for the investigation of liquid structure.

In the foregoing discussion, it has been assumed implicitly that the electric field applied to the liquid is purely static. The argument is, however, easily extended to harmonically varying fields. The electric intensities $\boldsymbol{E}_m$, $\boldsymbol{E}_0$, $\boldsymbol{E}_1$, etc., can then be represented as real parts of the complex vectors $\boldsymbol{e}_m\, e^{i\omega t}$, $\boldsymbol{e}_0\, e^{i\omega t}$, $\boldsymbol{e}_1\, e^{i\omega t}$, etc., where $\boldsymbol{e}_m$, $\boldsymbol{e}_0$ and $\boldsymbol{e}_1$ may be complex but do not depend on the time. The mean dipole moment $\boldsymbol{P}_1$ of a molecule is the real part of a complex vector $\boldsymbol{P}_1\, e^{i\omega t} = \alpha \boldsymbol{e}_1\, e^{i\omega t}$, where α may now be complex and depend on the angular frequency ω. The field at the point $\boldsymbol{x}^{(1)}$ due to a molecule at the point $\boldsymbol{x}^{(2)}$ with a dipole moment $\boldsymbol{P}_1^{(2)}$, which varies with angular frequency ω, is $\mathfrak{t}^{(12)} \cdot \boldsymbol{P}_1^{(2)}$, where

$$\mathfrak{t}^{(12)} = \frac{\omega^2 (r^2 \boldsymbol{\delta} - \boldsymbol{r}\,\boldsymbol{r})}{c^2 r^3}. \qquad (26.15)$$

The only modification required in the above discussion, for the variable field, is therefore the systematic replacement of $\boldsymbol{E}_m, \boldsymbol{E}_0, \boldsymbol{E}_1, \ldots, \boldsymbol{P}_1, \ldots$ and T by $\boldsymbol{e}_m, \boldsymbol{e}_0, \boldsymbol{e}_1, \ldots, \boldsymbol{p}_1, \ldots$ and $\mathfrak{t}$, keeping in mind the fact that α is now complex. The dielectric constant $\varkappa$ is also a complex quantity in general; the electric displacement is the real part of the vector $\varkappa\, \boldsymbol{e}\, e^{i\omega t}$ and is not normally in phase with the electric intensity.

β) Polar liquids. As might be expected, the theory of polar liquids is not quite so highly developed. Although the theory of the dielectric constant has received a good deal of attention, it has not, on the whole, provided much information on the structure of polar liquids, except from a qualitative point of view

[1] See footnote 1, p. 72.

(see Sect. 10α). Most of the progress which has been made is presented very well in FRÖHLICH's monograph[1]. But one simple consideration due to KIRKWOOD[2], is of great importance for the light it sheds on the relation between the dielectric constant and the structure of polar liquids, and will be discussed here.

In polar liquids each molecule has a dipole moment in the absence of an external electric field and though its moment is probably influenced somewhat in magnitude by the environment of the molecule in the liquid, the direction of the moment is determined primarily by the orientation of the molecule. The orientations of such molecules are strongly correlated, owing to the dipole interactions, even at quite large separations. The fact that the dipole interaction decreases only as the cube of the distance between two molecules, whereas the van der Waals forces decrease as the seventh power of the distance, ensures that the effects of a local anisotropy, such as exists near the surface of a liquid, are transmitted across a much greater distance, on the molecular scale, than in non-polar liquids. The problem of statistical mechanics which arises in connection with polar liquids, even in the absence of an applied electric field, is clearly one of great complexity, and has not yet been solved. KIRKWOOD, however, took the structure of the liquid in the absence of the field for granted, and examined the perturbation arising from the application of a *weak* electric field.

It is assumed, as before, that the polarization of the molecule has the same effect as an equivalent electric dipole at the mass centre. In a special configuration of the molecules, the dipole moment $\boldsymbol{P}$ of a molecule at any point $\boldsymbol{x}$ in the liquid is the resultant of two components, one independent of, and one proportional in magnitude to the applied field. Specifically

$$\boldsymbol{P} = \boldsymbol{\mu} + \alpha \boldsymbol{E}, \tag{26.16}$$

where $\boldsymbol{\mu}$ is independent of the applied field, though it may depend on the relative positions and orientations of surrounding molecules; α is the molecular polarizability and $\boldsymbol{E}$ is that part of the intensity from sources outside the molecule which is proportional in magnitude to the applied field. If $\boldsymbol{E}_0$ is the intensity due to sources outside the liquid,

$$\boldsymbol{E}^{(i)} = \boldsymbol{E}_0^{(i)} + \alpha \sum_j T^{(ij)} \cdot \boldsymbol{E}^{(j)}, \tag{26.17}$$

where the summation is over all molecules of the liquid. Obviously the average value of $\boldsymbol{E}$ for all configurations of the molecules is precisely analogous to $\boldsymbol{E}_1$ of Sect. 10α above, and is connected with $\boldsymbol{E}_0$ by the same relation. However, the average value of $\boldsymbol{P}$ is $\boldsymbol{P}_1 = \boldsymbol{\mu}_1 + \alpha \boldsymbol{E}_1$, where $\boldsymbol{\mu}_1$ does not vanish, because the most probable molecular configurations are changed by the applied field. The problem is, therefore, to determine the average value $\boldsymbol{\mu}_1$ of $\boldsymbol{\mu}$.

Now the potential energy Φ of the molecules of the liquid in a particular configuration includes the energy of their dipole interaction with the field $\boldsymbol{E}_0$ and also of the dipole-dipole interaction between the molecules, part of which depends on $\boldsymbol{E}_0$. If Φ_0 is the value of Φ for $\boldsymbol{E}_0 = 0$,

$$\Phi = \Phi_0 - \sum_j \boldsymbol{P}^{(j)} \cdot \boldsymbol{E}^{(j)}. \tag{26.18}$$

Also, the relative probability of any configuration is proportional to $\exp(-\beta\Phi)$, which reduces to $\exp(-\beta\Phi_0)\left\{1 + \beta \sum_j \boldsymbol{P}^{(j)} \cdot \boldsymbol{E}^{(j)}\right\}$, or $\exp(-\beta\Phi_0)\left\{1 + \beta \sum_j \boldsymbol{\mu}^{(j)} \cdot \boldsymbol{E}^{(j)}\right\}$ with neglect of terms of order $\boldsymbol{E}_0^2$. Hence the mean value of $\boldsymbol{\mu}^{(i)}$ for the existing

[1] H. FRÖHLICH: Theory of Dielectrics. Oxford 1949.
[2] J. G. KIRKWOOD: J. Chem. Phys. 7, 911 (1939).

field is the same as the mean value of $\beta\,\boldsymbol{\mu}^{(i)} \sum_j \boldsymbol{\mu}^{(j)} \cdot \boldsymbol{E}^{(j)}$, computed with the distribution functions for $\boldsymbol{E}_0 = 0$:

$$\mu_1^{(i)} = \beta \langle \boldsymbol{\mu}^{(i)} \sum_j \boldsymbol{\mu}^{(j)} \cdot \boldsymbol{E}^{(j)} \rangle_0 . \tag{26.19}$$

As no further progress is possible without approximation, one may examine the effect of replacing $\boldsymbol{E}^{(j)}$ by $\boldsymbol{E}_0^{(i)}$ (KIRKWOOD'S procedure[1]) or by its mean value $\boldsymbol{E}_1^{(j)}$. The first alternative gives

$$\mu_1^{(i)} \approx \tfrac{1}{3} \beta \sum_j \langle \boldsymbol{\mu}^{(i)} \cdot \boldsymbol{\mu}^{(j)} \rangle_0 \boldsymbol{E}_0^{(i)} , \tag{26.20}$$

while the second leads to

$$\boldsymbol{P}_1^{(i)} \approx \left\{ \alpha + \tfrac{1}{3} \beta \sum_j \langle \boldsymbol{\mu}^{(i)} \cdot \boldsymbol{\mu}^{(j)} \rangle_0 \right\} \boldsymbol{E}_1^{(i)} , \tag{26.21}$$

which is a simple generalization of the relation $\boldsymbol{P}_1^{(i)} = \alpha \boldsymbol{E}_1^{(i)}$ for non-polar liquids. Because α is usually small compared with $\frac{1}{3} \beta \sum_j \langle \boldsymbol{\mu}^{(i)} \cdot \boldsymbol{\mu}^{(j)} \rangle_0$, there is, in fact, little difference between $\boldsymbol{E}_0^{(i)}$, $\boldsymbol{E}_1^{(i)}$ and $\boldsymbol{E}^{(i)}$ in polar liquids.

It is still necessary to relate $\boldsymbol{E}_0^{(i)}$ to the macroscopic electric intensity. This has usually been done by regarding the liquid so far considered as contained within a sphere, immersed in a liquid of the same kind. It is known from macroscopic electrostatics that the field in a spherical cavity in a medium of dielectric constant $\varkappa$ is

$$\boldsymbol{E}_0 = \frac{3\varkappa}{2\varkappa + 1} \boldsymbol{E}_m , \tag{26.22}$$

where $\boldsymbol{E}_m$ is the macroscopic intensity in the dielectric KIRKWOOD'S final formula for the dielectric constant is therefore

$$\varkappa - 1 = 4\pi n \frac{3}{2\varkappa + 1} \left\{ \alpha + \frac{1}{3} \beta \sum_j \langle \boldsymbol{\mu}^{(i)} \cdot \boldsymbol{\mu}^{(j)} \rangle_0 \right\} . \tag{26.23}$$

Beyond, the theory is conserned mainly with the evaluation of the means $\langle \boldsymbol{\mu}^{(i)} \cdot \boldsymbol{\mu}^{(j)} \rangle_0$. Assuming that the magnitude of $\boldsymbol{\mu}$ does not vary much, as in fact it does not when α is small enough to justify the use of (26.20),

$$\langle \boldsymbol{\mu}^{(i)} \cdot \boldsymbol{\mu}^{(j)} \rangle_0 = \mu^2 \langle \cos \vartheta^{(ij)} \rangle_0 \tag{26.24}$$

where $\vartheta^{(ij)}$ is the angle between $\boldsymbol{\mu}^{(i)}$ and $\boldsymbol{\mu}^{(j)}$, and so measures the mutual orientation of the two molecules. One may therefore regard the dielectric constant in polar liquids as a quite sensitive function of the mutual orientations of the molecules. It is plausible to suppose that in most liquids the orientations are strongly correlated only when the molecules are near neighbours, and one can then restrict the summation in (26.23) to the first one or two co-ordination shells. For most polar liquids $\varkappa$ is considerably greater than 1, so that the ratio $3\varkappa : 2\varkappa + 1$ can be replaced by 1.5. The dielectric constant is therefore almost inversely proportional to the temperature, deviations being largely due to the temperature dependence of the molecular structure.

The formula (26.23) has been quite successful in the prediction of the dielectric constant. For water the predicted values are about 15% too low, even if one assumes that the relative orientations of neighbouring molecules are the same

---

[1] See previous reference. The same approximation was implicitly made by F. E. HARRIS and B. J. ALDER: J. Chem. Phys. **21**, 1031 (1953); the fact that their result differs from KIRKWOOD'S is due to a further approximation, which has been criticized by H. FRÖHLICH, J. Chem. Phys. **22**, 1804 (1954). Their method, however, secures good agreement with experiment.

as in ice; but this is not surprising in view of the nature of the model. The polarity of water and many other substances is located near the surface of the molecule and is only very roughly equivalent to a point dipole as the theory requires. The difference between the theoretical model and the actual molecule is not great enough, however, to impair confidence in the interpretation of the experimental data on the dielectric constant, which the model provides.

c) Structure in the surface zone.

27. Surface phenomena. Macroscopically, the interface between a liquid and its vapour or containing wall appears to be sharply defined, and the change of density and other properties across the surface is apparently discontinuous. On the molecular scale, the uniformity of density even within the liquid is, of course, a statistical abstraction, but it is clear that there can be no sharp division between the liquid and its environment. It is therefore necessary to think of the boundary as a zone rather than a mathematical surface, in which the structure of the fluid is slowly modified so as to change gradually from liquid to vapour type, when the boundary separates the two phases, or to effect the delimitation of the liquid by a solid wall. From a statistical point of view, the molecular number

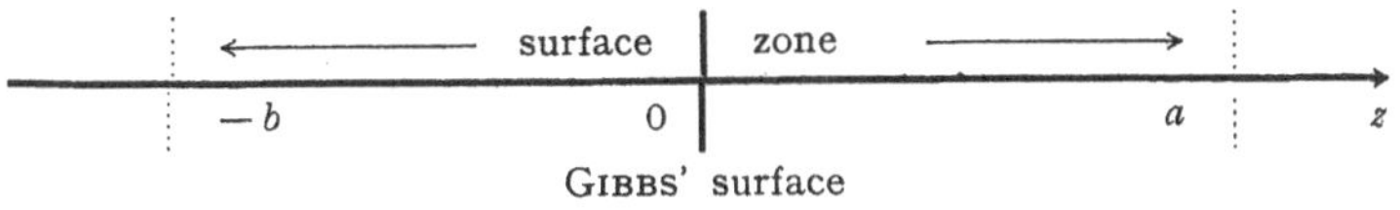

Fig. 7. Diagram of surface of liquid.

density, defined as in Sect. 11, will change continuously across the surface zone from its value in the liquid to its value outside, and the distribution function n_2 is also progressively modified. The structural changes at the surface of the liquid are, in fact, very complicated, and are exhibited macroscopically in the phenomenon of surface tension.

For a detailed discussion of the molecular theory of surface tension, the reader is referred to the companion work by Drs. ONO and KONDO in this volume. The theory outlined here is intended only to clarify the relationship which exists between the surface tension and the molecular structure. For this purpose it may be supposed that the radii of curvature of the liquid surface are large compared with the depth of the surface zone, at least in non-polar liquids which will be considered in this section. There will be no need to discuss very small drops of liquid, where the curvature of the surface is an important feature; the required modifications have been elaborated recently by HILL[1], BUFF[2] and, from the present point of view, by MCLELLAN[3].

Although the liquid boundary is not a geometrical surface, it has been found convenient to introduce a fictitious "GIBBS' surface" somewhere in the surface zone, where the transition from liquid to gas or solid may be imagined to take place. This surface will be defined here as one of constant density, under isothermal conditions, and drawn in such a way that, *if* the change of density across the surface were discontinuous (as it appears to be macroscopically), the mean number of molecules in the surface zone would be the same as it is in fact. To be more precise, let z be a coordinate along the normal to the GIBBS' surface, so that $z > 0$ in the liquid and $z < 0$ outside.

[1] T. L. HILL: J. Chem. Phys. **19**, 1203 (1951).
[2] F. P. BUFF: J. Chem. Phys. **19**, 1591 (1951); **23**, 419 (1955).
[3] A. G. MCLELLAN: Proc. Roy. Soc. Lond., Ser. A **217**, 92 (1953).

Then the surface is chosen so that

$$\int_{-b}^{a} n\,dz = a n_A + b n_B,$$

where n_A is the molecular density, assumed to be uniform, for $z > a$, and n_B the density (either that of the vapour, or zero in a solid wall) for $z < -b$. The above condition can also be written

$$\int_{-b}^{a} \frac{\partial n}{\partial z} z\,dz = 0. \tag{27.1}$$

One can then define the free energy f associated with this surface as the difference between the actual free energy of the fluid in the surface zone, and the free energy which the fluid within the same zone would have if the transition were discontinuous. If F_1 is the free energy per molecule, and σ the area of the surface,

$$\gamma = f/\sigma = \int_{-b}^{a} n F_1\,dz - a n_A F_{1A} - b n_B F_{1B} = \int_{-b}^{a} \frac{\partial}{\partial z}(n F_1)\,z\,dz. \tag{27.2}$$

The superficial internal energy u and the superficial entropy are connected with the internal energy per molecule U_1 and the entropy per molecule S_1 by similar relations.

The surface tension is defined as the free energy per unit area of surface, and is therefore the quantity γ defined by (27.2). The following thermodynamic relations hold:

$$\left.\begin{aligned} T\,ds &= du - \gamma\,d\sigma, \\ df &= \gamma\,d\sigma - s\,dT, \end{aligned}\right\} \tag{27.3}$$

showing that γ is also the work required to increase the surface area per unit area, under isothermal conditions. Anywhere except in the surface zone, the pressure is isotropic, provided there is no external force and the fluid is at rest. But in the surface zone, the change of density from layer to layer produces an asymmetry of the intermolecular forces, which in turn causes an asymmetry of the pressure tensor. Supposing that the surface is normal to the x_3-direction, it is easy to see that the equilibrium of the surface layer requires the component p_{33} of p to remain constant in passing from the liquid to the vapour, and to vary in such a way as to balance the forces exerted by the molecules of the wall, at a solid boundary. But the tangential components p_{11} and p_{22} may differ from p_{33} in the surface zone, and it is these components which do work when the surface area of the liquid is changed. If p is the hydrostatic pressure, one has in fact

$$\gamma = -\int_{-b}^{a} p_{11}\,dz + (a+b)\,p = \int_{-b}^{a} \frac{\partial p_{11}}{\partial z} z\,dz. \tag{27.4}$$

At this juncture the molecular structure enters, through the expression for the pressure tensor derived in Sect. 15. Some care is required, however, in using the result (15.11), since the variation of $\bar{n}_2(\boldsymbol{r}, \boldsymbol{x})$ with x_3 across the surface zone is so rapid that the truncation of the series (15.9) is not allowed there. McLELLAN[1] has derived an expression for the pressure tensor valid under these conditions,

[1] See footnote 3, p. 77.

which agrees with the prescription of Sect. 15, but refers to the point $\boldsymbol{X}$ through which the intermolecular forces are transmitted. It is necessary simply to substitute $\boldsymbol{x}^{(1)}=\boldsymbol{X}+\lambda\boldsymbol{r}$ and $\boldsymbol{x}^{(2)}=\boldsymbol{X}+(1-\lambda)\,\boldsymbol{r}$ in $n_2(\boldsymbol{x}^{(1)},\boldsymbol{x}^{(2)})$, and average with respect to $\lambda(0<\lambda<1)$, in order to obtain the expression

$$`n_2(\boldsymbol{r},\boldsymbol{X})' = \int_0^1 n_2(\boldsymbol{X}+\lambda\boldsymbol{r},\boldsymbol{X}+\boldsymbol{r}-\lambda\boldsymbol{r})\,d\lambda\,, \tag{27.5}$$

to be substituted for $\bar{n}_2(\boldsymbol{r},\boldsymbol{x})$ in (15.11). So

$$p_{11} = n/\beta - \tfrac{1}{2}\int `n_2'\,\phi'(r)\,(r_1^2/r)\,d^3r\,,$$
$$p_{33} = n/\beta - \tfrac{1}{2}\int `n_2'\,\phi'(r)\,(r_3^2/r)\,d^3r\,,$$

and, since $p_{33}=p$ at the interface of the liquid with its vapour, the surface tension there is

$$\left.\begin{aligned}\gamma &= \int_{-b}^{a}(p_{33}-p_{11})\,dz\\ &= \tfrac{1}{2}\iint `n_2'\,\phi'(r)\,(r_1^2-r_3^2)/r\,d^3r\,dX_3\,.\end{aligned}\right\} \tag{27.6}$$

Substituting from (27.5) and changing the variable of integration from X_3 to $x_3^{(1)}=X_3-\lambda r_3$, one has finally

$$\gamma = \tfrac{1}{2}\iint n_2\,\phi'(r)\,(r_3^2-r_1^2)/r\,d^3r\,d\,x_3^{(1)}. \tag{27.7}$$

This formula has also been derived by McLellan[1], using only the statistical mechanical expression for the partition function. The superficial internal energy can be obtained either by using Kelvin's formula

$$\frac{u}{\sigma} = \frac{d}{d\beta}(\beta\gamma) \tag{27.8}$$

which follows from (27.3), or directly from the formula analogous to (2.72).

These results are exact, but for computational purposes demand a knowledge of n_2 throughout the surface zone. Attempts to compute the surface tension have hitherto been based on rather crude assumptions concerning n_2. Probably the simplest approximation one could devise is to write

$$n_2 \approx n^{(1)}\,n^{(2)}\,g(r)\,, \tag{27.9}$$

where $g(r)$ is the correlation function for the liquid in bulk; this would lead rigorously to a non-linear integral equation for n as a function of z, of the type which arises in the theory of freezing due to Kirkwood and Monroe[2]. In the most recent calculation of Kirkwood and Buff[3], however, it was assumed that $n=n_A$ $z>0$ and $n=n_B$ for $z<0$; the integral expression in (27.6) can then be simplified, first using symmetry to replace $r_1^2-r_3^2$ by $\frac{1}{2}(r^2-3r_3^2)$ in the integrand. Replacing $x_3^{(1)}$ by z,

$$\left.\begin{aligned}\gamma &= \pi\Big[n_A\int_0^\infty dz\Big\{n_A\int_{-z}^{\infty}dr_3 + n_B\int_{-\infty}^{-z}dr_3\Big\} + \\ &\quad + n_B\int_{-\infty}^{0}dz\Big\{n_A\int_z^\infty dr_3 + n_B\int_{-\infty}^{z}dr_3\Big\}\Big]\int_{|r_3|}^{\infty}dr\,g(r)\,\phi'(r)\,(r^2-3r_3^2)\\ &= \tfrac{1}{8}\pi\int_0^\infty (n_A-n_B)^2\,g(r)\,\phi'(r)\,r^4\,dr\,,\end{aligned}\right\} \tag{27.10}$$

---

[1] A. G. McLellan: Proc. Roy. Soc. Lond., Ser. A **213**, 274 (1952).
[2] J. G. Kirkwood and E. Monroe: J. Chem. Phys. **9**, 514 (1941).
[3] J. G. Kirkwood and F. P. Buff: J. Chem. Phys. **17**, 338 (1948).

after some elementary manipulations. This results appears to be due to FOWLER[1], and has been verified by KIRKWOOD and BUFF for liquid argon at 90° K. They used a form of $g(r)$ based on the X-ray scattering determinations and thermodynamical data, and evaluated the integral of (27.10), obtaining for γ the value 14.9 dyne/cm. KIRKWOOD and BUFF also computed the superficial internal energy u from the formula

$$\frac{\mu}{\sigma} = - \int_{-b}^{a} \frac{\partial}{\partial z} (n\, U_1)\, z\, dz \tag{27.11}$$

which, with the same approximations, yields the result

$$\frac{\mu}{\sigma} = - \frac{1}{2} \pi \int_{0}^{\infty} (n_A - n_B)^2\, g(r)\, \phi(r)\, r^3\, dr. \tag{27.12}$$

This expression has the value 27.2 dyne/cm for liquid argon at 90 °K, compared with the experimental value 35 dyne/cm.

Such calculations as have been made on the basis of the rather crude approximation described are therefore in fairly good agreement with experiment. This might be regarded as indicating that in non-polar liquids like liquid argon, the transition from liquid to vapour, though not discontinuous as the approximation assumes, is in fact quite rapid: the thickness of the surface zone is not greater than a few molecular diameter. If so, the character of the surface zone is the feature which most decisively distinguishes non-polar from polar liquids.

28. The surface zone in polar and associated liquids. There is a considerable body of direct and indirect experimental evidence concerning the surface zone in liquids, which is the subject of an excellent review by HENNIKER[2]. HENNIKER does not attempt to classify the liquids to which he refers, but he is clearly concerned almost exclusively with polar or other types of associated liquids; since the theory of the last section does not apply to such liquids and our theoretical information concerning them is otherwise meagre, his discussion is of particular value.

Qualitatively the evidence can be understood by observing that, in the outer part of the surface zone, there is an anisotropy in the forces acting on a molecule, due to the variation in the density and other properties of the medium. At the interface between a liquid and its vapour, the comparative sparsity of the vapour molecules means that the attractive intermolecular forces are directed away from the surface. At an interface between a liquid and a solid or between two liquids there is a similar anisotropy, which is often augmented by the orientation of molecules of the solid or liquid opposite. The anisotropy of the intermolecular forces has the effect of imposing a preferential orientation on the surface molecules of the liquid, and this, in turn, produces an anisotropy in the forces affecting the molecules just below the liquid surface. It is possible in this way for co-operative orientation of the molecules to extend for considerable distances below the surface of the liquid, even where the range of the intermolecular forces is short. In polar liquids, however, the dipole-dipole interactions have a range which is unusually great, and there is an added reason to expect the anisotropy created by the surface to extend far into the liquid, on the molecular scale.

---

[1] R. H. FOWLER: Proc. Roy. Soc. Lond., Ser. A **159**, 229 (1937).

[2] J. C. HENNIKER: Rev. Mod. Phys. **21**, 322 (1949); this paper includes an extensive bibliography.

Under exceptional circumstances it is possible for the preferred orientation of the molecules to extend throughout the liquid, which then becomes what is called a liquid crystal. For example, this behaviour is found in some long-chain fatty acids and *p*-azoxyanisole above their melting point. Though the anisotropy is no longer confined to a surface zone in such liquids, they offer the opportunity to study on a macroscopic scale the conditions which are found near the surface of many liquids. There is no loss of fluidity, but birefringency and other optical phenomena characteristic of crystalline substances are found. Naturally co-operative orientation tends to be destroyed by the thermal agitation which accompanies increase in temperature, and the "orientation fusion" of a liquid crystal, which may be supposed to occasion the loss of such properties, is quite abrupt.

In many liquids other than liquid crystals, there is evidence for molecular orientation extending 1000 Å or more below the surface, the precise depth depending on the nature of the material with which the liquid is in contact. The orientation can be detected by X-ray diffraction techniques, and also by a wide range of macroscopic phenomena, some of which will be briefly mentioned. According to Sect. 26β, the dielectric constant of a polar liquid might be expected to be greater if the molecules are similarly oriented, as they are in the surface zone, than when they have various orientations, as they have to some extent within the liquid; there is some evidence that this expectation is realized. The electrical conductivity of certain liquids may also be very different in thin surface layers than in bulk, indicating a change of molecular structure. It has been shown that the viscosity of many liquids near their surface and in thin films is abnormally great, an effect which may be correlated with the increased resistance to lateral motion of similarly oriented molecules. The increased viscosity reveals itself also in the apparent rigidity of various liquids, including water, in their surface layers; in addition, there appears to be an unusually great resistance to compression normal to the surface[1].

Surface tension is due to changes in the structure of the liquid in the surface zone, and in polar liquids may be attributed only partly to the rapid change, accompanied by a change indensity, which occurs within a few molecular diameters of the interface. There is, in addition, a more gradual change of structure across the surface zone, in which the orientations of the molecules become more and more ordered as the interface is approached. The contribution to the surface tension from this zone might be expected to decrease, or at least to be considerably modified, in very thin films; and again, this expectation appears to be experimentally realized.

IV. Structure of non-uniform liquids.

a) Effect of irreversible processes.

29. Irreversible processes in general. The previous chapter has dealt almost exclusively with the structure of liquids at rest and in thermodynamic equilibrium. Of even greater interest are the modifications and changes of structure which accompany the approach to the equilibrium state.

It is known from macroscopic thermodynamics that the natural processes by which equilibrium is attained are irreversible, involving a spontaneous increase in the entropy of the liquid. It is known also that both the classical and quantum mechanical laws are time-reversible: that if one process with a given initial and final state is mechanically possible, so is the reverse process. The

[1] Cf. J. H. Sinfeld and H. G. Drickamer: J. Chem. Phys. **23**, 1095 (1955).

macroscopic irreversibility of natural processes must therefore be of a statistical character, and the result of a statistical principle as distinct from a mechanical one. The exact formulation of the principle in question has presented great difficulty in the past, though it was implicit in BOLTZMANN's equation in the kinetic theory of gases, and is closely related to the well known principle of molecular chaos. No attempt will be made at this stage to enunciate the principle, though it may be mentioned that it appears in molecular theory by way of the initial conditions which are required to solve a physical problem. The important point is that natural phenomena are irreversible only from a macroscopic point of view, so that only macroscopic processes can be described as irreversible.

In simple liquids, in which the molecules are all alike, the important irreversible processes are those associated with viscosity and thermal conduction. In mixtures it will be necessary to consider diffusion as well. But, in view of the remarks made at the beginning of Sect. 18, no special attention will be given to chemical reactions, and certain irreversible processes, involving the transition from a metastable to a stable phase of a mixture, will thus be excluded from consideration. For a macroscopic treatment applicable to such processes, the reader is referred to the paper by ECKART[1].

The various sources of irreversible phenomena can be analysed by computing the rate of increase of entropy in the liquid. The entropy per unit mass (S), which can vary from point to point in the liquid as well as with time, is defined macroscopically by the equation

$$T\,dS = dU + p^0\,d(1/\varrho) - \sum_j \mu_j\,d\alpha_j \tag{29.1}$$

where U is the internal energy per unit mass, ϱ is the mass density, μ_j is the chemical potential per unit mass of the j-th constituent of a mixture, and $\alpha_j = \varrho_j/\varrho$ is the mass fraction of this constituent. This definition is not restricted to quasistatic processes, but applies whenever the state of the liquid is specified by giving the values of a complete set of independent macroscopic variables, such as U, ϱ and α_j, throughout the liquid. To compute the rate of change of S, one requires the rates of change of U, ϱ and α_j. It has been assumed implicitly in (29.1) that there are no external forces causing diffusion, and for U one then has, according to (18.12),

$$\varrho\frac{dU}{dt} + \left(\mathsf{p}\cdot\frac{\partial}{\partial \boldsymbol{x}}\right)\cdot\boldsymbol{u} + \frac{\partial}{\partial \boldsymbol{x}}\cdot\boldsymbol{Q} = 0, \tag{29.2}$$

where p is the pressure tensor, and the energy flux vector $\boldsymbol{Q}$ can be expressed in the form

$$\boldsymbol{Q} = \boldsymbol{q} + \sum_j \mu_j\,\varrho_j\,\boldsymbol{w}_j. \tag{29.3}$$

Here $\boldsymbol{w}_j = \boldsymbol{u}_j - \boldsymbol{u}$ is the velocity of diffusion of the j-th constituent relative to the mass velocity $\boldsymbol{u}$, and $\boldsymbol{q}$ may be identified with the thermal flux vector. For ϱ and α one has, according to (18.2) and (18.6),

$$\left.\begin{aligned} \frac{d\varrho}{dt} + \varrho\frac{\partial}{\partial \boldsymbol{x}}\cdot\boldsymbol{u} &= 0;\\ \varrho\frac{d\alpha_j}{dt} + \frac{\partial}{\partial \boldsymbol{x}}\cdot(\varrho_j\,\boldsymbol{w}_j) &= 0. \end{aligned}\right\} \tag{29.4}$$

[1] C. ECKART: Phys. Rev. **58**, 269, 924 (1940).

On dividing (29.1) by the time differential dt, therefore, and substituting from (29.2) and (29.3), one obtains

$$\varrho \frac{dS}{dt} + \frac{\partial}{\partial \boldsymbol{x}} \cdot \left(\frac{\boldsymbol{q}}{T}\right) = -\frac{\boldsymbol{q}}{T^2} \frac{\partial T}{\partial \boldsymbol{x}} - \left(\frac{\mathsf{p}}{T} \cdot \frac{\partial}{\partial \boldsymbol{x}}\right) \cdot \boldsymbol{u} + \frac{p_0}{T} \frac{\partial}{\partial \boldsymbol{x}} \cdot \boldsymbol{u} - \sum_j \frac{\varrho_j \boldsymbol{w}_j}{T} \cdot \frac{\partial \mu_j}{\partial \boldsymbol{x}}. \quad (29.5)$$

Now, $\boldsymbol{q}/T$ is the flux of entropy, so the right-hand side of (29.5) represents the rate of entropy production per unit volume. Clearly the entropy can only change within the liquid if there is a temperature gradient, a velocity gradient or a gradient of the chemical potentials arising from the variation of the composition of the mixture. Further, the right-hand side of (29.5) must be essentially positive if the second law of thermodynamics is to be fulfilled. That it is positive definite can be proved only on the basis of the molecular theory, though an intuitive argument is sometimes available. For example, the empirical laws of viscosity are summarized in the formula

$$\mathsf{p} = \left(p^0 - \eta^0 \frac{\partial}{\partial \boldsymbol{x}} \cdot \boldsymbol{u}\right) \boldsymbol{\delta} - 2\eta \frac{\partial_s \boldsymbol{u}}{\partial \boldsymbol{x}}, \quad (29.6)$$

where η^0 and η are the coefficients of volume and shear viscosity, and $\partial_s \boldsymbol{u}/\partial \boldsymbol{x}$ is the rate of strain tensor (see Sect. 33). Hence

$$-\left(\frac{\mathsf{p}}{T} \cdot \frac{\partial}{\partial \boldsymbol{x}}\right) \cdot \boldsymbol{u} + \frac{p^0}{T} \frac{\partial}{\partial \boldsymbol{x}} \cdot \boldsymbol{u} = \frac{\eta^0}{T} \left(\frac{\partial}{\partial \boldsymbol{x}} \cdot \boldsymbol{u}\right)^2 + \frac{2\eta}{T} \left(\frac{\partial_s \boldsymbol{u}}{\partial \boldsymbol{x}}\right)^2 \quad (29.7)$$

is the rate of entropy production due to viscosity, and this is seen to be positive definite if one accepts the intuitive fact that the coefficients of viscosity must be positive. For $\boldsymbol{q}$ and $\boldsymbol{w}_j$ there are similar empirical laws, summarized in the formulae

$$\boldsymbol{q} = -\frac{\lambda}{T} \frac{\partial T}{\partial \boldsymbol{x}} - \sum_j \nu_j \frac{\partial \mu_j}{\partial \boldsymbol{x}} \quad (29.8)$$

and

$$\varrho_j \boldsymbol{w}_j = -\frac{\nu_j}{T} \frac{\partial T}{\partial \boldsymbol{x}} - \sum_k \nu_{jk} \frac{\partial \mu_k}{\partial \boldsymbol{x}} \quad (29.9)$$

where $\sum_j \nu_j = 0$ and $\sum_j \nu_{jk} = 0$. It is intuitively obvious that λ must be positive, so that the rate of entropy production due to thermal conduction alone, $\frac{\lambda}{T^3}\left(\frac{\partial T}{\partial \boldsymbol{x}}\right)^2$, is also positive; but this can only be proved on the basis of the molecular theory. Similarly the fact that the coefficients ν_j in (29.8) and (29.9) are identical, and that $\nu_{jk} = \nu_{kj}$ can only be proved by using the "principle of microscopic reversibility"[1], which is abstracted from the molecular theory, together with certain other assumptions.

The theory of the molecular constitution of liquids therefore has the task of explaining a large number of empirical facts, some of which are intuitively "obvious" but all of which nevertheless have to be assumed by the macroscopic theories. In addition, it should be possible to establish quantitative relations between the molecular structure and the coefficients λ, μ^0, μ, ν_j and ν_{jk} which appear in the empirical laws of thermal conduction, viscosity and diffusion. These are the practical problems which have stimulated most of the work described in the present chapter. Some of them are at the time of writing only partially solved, but their influence on the concept of liquid structure today is profound.

[1] See S. R. de Groot: Thermodynamics of Irreversible Processes. Amsterdam: North-Holland Publishing Co. 1951.

30. Diffusion and thermal diffusion. In this section the discussion will, for the sake of simplicity, be limited to binary mixtures, *i.e.*, to liquids containing molecules of two different kinds. The molecular number densities of the two components will be denoted by n_a and n_b, and the mean velocities of molecules of the two different types by $\boldsymbol{u}_a$ and $\boldsymbol{u}_b$. In a steady state of the liquid, $\boldsymbol{u}_a$ and $\boldsymbol{u}_b$ can differ because of inhomogeneities of $\vartheta = n_a/n_b$, or of temperature or pressure. One may therefore write

$$\boldsymbol{u}_a - \boldsymbol{u}_b = \frac{-D_0}{\vartheta}\frac{\partial\vartheta}{\partial\boldsymbol{x}} - \frac{(n_a+n_b)^2}{n_a n_b}\left(\frac{D_T}{T}\frac{\partial T}{\partial\boldsymbol{x}} + \frac{D_p}{p}\frac{\partial p}{\partial\boldsymbol{x}}\right) \tag{30.1}$$

where D_0 is called simply the coefficient of diffusion, and D_T and D_p are the coefficients of thermal and pressure diffusion respectively. The coefficiently must obviously be related to those appearing on the right-hand side of (29.9), of which there are essentially only two for binary mixtures, since $\nu_b = -\nu_a$ and $\nu_{bb} = -\nu_{ab} = -\nu_{ba} = \nu_{aa}$. If one regards the chemical potential difference $\mu = \mu_a - \mu_b$ as a function of $\vartheta = n_a/n_b$, T and p one has $\frac{\partial\mu}{\partial\boldsymbol{x}} = \frac{\partial\mu}{\partial\vartheta}\frac{\partial\vartheta}{\partial\boldsymbol{x}} + \frac{\partial\mu}{\partial T}\frac{\partial T}{\partial\boldsymbol{x}} + \frac{\partial\mu}{\partial p}\frac{\partial p}{\partial\boldsymbol{x}}$, so, comparing the coefficients of $\frac{\partial\vartheta}{\partial\boldsymbol{x}}$ and $\frac{\partial p}{\partial\boldsymbol{x}}$ in (30.1) and the corresponding formula derived from (29.9),

$$\frac{(n_a+n_b)^2}{n_a n_b p}\,\frac{D_p}{D_0}\,\frac{\partial\mu}{\partial\vartheta} = \frac{\partial\mu}{\partial p}. \tag{30.2}$$

Only two coefficients of diffusion are therefore unrelated.

On account of thermal diffusion, it is possible to effect a partial separation of two different molecular species from a mixture, by applying a temperature gradient. In liquids this is known as the Soret effect; the reader is referred to DE GROOT's dissertation[1] for an excellent discussion based on macroscopic theories. Molecular interpretations, based on more or less plausible but unverified assumption, are due to WIRTZ and HIBY[2], DENBIGH[3], PRIGOGINE, DE BROUKERE and AMAND[4], and RUTHERFORD and DRICKAMER[5]. The rigorous treatment in terms of molecular distribution is due to YANG[6], on whose work the present discussion will be based.

It should be observed first that since

$$\boldsymbol{u}_a - \boldsymbol{u}_b = \int (f_a/n_a)\,\xi_a\,d^3\xi_a - \int (f_b/n_b)\,\xi_b\,d^3\xi_b \tag{30.3}$$

diffusion of any kind may be attributed to deviation of the simple velocity distribution functions from their equilibrium form. In the kinetic theory of gases, indeed, to determine the coefficients of diffusion and thermal diffusion there is no other task than the computation of f_a and f_b under non-uniform conditions. It might appear, therefore, that the connection between diffusion and the actual molecular structure of the fluid must be slight. However, the mechanism of diffusion in gases is very profoundly modified as the free path contracts, and is fundamentally changed in liquids. The rate of progress of a molecule through the liquid is much smaller than in gases and is essentially determined by the resistance it encounters from its immediate environment, i.e., by the molecular structure. Although (30.3) is formally correct, it fails to exhibit the close connection which

[1] S. R. DE GROOT: L'Effet Soret. Amsterdam: North-Holland Publishing Co. 1945.
[2] K. WIRTZ and J. W. HIBY: Phys. Z. **44**, 369 (1943).
[3] K. G. DENBIGH: Trans. Faraday Soc. **48**, 1 (1952).
[4] I. PRIGOGINE, L. DE BROUKERE and R. AMAND: Physica, Haag **16**, 577, 851 (1950).
[5] W. M. RUTHERFORD and H. G. DRICKAMER: J. Chem. Phys. **22**, 1157 (1954).
[6] L. M. YANG: Proc. Roy. Soc. Lond., Ser. A **198**, 94, 471 (1949).

undoubtedly exists between the distribution of velocities and the relative positions of the molecules in a liquid. Following YANG, it will be shown how the coefficients of diffusion are connected with the so-called "friction constants" of the two molecular species.

Suppose that a molecule of the first species is situated at the point $\boldsymbol{x}_0$ in the liquid at some initial time t_0, and let $\psi_a(\boldsymbol{x}_0, t_0; \boldsymbol{x}, t)\, d^3x$ be the probability that this molecule will be found at the point $\boldsymbol{x}$ at time t. The function ψ_a connects the number density distributions at time t_0 and t:

$$n_a(\boldsymbol{x}, t) = \int n_a(\boldsymbol{x}_0, t_0)\, \psi_a(\boldsymbol{x}_0, t_0; \boldsymbol{x}, t)\, d^3x_0. \tag{30.4}$$

The mean displacement suffered in the time $t - t_0$ by a molecule at the point $\boldsymbol{x}$ at time t is

$$\langle \boldsymbol{r}_a \rangle = \int (\boldsymbol{x} - \boldsymbol{x}_0)\, \psi_a(\boldsymbol{x}_0, t_0; \boldsymbol{x}, t)\, d^3x_0 \tag{30.5}$$

and the mean square deviation from the mean is given by the tensor

$$\left.\begin{aligned} &\langle \boldsymbol{r}_a \boldsymbol{r}_a \rangle - \langle \boldsymbol{r}_a \rangle \langle \boldsymbol{r}_a \rangle \\ &\quad = \int (\boldsymbol{x} - \boldsymbol{x}_0 - \langle \boldsymbol{r}_a \rangle)\, (\boldsymbol{x} - \boldsymbol{x}_0 - \langle \boldsymbol{r}_a \rangle)\, \psi_a(\boldsymbol{x}_0, t_0; \boldsymbol{x}, t)\, d^3x_0. \end{aligned}\right\} \tag{30.6}$$

It will be found convenient to write

$$\left.\begin{aligned} \langle \boldsymbol{r}_a \rangle &= \boldsymbol{u}_a^* (t - t_0); \\ \langle \boldsymbol{r}_a \boldsymbol{r}_a \rangle - \langle \boldsymbol{r}_a \rangle \langle \boldsymbol{r}_a \rangle &= \tfrac{1}{3} \mathsf{D}_a (t - t_0). \end{aligned}\right\} \tag{30.7}$$

If there are no inhomogeneities of the type causing diffusion, $\boldsymbol{u}_a^*$ reduces to the mass velocity $\boldsymbol{u}$, and D_a is a multiple of the unit tensor $\boldsymbol{\delta}$:

$$\mathsf{D}_a = D_a \boldsymbol{\delta}. \tag{30.8}$$

Provided the time interval $t - t_0$ is long enough for the velocity of the molecule at time t to be nearly uncorrelated with its velocity at time t_0, it is known from the theory of Brownian motion[1] that the coefficient D_a is independent of $t - t_0$; it is called the coefficient of self-diffusion of the molecular species concerned. Obviously D_a is inversely proportional to the resistance encountered by a molecule in its progress through the liquid, and a "friction constant" ζ_a can be defined by the relation

$$\beta \zeta_a D_a = 1, \tag{30.9}$$

where β is, as usual, inversely proportional to the temperature. This friction constant is identical with that which appears in LANGEVIN'S equation[1] in the theory of Brownian motion. A discussion of methods by which it can be computed in relation to the molecular structure of the liquid will be deferred to Sect. 41.

From its definition, the function ψ_a appearing in (30.4) clearly satisfies

$$\int \psi_a(\boldsymbol{x}_0, t_0; \boldsymbol{x}, t)\, d^3x_0 = 1. \tag{30.10}$$

Hence, if a TAYLOR'S expansion is made of $n_a(\boldsymbol{x}_0, t_0)$ about the point $\boldsymbol{x}$ and the time t:

$$\left.\begin{aligned} n_a(\boldsymbol{x}_0, t_0) = \Big\{1 &- (t - t_0) \frac{\partial}{\partial t} - (\boldsymbol{x} - \boldsymbol{x}_0) \cdot \frac{\partial}{\partial \boldsymbol{x}} + \\ &+ \frac{1}{2} (\boldsymbol{x} - \boldsymbol{x}_0)\, (\boldsymbol{x} - \boldsymbol{x}_0) : \frac{\partial^2}{\partial \boldsymbol{x}\, \partial \boldsymbol{x}} \Big\}\, n_a(\boldsymbol{x}, t) + \cdots \end{aligned}\right\} \tag{30.11}$$

[1] See, for example, S. CHANDRASEKHAR: Rev. Mod. Phys. 15, 20 (1943).

and the terms of order $(t-t_0)^2$ are neglected, one obtains by substitution in (30.4)

$$\frac{\partial n_a}{\partial t} + \frac{\partial}{\partial \boldsymbol{x}} \cdot (n_a \boldsymbol{u}_a) = 0,$$

where

$$\boldsymbol{u}_a = \boldsymbol{u}_a^* - \frac{1}{6} \frac{\partial}{\partial \boldsymbol{x}} \cdot \mathsf{D}_a. \tag{30.12}$$

It is presumably possible to choose $t-t_0$ small enough to permit the neglect of terms of order $(t-t_0)^2$ in (30.11), yet large enough to make D_a independent of $t-t_0$; this is one of the basic assumptions of the theory of the Brownian motion. A similar assumption appears in KIRKWOOD's theory of the transport processes, which will be described in Sects. 40 to 42. It is worth noticing, however, that the neglected terms involve second or higher derivatives of n_a with respect to time, and third or higher spatial derivatives, all of which would be negligible under steady state conditions, not too far removed from equilibrium. The above procedure will therefore not lead to any error in the computed coefficients of diffusion. For a similar reason it is permissible to substitute for D_a in (30.12) its equilibrium value, given by (30.8); any correction to D_a, due to departures from equilibrium, proportional to the gradient of concentration, temperature or pressure, and can be omitted from (30.12) under the conditions described.

According to (30.12), therefore, the relative velocity of diffusion of the two molecular species is

$$\boldsymbol{u}_a - \boldsymbol{u}_b = \boldsymbol{u}_a^* - \boldsymbol{u}_b^* - \frac{1}{6} \frac{\partial}{\partial \boldsymbol{x}} (D_a - D_b). \tag{30.13}$$

Now there seems to be no easy way of calculating $\boldsymbol{u}_a^* - \boldsymbol{u}_b^*$, but there is reason to suppose it is small, compared with the remaining term on the right-hand side of (30.13), in liquids. For the velocities of a molecule at the times t_0 and t are nearly uncorrelated, as the result of its continual interchange of momentum with surrounding molecules; its mean velocity in the interval will therefore approximate to the mass velocity $\boldsymbol{u}$, irrespective of the species to which it belongs. No estimate is available of the error entailed in neglecting $\boldsymbol{u}_a^* - \boldsymbol{u}_b^*$, but if this term is omitted from (30.13), the very reasonable conclusion can be drawn that diffusion in liquids is due primarily to differences in the rates of self-diffusion of the different molecular species. Comparing (30.1) and (30.13), one has finally

$$\left.\begin{aligned} D_0 &\approx \frac{\vartheta}{6} \frac{\partial}{\partial \vartheta} (D_a - D_b); \\ D_T &\approx \frac{\vartheta T}{6(1+\vartheta)^2} \frac{\partial}{\partial T} (D_a - D_b); \\ D_p &\approx \frac{\vartheta p}{6(1+\vartheta)^2} \frac{\partial}{\partial p} (D_a - D_b). \end{aligned}\right\} \tag{30.14}$$

The theory of the friction constant will determine $D_a - D_b$ as a function of the required thermodynamic variables, but no reliable numerical calculations for diffusion in liquids have been made up to the time of writing. Calculations of the coefficients of viscosity and thermal conduction based on similar reasoning, however, are sufficiently successful to justify the method.

31. Thermal conduction. In a gas, the transfer of heat from one part of the fluid to another is accomplished mainly by the motion of the molecules themselves. The faster molecules in the hotter part of the gas carry their excess kinetic energy with them as they diffuse into regions of lower temperature and, conversely, the slower molecules in the cooler part of the gas diffuse into regions of higher temperature. There is thus a tendency for thermal energy to be distributed more evenly in the fluid. While the process described undoubtedly takes place also in liquids, its efficacy depends on a lack of resistance to the motion of the molecules. It happens, therefore, that though kinetic transfer of energy accounts very well for the thermal conduction of rare gases, it is responsible for only a small and almost negligible contribution to the coefficient of thermal conduction in liquids. A molecule in a liquid possesses not only kinetic energy but also potential energy due to its continual interaction with its neighbours, and this energy can be transferred from molecule to molecule by a process in which the physical motion of the molecules plays a secondary albeit indispensible role. Before the detailed mathematical description of this process is attempted, it will first be analysed in physical terms[1].

The coefficient of thermal conductivity is normally measured in liquids which are at rest, and in which the pressure is therefore uniform, apart from the unimportant variation due to gravity. Although there is no pressure gradient in the liquid, there may be a mean resultant force on a molecule due to the temperature gradient. It was seen in Sect. 15 that the mean force acting on a molecule is

$$-\frac{1}{n}\int n_2 \frac{\partial\phi^{(12)}}{\partial \boldsymbol{x}^{(1)}}\, d\boldsymbol{x}^{(2)}, \qquad \text{or} \qquad \frac{1}{n}\frac{\partial}{\partial \boldsymbol{x}}(n\,kT - p).$$

When the pressure is uniform, the rate of working of this force is just sufficient, in the steady state, to give the molecule a mean kinetic energy appropriate to its changing environment. Now consider the same molecule, and also a second molecule at the point $\boldsymbol{x}^{(2)}$. The mean force on the first molecule is

$$-\frac{\partial\phi^{(12)}}{\partial \boldsymbol{x}^{(1)}} - \frac{1}{n_2}\int n_3 \frac{\partial\phi^{(13)}}{\partial \boldsymbol{x}^{(1)}}\, d^3x^{(3)}.$$

In the steady state, the rate of working of this force is also just sufficient to give the molecule a mean kinetic energy appropriate to its changing environment; it depends on the local temperature distribution, and is therefore not quite balanced by the mean force acting on the molecule at $\boldsymbol{x}^{(2)}$. There is, therefore, a mean resultant force acting on the pair of molecules, and there is a tendency for them to move together in a direction determined by their relative displacement and the temperature gradient.

The mean resultant force on the pair of molecules changes, of course, with their known velocities. In addition to the forces independent of velocity, there are frictional forces which inhibit the motion of the faster molecules, and effectively determine the mean velocity of diffusion of the pair of molecules through the liquid. It should, perhaps, be emphasised that this velocity of diffusion depends on the relative displacement of the two molecules as well as the temperature gradient, and varies with the displacement in such a way that there is no motion of the liquid as a whole.

The diffusion of pairs of molecules with velocities depending on their distance apart is clearly a very effective means of energy transport. For the interaction energy of any pair is positive when the two are very close together, and negative when they are separated by a somewhat greater distance. If, on the average, pairs with positive interaction energy move down the temperature gradient, and pairs with negative interaction energy move up the temperature gradient, a large contribution to the coefficient of thermal conduction will result.

[1] This analysis was first given in substance by the author in § 4.2, Chap. 5 of Molecular Theory of Fluids. Amsterdam: North-Holland Publishing Co. 1952.

Referring now to the formula (16.13) for the thermal flux, it will be seen that there are two contributions

$$q_k = \int \tfrac{1}{2} m (\xi - u)^2 (\xi - u) f\, d^3\xi \tag{31.1}$$

and

$$q_P = \frac{1}{2}\int \phi(r)\,(u_2^{(1)} - u)\, n_2\, d^3r - \frac{1}{4}\int r\,\frac{\partial\phi}{\partial r}\cdot(\bar{u}_2^{(1)} + \bar{u}_2^{(2)} - 2u)\,\bar{n}_2\, d^3r. \tag{31.2}$$

The first is that corresponding to kinetic transfer, which is important in gases but not in liquids. The second corresponds to the process just described in detail. To make use of these results, it is necessary to know the effect of a temperature gradient on the velocity distribution function f, and also on the diffusion velocity $u_2^{(1)} - u$, regarded as a function of r. No attempt will be made to determine these functions at present, but attention will be drawn to some general conditions which they must satisfy.

The functions f and $n_2(u_2^{(1)} - u)$, since they depend on the temperature distribution, may be regarded as functions of the temperature T and its derivatives $\frac{\partial T}{\partial x}, \frac{\partial^2 T}{\partial x\,\partial x}, \ldots$ at any convenient point x. They may therefore be developed in powers of the components of $\frac{\partial T}{\partial x}, \frac{\partial^2 T}{\partial x\,\partial x} \ldots$, and if the temperature varies only slowly from point to point, one has very nearly $f = f^0 + f'$ where f^0 depends on T but not the temperature gradient, and f' is linear in the temperature gradient. Similarly $u_2^{(1)} = u_2^{(1)0} + u_2^{(1)\prime}$, where $u_2^{(1)0}$ is independent of, and $u_2^{(1)\prime}$ linear in, the temperature gradient. The functions f^0 and $u_2^{(1)0}$ are those to which f and $u_2^{(1)}$ would reduce if the temperature had everywhere the value T at the point x; they are therefore the functions appropriate to thermodynamic equilibrium, and

$$f^0 = n(\beta m/2\pi)^{\frac{3}{2}} \exp\{-\tfrac{1}{2}\beta m(\xi - u)^2\}, \qquad u_2^{(1)0} = u^{(1)}, \tag{31.3}$$

where $\beta k T = 1$. Setting as usual $v = \xi - u$, f' must clearly be of the form

$$f' = -\vartheta\, v\cdot\frac{\partial T}{\partial x}, \tag{31.4}$$

where ϑ depends on the magnitude of the vector $v = \xi - u$, but not on its direction. On substitution from (31.3) and (31.4), one obtains

$$q_k = -\left\{\frac{1}{6} m \int \vartheta\, v^4\, dv\right\}\frac{\partial T}{\partial x}. \tag{31.5}$$

Similarly $u_2^{(1)\prime}$ must have the form

$$u_2^{(1)\prime} = \left(\frac{\sigma}{r^2}\, r\cdot\frac{\partial T}{\partial x}\, r + \sigma_0\frac{\partial T}{\partial x}\right) \tag{31.6}$$

where σ and σ_0 depend on the magnitude of the displacement r, but not on its direction. It is easy to see, further, that n_2 is unaffected by the temperature gradient to this order of approximation, for scalar terms linear in the temperature gradient would involve a factor $r\cdot\frac{\partial T}{\partial x}$ which changes sign with r, whereas n_2 is unaltered when the sign of r is changed, leaving the mean centre $x = \frac{1}{2}(x^{(1)} + x^{(2)})$ unchanged. The temperature gradient cannot, in fact, alter the character of the radial distribution of thermodynamic equilibrium, except to the extent of terms quadratic in the temperature gradient or linear in the second spatial derivatives of the temperature. Substituting from (31.6) into (31.2), one has, therefore,

$$q_P = \frac{1}{6}\left[\int\{(\sigma + 3\sigma_0)\,\phi(r) - (\sigma + \sigma_0)\, r\,\phi'(r)\}\, n_2\, d^3r\right]\frac{\partial T}{\partial x}. \tag{31.7}$$

By combining the formulae (31.5) and (31.7) and comparing with (29.8), it is found that the coefficient of thermal conduction is given by

$$\lambda = \tfrac{1}{6} m \int \vartheta v^4 \, d^3 \boldsymbol{v} + \tfrac{1}{6} \int \{(\sigma + \sigma_0)\, r \phi'(r) - (\sigma + 3\sigma_0)\, \phi(r)\}\, n_2 \, d^3 r. \qquad (31.8)$$

Some consideration will be given in Sects. 40 and 42 to the problem of finding explicit expressions for ϑ, σ and σ_0. Here, however, it is worth noticing a condition which must be satisfied by σ and σ_0. The mean velocity of a molecule in the liquid, irrespective of its position, cannot be influenced by the presence of a molecule at a given point $\boldsymbol{x}^{(2)}$; so

$$\int n_2 (\boldsymbol{u}_2^{(1)} - \boldsymbol{u}^{(1)})\, d^3 r = 0. \qquad (31.9)$$

On substitution from (31.6) one obtains the condition

$$\int n_2 (\sigma + 3\sigma_0)\, d^3 r = 0. \qquad (31.10)$$

There are, of course, various ways in which this condition can be satisfied identically: the simplest is to take $\sigma_0 = -\tfrac{1}{3}\sigma$, and the most general is make $r^2 n_2 (\sigma + 3\sigma_0)$ the derivative of any function which vanishes at both $r = 0$ and $r = \infty$. In practice the condition can be used to check the consistency of any approximation used to determine $\boldsymbol{u}_2^{(1)\prime}$.

b) Deformation of structure by viscosity.

32. Theory of viscosity in liquids. Viscosity is a phenomenon fundamentally very similar to thermal conduction, which has been considered in the previous section; the difference is that whereas thermal conduction is the means of energy transfer, viscosity is the means of momentum transfer in the liquid. In a gas, momentum is transferred directly by the motion of the molecules, which carry their own momentum with them in passing between two regions with different macroscopic velocities. In a liquid, however, this process is not very effective, owing to the resistance to the motion of a molecule, created by the surrounding molecules. There is instead, a very effective process depending on the continual interaction between neighbouring molecules. Pairs of molecules in neighbouring regions of the liquid moving with different mean velocities tend to drag one another in such a way as to reduce their relative velocity. The mechanism of this process is now to be discussed in some detail.

A distinction should be drawn between shearing viscosity, which requires no change in the density of the liquid, and volume viscosity, which occurs when the liquid suffers a dilatation. In gases volume viscosity can hardly be detected, since it depends exclusively on molecular interactions. In liquids, on the other hand, its effects are not negligible, but are still difficult to measure since a steady state of uniform dilatation is necessarily of short duration.

Confining attention for the moment to shearing viscosity, consider a liquid in a state of non-uniform motion but constant density. The mean velocities of two molecules at the points $\boldsymbol{x}^{(1)}$ and $\boldsymbol{x}^{(2)}$ are $\boldsymbol{u}_2^{(1)}$ and $\boldsymbol{u}_2^{(2)}$ respectively, and their relative velocity is $\boldsymbol{u}_2^{(2)} - \boldsymbol{u}_2^{(1)}$, which is approximately $\boldsymbol{u}^{(2)} - \boldsymbol{u}^{(1)}$ or $\boldsymbol{r} \cdot \frac{\partial}{\partial \boldsymbol{x}} \boldsymbol{u}$ if they are not too near together. The fact that the mean relative velocity of a pair of molecules does not generally vanish in a liquid where there is a velocity gradient, but depends on the angle between the line joining them and the direction of the velocity gradient implies that the distribution of molecules about a given molecule can no longer be purely radial. The co-ordination shells, instead of being spherical, are somewhat ellipsoidal, with principal axes coinciding with those of the rate of strain tensor $\frac{\partial_s \boldsymbol{u}}{\partial \boldsymbol{x}}$ whose components are

$$\left(\frac{\partial_s \boldsymbol{u}}{\partial \boldsymbol{x}}\right)_{\alpha\beta} = \frac{1}{2}\left(\frac{\partial u_\alpha}{\partial x_\beta} + \frac{\partial u_\beta}{\partial x_\alpha} - \frac{2}{3}\frac{\partial}{\partial \boldsymbol{x}} \cdot \boldsymbol{u}\,\delta_{\alpha\beta}\right), \tag{32.1}$$

where the suffixes α and β denote cartesian components and

$$\delta_{\alpha\beta} = 1 \quad \text{if} \quad \alpha = \beta \quad \text{but} \quad \delta_{\alpha\beta} = 0 \quad \text{if} \quad \alpha \neq \beta .$$

Owing to this deformation of the co-ordination shells by the velocity gradient, the mean resultant force exerted on a molecule at $\boldsymbol{x}^{(1)}$, when there is another molecule at $\boldsymbol{x}^{(2)}$, is not what it would be if the liquid were at rest. There is no acceleration of the mass centre of the pair of molecules, as there would be if, for example, the temperature were not uniform, but there is an acceleration of one molecule relative to the other. In so far as it differs from its value in the liquid at rest, this relative acceleration is caused by the velocity-dependent forces which inhibit the relative motion of the pair of molecules. There is therefore a precise relation between the deformation of the radial distribution function and the mean relative velocity of a pair of molecules, or hence the velocity gradient.

It has been seen that shearing viscosity is accompanied by a deformation of liquid structure. Volume viscosity, which will now receive consideration, is accompanied by a progressive change in the liquid structure, which remains isotropic but nevertheless departs from the normal structure for the density instantaneously achieved. Otherwise the mechanism is very similar to that which has already been described for shearing viscosity. The relative velocity $\boldsymbol{u}_2^{(2)} - \boldsymbol{u}_2^{(1)}$ of a pair of molecules is directed along the line joining them, and is determined by the rate of change of the radial distribution with density. The relative acceleration of the pair of molecules is not quite what one would expect in a liquid at rest, because the adaptation of the liquid structure to changes in density is not quite instantaneous. The additional forces which account for the difference depend on the relative velocity of the molecules, and there is therefore a direct connection between the lag in the structural changes and the relative velocity of a pair of molecules, or hence the rate of dilatation.

Knowing that the distribution function n_2 is affected by the presence of a non-uniform velocity field, it may be regarded as a function of $\frac{\partial \boldsymbol{u}}{\partial \boldsymbol{x}}, \frac{\partial^2 \boldsymbol{u}}{\partial \boldsymbol{x}\,\partial \boldsymbol{x}}, \ldots$ at any fixed point $\boldsymbol{x}$ in the liquid, which may be conveniently identified with the mean centre $\frac{1}{2}(\boldsymbol{x}^{(1)} + \boldsymbol{x}^{(2)})$ of the two positions concerned. When the variation of the velocity $\boldsymbol{u}$ from point to point in the liquid is slow, n_2 may be expanded in powers of the spatial derivatives of $\boldsymbol{u}$, and all except the terms independent of or linear in the velocity gradient may be discarded. Then n_2 is necessarily of the form

$$n_2 = n_2^0 + n_2' \tag{32.2}$$

where n_2^0 is the distribution function for the liquid at rest, and

$$n_2' = \frac{\nu}{r^2}\boldsymbol{r} \cdot \left(\frac{\partial_s \boldsymbol{u}}{\partial \boldsymbol{x}}\right) \cdot \boldsymbol{r} + \nu_0 \frac{\partial}{\partial \boldsymbol{x}} \cdot \boldsymbol{u}. \tag{32.3}$$

Similarly the velocity distribution function can be expressed in the form $f = f^0 + f'$, where f^0 is the function for the liquid in a state of uniform motion, given by (31.3), and

$$f' = -\left\{\frac{\varkappa}{v^2}\boldsymbol{v} \cdot \left(\frac{\partial_s \boldsymbol{u}}{\partial \boldsymbol{x}}\right) \cdot \boldsymbol{v} + \varkappa_0 \frac{\partial}{\partial \boldsymbol{x}} \cdot \boldsymbol{u}\right\}. \tag{32.4}$$

If there were a gradient of temperature as well as velocity in the liquid, it would be necessary to add the term on the right-hand side of (31.4).

Now the pressure tensor, according to the theory leading to (15.11), consists of two terms,

$$p_k = m \int \boldsymbol{v}\,\boldsymbol{v}\,f\,d^3v \tag{32.5}$$

and

$$p_p = -\tfrac{1}{2}\int \{\phi'(r)/r\}\,\boldsymbol{r}\,\boldsymbol{r}\,\bar{n}_2\,d^3r. \tag{32.6}$$

The first, on substitution from (31.3) and (32.4), gives

$$p_k = n\,k\,T\,\boldsymbol{\delta} - \left(\frac{2}{15}\int \varkappa\,v^2\,d^3v\right)\frac{\partial_s \boldsymbol{u}}{\partial \boldsymbol{x}} - \left(\frac{1}{3}\int \varkappa_0\,v^2\,d^3v\right)\frac{\partial}{\partial \boldsymbol{x}}\cdot \boldsymbol{u}\,\boldsymbol{\delta}, \tag{32.7}$$

while the second, on substitution from (32.2) and (32.3), gives

$$\left.\begin{aligned} p_p = &-\left\{\frac{1}{6}\int r\,\phi'(r)\,n_2^0\,d^3r\right\}\boldsymbol{\delta} - \\ &-\left\{\frac{1}{15}\int r\,\phi'(r)\,\nu\,d^3r\right\}\frac{\partial_s \boldsymbol{u}}{\partial \boldsymbol{x}} - \left\{\frac{1}{6}\int r\,\phi'(r)\,\nu_0\,d^3r\right\}\frac{\partial}{\partial \boldsymbol{x}}\cdot \boldsymbol{u}\,\boldsymbol{\delta}. \end{aligned}\right\} \tag{32.8}$$

By comparison of (32.7) and (32.8) with (29.6), it will be seen that the coefficient of shearing viscosity is

$$2\eta = \tfrac{2}{15}\int \varkappa\,v^2\,d^3v + \tfrac{1}{15}\int r\,\phi'(r)\,\nu\,d^3r \tag{32.9}$$

and the coefficient of volume viscosity

$$\eta_0 = \tfrac{1}{3}\int \varkappa_0\,v^2\,d^3v + \tfrac{1}{6}\int r\,\phi'(r)\,\nu_0\,d^3r. \tag{32.10}$$

There are certain restrictions on $\varkappa_0$ which are worthy of attention. Since

$$\int f\,d^3v = \int f^0\,d^3\boldsymbol{v} = n,$$

one finds from (32.4) that

$$\int \varkappa_0\,d^3v = 0. \tag{32.11}$$

Also, since

$$\int m\,\boldsymbol{v}^2 f\,d^3v = \int m\,\boldsymbol{v}^2 f^0\,d^3v = 3\,kT,$$

it is necessary that $\varkappa_0$ should satisfy

$$\int \varkappa_0\,\boldsymbol{v}^2\,d^3v = 0. \tag{32.12}$$

The simplest way of satisfying these conditions is to set $\varkappa_0 = 0$; and although there are of course other ways in which they can be satisfied, the evidence of gas-theory suggests that $\varkappa_0$ does really vanish. However this may be, it can be seen from (32.12) that the first term on the right-hand side of (32.10) vanishes. This confirms the statement already made that volume viscosity is due to the intermolecular interaction alone.

Methods for the explicit determination of the functions $\varkappa$, ν and ν_0 will be found in Sects. 40 and 42; they are necessarily approximate in character. The results expressed by (32.9) and (32.10), on the other hand, are exact. They show that the coefficients of viscosity are determined by the deformation of the distribution of molecules about a given molecule which is produced by the velocity gradient. This fact will be seen to enhance the significance for molecular theory of the considerable body of experimental data which has been accumulated by the study of sonic and ultrasonic propagation in liquids.

33. Sonic and ultrasonic absorption in fluids. In the reaction of a fluid to a shearing stress, the reversible effect, elastic deformation, is masked by the irreversible effect, viscous flow. When an isotropic stress system is set up, on the other hand, so that the fluid dilates or is compressed, the reverse occurs: the elastic deformation is permanent and therefore easily observed, but any process of a viscous nature is transitory and difficult to detect. Techniques similar to those devised to measure the shear modulus of elasticity (Sect. 25) can, however, be applied to measure the volume viscosity of a fluid. A stress system which varies harmonically, instead of changing suddenly and then remaining constant, is applied. Ideally the stress should have no shearing components, but it is much simpler experimentally to study the transmission of sound through the fluid. The deformation caused by a sound wave is partly a shear and partly a change of volume, but as the effects of shearing viscosity and other irreversible processes can be investigated by other means, it is possible to isolate the specific effect of volume viscosity on the propagation of sound in the fluid. In view of the connection between the volume viscosity and the molecular structure of liquids, the study of sonic absorption is of interest and importance to molecular theory on these grounds alone.

However, there may be another reason for expecting the absorption of sound in liquids, especially at high frequencies, to provide information concerning molecular structure. There is some evidence that the coefficients of viscosity and thermal conduction exhibited in the absorption of sound by at least some liquids depends on the frequency, in such a way that the absorption is less than one would expect from their values at low frequencies. The experimental results are sometimes accounted for in terms of "relaxation" processes at the molecular level. The reader should understand that many of the relaxation processes which have been described are at best naive representations of the complex molecular phenomena which accompany ordinary irreversible processes. There is little doubt, however, that a careful interpretation of the experiments on supersonic absorption could add much to our understanding of these phenomena.

The amount of experimental data available has increased very rapidly in recent years. Data accumulated up to the year 1951 are admirably summarized in a review article by MARKHAM, BEYER and LINDSAY[1], together with the various theories. A somewhat more recent but also more specialized review is due to LAMB[2].

On the theoretical side, it is clear that sonic absorption is an irreversible effect, and that the only possible irreversible processes are viscosity and thermal conduction (including conduction by radiation) in simple liquids, and diffusion and certain types of chemical reaction (including association or dissociation) in mixtures. Evidence is available[3] that abnormal diffusion is not involved in sonic propagation, so that this possibility can be disregarded. The effect of shear viscosity on sonic absorption was originally discussed by STOKES, who on the strength of the theory of rare gases, mistakenly thought that volume viscosity could probably not exist. As a result, any effect which could not be ascribed to shear viscosity or thermal conduction has been classified as anomalous. It is, however, widely accepted today that most "anomalous" contributions to the absorption coefficient in liquids are due to volume viscosity[4]. It will be shown in the next section that no contribution to the absorption which increases as the square of the frequency need be regarded as truly anomalous. The experimental data for the absorption coefficient α in simple liquids can be analysed by decomposition into three terms

$$\alpha = \alpha_{\text{th}} + \alpha_{\text{vis}} + \alpha'_{\text{vis}} \tag{33.1}$$

[1] J. J. MARKHAM, R. T. BEYER and R. B. LINDSAY: Rev. Mod. Phys. **23**, 353 (1951).
[2] J. LAMB: Research. **5**, 553 (1952).
[3] K. ALTENBURG: Z. phys. Chem. **202**, 460 (1953).
[4] See L. TISZA: Phys. Rev. **61**, 531 (1942).

where α_{vis} is the viscous absorption, due to shear viscosity alone, α_{th} is due to thermal conduction alone, and α'_{vis} is due to volume viscosity. The three contributions are strictly additive only if the total absorption is not large enough to affect the velocity of sound in the liquid; this analysis is therefore liable to fail at very high frequencies, and in very viscous fluids. In mixtures there may be an additional term associated with chemical reaction. But no other contribution to the absorption need be envisaged.

In gases, especially in gases at low densities, the "anomalous" contribution α'_{vis} is negligible. The thermal conductive contribution α_{th} decreases from about 70% of the viscous contribution α_{vis} in monatomic gases to about 40% in diatomic gases and still less in polyatomic gases. These facts are readily understood on the basis of the kinetic theory of gases which, in fact, allows accurate predictions to be made. In liquids, the situation is much more complex, because of the complicated and diverse behaviour of the coefficients of viscosity and thermal conduction. Usually the thermal conductive contributions to the absorption can be neglected, though exceptions can be found in liquid metals, such as mercury. On the relative contributions of shear and volume viscosity, significant regularities are most easily discerned by studying groups of liquids with similar molecular constitution.

At least in certain liquids, there is found, for example, a similarity in the behaviour of the volume/shear viscosity ratio and the volume/shear elasticity ratio. Litovitz, Higgs and Meister[1], by measuring sonic absorption in the diols (ethanediol, propanediol and butanediol), found that the effect of adding CH_2 groups was to increase the ratio of volume to shear viscosity, though the shear viscosity naturally increases with increasing molecular complexity. This contrast with the behaviour of the alcohols, determined by Pellam and Galt[2] and Pinkerton[3], from whose work it can be inferred that the volume/shear viscosity ratio decreases as CH_2 groups are added. These facts are correlated by the observation that the adiabatic compressibility also increases with the addition of CH_2 groups in the diols, but increases in the alcohols. Taking account of the relation between the coefficients of elasticity and viscosity and the molecular structure, one can draw the conclusion that the deformation of the molecular structure in viscous processes must be at least roughly proportional to the known deformation in elastic processes. This may be regarded as experimental evidence in favour of a subsequent approximation expressed by (40.4) or (40.5).

A few experimenters have detected the dispersion as well as absorption of acoustic waves. Dispersion can arise through the use of waves of too great an amplitude at the source and also in very viscous media, without any important significance. Lamb and Pinkerton[4], however, attributed the dispersion which they found in acetic acid to the known tendency of the molecules to associate in pairs. Association and dissociation are irreversible processes which can play an important part in absorption on their own account, as well as through their effect on the viscosity of liquids. The evidence seems to be that these processes may sometimes proceed rather slowly, and therefore not be fully effective at high frequencies.

Quantum effects are very clearly shown in the temperature dependence of the absorption of sound in liquid helium, which was investigated by Pellam

[1] T. A. Litovitz, R. Higgs and R. Meister: J. Chem. Phys. **22**, 1281 (1954).
[2] J. R. Pellam and J. K. Galt: J. Chem. Phys. **14**, 608 (1946).
[3] J. M. M. Pinkerton: Proc. Phys. Soc. Lond. B **62**, 129 (1949).
[4] J. Lamb and J. J. Pinkerton: Proc. Roy. Soc. Lond., Ser. A **199**, 114 (1949).

and SQUIRE[1] and ATKINS and CHASE[2]. There is a λ-phenomenon in the absorption which coincides with the transition from He I to He II; the sonic velocity, however, is little affected, at the pressure of the saturated vapour. In He I the data can be explained in terms of normal thermal conduction and shear viscosity, without and appreciable contribution from volume viscosity. In He II, on the other hand, it is difficult to draw any sure conclusion, since no consistent measurements of the coefficient of thermal conduction can be made. The experiments mentioned have been restricted to the absorption of ordinary sound, and though absorption of "second sound" does occur in the liquid[3], its significance is not yet clear.

34. Theory of sonic absorption. For the correct interpretation of the experimental data, it is important to have available a complete quantitative theory of sonic absorption. A brief derivation will therefore be given of a general result which shows the effect of thermal conduction, and both shearing and volume viscosity. Radiation, and diffusion in liquid mixtures, will be left out of account, since they are known to be unimportant. Also, to avoid irrelevant complications it will be supposed that the amplitude of the sonic waves is small; then, if ϑ represents any macroscopic variable, such as the mean density of the fluid velocity $\boldsymbol{u}$, one can write

$$\vartheta = \vartheta_0 + \mathrm{Re}\,(\delta\vartheta) \tag{34.1}$$

where ϑ_0 is the value ϑ would have if there were no sonic disturbance, and $\mathrm{Re}\,(\delta\vartheta)$ represents the real part of the harmonically varying complex amplitude

$$\delta\vartheta = \delta\vartheta_0 \exp\{i(\omega t - \boldsymbol{k}\cdot\boldsymbol{x})\}. \tag{34.2}$$

Here t is the time and $\boldsymbol{x}$ the position in the liquid; $\delta\vartheta_0$ and $\boldsymbol{k}$ can be complex and $\delta\vartheta_0$ must be small enough to permit the neglect of quantities quadratic in the δ's. For example, since $\boldsymbol{u}_0 = 0$, $\boldsymbol{u}\cdot\frac{\partial\vartheta}{\partial\boldsymbol{x}} = \delta\boldsymbol{u}\cdot\frac{\partial(\delta\vartheta)}{\partial\boldsymbol{x}}$ will be negligible, and

$$\frac{d\vartheta}{dt} = \frac{\partial\vartheta}{\partial t} = \mathrm{Re}\,(i\,\omega\,\delta\vartheta); \qquad \frac{\partial\vartheta}{\partial\boldsymbol{x}} = \mathrm{Re}\,(-\,i\,\boldsymbol{k}\,\delta\vartheta). \tag{34.3}$$

Using (34.3), the equation of conservation of mass (29.4) reduces to

$$\omega\,\delta\varrho = \varrho\,\boldsymbol{k}\cdot\delta\boldsymbol{u}, \tag{34.4}$$

and the hydrodynamical equation of motion (15.3) to

$$\varrho\,\omega\,\delta\boldsymbol{u} = \boldsymbol{k}\cdot\boldsymbol{\delta}\mathsf{p}; \tag{34.5}$$

also, according to (29.6),

$$\boldsymbol{\delta}\mathsf{p} = \delta p\,\boldsymbol{\delta} + i\,\eta\,(\boldsymbol{k}\,\delta\boldsymbol{u} + \delta\boldsymbol{u}\,\boldsymbol{k} - \tfrac{2}{3}\,\boldsymbol{k}\cdot\delta\boldsymbol{u}\,\boldsymbol{\delta}) + i\,\eta_0\,\boldsymbol{k}\cdot\delta\boldsymbol{u}\,\boldsymbol{\delta} \tag{34.6}$$

where p is the pressure, and η and η_0 are the coefficients of shearing volume viscosity. Combining (34.5) and (34.6), one finds that $\boldsymbol{k}$ and $\delta\boldsymbol{u}$ must be parallel and

$$k^2\,\delta p = \{\varrho\,\omega - i\,(\tfrac{4}{3}\,\eta + \eta_0)\,k^2\}\,\boldsymbol{k}\cdot\delta\boldsymbol{u}. \tag{34.7}$$

To determine the ratio $\delta p/\delta\varrho$ when there is thermal conduction, one needs the equation of energy transfer (16.4), which reduces to

$$\omega\,\varrho\,\delta U = \boldsymbol{k}\cdot\delta\boldsymbol{q} + p\,\boldsymbol{k}\cdot\delta\boldsymbol{u}, \tag{34.8}$$

[1] J. R. PELLAM and C. F. SQUIRE: Phys. Rev. **72**, 1245 (1947).
[2] K. R. ATKINS and C. E. CHASE: Proc. Phys. Soc. Lond. A **64**, 826 (1951).
[3] See J. PELLAM: Phys. Rev. **75**, 1183 (1949). — D. V. OSBORNE: Proc. Phys. Soc. Lond. A **64**, 114 (1951).

where $\delta \boldsymbol{q}$, according to (29.8), is given by

$$\delta \boldsymbol{q} = i\,\lambda\, \boldsymbol{k}\, \delta T, \tag{34.9}$$

and λ is the coefficient of thermal conduction. If a is the coefficient of volume expansion, c_p and c_v the specific heats at constant pressure and volume respectively, and v_s^2 the adiabatic compressibility,

$$\left.\begin{aligned} \delta U &= \frac{c_p}{a\varrho}\left(\frac{\delta p}{v_s^2} - \delta\varrho\right) + \frac{p\,\delta\varrho}{\varrho^2} \\ \delta T &= \frac{1}{a\varrho}\left(\frac{c_p\,\delta p}{c_v\, v_s^2} - \delta\varrho\right). \end{aligned}\right\} \tag{34.10}$$

(34.8) and (34.9) yield

$$\delta p\left(1 - \frac{i\lambda k^2}{\omega c_v \varrho}\right) = v_s^2\,\delta\varrho\left(1 - \frac{i\lambda k^2}{\omega c_p \varrho}\right). \tag{34.11}$$

Eliminating δp, $\delta\varrho$ and $\delta\boldsymbol{u}$ between (34.4), (34.7) and (34.11), one obtains

$$\left\{\frac{\omega^2}{k^2} - \frac{i(4\eta + 3\eta_0)\,\omega}{3}\right\}\left(1 - \frac{i\lambda k^2}{\omega c_v \varrho}\right) = v_s^2\left(1 - \frac{i\lambda k}{\omega c_p \varrho}\right). \tag{34.12}$$

This relation determines the complex wave number k, whose imaginary part is the coefficient of absorption. If the coefficients of viscosity and thermal conduction are not so large that the damping is appreciable in a single wave length, the absorption coefficient is

$$\alpha = \frac{\omega^2}{2 v_s^2 \varrho}\left\{\left(\frac{1}{c_v} - \frac{1}{c_p}\right)\lambda + \frac{4}{3}\eta + \eta_0\right\}. \tag{34.13}$$

This formula may be compared with (33.1); v_s is, of course, the velocity of sound in the absence of absorption. One sees that, so long as the absorption is proportional to the square of the angular frequency ω, there is no reason to suspect any anomalous effect which cannot be explained in terms of volume viscosity. The formula indeed permits the determination of η_0 from experimental data on the sonic absorption, provided λ and η are already known with sufficient accuracy. If, however, α/ω^2 is not a constant, but varies with ω, this is an indication of the existence of "relaxation frequencies" in the range under investigation. Phenomenologically one may explain such anomalies by supposing that λ, η and η_0 vary with frequency in accordance with laws of the type

$$\lambda = \lambda_0 + \int_0^\infty \frac{\lambda'(\tau)\, d\tau}{1 + \tau^2\omega^2}$$

where $\lambda'(\tau)$ is some function of τ. The interpretation given to a formula of this type is that irreversible processes are compounded of separate processes, each with its own "relaxation time" τ. In order that such processes should be detected, it is necessary to experiment with frequencies of order $1/\tau$, which are usually very high. While neither the experimental or theoretical position with regard to relaxation processes is very satisfactory at present, it seems likely that a detailed investigation of them could provide valuable information concerning liquid structure.

For an account of some relaxation processes which have been suggested, the reader is referred to the review article already mentioned[1]. Only two of the more appealing instances will be mentioned here. The first affects the coefficient of thermal conduction. Most molecules have internal modes of rotation or

[1] J. J. MARKHAM, R. T. BEYER and R. B. LINDSAY: Rev. Mod. Phys. **23**, 353 (1951).

vibration with energy which is quantized and only weakly coupled with the external modes of the molecule, in interaction with its environment. The energy of these internal modes normally contributes to the energy flux, but time is required for this energy to be made available. If, therefore, the energy flux is alternating with a frequency greater than the inverse relaxation time, the internal modes are unable to contribute and the result is a decline in the thermal conductivity of the liquid. The second relaxation process which deserves mention is associated with viscosity. Suppose, as envisaged in Sect. 25, a strain is applied almost instantaneously to a liquid at rest. The deformation of the isotropic structure is accomplished gradually, and a short time is required before the steady state can be reached. If the strain is varying with a period less than this relaxation time, the structure characteristic of the steady state will never be attained, and consequently the response of the liquid to the strain may be elastic rather than viscous. The result will be a decrease in the coefficients of viscosity at high frequencies, which is manifest in a decrease in the expected ultrasonic absorption.

It is not likely that viscous relaxation processes always have a single characteristic time, or even a discrete spectrum of relaxation times, as is sometimes assumed. More probably there is a continuous spectrum of relaxation times, which may however, contribute to the viscosity mainly in distinguished frequency bands. The mathematical treatment of unsteady states of a liquid, in which the relaxation frequencies enter, is not yet far developed, but such progress as has already been made will be discussed in Sect. 37 below.

c) Theory of non-uniform liquids.

35. Changes in molecular distribution. In a non-uniform liquid the distribution functions which specify the molecular structure are continually changing, and irreversible processes are associated with these changes. In this section the fundamental equations, which determine the time dependence of the distribution functions will be derived.

It is necessary for the moment to introduce an acceleration distribution function $G(t, \boldsymbol{x}, \boldsymbol{\xi}, \dot{\boldsymbol{\xi}})$, defined in such a way that $G(t, \boldsymbol{x}, \boldsymbol{\xi}, \dot{\boldsymbol{\xi}})\, d^3\boldsymbol{x}\, d^3\boldsymbol{\xi}\, d^3\dot{\boldsymbol{\xi}}$ is the probability at time t of finding a molecule in the volume element $d^3\boldsymbol{x}$, with velocity in the range $d^3\boldsymbol{\xi}$ and acceleration in the range $d^3\dot{\boldsymbol{\xi}}$. The velocity distribution function f of Sect. 12 is, of course, given by

$$f(t, \boldsymbol{x}, \boldsymbol{\xi}) = \int G(t, \boldsymbol{x}, \boldsymbol{\xi}, \dot{\boldsymbol{\xi}})\, d^3\dot{\xi}. \tag{35.1}$$

Also, any molecule with velocity $\boldsymbol{\xi}$, at the point $\boldsymbol{x}$ at time t, must have been at the point $\boldsymbol{x} - \boldsymbol{\xi}\,dt$ and had velocity $\boldsymbol{\xi} - \dot{\boldsymbol{\xi}}\,dt$, if $\dot{\boldsymbol{\xi}}$ was its acceleration, at time $t - dt$. Hence

$$f(t, \boldsymbol{x}, \boldsymbol{\xi}) = \int G(t - dt, \boldsymbol{x} - \boldsymbol{\xi}\,dt, \boldsymbol{\xi} - \dot{\boldsymbol{\xi}}\,dt, \dot{\boldsymbol{\xi}})\, d^3\dot{\xi}. \tag{35.2}$$

Subtracting (35.2) from (35.1), one obtains

$$0 = dt \int \left(\frac{\partial G}{\partial t} + \boldsymbol{\xi} \cdot \frac{\partial G}{\partial \boldsymbol{x}} + \dot{\boldsymbol{\xi}} \cdot \frac{\partial G}{\partial \boldsymbol{\xi}} \right) d^3\dot{\xi}$$

or

$$\frac{\partial f}{\partial t} + \boldsymbol{\xi} \cdot \frac{\partial f}{\partial \boldsymbol{x}} + \frac{\partial}{\partial \boldsymbol{\xi}} \cdot (\eta f) = 0 \tag{35.3}$$

where η is the mean acceleration defined by

$$f\eta = \int G\, \dot{\boldsymbol{\xi}}\, d^3\dot{\xi}. \tag{35.4}$$

The mean acceleration, can, however, be inferred from the forces which act on the molecule, including the mean force due to the presence of the other molecules. Since the probability of finding another molecule in the volume element $d^3\boldsymbol{x}^{(2)}$, with velocity in the range $d^3\xi^{(2)}$, is $f_2\, d^3\boldsymbol{x}^{(2)}\, d^3\xi^{(2)}/f$, one has

$$m\eta = \boldsymbol{F} - \frac{1}{f}\iint \frac{\partial\phi^{(12)}}{\partial\boldsymbol{x}^{(1)}} f_2\, d^3x^{(2)}\, d^3\xi^{(2)}, \tag{35.5}$$

where $\boldsymbol{F}$ is the force due to external forces.

Combining (35.3) and (35.5) one has

$$\frac{\partial f}{\partial t} + \xi\cdot\frac{\partial f}{\partial\boldsymbol{x}} + \frac{1}{m}\frac{\partial}{\partial\xi}\cdot\left(f\boldsymbol{F} - \iint \frac{\partial\phi^{(12)}}{\partial\boldsymbol{x}^{(1)}} f_2 d^3x^{(2)} d^3\xi^{(2)}\right) = 0. \tag{35.6}$$

This equation can also be obtained by applying an averaging procedure to Liouville's equation for the distribution in the phase space of the whole system of molecules which comprise the liquid.

For future convenience, some elementary consequences of (35.3) will be noticed here. If this equation is multiplied by any function $\varphi(\xi)$ of the molecular velocity ξ, and the result integrated with respect to ξ, one obtains

$$\frac{\partial}{\partial t}(n\langle\varphi\rangle) + \frac{\partial}{\partial\boldsymbol{x}}\cdot(n\langle\varphi\rangle) = n\left\langle\eta\cdot\frac{\partial\varphi}{\partial\xi}\right\rangle \tag{35.7}$$

where $\langle\varphi\rangle$ denotes the average value of φ, defined by

$$n\langle\varphi\rangle = \int f\varphi\, d^3\xi. \tag{35.8}$$

For the particular substitution $\varphi = 1$, (35.7) reduces the equation of continuity

$$\frac{\partial n}{\partial t} + \frac{\partial}{\partial\boldsymbol{x}}\cdot(n\boldsymbol{u}) = 0, \tag{35.9}$$

and this can be used to re-write (35.7) in the form

$$n\frac{d\langle\varphi\rangle}{dt} + \frac{\partial}{\partial\boldsymbol{x}}\cdot(n\langle\boldsymbol{v}\varphi\rangle) = n\left\langle\eta\cdot\frac{\partial\varphi}{\partial\xi}\right\rangle \tag{35.10}$$

where

$$\boldsymbol{v} = \xi - \boldsymbol{u};\qquad \frac{d}{dt} = \frac{\partial}{\partial t} + \boldsymbol{u}\cdot\frac{\partial}{\partial\boldsymbol{x}}. \tag{35.11}$$

If one substitutes $\varphi = \xi$ into (35.10), one obtains the equation of motion in the form

$$n\frac{d\boldsymbol{u}}{dt} + \frac{\partial}{\partial\boldsymbol{x}}\cdot(n\langle\boldsymbol{v}\xi\rangle) = n\langle\eta\rangle. \tag{35.12}$$

If, on the other hand, one substitutes $\varphi = m\xi^2$, and uses the classical formula (12.3) for the temperature, one has

$$n\frac{d}{dt}(3kT + m\boldsymbol{u}^2) + m\frac{\partial}{\partial\boldsymbol{x}}\cdot(n\langle\boldsymbol{v}\xi^2\rangle) = 2n\langle\eta\cdot\xi\rangle. \tag{35.13}$$

An equation analogous to (35.3) is readily derived for the distribution function f_2:

$$\frac{\partial f_2}{\partial t} + \xi^{(1)}\cdot\frac{\partial f_2}{\partial\boldsymbol{x}^{(1)}} + \xi^{(2)}\cdot\frac{\partial f_2}{\partial\boldsymbol{x}^{(2)}} + \frac{\partial}{\partial\xi^{(1)}}\cdot(\eta_2^{(1)} f_2) + \frac{\partial}{\partial\xi^{(2)}}\cdot(\eta_2^{(2)} f_2) = 0 \tag{35.14}$$

where

$$m\eta_2^{(i)} = F^{(i)} - \frac{\partial\phi^{(12)}}{\partial\boldsymbol{x}^{(i)}} - \frac{1}{f_2}\iint \frac{\partial\phi^{(i3)}}{\partial\boldsymbol{x}^{(i)}} f_3\, d^3x^{(3)}\, d^3\xi^{(3)}. \tag{35.15}$$

If $\langle\varphi\rangle$ now denotes the average of a function $\varphi(\xi^{(1)}, \xi^{(2)})$ of the two velocities, the analogue of (35.7) is

$$\left.\begin{aligned}\frac{\partial}{\partial t}(n_2\langle\varphi\rangle)+\frac{\partial}{\partial \boldsymbol{x}^{(1)}}\cdot(n_2\langle\xi^{(1)}\varphi\rangle)+\frac{\partial}{\partial \boldsymbol{x}^{(2)}}\cdot(n_2\langle\xi^{(2)}\varphi\rangle)\\ =n_2\left\langle\eta_2^{(1)}\cdot\frac{\partial\varphi}{\partial\xi^{(1)}}+\eta_2^{(2)}\cdot\frac{\partial\varphi}{\partial\xi^{(2)}}\right\rangle.\end{aligned}\right\} \tag{35.16}$$

Substituting $\varphi=1$ in this equation, one has

$$\frac{\partial n_2}{\partial t}+\frac{\partial}{\partial \boldsymbol{x}^{(1)}}\cdot(n_2\boldsymbol{u}_2^{(1)})+\frac{\partial}{\partial \boldsymbol{x}^{(2)}}\cdot(n_2\boldsymbol{u}_2^{(2)})=0, \tag{35.17}$$

with $\boldsymbol{u}_2^{(1)}$ and $\boldsymbol{u}_2^{(2)}$ defined by (12.6). Making use of this result, and setting

$$\left.\begin{aligned}\boldsymbol{v}_2^{(i)}&=\xi^{(i)}-\boldsymbol{u}_2^{(i)},\\ \frac{d_2}{dt_2}&=\frac{\partial}{\partial t}+\boldsymbol{u}_2^{(1)}\cdot\frac{\partial}{\partial \boldsymbol{x}^{(1)}}+\boldsymbol{u}_2^{(2)}\cdot\frac{\partial}{\partial \boldsymbol{x}^{(2)}},\end{aligned}\right\} \tag{35.18}$$

(35.16) can be written

$$\left.\begin{aligned}n_2\frac{d_2\langle\varphi\rangle}{dt_2}+\frac{\partial}{\partial \boldsymbol{x}^{(1)}}\cdot(n_2\langle\boldsymbol{v}_2^{(1)}\varphi\rangle)+\frac{\partial}{\partial \boldsymbol{x}^{(2)}}\cdot(n_2\langle\boldsymbol{v}_2^{(2)}\varphi\rangle)\\ =n_2\left\langle\eta_2^{(1)}\cdot\frac{\partial\varphi}{\partial\xi^{(1)}}+\eta_2^{(2)}\cdot\frac{\partial\varphi}{\partial\xi^{(2)}}\right\rangle.\end{aligned}\right\} \tag{35.19}$$

The generalization of these results for the function f_q of Sect. 12 will now be sufficiently obvious; the detailed formulae have been given by BORN and GREEN[1]. For subsequent purposes only the equation satisfied by a distribution function F_q, defined by analogy with the function N_q in Sect. 13, will be required, and this will now be obtained. It will be recalled that N_q specifies the distribution of q particles, within a finite region of volume V, when there are no other molecules in the region. The function F_q is defined so that $F_q\, d^3x^{(1)}\, d^3\xi^{(1)} \ldots d^3x^{(q)}\, d^3\xi^{(q)}$ is the probability of finding q volume elements $d^3x^{(1)} \ldots d^3x^{(q)}$, in the region, occupied by molecules with velocities in the ranges $d^3\xi^{(1)} \ldots d^3\xi^{(q)}$ respectively, and no other molecule anywhere in the region. If no molecule could enter or leave the region, the equation satisfied by F_q would be LIOUVILLE's equation, i.e., if Φ_q is the total potential energy of the system of q molecules,

$$\frac{D_q F_q}{D t_q}\equiv\frac{\partial F_q}{\partial t}+\sum_{i=1}^{q}\left\{\xi^{(i)}\cdot\frac{\partial F_q}{\partial \boldsymbol{x}^{(i)}}-\frac{1}{m}\frac{\partial\Phi_q}{\partial \boldsymbol{x}^{(i)}}\cdot\frac{\partial F_q}{\partial\xi^{(i)}}\right\} \tag{35.20}$$

would vanish. Actually F_q can increase on account of the departure of a molecule from the region when $q+1$ molecules are present, or decrease due to the entry of an additional molecule when q molecules are present. So

$$\frac{D_q F_q}{D t_q}=\iint F_{q+1}\,(\xi^{(q+1)}-\boldsymbol{u}_0^{(q+1)})\cdot d\boldsymbol{S}^{(q+1)}\, d^3\xi^{(q+1)}, \tag{35.21}$$

where $\boldsymbol{u}_0^{(q+1)}$ is the velocity of the boundary of the region, if it moves, and $d\boldsymbol{S}^{(q+1)}$ is a vector element of the bounding surface at the point $\boldsymbol{x}^{(q+1)}$. Using GAUSS's theorem, this can also be written

$$\frac{D_q F_q}{D t_q}=\iint\frac{\partial}{\partial \boldsymbol{x}^{(q+1)}}\cdot(F_{q+1}\boldsymbol{v}_0^{(q+1)})\, d^3x^{(q+1)}\, d^3\xi^{(q+1)} \tag{35.22}$$

where $\boldsymbol{v}_0^{(q+1)}=\xi^{(q+1)}-\boldsymbol{u}_0^{(q+1)}$.

[1] M. BORN and H. S. GREEN: Proc. Roy. Soc. Lond., Ser. A **190**, 455 (1947).

The distribution functions f_q and F_q are related by

$$f_q = F_{q,0} + F_{q,1}/1! + F_{q,2}/2! + \cdots, \tag{35.23}$$

where $F_{q,0}$ is the same as F_q, and

$$F_{q,j} = \overset{(2j)}{\int \cdots \int} F_{q+j}\, d^3x^{(q+1)}\, d^3\xi^{(q+1)} \ldots d^3x^{(q+j)}\, d^3\xi^{(q+j)}. \tag{35.24}$$

Using (35.20) and (35.21), it is a straight-forward matter to prove the relation

$$\left.\begin{aligned} &\frac{\partial f_q}{\partial t} + \sum_{i=1}^{q}\Bigg[\xi^{(i)}\cdot\frac{\partial f_q}{\partial x^{(i)}} - \frac{1}{m}\frac{\partial}{\partial \xi^{(i)}}\times \\ &\times\left\{\frac{\partial \Phi_q}{\partial x^{(i)}} f_q + \iint \frac{\partial(\Phi_{q+1}-\Phi_q)}{\partial x^{(i)}} f_{q+1}\, d^3x^{(q+1)}\, d^3\xi^{(q+1)}\right\}\Bigg] = 0. \end{aligned}\right\} \tag{35.25}$$

This is the generalization of the relation (35.6) above.

36. The steady state of a non-uniform liquid. In nearly all experiments, the positions and velocities of individual molecules are completely unknown, and even the distribution functions, which provide a statistical description of the molecular configurations and velocities, have to be calculated from purely macroscopic data. The data immediately available concern the density, temperature and velocity of flow of the liquid, which may ideally be regarded as known at every point in the liquid. The distribution functions, determined from this data, may therefore be regarded as functions of n, T and $\boldsymbol{u}$, together with their spatial derivatives, at any conveniently chosen point $\boldsymbol{x}$ within the liquid. If an external force field should be specified, the distribution functions may be regarded as depending also on the interaction energy $\Psi(\boldsymbol{x})$ of a molecule at the point $\boldsymbol{x}$ with this field, together with the spatial derivatives of $\Psi(\boldsymbol{x})$. But for the purposes of this section, any external field of force will be disregarded.

Accordingly, the velocity distribution function $f(t, \boldsymbol{x}, \boldsymbol{\xi})$ may be expressed as a new function of the velocity $\boldsymbol{\xi}$, the values of n, T and $\boldsymbol{u}$ at the point $\boldsymbol{x}$, and the spatial derivatives of n, T and $\boldsymbol{u}$ at the same point:

$$f(t, \boldsymbol{x}, \boldsymbol{\xi}) = \bar{f}\left(\boldsymbol{\xi}, \lambda, \frac{\partial \lambda}{\partial \boldsymbol{x}}, \frac{\partial^2 \lambda}{\partial \boldsymbol{x}\,\partial \boldsymbol{x}}, \ldots\right) \tag{36.1}$$

where λ represents the group of variables n, T and $\boldsymbol{u}$. The new function $\bar{f}$ does not depend explicitly on the time t or the position $\boldsymbol{x}$, though it does of course depend on them implicitly through the parameters λ. It is, moreover, redundant to suppose that $\bar{f}$ might depend on the time derivatives of these parameters, since, according to (35.9), (35.12) and (35.13)

$$\left.\begin{aligned} \frac{\partial n}{\partial t} &= -\boldsymbol{u}\cdot\frac{\partial n}{\partial \boldsymbol{x}} - n\frac{\partial}{\partial \boldsymbol{x}}\cdot\boldsymbol{u}, \\ \frac{\partial \boldsymbol{u}}{\partial t} &= -\boldsymbol{u}\cdot\frac{\partial \boldsymbol{u}}{\partial \boldsymbol{x}} - \frac{1}{n}\frac{\partial}{\partial \boldsymbol{x}}\cdot(n\langle \boldsymbol{v}\boldsymbol{v}\rangle) + \langle \boldsymbol{\eta}\rangle, \\ \frac{\partial T}{\partial t} &= -\boldsymbol{u}\cdot\frac{\partial T}{\partial \boldsymbol{x}} - \frac{m}{3kn}\frac{\partial}{\partial \boldsymbol{x}}\cdot(n\langle \boldsymbol{v}\,\boldsymbol{v}^2\rangle) + \frac{2}{3k}\langle \boldsymbol{\eta}\cdot\boldsymbol{v}\rangle, \end{aligned}\right\} \tag{36.2}$$

and the right-hand sides of these equations can be expressed entirely in terms of spatial derivatives. It should be noticed also that $\bar{f}$ does not depend on $\boldsymbol{\xi}$ and $\boldsymbol{u}$ independently, but only on their difference $\boldsymbol{v}$; if the velocity of every molecule is changed by an amount $\delta\boldsymbol{u}$, so is the velocity $\boldsymbol{u}$ of the liquid as a whole.

Not only the function f, but also f_2 and the functions of higher order, can be expressed in a form analogous to (36.1). Thus one may write

$$f_2 = \bar{f}_2\left(\boldsymbol{r}, \boldsymbol{\xi}^{(1)}, \boldsymbol{\xi}^{(2)}, \lambda, \frac{\partial \lambda}{\partial \boldsymbol{x}}, \frac{\partial^2 \lambda}{\partial \boldsymbol{x}\, \partial \boldsymbol{x}}, \ldots\right), \tag{36.3}$$

where $\boldsymbol{r} = \boldsymbol{x}^{(2)} - \boldsymbol{x}^{(1)}$, and the point $\boldsymbol{x}$ at which the values of the parameters λ and derivatives are specified may be conveniently identified with the mean centre $\frac{1}{2}(\boldsymbol{x}^{(1)} + \boldsymbol{x}^{(2)})$.

Though the dependence of $\bar{f}$ on the second spatial derivatives $\frac{\partial^2 \lambda}{\partial \boldsymbol{x}\, \partial \boldsymbol{x}}$ has been investigated in the kinetic theory of gases[1], little is yet known in relation to liquids. In practice it is necessary to suppose that f and f_2 are expanded in powers of the derivates of the λ, and so far attention has been restricted to the coefficient of $\partial \lambda / \partial \boldsymbol{x}$. Writing

$$\left.\begin{aligned} \bar{f} &= f^0 + f' + f'' + \cdots, \\ f^0 &= \bar{f}\,(\boldsymbol{\xi}, \lambda, 0, 0, \ldots), \\ f' &= \frac{\partial f}{\partial(\partial \lambda / \partial \boldsymbol{x})}\,(\boldsymbol{\xi}, \lambda, 0, 0, \ldots) \cdot \frac{\partial \lambda}{\partial \boldsymbol{x}}, \end{aligned}\right\} \tag{36.4}$$

etc., it is clear that f^0 must be the function appropriate to mechanical and thermodynamical equilibrium, and the essential problem at present is to obtain f'. It has already been seen in Sects. 31 and 32 that f' determines certain contributions to the coefficients of shearing viscosity and thermal conduction. Similarly, if

$$\bar{f}_2 = f_2^0 + f_2' + f_2'' + \cdots, \tag{36.5}$$

where f_2^0 is the function appropriate to equilibrium and f_2' is linear in the spatial gradients etc., the essential problem at present is to obtain f_2'; this also determines certain contributions—in fact, the major contributions—to the coefficients of viscosity and thermal conduction. For f_2' determines both n_2' and $\boldsymbol{u}_2^{(1)\prime}$, which appear in the formulae (32.3) and (31.6) respectively.

Some equations connecting f', f_2', n_2' and $\boldsymbol{u}_2'$ will be needed subsequently and will now be derived. Since only terms linear in the first spatial derivatives of λ are required, (36.2) may be used in the approximation

$$\left.\begin{aligned} \frac{dn}{dt} &= -n \frac{\partial}{\partial \boldsymbol{x}} \cdot \boldsymbol{u}, \\ \frac{d\boldsymbol{u}}{dt} &\approx -\frac{1}{mn} \frac{\partial p^0}{\partial \boldsymbol{x}} = -\frac{1}{mn}\left(\frac{\partial p^0}{\partial n} \frac{\partial n}{\partial \boldsymbol{x}} + \frac{\partial p^0}{\partial T} \frac{\partial T}{\partial \boldsymbol{x}}\right), \\ \frac{dT}{dt} &\approx -A \frac{\partial}{\partial \boldsymbol{x}} \cdot \boldsymbol{u}, \end{aligned}\right\} \tag{36.6}$$

where

$$A \frac{\partial}{\partial \boldsymbol{x}} \cdot \boldsymbol{u} \approx -\frac{2}{3k} \langle \boldsymbol{\eta} \cdot \boldsymbol{v} \rangle.$$

The coefficient A can be determined more explicitly by noticing that, according to (29.5),

$$T \frac{dS}{dt} = \frac{dU}{dt} - \frac{p^0}{mn^2} \frac{dn}{dt} \approx 0,$$

to the required approximation, so that

$$A \frac{\partial U}{\partial T} + n \frac{\partial U}{\partial n} = \frac{p^0}{mn}. \tag{36.7}$$

[1] See Vol. XII, article of GRAD.

Now, remembering that for the function n_2, $\boldsymbol{x} = \frac{1}{2}(\boldsymbol{x}^{(1)} + \boldsymbol{x}^{(2)})$, one has

$$\left.\begin{aligned} \frac{d n_2}{dt} &= \frac{\partial \bar{n}_2}{\partial n}\frac{dn}{dt} + \frac{\partial \bar{n}_2}{\partial T}\frac{dT}{dt}, \\ \frac{\partial n_2}{\partial \boldsymbol{x}} &= \frac{\partial \bar{n}_2}{\partial n}\frac{\partial n}{\partial \boldsymbol{x}} + \frac{\partial \bar{n}_2}{\partial T}\frac{\partial T}{\partial \boldsymbol{x}}, \end{aligned}\right\} \tag{36.8}$$

so (35.17) yields[1]

$$\left(n_2^0 - n\frac{\partial n_2^0}{\partial n} - A\frac{\partial n_2^0}{\partial T}\right)\frac{\partial}{\partial \boldsymbol{x}}\cdot \boldsymbol{u} + \frac{\partial}{\partial \boldsymbol{r}}\cdot\{n_2^0(\boldsymbol{u}_2^{(2)\prime} - \boldsymbol{u}_2^{(1)\prime})\} = 0. \tag{36.9}$$

A similar procedure can be applied to (35.19), with $\xi^{(2)} - \xi^{(1)}$ substituted for φ; the result is simply

$$\left.\begin{aligned} \frac{\partial}{\partial \boldsymbol{r}}\cdot\{n_2\langle \boldsymbol{v}_2^{(2)} - \boldsymbol{v}_2^{(1)}\,\boldsymbol{v}_2^{(2)} - \boldsymbol{v}_2^{(1)}\rangle\}' &= \{n_2\langle \eta_2^{(2)} - \eta_1^{(1)}\rangle\}' \\ &= -\frac{2}{m}\left\{n_2'\frac{\partial\phi^{(12)}}{\partial \boldsymbol{x}^{(2)}} + \int n_3'\frac{\partial\phi^{(23)}}{\partial \boldsymbol{x}^{(2)}}\,d^3x^{(3)}\right\}. \end{aligned}\right\} \tag{36.10}$$

Also, if $\xi^{(1)} + \xi^{(2)}$ is substituted for φ, the result is

$$\left.\begin{aligned} -\frac{2n_2}{mn}\frac{\partial p^0}{\partial \boldsymbol{x}} + \frac{\partial}{\partial \boldsymbol{x}}(n_2 kT) + \frac{\partial}{\partial \boldsymbol{r}}\cdot\{n_2\langle \boldsymbol{v}_2^{(2)} - \boldsymbol{v}_2^{(1)}\,\boldsymbol{v}_2^{(1)} + \boldsymbol{v}_2^{(2)}\rangle\} \\ = \{n_2(\eta_2^{(1)} + \eta_2^{(2)})\}' \\ = -\frac{1}{m}\left\{\int n_3\left(\frac{\partial\phi^{(13)}}{\partial \boldsymbol{x}^{(1)}} + \frac{\partial\phi^{(23)}}{\partial \boldsymbol{x}^{(2)}}\right)d^3x^{(3)}\right\}'. \end{aligned}\right\} \tag{36.11}$$

Finally, terms linear in the spatial gradients can be extracted from (35.6) in a similar way. When substituting $f_2 = f_2^0 + f_2'$ in the integral, it is important to observe that the centre of mass $\boldsymbol{x} = \frac{1}{2}(\boldsymbol{x}^{(1)} + \boldsymbol{x}^{(2)})$ is no longer the point at which the parameters λ and their derivatives are specified; it is therefore necessary to set

$$f_2^0(\boldsymbol{x}) = f_2^0(\boldsymbol{x}^{(1)}) + \tfrac{1}{2}\boldsymbol{r}\cdot\frac{\partial}{\partial \boldsymbol{x}} f_2^0(\boldsymbol{x}^{(1)}).$$

The result obtained is

$$\left.\begin{aligned} &\frac{p^0 f^0}{nkT}\left\{\left(\frac{1}{2}\frac{m\boldsymbol{v}^2}{kT} - \frac{5}{2}\right)\frac{\boldsymbol{v}}{T}\cdot\frac{\partial T}{\partial \boldsymbol{x}} + \frac{m}{kT}\boldsymbol{v}\cdot\left(\frac{\partial_s \boldsymbol{u}}{\partial \boldsymbol{x}}\right)\cdot\boldsymbol{v}\right\} \\ &+\left(\frac{2}{3}\frac{p^0}{nkT} - \frac{A}{T}\right)f^0\left(\frac{1}{2}\frac{m\boldsymbol{v}^2}{kT} - \frac{3}{2}\right)\frac{\partial}{\partial \boldsymbol{x}}\cdot\boldsymbol{u} = \frac{1}{m}\iint\frac{\partial f_2'}{\partial \xi^{(1)}}\cdot\frac{\partial\phi^{(12)}}{\partial \boldsymbol{x}^{(1)}}\,d^3x^{(2)}\,d^3\xi^{(2)}. \end{aligned}\right\} \tag{36.12}$$

These are all exact relations, which do not, however, suffice for the complete determination of f_2', n_2' and $\boldsymbol{u}_2^{(i)\prime}$. One can, however by adopting plausible approximations, obtain sufficient information to calculate the coefficients of viscosity and thermal conduction. This will be done in Sects. 40 to 42 below. In Sects. 38 and 39, some developments will be described which, in principle, eliminate the need for approximation. But first some consideration will be given to the problems associated with a liquid in an unsteady state.

37. Unsteady states. The formalism of the previous section, though adequate for the discussion of most natural states of a non uniform liquid, in which the macroscopic parameters (density, temperature and fluid velocity) change rather slowly from point to point and from time to time, leaves out of account the possibility of rapid fluctuations. A universal method for the treatment of unsteady states has yet to be devised, but will probably be founded on the general theory

[1] This result, in an integrated form, first appears as Eq. (3.30) of M. Born and H. S. Green: Proc. Roy. Soc. Lond., Ser. A **190**, 455.

of stochastic processes. The method which will now be described is specially adapted to unsteady states characterized by small, though possibly very rapid oscillations in the vicinity of thermal and mechanical equilibrium. Such conditions are present, for example, in experiments on ultrasonic propagation and absorption of the type discussed in Sect. 34, and also in the experiments, described in Sect. 25, designed to detect and measure the elasticity of liquids.

It will be assumed that the distribution function f_q for q particles in the liquid can be expressed in the form

$$f_q = f_q^0(\boldsymbol{x} - \mathbf{s}_q, \boldsymbol{\xi} - \sigma_q) \prod_{i=1}^{q} \left(1 - \frac{\partial}{\partial \boldsymbol{x}^{(i)}} \cdot \mathbf{s}_q^{(i)}\right) \left(1 - \frac{\partial}{\partial \boldsymbol{\xi}^{(i)}} \cdot \sigma_q^{(i)}\right) \tag{37.1}$$

where f_q^0 is the function appropriate to equilibrium, and $\mathbf{s}_q^{(1)}, \ldots, \mathbf{s}_q^{(q)}$ and $\sigma_q^{(1)}, \ldots, \sigma_q^{(q)}$ are time dependent functions of the positions and velocities, which may be regarded as the mean departures from $\boldsymbol{x}^{(1)}, \ldots, \boldsymbol{x}_q^{(q)}$ and $\boldsymbol{\xi}_q^{(1)}, \ldots, \boldsymbol{\xi}_q^{(q)}$ respectively in the unsteady state considered. Since the $\mathbf{s}_q^{(i)}$ and $\sigma_q^{(i)}$ are small, (37.1) may also be written

$$f_q = f_q^0 - \sum_{i=1}^{q} \left\{\frac{\partial}{\partial \boldsymbol{x}^{(i)}} \cdot (f_q^0\, \mathbf{s}_q^{(i)}) + \frac{\partial}{\partial \boldsymbol{\xi}^{(i)}} \cdot (f_q^0\, \sigma_q^{(i)})\right\}. \tag{37.2}$$

The equations satisfied by $\mathbf{s}_q^{(i)}$ and $\sigma_q^{(i)}$ may be obtained by substitution from (37.2) into (35.25), which can be written

$$\frac{D_q f_q}{D t_q} = \frac{1}{m} \sum_{i=1}^{q} \frac{\partial}{\partial \boldsymbol{\xi}^{(i)}} \cdot \iint \frac{\partial \phi^{(i\,q+1)}}{\partial \boldsymbol{x}^{(i)}}\, f_{q+1}\, d^3x^{(q+1)}\, d^3\xi^{(q+1)} \tag{37.3}$$

where the left-hand side is interpreted in the same sense as in (35.20). It is then found that this equation is satisfied identically, provided

$$f_q^0\, \sigma_q^{(i)} = \frac{D_q}{D t_q} (f_q^0\, \mathbf{s}_q^{(i)}) - \frac{1}{m} \sum_{i=1}^{q} \frac{\partial}{\partial \boldsymbol{\xi}^{(i)}} \cdot \iint \frac{\partial \phi^{(i\,q+1)}}{\partial \boldsymbol{x}^{(i)}}\, f_{q+1}^0\, \mathbf{s}_{q+1}^{(i)}\, d^3x^{(q+1)}\, d^3\xi^{(q+1)} \tag{37.4}$$

and

$$\left.\begin{aligned} &\frac{D_q}{D t_q} (f_q^0\, \sigma_q^{(i)}) - \frac{1}{m} \sum_{i=1}^{q} \frac{\partial}{\partial \boldsymbol{\xi}^{(i)}} \cdot \iint \frac{\partial \phi^{(i\,q+1)}}{\partial \boldsymbol{x}^{(i)}}\, f_{q+1}^0\, \sigma_{q+1}^{(i)}\, d^3x^{(q+1)}\, d^3\xi^{(q+1)} + \\ &\quad + \frac{1}{m} \sum_{j=1}^{q} (f_q^0\, \mathbf{s}_q^{(j)}) \cdot \frac{\partial^2 \Phi_q}{\partial \boldsymbol{x}^{(j)}\, \partial \boldsymbol{x}^{(i)}} + \\ &\quad + \frac{1}{m} \iint \frac{\partial^2 \phi^{(i\,q+1)}}{\partial \boldsymbol{x}^{(i)}\, \partial \boldsymbol{x}^{(i)}} \cdot (\mathbf{s}_{q+1}^{(i)} - \mathbf{s}_{q+1}^{(q+1)})\, f_{q+1}^0\, d^3x^{(q+1)}\, d^3\xi^{(q+1)} = 0. \end{aligned}\right\} \tag{37.5}$$

It is convenient at this stage to introduce a mean displacement $\mathfrak{s}_q^{(i)}$, corresponding to $\mathbf{s}_q^{(i)}$, which depends on the molecular coordinates but not on their velocities, and defined by

$$n_q^0\, \mathfrak{s}_q^{(i)} = \int f_q^0\, \mathbf{s}_q^{(i)}\, d\xi_q, \tag{37.6}$$

where $d\xi_q$ stands for $d^3\xi^{(1)} \ldots d^3\xi^{(q)}$. Then, if $\mathfrak{s} = \mathfrak{s}_1^{(1)}$ is prescribed for some initial time at every point in the liquid, it should be possible to obtain a unique solution of the series of Eqs. (37.4) and (37.5) above, corresponding to a particular set of boundary conditions. If the liquid is subject to no thermal or mechanical disturbance, it is natural to suppose that it will approach equilibrium: $\mathfrak{s}$, regarded as a function of the time t, will tend to zero as $t \to \infty$. Suppose, therefore, $\mathfrak{s}$ is resolved into its Fourier components:

$$\mathfrak{s} = \sum_{\varkappa} \mathfrak{a}_{\varkappa}\, e^{-i\varkappa \cdot \boldsymbol{x}} \tag{37.7}$$

and a solution is obtained which fulfils

$$\frac{d\mathfrak{a}_\varkappa}{dt} = i\,\omega\,\mathfrak{a}_\varkappa\,; \tag{37.8}$$

then the imaginary part of the complex angular frequency ω must be positive. There will, of course, be another solution, with the sign of ω reversed; but this represents anti-causal processes, and must be rejected as unphysical.

One procedure will now be described for the solution of the coupled Eqs. (37.4) and (37.5). As a first approximation, the unknowns $\mathbf{s}_{q+1}^{(i)}$ and $\sigma_{q+1}^{(i)}$ in the integrals of these equations are replaced by $\mathbf{s}_1^{(i)}$ and $\sigma_1^{(i)}$. This approximation amounts to neglecting the correlations between the displacements of the molecules from their positions in thermal and mechanical equilibrium, and can be expected to give good results only for frequencies of the order of, or greater than, the lowest "relaxation" frequency. However, the equations can then be solved in a straightforward way, and if a better approximation should be required, the solutions $\mathbf{s}_{q+1}^{(i)}$ and $\sigma_{q+1}^{(i)}$ obtained with the help of the first approximation could be substituted in the integrals, and the equations solved again. By iteration, the exact solution of the equations could be approached as nearly as desired. Only the results of the first approximation will be considered here.

It is required to integrate the equations

$$\left.\begin{aligned} f^0\,\sigma &= \frac{D(f^0\mathbf{s})}{Dt}\,;\\ \frac{D(f^0\sigma)}{Dt} &= \frac{1}{m}\iint \frac{\partial^2\phi^{(12)}}{\partial\boldsymbol{x}^{(1)}\,\partial\boldsymbol{x}^{(1)}}\cdot(\mathbf{s}^{(2)}-\mathbf{s})\,f_2^0\,d^3x^{(2)}\,d^3\xi^{(2)} \end{aligned}\right\} \tag{37.9}$$

obtained by applying the approximation just described to the Eqs. (37.4) and (37.5), with $q=1$, and suppressing the suffix 1 and the superfix $^{(1)}$ where these are inessential. A solution of (37.9) will be sought proportional to $\exp.\{i(\omega t-\boldsymbol{\varkappa}\cdot\boldsymbol{x})\}$, i.e., corresponding to the existence of only a single Fourier component in the expansion of (37.7). For this particular solution, one has

$$\frac{D(f^0\sigma)}{Dt} = f^0(\omega-\boldsymbol{\varkappa}\cdot\boldsymbol{\xi})^2\,\mathbf{s} \tag{37.10}$$

and (37.9) reduces to

$$m\,n\,\{(\omega-\boldsymbol{\varkappa}\cdot\boldsymbol{\xi})^2-\omega_r^2\}\,\mathbf{s} = \hat{\mathbf{s}}\cdot\int\frac{\partial^2\phi}{\partial\boldsymbol{r}\,\partial\boldsymbol{r}}\cos(\boldsymbol{\varkappa}\cdot\boldsymbol{r})\,n_2^0\,d^3r, \tag{37.11}$$

where ω_r is a relaxation frequency, given by

$$m\,n\,\omega_r^2 = -\frac{1}{3}\int\frac{\partial^2\phi}{\partial\boldsymbol{r}\cdot\partial\boldsymbol{r}}\,n_2^0\,d^3r. \tag{37.12}$$

In the liquid state, in contradistinction to the solid state, ω_r is real.

The result (37.11) shows the dependence of $\mathbf{s}$ on the molecular velocity $\boldsymbol{\xi}$. An equation to determine ω is obtained by dividing through by the factor $\{(\omega-\boldsymbol{\varkappa}\cdot\boldsymbol{\xi})^2-\omega_r^2\}$ and integrating with respect to $\boldsymbol{\xi}$. The fact that the imaginary part of ω must be positive is of importance in evaluating the integral. If $\varphi(z)$ denotes the complex function

$$\varphi(z) = \mathrm{e}^{-\frac{1}{2}z^2}\Big\{\int_0^z \mathrm{e}^{\frac{1}{2}\zeta^2}\,d\zeta - i\sqrt{\tfrac{1}{2}\pi}\Big\}, \tag{37.13}$$

then

$$\left.\begin{aligned} 2mn\,\omega_r\,\varkappa\,\hat{\mathbf{s}} = (m\beta)^{\frac{1}{2}}\,[\varphi\{(m\beta)^{\frac{1}{2}}(\omega-\omega_r)/\varkappa\}-\varphi\{(m\beta)^{\frac{1}{2}}(\omega+\omega_r)/\varkappa\}]\,\hat{\mathbf{s}}\times\\ \times\int\frac{\partial^2\phi}{\partial\boldsymbol{r}\,\partial\boldsymbol{r}}\cos(\boldsymbol{\varkappa}\cdot\boldsymbol{r})\,n_2^0\,d^3r. \end{aligned}\right\} \tag{37.14}$$

If ω is of the same order of magnitude as ω_r, so that $(m\beta)^{\frac{1}{2}}(\omega-\omega_r)/\varkappa$ is small, and yet $1/\varkappa$ is large compared with the intermolecular distance, the approximation

$$\omega \approx \omega_r - \frac{2\varkappa^2}{m\beta\omega_r} + i\left(\frac{\pi}{2m\beta}\right)^{\frac{1}{2}}\varkappa \tag{37.15}$$

holds. The rate at which the disturbance is damped is then inversely proportional to the wave-length, and there is much dispersion. At still higher frequencies, both transverse and longitudinal waves are almost undamped, and the dispersion is negligible, so long as $1/\varkappa$ remains large compared with the distance between nearest molecules. This is the region in which the liquid may be expected to display nearly elastic behaviour. The velocity of propagation of elastic waves is $v_s = \omega/\varkappa$, where

$$m\,n\,v_s^2 = -\frac{c}{30}\int \frac{\partial^2 \phi}{\partial \boldsymbol{r}\cdot\partial \boldsymbol{r}}\, n_2^0\, r^2\, d^3 r, \tag{37.16}$$

and $c=1$ for transverse waves and $c=3$ for longitudinal waves.

In the theory of the elasticity of liquids presented in Sect. 25, it was required to evaluate integrals containing σ. For example,

$$\left.\begin{aligned} \boldsymbol{u} &= \frac{1}{n_0}\int f^0 \sigma\, d^3\xi \\ &= \frac{i\omega_r\,[\varphi\{(m\beta)^{\frac{1}{2}}(\omega-\omega_r)/\varkappa\} + \varphi\{(m\beta)^{\frac{1}{2}}(\omega+\omega_r)/\varkappa\}]\,\boldsymbol{s}}{\varphi\{(m\beta)^{\frac{1}{2}}(\omega-\omega_r)/\varkappa\} - \varphi\{(m\beta)^{\frac{1}{2}}(\omega+\omega_r)/\varkappa\}}\,. \end{aligned}\right\} \tag{37.17}$$

It can now be verified, using the asymptotic development of $\varphi(z)$, that $\boldsymbol{u}$ differs from $\frac{\partial \boldsymbol{s}}{\partial t} = i\omega\boldsymbol{s}$ only by terms of relative order $\frac{\varkappa}{(m\beta)^{\frac{1}{2}}\omega}$.

The theory of the present section is of interest in showing how it is possible for irreversible effects to be predicted on the basis of equations which are time-reversible, i.e., invariant with respect to a change of the sign of t and all velocities. It has been contended in the past[1] that this was impossible; but it is now clear that all that may be required is the choice between two solutions of the equations, one of which is clearly inadmissible on physical grounds. Nevertheless it is desirable that the physically significant solution should result from the application of a more fundamental principle than the rejection of those solutions which are obviously wrong. The fundamental principle required for this purpose will be formulated in Sect. 38 which follows.

38. Generalization of BOLTZMANN's equation. The equations to determine the distribution functions which were developed in Sect. 35 differ from BOLTZMANN's equation in gas theory in that they are time-reversible, and need to be supplemented by a "boundary condition" of some type if it is desired to predict irreversible processes. The condition that the equilibrium state should be approached after a long time was used in Sect. 37, but this condition is difficult to apply in general, and it provides no indication of how the statistical preponderances, which are supposed to favour the approach to equilibrium, are effective. In this section, a condition of more general applicability, which has been proposed by the author[2], will be formulated and used to derive a generalization of BOLTZMANN's equation for liquids.

Attention will be restricted to a finite region R, contained in a liquid of greater extent, and bounded by a surface S moving everywhere with the macroscopic

[1] e.g. G. KLEIN and I. PRIGOGINE: Physica, Haag 19, 74, 89, 1053 (1953).
[2] H. S. GREEN: Proc. Phys. Soc. Lond. B 69, 269 (1956).

fluid velocity $\boldsymbol{u}$. The equation (35.21) can then be written in the form

$$\frac{D_q F_q}{D t_q} = \iint F_{q+1} \boldsymbol{v}^{(q+1)} \cdot d\boldsymbol{S}^{(q+1)} \, d^3 \xi^{(q+1)}, \tag{38.1}$$

where $\boldsymbol{v}^{(q+1)} = \xi^{(q+1)} - \boldsymbol{u}^{(q+1)}$. The attempt will be made to expresse F_{q+1} in terms of F_q.

It should be noticed that if the velocity of a molecule at the point $\boldsymbol{x}^{(q+1)}$ on S is such as to carry it outward across the surface, i.e., if $\boldsymbol{v}^{(q+1)} \cdot d\boldsymbol{S}^{(q+1)} > 0$, the molecule has recently been in interaction with the molecules within S, and there can be no question of neglecting the correlation between its position and velocity, and those of the other q molecules. However, if $\boldsymbol{v}^{(q+1)} \cdot d\boldsymbol{S}^{(q+1)} < 0$, so that it is entering the region, it can have interacted only very weakly with the molecules inside S in the recent past, and the correlation between its position and velocity and those of the other q molecules is really very small. Thus, in writing

$$F_{q+1} = F_q \vartheta(\boldsymbol{x}^{(q+1)}, \xi^{(q+1)}) \quad \text{for} \quad \boldsymbol{v}^{(q+1)} \cdot d\boldsymbol{S}^{(q+1)} < 0, \tag{38.2}$$

one is merely neglecting the correlations arising from interactions across the surface, and the error will truly be negligible, provided the region R is sufficiently large to accommodate many molecules. Even if the volume of R were comparable with the molecular volume, the error should not be worse than that involved in the statistical mechanics of the cell model (Sect. 20). But it will be supposed here that R is of macroscopic extent.

The function $\vartheta(\boldsymbol{x}^{(q+1)}, \xi^{(q+1)})$ in (38.2) measures the probability that a molecule will enter the region with velocity $\xi^{(q+1)}$, through the point $\boldsymbol{x}^{(q+1)}$ on the surface. It might be expected that $\vartheta = f$, and this can be proved by using the formula (35.23) with $q = 1$ in the form

$$f = F_{1,0}^{(1)} + F_{1,1}^{(1)}/1! + F_{1,2}^{(1)}/2! + \cdots. \tag{38.3}$$

On introducing (38.2) into the integral $F_{1,q}^{(1)}$ defined by (35.24), one finds that, provided $\boldsymbol{x}^{(1)}$ is on the surface S, and $\boldsymbol{v}^{(1)} \cdot d\boldsymbol{S}^{(1)} < 0$, $F_{1,q}^{(1)} = \vartheta(\boldsymbol{x}^{(1)}, \xi^{(1)}) F_{0,q}$; so that by using (35.23) again, this time with $q = 0$ and remembering that $f_0 = 1$, one finds that (38.3) reduces to $f = \vartheta$. So (38.2) can be re-written in the form

$$F_{q+1} = F_q f^{(q+1)} \quad \text{for} \quad \boldsymbol{v}^{(q+1)} \cdot d\boldsymbol{S}^{(q+1)} < 0. \tag{38.4}$$

This condition is not time-reversible and is in fact the sole premise of the theory which permits the prediction of natural irreversible phenomena.

The Eq. (38.1) can be written

$$\frac{D_q F_q}{D t_q} = \iint (F_{q+1} - F_q f^{(q+1)}) \, \boldsymbol{v}^{(q+1)} \cdot d\boldsymbol{S}^{(q+1)} \, d^3 \xi^{(q+1)} \tag{38.5}$$

since the additional term vanishes. The point of writing it in this way is that, according to (38.4), the integrand vanishes for $\boldsymbol{v}^{(q+1)} \cdot d\boldsymbol{S}^{(q+1)} < 0$, and *the integration* therefore may and *will be restricted to velocities for which* $\boldsymbol{v}^{(q+1)} \cdot d\boldsymbol{S}^{(q+1)}$ *is positive.*

The next step is to express F_{q+1} in terms of F_q; this can be done as follows. One writes (38.5), with the value of q advanced by 1, in the form

$$\frac{D_q F_{q+1}}{D t_q} + \alpha F_{q+1} = \iint F_{q+2} \boldsymbol{v}^{(q+1)} \cdot d\boldsymbol{S}^{(q+1)} \, d^3 \xi^{(q+1)}, \tag{38.6}$$

where

$$\alpha = \iint f \cdot \boldsymbol{v} \cdot d\boldsymbol{S} \, d^3 \xi \tag{38.7}$$

does not vanish, since it represents the flux of molecules from one side of the surface to the other, the integration being limited to values of ξ such that

$\boldsymbol{v}\cdot d\boldsymbol{S}>0$. Let t_1 be any time previous to t, and suppose that a set of $q+1$ molecules, situated at the points $\boldsymbol{x}_1^{(i)}$ and having velocities $\boldsymbol{\xi}_1^{(i)}$ $(i=1, 2, \ldots, q+1)$, would, by moving freely under their mutual interactions alone, reach the points $\boldsymbol{x}^{(i)}$ with velocities $\boldsymbol{\xi}^{(i)}$ at time t. Then the $\boldsymbol{x}_1^{(i)}$ and $\boldsymbol{\xi}_1^{(i)}$ are functions of the $\boldsymbol{x}^{(i)}$ and $\boldsymbol{\xi}^{(i)}$ and the interval $t-t_1$, which can in principle be determined by integrating the equations of motion of the $q+1$ molecules. Also, let $F_{q+1}(t_1)$ be the value of the function F_{q+1} with $\boldsymbol{x}_1^{(i)}$ and $\boldsymbol{\xi}_1^{(i)}$ substituted for $\boldsymbol{x}^{(i)}$ and $\boldsymbol{\xi}^{(i)}$ $(i=1, \ldots, q+1)$; and let $F_{q+2}(t_1)$ be the value of F_{q+2} with $\boldsymbol{x}_1^{(i)}$ and $\boldsymbol{\xi}_1^{(i)}$ substituted for $\boldsymbol{x}^{(i)}$ and $\boldsymbol{\xi}^{(i)}$ $(i=1, \ldots, q+2)$, even though $\boldsymbol{x}_1^{(q+2)}$ and $\boldsymbol{\xi}_1^{(q+2)}$ do not depend on $t-t_1$. Then (38.6) has the integral

$$F_{q+1}=I(t_1,t)\,F_{q+1}(t_1)+\int_{t_1}^{t} I(t',t)\,F_{q+2}(t')\,d\tau^{(q+2)}(t'), \tag{38.8}$$

where $\int_{t_1}^{t}\ldots d\tau^{(q+2)}(t')$ means $\int_{t_1}^{t}\iint\ldots \boldsymbol{v}^{(q+2)}\cdot d\boldsymbol{S}^{(q+2)}\,d^3\xi^{(q+2)}\,dt'$ and $F_{q+2}(t')$ is simply $F_{q+2}(t_1)$ with t' instead of t_1, also

$$I(t_1,t)=\exp\Big\{-\int_{t_1}^{t}\alpha(t')\,dt'\Big\}. \tag{38.9}$$

Iteration of the result (38.8) gives

$$\left.\begin{aligned} F_{q+1}&=I(t_1,t)\,F_{q+1}(t_1)+\int_{t_1}^{t} I(t_2,t)\,F_{q+2}(t_2)\,d\tau^{(q+2)}(t')+\\ &\quad+\int_{t_1}^{t}\int_{t_2}^{t'} I(t_3,t)\,F_{q+3}(t_3)\,d\tau^{(q+3)}(t'')\,d\tau^{(q+2)}(t')+\cdots,\end{aligned}\right\} \tag{38.10}$$

where $F_{q+2}(t_2)$ is derived from $F_{q+2}(t')$ in the same way as $F_{q+1}(t_1)$ from F_{q+1}, and in general $F_{q+j+1}(t_{j+1})$ is derived from $F_{q+j+1}(t^{(j)})$ in the same way. Thus $\boldsymbol{x}_{j+1}^{(i)}$ and $\xi_{j+1}^{(i)}$ depend implicitly on $t^{(j)}, t^{(j-1)}, \ldots$ and t' as well as $\boldsymbol{x}^{(i)}$, $\boldsymbol{\xi}^{(i)}$ and t. The values of $t_1, t_2, \ldots$ are arbitrary, but will now be specified. The first, t_1, is chosen so that a molecule is entering the region, at the point $\boldsymbol{x}_1^{(a)}$, say, on S with velocity $\boldsymbol{\xi}_1^{(a)}$ at time t_1, and no molecule crosses the surface between the times t_1 and t. Then one will have

$$F_{q+1}(t_1)=F_q^{[a]}(t_1)\,f^{(a)}(t_1), \tag{38.11}$$

where $F_q^{[a]}$ is the function F_q of the variables $\boldsymbol{x}_1^{(i)}$ and $\xi_1^{(i)}$ $(i=1, 2, \ldots, q+1)$, with $\boldsymbol{x}_1^{(a)}$ and $\xi_1^{(a)}$ omitted. Similarly t_2 is chosen so that a molecule entered the region at the point $\boldsymbol{x}_2^{(b)}$ on the boundary, with velocity $\boldsymbol{\xi}_2^{(b)}$ at time t_2, and a second molecule entered the region at a later time t_2', but no other molecule entered the region between the times t_2 and t. Then

$$F_{q+2}(t_2)=F_q^{[ab]}(t_2)\,f^{(b)}(t_2)\,f^{(a)}(t_2'), \tag{38.12}$$

where $t_2<t_2'<t'$; and in general

$$F_{q+j+1}(t_{j+1})=F_q^{[ab\ldots z]}(t_{j+1})\,f^{(z)}(t_{j+1})\ldots f^{(a)}(t_{j+1}^{(j)}), \tag{38.13}$$

where $t_{j+1}<t_{j+1}'<\cdots<t_{j+1}^{(j)}<t^{(j)}$. By substituting (38.11), (38.12) and (38.13) into the right-hand side of (38.10), one obtains the required expression for F_{q+1} in terms of F_q. If this is again substituted into (38.5), the result is

$$\left.\begin{aligned} \frac{D_qF_q}{Dt_q}&=\sum_{j=0}^{\infty}\overbrace{\int\cdots\int}^{(j+1)} I(t_{j+1},t)\,[F_q^{[ab\ldots z]}(t_{j+1})\,f^{(z)}(t_{j+1})\ldots f^{(a)}(t_{j+1}^{(j)})-\\ &\quad-F_q(t)\,f_1^{(q+1)}(t)\ldots f_1^{(q+1)}(t^{(j)})]\,d\tau^{(q+j+1)}(t^{(j)})\ldots d\tau^{(q+2)}(t)\,d\sigma^{(q+1)},\end{aligned}\right\} \tag{38.14}$$

where $\int\ldots d\sigma^{(q+1)}$ means $\iint\ldots \boldsymbol{v}^{(q+1)}\cdot d\boldsymbol{S}^{(q+1)}\,d^3\xi^{(q+1)}$.

The final result (36.14) is a formal generalization of BOLTZMANN'S equation. In gas-theory, since the molecular density is small, F_1 can be approximated by f_1 if the region considered is not too large, and $I(t_1, t)$ is not very different from 1; also, the series on the right-hand side of (38.14) is sufficiently well represented by its first term. But these approximations are no longer possible at liquid densities, and the behaviour of F_1 is of little interest anyway. It is easy to prove, at any rate for thermal and mechanical equilibrium, that the major contribution to the series (38.3) for f, or the similar series for f_q, comes from the terms involving $F_{\bar{q}\pm j}$, where $\bar{q} = \int n\, d^3x$ is the mean number of molecules in the region considered, and $j/\bar{q}$ is very small, if the region is of macroscopic extent. If, in fact, only $F_{\bar{q}}$ were determined, most properties of the liquid could be predicted with sufficient accuracy.

39. The *H*-theorem for liquids. An important feature of Boltzmann's equation in gas-theory is the proof which it affords of the existence of a function of state (the entropy) which spontaneously increases in natural irreversible processes. For a long time this was the only derivation of the second law of thermodynamics founded on molecular theory. Even up to quite recently there has been no precise generalization of BOLTZMANN'S H-theorem, as it is usually called, for liquids; though there are various general theorems of a similar nature[1], the attempt to apply them to a particular system such as a liquid has led to difficulties. It might be expected, however, that the exact analogue of BOLTZMANN'S equation would enable the proof of a generalized H-theorem; and it will indeed be shown that the Eq. (38.14) is sufficient to establish the second law of thermodynamics.

The total entropy S of the liquid within the region R considered will be defined by

$$S = -k \sum_{q=0}^{\infty} \int F_q \log F_q \, d\tau_q / q! \tag{39.1}$$

where $\int \ldots d\tau_q$ means $\overset{(2q)}{\int \cdots \int} \cdots d^3x^{(1)}\, d^3\xi^{(1)} \ldots d^3x^{(q)}\, d^3\xi^{(q)}$, and k is BOLTZMANN'S constant. To justify this definition, it must first be verified that S has the properties ascribed to the entropy in macroscopic thermodynamics. It must be shown that (a) in a state of thermal and mechanical equilibrium, S reduces to $S^0 = (U + p^0 V - G)/T$, where U is the internal energy, p^0 the pressure, and G is the mean GIBB'S thermodynamic potential; (b) in any other state with the same volume and interval energy, $S < S^0$; and (c) the total entropy of two or more separate regions of the liquids is the sum of their separate total entropies. To prove (a) and (b), it should be noticed that the internal energy associated with the region considered is

$$U = \sum_{q=0}^{\infty} \int F_q H_q \, d\tau_q / q! \tag{39.2}$$

where H_q is the energy of the system of q molecules, regarded as a function of their positions and velocities; and also that the mean number of molecules in the region is

$$\bar{q} = \sum_{q=0}^{\infty} q \int F_q \, d\tau_q / q!. \tag{39.3}$$

It can be seen from (12.7) and (13.5) that, in equilibrium, F_q reduces to

$$\left.\begin{aligned} F_q^0 &= \left(\frac{\beta m}{2\pi}\right)^{3q/2} \exp\left\{-\beta \sum_{i=1}^{q} \left(\frac{1}{2} m\, \xi^{(i)2}\right)\right\} N_q^0 \\ &= \exp\{-\beta (H_q + p^0 V - q G_1)\} \end{aligned}\right\} \tag{39.4}$$

[1] e.g. M. BORN and H. S. GREEN: Proc. Roy. Soc. Lond., Ser. A **192**, 166 (1948).

where G_1 is GIBB's thermodynamic potential per molecule. Let

$$F_q = \alpha_q F_q^0; \tag{39.5}$$

then, provided the distribution functions F_q and F_q^0 yield the same values of U and $\bar{q}$, one has by substitution in (39.1)

$$\left.\begin{aligned} S = \beta k \sum_{q=0}^{\infty} \int F_q (H_q + p^0 V - q G_1 - \log \alpha_q/\beta)\, d\tau_q/q! = (U + p^0 V - G)/T + \\ + k \sum_{q=0}^{\infty} \int F_q^0 (\alpha_q - 1 - \alpha_q \log \alpha_q)\, d\tau_q/q!. \end{aligned}\right\} \tag{39.6}$$

The expression $\alpha_q - 1\alpha_q \log \alpha_q$ vanishes in equilibrium when $\alpha_q = 1$ and is negative otherwise, so that (a) and (b) are fulfilled. To prove (c), it need merely be noticed that the configurations in two or more separate regions are practically uncorrelated, so that their distribution functions are multiplicative.

Although the entropy defined by (39.1) has the required properties, it still remains to be proved that, except in the state of equilibrium, the entropy will spontaneously increase and so approach its maximum value. It has been seen in Sect. 29 that this is sufficient to ensure positive values of the coefficients of viscosity and thermal conduction, and indeed that these coefficients are directly related to the rate of increase of entropy. The problem of the calculation of the coefficients of viscosity and thermal conduction is therefore solved in principle, though of course not explicitly, by the proof of the H-theorem.

Before outlining the proof of the theorem itself, it is convenient to notice a series of identities, which can be written together in the form

$$\left.\begin{aligned} &\sum_{q=0}^{\infty} \int \overset{(q-1)}{\cdots} \int I(t^{(j)}, t)\, F_{q+j+1}(t^{(j)})\, d\tau_q^{[1,\ldots j+1]}\, d\tau^{(1)}(t) \ldots d\tau^{(j)}(t^{(j-1)})/q! \\ &\quad = \sum_{q=0}^{\infty} \int \overset{(q-1)}{\cdots} \int I(t^{(j)}, t)\, F_q^{[1,\ldots j+1]} f^{(1)}(t) \ldots f^{(j)}(t^{(j-1)}) \times \\ &\quad \times d\tau_q^{[1,\ldots j+1]}\, d\tau^{(1)}(t) \ldots d\tau^{(j)}(t^{(j-1)})/q!, \end{aligned}\right\} \tag{39.7}$$

where $d\tau_q^{[1,\ldots j+1]} = d\tau^{(j+2)}\, d\tau^{(j+3)} \ldots d\tau^{(q)}$, $F_q^{[1,\ldots j+1]}$ is the function F_q of the variables $t, \boldsymbol{x}^{(j+2)}, \boldsymbol{\xi}^{(j+2)}, \ldots, \boldsymbol{x}^{(q+j+1)}, \boldsymbol{\xi}^{(q+j+1)}$, and $\int \ldots d\tau^{(1)}(t)$ means $\int_t^{t+\Delta t} \iint \ldots \boldsymbol{v}^{(1)} \times \times d\boldsymbol{S}^{(1)}\, d^3\xi^{(1)}\, dt$; otherwise the notation is that explained in Sect. 38. The correctness of the identity can be seen from the fact that each side represents the probability density of finding a molecule (superfix $j+1$) in an assigned position and with an assigned velocity at time $t^{(j)}$, given that j molecules will leave the region (the last between times t and $t + \Delta t$) before one enters.

The spontaneous increase of entropy within the region between times t and $t + \Delta t$ is

$$\left.\begin{aligned} \Delta S &= -k \sum_{q=0}^{\infty} \iint \overset{(q)}{\cdots} \int \frac{D_q}{D t_q} \{F_q \log F_q\}\, d\tau_q\, dt/q! \\ &= -k \sum_{q=0}^{\infty} \int \overset{(q+1)}{\cdots} \int \log F_q \{F_{q+1} - F_q f^{(q+1)}\}\, d\tau_q\, d\tau^{(q+1)}(t)/q!. \end{aligned}\right\} \tag{39.8}$$

To the right-hand side of this equation one first adds the expression

$$-k \sum_{q=0}^{\infty} \int \overset{(q+1)}{\cdots} \int \log f^{(q+1)} \{F_{q+1} - F_y f^{(q+1)}\}\, d\tau_q\, d\tau^{(q+1)}(t)/q!$$

which vanishes, according to the identity (39.7) with $j=0$. One next substitutes (38.8) for F_{q+1} and adds a term

$$\left.\begin{aligned} &-k\sum_{q=0}^{\infty}\int\overset{(q+2)}{\cdots}\int I(t',t)\log f^{(q+2)}(t')\times\\ &\qquad\times\{F_{q+2}(t')-F_q f_1^{(q+1)} f_1^{(q+2)}(t')\}\,d\tau^{(q+2)}(t')\,d\tau_q\,d\tau^{(q+1)}(t)\end{aligned}\right\}\qquad(39.9)$$

which also vanishes according to (39.7) with $j=1$. Proceeding indefinitely in this way, one obtains finally

$$\left.\begin{aligned} \Delta S &= -k\sum_{q=0}^{\infty}\sum_{j=0}^{\infty}\int\overset{(q+j+1)}{\cdots}\int I(t_{j+1},t)\,(\log A_{q,j})\,(A'_{q,j}-A_{q,j})\times\\ &\qquad\times d\tau^{(q+j+1)}(t^{(j)})\ldots d\tau^{(q+1)}(t)\,d\tau_q/q!\\ A_{q,j} &= F_q(t)\,f^{(q+1)}(t)\ldots f^{(q+j+1)}(t^{(j)}),\\ A'_{q,j} &= F_q^{[ab\ldots z]}(t_{j+1})f^{(z)}(t_{j+1})\ldots f^{(a)}(t_{j+1}^{(j)}).\end{aligned}\right\}\qquad(39.10)$$

It now has to be shown that the value of each integral in (39.10) remains unchanged if one interchanges $A_{q,j}$ with $A'_{q,j}$ in the integrand. This is achieved by a succession of three transformations:

(i) The variables $\boldsymbol{x}_q$ (*i.e.*, $\boldsymbol{x}^{(1)},\ldots,\boldsymbol{x}^{(q)}$), $\boldsymbol{\xi}_q$, $\boldsymbol{x}^{(q+i+1)}(t^{(i)})$ and $\boldsymbol{\xi}^{(q+i+1)}(t^{(i)})$ in (39.11) are replaced by $-\boldsymbol{x}_q^{[ab\ldots z]}(-t_{j+1})$, $\boldsymbol{\xi}_q^{[ab\ldots z]}(-t_{j+1})$, $-\boldsymbol{x}^{(p)}(-t_{j+1}^{(i)})$ and $\boldsymbol{\xi}^{(p)}(-t_{j+1}^{(i)})$, where p is the $(i+1)$-th of the superfixes $z,\ldots,b,a$, and the limits of the spatial integrations are similarly changed so that the value of the integral is unchanged. Then, since the motion of the set of molecules within the region is reversible, the variables $\boldsymbol{x}_q^{[ab\ldots z]}(t_{j+1})$, $\boldsymbol{\xi}_q^{[ab\ldots z]}(t_{q+1})$, $\boldsymbol{x}^{(p)}(t_{j+1}^{(i)})$ and $\boldsymbol{\xi}^{(p)}(t_{j+1}^{(i)})$ must be replaced by $-\boldsymbol{x}_q(-t)$, $\boldsymbol{\xi}_q(-t)$, $-\boldsymbol{x}^{(q+i+1)}(-t^{(i)})$ and $\boldsymbol{\xi}^{(q+i+1)}(-t^{(i)})$.

(ii) The independent variables of integration are now changed from $-\boldsymbol{x}_q^{[ab\ldots z]}\times(-t_{j+1})$, $\boldsymbol{\xi}_q^{[ab\ldots z]}(-t_{j+1})$, $-\boldsymbol{x}^{(p)}(-t_{j+1}^{(i)})$ and $\boldsymbol{\xi}^{(p)}(-t_{j+1}^{(i)})$ to $-\boldsymbol{x}_q(-t)$, $\boldsymbol{\xi}_q(-t)$, $-\boldsymbol{x}^{(q+i+1)}(-t^{(i)})$ and $\boldsymbol{\xi}^{(q+i+1)}(-t^{(i)})$. Since each volume element of phase space is unchanged by motion in accordance with mechanical laws, the integrand is not affected, though the limits of integration are changed.

(iii) The signs of all co-ordinate variables, and t and $(t^{(i)})$ are changed, so that the limits of integration return to their original values.

The combined effect of these transformations is to interchange $A_{q,j}$ and $A'_{q,j}$, provided F_q and f do not depend explicitly on the time. Actually it may be supposed that F_q and f depend on the time only through the variation with time of the density, temperature and fluid velocity of which F_q and f are functionals. Also, since n, $\boldsymbol{u}$ and T satisfy the time-reversible equations (36.6) to a sufficient degree of approximation, the sequence of transformations described will not have any resultant effect on their values. It follows, therefore, from (39.10) that

$$\left.\begin{aligned}\Delta S = \tfrac{1}{2}k\sum_{q=0}^{\infty}\sum_{j=0}^{\infty}\int\overset{(q+j+1)}{\cdots}\int I(t_{j+1},t)\log(A'_{q,j}/A_{q,j})\,(A'_{q,j}-A_{q,j})\\ d\tau^{(q+j+1)}(t^{(j)})\ldots d\tau^{(q+1)}(t)\,d\tau_q/q!,\end{aligned}\right\}\qquad(39.11)$$

and this is an essentially positive expression. Physically, it is now clear that every cycle beginning and ending with the same number of molecules in the region considered makes its own contribution to the spontaneous increase in entropy. The proof of the H-theorem outlined above is a natural generalization of the corresponding theorem for rare gases. It is fairly obvious how a similar argument could be developed for liquid mixtures. Thus one might establish, for example, that the coefficient of diffusion is positive, and also derive the "Onsager relations" [the identity of the ν_j appearing in (29.8) and (29.9) and the symmetry of the matrix ν_{jk}] from a purely molecular theory.

40. Applications of the theory. The theory which has been developed has not yet led to exact determinations of the coefficients of viscosity and thermal conduction. Considerable progress has, however, been made with the aid of reasonable approximations, and the results thereby obtained will now be reviewed.

First it should be mentioned that the implications of (36.12) have been analysed by BORN and GREEN[1]. It is obvious that this equation cannot be made to determine f_2' without approximation, but it does partially determine the dependence of f_2' on one of the velocity variables. Also, with the help of an approximation suggested by the author elsewhere[2], the functions f_1' and f_2' can both be determined, in principle. The analysis suggests that the form of f_1' is not very different in liquids and gases, and the methods of gas-theory can be used to obtain a first approximation to f_1'. But since the viscosity and thermal conduction of liquids is more associated with the deformation of the molecular structure than with small changes in the velocity distribution, it is unnecessary to enter into a discussion of these developments here.

The quantitative theory of viscosity in liquids is at present founded on the Eq. (36.10), which was first stated explicitly by the author[2]. Since f_2' is not known, it is necessary to approximate $\langle \boldsymbol{v}_2^{(i)} \boldsymbol{v}_2^{(j)} \rangle$ by $\langle \boldsymbol{v}^{(i)} \boldsymbol{v}^{(j)} \rangle$, i.e., to write

$$\left.\begin{aligned} &\langle \boldsymbol{v}_2^{(1)} \boldsymbol{v}_2^{(2)} \rangle \approx 0 \qquad \langle \boldsymbol{v}_2^{(2)} \boldsymbol{v}_2^{(1)} \rangle \approx 0\,, \\ &m \langle \boldsymbol{v}_2^{(1)} \boldsymbol{v}_2^{(1)} \rangle \approx \mathsf{p}_{\mathrm{K}}^{(1)}/n = kT^{(1)}\boldsymbol{\delta} + \mathsf{p}_{\mathrm{K}}'/n\,, \\ &m \langle \boldsymbol{v}_2^{(2)} \boldsymbol{v}_2^{(2)} \rangle \approx \mathsf{p}_{\mathrm{K}}^{(2)}/n = kT^{(2)}\boldsymbol{\delta} + \mathsf{p}_{\mathrm{K}}'/n\,. \end{aligned}\right\} \tag{40.1}$$

Then one has

$$kT \frac{\partial n_2'}{\partial \boldsymbol{r}} + n_2' \frac{\partial \phi}{\partial \boldsymbol{r}} + \int n_3' \frac{\partial \phi^{(23)}}{\partial \boldsymbol{x}^{(2)}} d^3 x^{(2)} = - \frac{\mathsf{p}_{\mathrm{K}}'}{n} \frac{\partial n_2^0}{\partial \boldsymbol{r}}\,. \tag{40.2}$$

To solve this equation, it is necessary to make use of the superposition approximation (21.3), from which it follows that

$$\frac{n_3'}{n_3^0} \approx \frac{n_2^{(12)'}}{n_2^{(12)0}} + \frac{n_2^{(23)'}}{n_2^{(23)0}} + \frac{n_2^{(31)'}}{n_2^{(31)0}}\,. \tag{40.3}$$

On substitution for n_3' from (40.3) into (40.2), one obtains an integro-differential equation for the determination of n_2'. The inhomogeneous term, involving p', can be determined by using the methods of gas-theory. Only approximate solutions have as yet been obtained, however. An indirect method developed by KIRKWOOD and his associates will be described in Sect. 42. The approximation considered here is suggested by the fact that the right-hand side of (40.2) is small, and also, when (40.3) is substituted, the last two terms, involving $n_2^{(23)'}$ and $n_2^{(31)'}$, are found to give an appreciable smaller contribution to the integral of (40.2) than the first, which involves $n_2^{(12)'}$. It may be concluded from this that n_2'/n_2^0 is a slowly varying function of r, compared with n_2^0 itself.

BORN and the author proposed to express the function ν of (32.3) in the form

$$\nu(r) = r\, c(r)\, (\beta m)^{\frac{1}{2}}\, n_2^0(r) \tag{40.4}$$

[1] See M. BORN and H. S. GREEN: Proc. Soc. Roy. Lond., Ser. A **190**, 455 (1947), especially 5–6; also Ch. 8·3 of H. S. GREEN: Molecular Theory of Liquids. Amsterdam 1952.

[2] H. S. GREEN: Proceedings of the International Congress on Rheology (1948) (1–12, Appendix II). Amsterdam: North-Holland Publishing Co. 1949.

and treat the variation of the function $c(r)$, over the interval in which $n_2^0(r)\,\phi'(r)$ differs appreciably from zero, as small. On substitution from (40.4) into (32.9) it then follows that

$$\eta = \frac{c(r_1)}{30}\left(\frac{m}{kT}\right)^{\frac{1}{2}} \int n_2^0(r)\,\phi'(r)\,r^2\,dr, \tag{40.5}$$

where $c(r_1)$ is a weighted average of $c(r)$, or, sufficiently nearly, the value of $c(r)$ at the first maximum $r=r_1$ of $n_2^0(r)$. The value of the integral obviously depends rather sensitively on the distance r_1 and magnitude $n_2^0(r_1)$ of the first "co-ordination peak". If r_1 is significantly greater than the distance r_0 at which the force $\phi'(r)$ vanishes in changing sign, and $\Psi(r_1)$ is the mean potential energy of a pair of molecules separated by the distance r_1, then the temperature dependence of η will be similar to that of $n_2(r_1)$, or $\exp\{-\beta\Psi(r_1)\}$. It has in fact been shown by ANDRADE[1] that the viscosity of many liquids has a temperature dependence of this type. It is, of course important to observe that $\Psi(r_1)$ varies slowly with temperature. Even so, it is now known that a number of organic liquids do not closely obey ANDRADE's rule; it may be that r_1 does not exceed r_0 in these liquids, and in that event the discrepancies would be explained. On the whole, it seems that the formula (40.4) provides a fairly adequate account of the viscosity of most liquids. Moreover, where ANDRADE's rule is followed, the experimental data provide a useful estimate of, or check on, the mean potential energy $\Psi(r_1)$ of nearest neighbours in the liquid.

The theory of thermal conduction in liquids can be developed in a similar way. The analogue of (40.2) is obtained by substituting $\xi^{(2)2}-\xi^{(1)2}$ for φ in (35.19). Separating terms linear in the temperature gradient, one has

$$\left.\begin{aligned}&\frac{\partial}{\partial r}\cdot\left(n_2\langle(\boldsymbol{v}_2^{(2)}-\boldsymbol{v}_2^{(1)})\{(\boldsymbol{v}_2^{(2)}+\boldsymbol{u}_2^{(2)})^2-(\boldsymbol{v}_2^{(1)}+\boldsymbol{u}_2^{(1)})^2\}\rangle\right)'=n_2\langle\eta_2^{(2)}\cdot\xi^{(2)}-\eta_2^{(1)}\cdot\xi^{(1)}\rangle\\&=-\frac{n_2}{m}(\boldsymbol{u}_2^{(1)}+\boldsymbol{u}_2^{(2)})\cdot\frac{\partial\phi}{\partial r}-\frac{1}{2m}\int n_3(\boldsymbol{u}_3^{(1)}+\boldsymbol{u}_3^{(2)})\cdot\left(\frac{\partial\phi^{(23)}}{\partial\boldsymbol{x}^{(2)}}-\frac{\partial\phi^{(13)}}{\partial\boldsymbol{x}^{(1)}}\right)d^3x^{(3)}.\end{aligned}\right\} \tag{40.6}$$

Using the approximations of (40.1) and also

$$\left.\begin{aligned}&\langle\boldsymbol{v}_2^{(2)}\boldsymbol{v}_2^{(1)2}\rangle\approx 0, \qquad \langle\boldsymbol{v}_2^{(1)}\boldsymbol{v}_2^{(2)2}\rangle\approx 0,\\&\tfrac{1}{2}m\langle\boldsymbol{v}_2^{(1)}\boldsymbol{v}_2^{(1)2}\rangle\approx \boldsymbol{q}_{\mathrm{k}}/n,\\&\tfrac{1}{2}m\langle\boldsymbol{v}_2^{(2)}\boldsymbol{v}_2^{(2)2}\rangle\approx \boldsymbol{q}_{\mathrm{k}}/n,\\&\boldsymbol{u}_3^{(1)}\approx\tfrac{1}{2}(\boldsymbol{u}_2^{(1,2)}+\boldsymbol{u}_2^{(1,3)}),\\&\boldsymbol{u}_3^{(2)}\approx\tfrac{1}{2}(\boldsymbol{u}_2^{(2,1)}+\boldsymbol{u}_2^{(2,3)}),\end{aligned}\right\} \tag{40.7}$$

the fundamental equation reduces to

$$\left.\begin{aligned}&kT\frac{\partial}{\partial r}\cdot\{n_2(\boldsymbol{u}_2^{(1)}+\boldsymbol{u}_2^{(2)})\}+n_2\frac{\partial\phi}{\partial r}\cdot(\boldsymbol{u}_2^{(1)}+\boldsymbol{u}_2^{(2)})+\\&+\frac{1}{4}\int n_3(\boldsymbol{u}_2^{(1)}+\boldsymbol{u}_2^{(2)}+\boldsymbol{u}_2^{(1,3)}+\boldsymbol{u}_2^{(2,3)})\cdot\left(\frac{\partial\phi^{(23)}}{\partial\boldsymbol{x}^{(2)}}-\frac{\partial\phi^{(13)}}{\partial\boldsymbol{x}^{(1)}}\right)d^3x^{(3)}\\&=-2\frac{\boldsymbol{q}_{\mathrm{k}}}{n}\cdot\frac{\partial n_2}{\partial r}.\end{aligned}\right\} \tag{40.8}$$

Now, the right-hand side of this equation is small, and the mean value of $\boldsymbol{u}_2^{(1,3)}-\boldsymbol{u}_2^{(1)}$ can also be neglected without much error. BORN and GREEN[2] inferred that $\boldsymbol{u}_2^{(1)}+\boldsymbol{u}_2^{(2)}$ must be a slowly varying function compared with n_2, and suggested,

---

[1] E. N. DA C. ANDRADE: Phil. Mag. (7) **17**, 497, 698 (1934).

[2] See footnote 1, p. 110.

in effect, that the functions σ_0 and σ defined in (31.6) should be expressed in the form

$$\sigma_0 = k\,r\,c_0(r)\,(\beta/m)^{\frac{1}{2}}, \qquad \sigma = k\,r\,c_1(r)\,(\beta/m)^{\frac{1}{2}} \tag{40.9}$$

where $c_0(r)$ and $c_1(r)$ are slowly varying functions determined by (40.8). The simplest way to satisfy the condition (31.10) is to take $c_0(r) = -\frac{1}{3}c_1(r)$. The coefficient of thermal conduction can then be obtained in a form analogous to (40.5), by substitution in (31.8). However, the coefficient of thermal conduction is measured at constant pressure, and the coefficient $c_1(r_1)$ probably depends on the pressure. There is obviously a need for a more exacting treatment of the fundamental equations (40.3) and (40.8) if detailed calculations of the coefficients of viscosity and thermal conduction are to be made.

KIRKWOOD's approach to the calculation of the thermal conductivity of liquids will be described in Sect. 42. It is based not on (40.8) but on the somewhat simpler Eq. (36.11) which determines the mean resultant force on a pair of molecules at distance r. If one is prepared to assume that the coefficient of diffusion under pressure is the same for a pair of molecules at any separation as for a single molecule, the drift velocity $\frac{1}{2}(\boldsymbol{u}_2^{(1)} + \boldsymbol{u}_2^{(2)})$ of the pair of molecules can be deduced, and the coefficient of thermal conduction calculated as before. The Eq. (40.8), however, eliminates the need for any further assumption, and might be expected to yield a more accurate result.

d) KIRKWOOD's theory of dissipative processes.

41. Molecular theory of the friction constant. The distinguishing feature of the various approaches to the molecular theory of irreversible processes, is the way in which the time-reversible Eqs. (35.6) and (35.25) are converted into time-irreversible equations. The method, already considered, of BORN and GREEN, is strongly influenced by analogies with gas-theory. Another method, due to KIRKWOOD[1], derives its inspiration from the theory of the Brownian motion. The well-known Fokker-Planck equation, or rather a slight generalization of it given and solved by CHANDRASEKHAR[2], is written in the form

$$\frac{\partial f}{\partial t} + \boldsymbol{\xi}\cdot\frac{\partial f}{\partial \boldsymbol{x}} = \frac{\zeta}{m}\frac{\partial}{\partial \boldsymbol{\xi}}\cdot\left(\boldsymbol{\xi} f + \frac{kT}{m}\frac{\partial f}{\partial \boldsymbol{\xi}}\right). \tag{41.1}$$

This time-irreversible equation is supposed to determine the distribution function f for a Brownian particle with mass m, where T is the absolute temperature, and ζ is the friction constant. KIRKWOOD has conjectured that an equation of the type (41.1) may be satisfied approximately by the molecular velocity distribution function, and shown how to derive such an equation from (35.6) by introducing approximations suggested by the theory of the Brownian motion. In addition he has derived an explicit, though rather intractable, formula for ζ from the molecular theory.

Consider a set of q molecules within a macroscopic region with a fixed boundary, and let Φ_q be their total potential energy, regarded as a function of their positions $\boldsymbol{x}^{(1)}, \ldots, \boldsymbol{x}^{(q)}$. Suppose that at some initial time t_0, the molecules have positions $\boldsymbol{x}_0^{(1)}, \ldots, \boldsymbol{x}_0^{(q)}$ and velocities $\boldsymbol{\xi}_0^{(1)}, \ldots, \boldsymbol{\xi}_0^{(q)}$, and that by moving under their mutual interactions they reach the positions $\boldsymbol{x}^{(1)}, \ldots, \boldsymbol{x}^{(q)}$ with velocities $\boldsymbol{\xi}^{(1)}, \ldots, \boldsymbol{\xi}^{(q)}$ at time $t = t_0 + \delta t$. The relation between the two sets of co-ordinates and velocities could be determined in principle by integrating the equations of motion of the

[1] J. G. KIRKWOOD: J. Chem. Phys. **14**, 180 (1946).
[2] S. CHANDRASEKHAR: Rev. Mod. Phys. **15**, 20 (1943).

system of molecules. If g_q is any function of the $\boldsymbol{x}^{(i)}$ and $\boldsymbol{\xi}^{(i)}$, and t, the same function of the $\boldsymbol{x}_0^{(i)}$ and $\boldsymbol{\xi}_0^{(i)}$, and t_0 will be denoted by $g_q(t_0)$. Since the configurations at the two times are causally related, they must have the same probability; hence, if F_q is the function introduced in Sect. 35 to specify the probability of finding precisely q molecules, in a given configuration, in the region considered, one will have

$$F_q(t_0)\, d\tau_q(t_0) = F_q\, d\tau_q, \tag{41.2}$$

where $d\tau_q(t_0) = d^3x_0^{(1)}\, d^3\xi_0^{(1)} \ldots d^3x_0^{(q)}\, d^3\xi_0^{(q)}$ and $d\tau_q = d^3x^{(1)} d^3\xi^{(1)} \ldots d^3x^{(q)}\, d^3\xi^{(q)}$.

The change of velocity of the j-th molecule in the interval between times $t_0 = t - \delta t$ and t is $\varrho_0^{(j)} = \boldsymbol{\xi}^{(j)} - \boldsymbol{\xi}_0^{(j)}$; the average value $\langle \varrho_0^{(1)} \rangle$ of $\varrho_0^{(1)}$ when only $\boldsymbol{x}_0^{(1)}$ and $\boldsymbol{\xi}_0^{(1)}$ are fixed is given by

$$f_1(t_0) \langle \varrho_0^{(1)} \rangle = \sum_{q=0}^{\infty} \int F_{q+1}(t_0)\, \varrho_0^{(1)}\, d\tau_{q+1}^{[1]}(t_0)/q! \tag{41.3}$$

where $f_1(t_0)$ is the velocity distribution function f_1 with the arguments t_0, $\boldsymbol{x}_0^{(1)}$ and $\boldsymbol{\xi}_0^{(1)}$, and $d\tau_{q+1}^{[1]}(t_0) = d^3x_0^{(2)}\, d^3\xi_0^{(2)} \ldots d^3x^{(q+1)}\, d^3\xi^{(q+1)}$. The average change of velocity between the times t and $t + \delta t$, when only $\boldsymbol{x}^{(1)}$ and $\boldsymbol{\xi}^{(1)}$ are fixed, is given similarly by

$$f_1 \langle \varrho^{(1)} \rangle = \sum_{q=0}^{\infty} \int F_{q+1}\, \varrho^{(1)}\, d\tau_{q+1}^{[1]}/q!\,. \tag{41.4}$$

Also, the average acceleration η_1 of a molecule during the interval between the time $t - \delta t$ and $t + \delta t$, when only $\boldsymbol{x}^{(1)}$ and $\boldsymbol{\xi}^{(1)}$ are fixed, is given by

$$2\delta t\, f_1\, \eta_1 = \sum_{q=0}^{\infty} \int F_{q+1} (\varrho_0^{(1)} + \varrho^{(1)})\, d\tau_{q+1}^{[1]}/q!\,. \tag{41.5}$$

The object of the present calculation is to find a serviceable expression for η_1.

Let g_1 be an *arbitrary* function of t, $\boldsymbol{x}^{(1)}$ and $\boldsymbol{\xi}^{(1)}$. Then it follows from (41.3), with the help of (41.2), that

$$\left.\begin{aligned} \int f_1(t_0)\, g_1(t_0) \langle \varrho_0^{(1)} \rangle\, d\tau_0^{(1)} &= \sum_{q=0}^{\infty} \int F_{q+1}(t_0)\, g_1(t_0)\, \varrho_0^{(1)}\, d\tau_{q+1}(t_0)/q! \\ &= \sum_{q=0}^{\infty} \int F_{q+1}\, g_1(t_0)\, \varrho_0^{(1)}\, d\tau_{q+1}/q!\,. \end{aligned}\right\} \tag{41.6}$$

At this stage an approximation is introduced, valid if δt is sufficiently small. One writes

$$g_1(t_0) \approx g_1 - \varrho_0^{(1)} \cdot \frac{\partial g_1}{\partial \boldsymbol{\xi}^{(1)}}, \tag{41.7}$$

neglecting the differences between t and t_0 and between $\boldsymbol{x}^{(1)}$ and $\boldsymbol{x}_0^{(1)}$, but taking account of the change of velocity $\varrho_0^{(1)}$ so far as terms linear in $\varrho_0^{(1)}$ are concerned. When (41.7) is substituted into (41.6), and an integration by parts is performed, one has

$$\int f_1(t_0)\, g_1(t_0) \langle \varrho_0^{(1)} \rangle\, d\tau_0^{(1)} = \sum_{q=0}^{\infty} \int g_1(t) \left\{ F_{q+1} \varrho_0^{(1)} + \frac{\partial}{\partial \boldsymbol{\xi}^{(1)}} (F_{q+1} \varrho_0^{(1)} \varrho_0^{(1)}) \right\} d\tau_{q+1}/q!\,. \tag{41.8}$$

The integral on the left-hand side of this equation depends only weakly, if at all, on the time, so t_0 may be increased to t. Since $g_1(t)$ is arbitrary, the integrations with respect to $\boldsymbol{x}^{(1)}$ and $\boldsymbol{\xi}^{(1)}$ may then be removed, with the result

$$f_1 \langle \varrho_0^{(1)} \rangle = \sum_{q=0}^{\infty} \int \left\{ F_{q+1} \varrho_0^{(1)} + \frac{\partial}{\partial \boldsymbol{\xi}^{(1)}} \cdot (F_{q+1} \varrho_0^{(1)} \varrho_0^{(1)}) \right\} d\tau_{q+1}^{[1]}/q!\,. \tag{41.9}$$

Adding (41.9) to (41.4) and making use of (41.5), one has finally

$$\eta_{1r} = \frac{\langle \varrho^{(1)} \rangle}{\delta t} = \eta_1 + \frac{kT}{m f_1} \frac{\partial}{\partial \boldsymbol{\xi}^{(1)}} \cdot (f_1 \boldsymbol{\zeta}_1) \tag{41.10}$$

where

$$kT f_1 \boldsymbol{\zeta}_1 = \frac{m}{2\delta t} \sum_{q=0}^{\infty} \int F_{q+1} \varrho_0^{(1)} \varrho_0^{(1)} d\tau_{q+1}^{[1]}/q!\,. \tag{41.11}$$

The tensor $\boldsymbol{\zeta}_1$ is called the friction tensor; for thermal and mechanical equilibrium, it reduces to a numerical multiple ζ of the unit tensor $\boldsymbol{\delta}$:

$$\boldsymbol{\zeta}_1^0 = \zeta \boldsymbol{\delta}\,. \tag{41.12}$$

Since $\varrho_0^{(1)}$ is proportional to δt when δt is very small, it can be seen from (41.11) that $\boldsymbol{\zeta}_1$ and ζ must approach zero with δt. But KIRKWOOD supposes that there is a fairly wide range of δt, below the value where the approximate (41.7) could no longer be sustained, where ζ is almost constant in value. He also assumes that this "plateau" value is almost independent of the velocity, by analogy with the friction constant which appears in the theory of Brownian motion. Unfortunately it appears to be very difficult to determine the plateau value; no explicit method of calculation has yet been described though an estimate

$$\zeta = \left\{\frac{1}{3mn} \int \frac{\partial^2 \phi(r)}{\partial \boldsymbol{r} \cdot \partial \boldsymbol{r}} n_2 d^3 r\right\}^{\frac{1}{2}} \tag{41.13}$$

has been published without derivation[1]. This would identify ζ as the relaxation frequency ω_r defined by (37.12).

In thermal and mechanical equilibrium, the mean acceleration η_1 vanishes, irrespective of the velocity of a molecule, so that (41.10) reduces to

$$\eta_{1r}^0 = \left\{-(\boldsymbol{\xi}^{(1)} - \boldsymbol{u}) + \frac{kT}{m} \frac{\partial}{\partial \boldsymbol{\xi}}\right\} \cdot \boldsymbol{\zeta}_1\,. \tag{41.14}$$

Subtracting this result from (41.10), one has

$$\eta'_{1r} = \eta_1 + \left\{\frac{kT}{m f_1} \frac{\partial f_1}{\partial \boldsymbol{\xi}^{(1)}} + \boldsymbol{v}^{(1)}\right\} \cdot \boldsymbol{\zeta}_1\,. \tag{41.15}$$

Substituting for η_1 from this equation into (35.3) one obtains finally

$$\frac{\partial f_1}{\partial t} + \boldsymbol{\xi}^{(1)} \cdot \frac{\partial f_1}{\partial \boldsymbol{x}^{(1)}} + \frac{\partial}{\partial \boldsymbol{\xi}^{(1)}} \cdot \left[f_1 \eta'_{1r} - \left\{\frac{kT}{m} \frac{\partial f_1}{\partial \boldsymbol{\xi}^{(1)}} + f_1 \boldsymbol{v}^{(1)}\right\} \cdot \boldsymbol{\zeta}_1\right] = 0\,. \tag{41.16}$$

This may be compared with (41.1), to which it reduces if η'_{1r} is neglected.

The above argument may be repeated, keeping two sets of molecular coordinates and velocities fixed instead of one. Then the result (41.15) is replaced by

$$\eta_{2r}^{(i)\prime} = \eta_2^{(i)\prime} + \left\{\frac{kT^{(i)}}{m f_2} \frac{\partial f_2}{\partial \boldsymbol{\xi}^{(i)}} + \boldsymbol{v}^{(i)}\right\} \cdot \boldsymbol{\zeta}_2^{(i)} \tag{41.17}$$

where the tensor $\boldsymbol{\zeta}_2^{(i)}$ is given by

$$kT^{(i)} f_2 \boldsymbol{\zeta}_2^{(i)} = \frac{m}{2\delta t} \sum_{q=0}^{\infty} \int F_{q+2} \varrho_0^{(i)} \varrho_0^{(i)} d\tau_{q+2}^{[12]}/q!\,. \tag{41.18}$$

It might appear from the last equation that $\boldsymbol{\zeta}_2^{(i)}$ should depend on the distance $r = |\boldsymbol{x}^{(2)} - \boldsymbol{x}^{(1)}|$, and possibly on $\boldsymbol{\xi}^{(1)}$ and $\boldsymbol{\xi}^{(2)}$ as well. In practice, however, KIRKWOOD treats it as constant and in fact identifies it with the friction tensor $\boldsymbol{\zeta}_1$.

[1] J. G. KIRKWOOD: Nuovo Cim. (9) 6 (Suppl. VIII) (1949).

When $\eta_2^{(i)\prime}$ is substituted from (41.17) into (35.14), the result is

$$\left.\begin{aligned}&\frac{\partial f_2}{\partial t}+\xi^{(1)}\cdot\frac{\partial f_2}{\partial \boldsymbol{x}^{(1)}}+\xi^{(2)}\cdot\frac{\partial f_2}{\partial \boldsymbol{x}^{(2)}}+\\&+\frac{\partial}{\partial \xi^{(1)}}\cdot\left[f_2(\eta_2^{(1)0}+\eta_{2r}^{(1)})-\left\{\frac{kT^{(1)}}{m}\frac{\partial f_2}{\partial \xi^{(1)}}+f_2\boldsymbol{v}^{(1)}\right\}\vartheta\cdot\zeta_2^{(1)}\right]+\\&+\frac{\partial}{\partial \xi^{(2)}}\cdot\left[f_2(\eta_2^{(2)0}+\eta_{2r}^{(2)})-\left\{\frac{kT^{(2)}}{m}\frac{\partial f_2}{\partial \xi^{(2)}}+f_2\boldsymbol{v}^{(2)}\right\}\vartheta\cdot\zeta_2^{(2)}\right]=0\end{aligned}\right\}\tag{41.19}$$

where

$$m\,\eta_2^{(i)0}=-\frac{\partial\phi^{(12)}}{\partial \boldsymbol{x}^{(i)}}-\int\frac{n_3^0}{n_2^0}\frac{\partial\phi^{(i3)}}{\partial \boldsymbol{x}^{(i)}}\,d^3x^{(3)}.\tag{41.20}$$

A similar equation can of course be derived for f_q. However, (41.16) and (41.19) above are presumably enough to determine the functions f_1 and f_2, which are of paramount importance in the applications of the theory.

42. Theory of viscosity and thermal conduction. Before discussing KIRKWOOD's theory of viscosity and thermal conduction, it is convenient to summarize the important hypotheses which have appeared in the theory of the friction constant.

The mean force acting on a molecule with velocity $\xi^{(1)}$, situated at the point $\boldsymbol{x}^{(1)}$ in the liquid, is rigorously

$$m\,\eta_1=-\iint\frac{f_2}{f_1}\frac{\partial\phi^{(12)}}{\partial \boldsymbol{x}^{(1)}}\,d^3x^{(2)}\,d^3\xi^{(2)},$$

according to (35.5). In KIRKWOOD's theory, this expression is approximated by

$$m\,\eta_1=\frac{\partial}{\partial \boldsymbol{x}}(n\,kT-p)-\zeta\left(\frac{kT}{f_1}\frac{\partial f_1}{\partial \xi^{(1)}}+m\,\boldsymbol{v}^{(1)}\right).\tag{42.1}$$

Thus the part of the acceleration representd by η_{1r}' in (41.15) is assumed to be independent of the molecular velocity; the velocity dependent forces, proportional to ζ, vanish on averaging over all velocities, and the acceleration η_{1r}' must therefore be produced by that part of the pressure associated with the intermolecular forces $(p-n\,kT)$. The velocity-dependent forces, on the other hand, are precisely those assumed in the theory of the Brownian motion.

The mean force acting on a molecule is influenced by the presence of a molecule known to be in the vicinity, and, according to (35.15), is rigorously

$$m\,\eta_2^{(1)}=-\frac{\partial\phi^{(12)}}{\partial \boldsymbol{x}^{(1)}}-\iint\frac{f_3}{f_2}\frac{\partial\phi^{(13)}}{\partial \boldsymbol{x}^{(1)}}\,d^3x^{(3)}\,d^3\xi^{(3)}.$$

In KIRKWOOD's theory, this is approximated by

$$m\,\eta_2^{(1)}=m\,\eta_2^{(1)0}+\frac{\partial}{\partial \boldsymbol{x}^{(1)}}(n^{(1)}kT^{(1)}-p^{(1)})-\zeta\left(\frac{kT^{(1)}}{f_2}\frac{\partial f_2}{\partial \xi^{(1)}}+m\,\boldsymbol{v}^{(1)}\right).\tag{42.2}$$

According to (41.20) and (21.4), $\eta_0^{(1)0}$ may be expressed in the form

$$\eta_2^{(1)0}=-\frac{kT^{(1)}}{m}\frac{\partial}{\partial r}\{\log g^0(r,\boldsymbol{x}^{(1)})\}\tag{42.3}$$

where $g^0(r,\boldsymbol{x}^{(1)})$ is the equilibrium value of $g=n_2/n^2$, for the density and temperature at the point $\boldsymbol{x}^{(1)}$. It will be noticed that the acceleration $\eta_{2r}^{(1)\prime}$ in (41.17) has been approximated by the same expression as η_{1r}'.

It should be remarked, perhaps, that since (42.1) and (42.2) are approximations, their indiscriminate use may lead to contradictions. For example, the mean value of $(\boldsymbol{v}^2)^2$ in thermal equilibrium is $15\,(kT/m)^2$, but a calculation made with (42.1) above gave the value $12\,(kT/m)^2$. Again, KIRKWOOD and his associates have obtained an expression for the coefficient of thermal conduction, (42.16) below, which differs from what they would have obtained had they proceeded, using the same approximations, from the Eq. (40.6). It is, however, in favour of the approximations that no very serious discrepancies have yet been uncovered in this way.

First the direct contributions from the thermal motion to the pressure tensor and thermal flux vector will be calculated, using (42.1). If the tensor $\boldsymbol{\xi}\boldsymbol{\xi}$ is substituted for φ in (35.10), the result is

$$n\frac{d}{dt}\langle\boldsymbol{\xi}\boldsymbol{\xi}\rangle+\frac{\partial}{\partial\boldsymbol{x}}\cdot(n\langle\boldsymbol{v}\boldsymbol{\xi}\boldsymbol{\xi}\rangle)=n\langle\boldsymbol{\eta}\boldsymbol{\xi}+\boldsymbol{\xi}\boldsymbol{\eta}\rangle,$$

which may also be written

$$\left.\begin{aligned}n\frac{d}{dt}\langle\boldsymbol{v}\boldsymbol{v}\rangle+\frac{\partial}{\partial\boldsymbol{x}}\cdot(n\langle\boldsymbol{v}\boldsymbol{v}\boldsymbol{v}\rangle)&=n\langle\boldsymbol{\eta}\boldsymbol{v}+\boldsymbol{v}\boldsymbol{\eta}\rangle+\\&+n\left\{\left(\frac{\partial_s\boldsymbol{u}}{\partial\boldsymbol{x}}\right)\cdot\langle\boldsymbol{v}\boldsymbol{v}\rangle+\langle\boldsymbol{v}\boldsymbol{v}\rangle\cdot\left(\frac{\partial_s\boldsymbol{u}}{\partial\boldsymbol{x}}\right)\right\}\end{aligned}\right\}\qquad(42.4)$$

if[1] $\frac{\partial}{\partial\boldsymbol{x}}\cdot\boldsymbol{u}=0$. Substituting $mn\langle\boldsymbol{v}\boldsymbol{v}\rangle=nkT+p'_{\mathrm{k}}$, and discarding all except terms linear in the gradients of $\boldsymbol{u}$, n and T, one finds that the left-hand side of the above equation disappears, and one is left with

$$\zeta p'_{\mathrm{k}}=-nkT\frac{\partial_s\boldsymbol{u}}{\partial\boldsymbol{x}}.\qquad(42.5)$$

Similarly, if the vector $\xi^2\boldsymbol{\xi}$ is substituted for φ in (35.10), the result is

$$\begin{aligned}n\frac{d}{dt}\langle\xi^2\boldsymbol{\xi}\rangle+\frac{\partial}{\partial\boldsymbol{x}}\cdot(n\langle\boldsymbol{v}\xi^2\boldsymbol{\xi}\rangle)&=n\langle2\boldsymbol{\eta}\cdot\boldsymbol{\xi}\boldsymbol{\xi}+\xi^2\boldsymbol{\eta}\rangle\\&=n(2\langle\boldsymbol{\xi}\boldsymbol{\xi}\rangle+\langle\xi^2\rangle\boldsymbol{\delta})\cdot\frac{\partial}{\partial\boldsymbol{x}}\frac{(nkT-p)}{m}-\\&\quad-n\zeta\langle2\boldsymbol{v}\cdot\boldsymbol{\xi}\boldsymbol{\xi}+\xi^2\boldsymbol{v}-10kT\boldsymbol{\xi}/m\rangle.\end{aligned}$$

Therefore, setting $\boldsymbol{u}=0$ so that $\boldsymbol{\xi}=\boldsymbol{v}$, and $\frac{1}{2}mn\langle v^2\boldsymbol{v}\rangle=\boldsymbol{q}_{\mathrm{k}}$, one has[2]

$$\left.\begin{aligned}6\zeta\boldsymbol{q}_{\mathrm{k}}&=\frac{5kT}{m}\frac{\partial}{\partial\boldsymbol{x}}(nkT-p)-5\frac{\partial}{\partial\boldsymbol{x}}(nk^2T^2)\\&=-\frac{5kT}{m}\left(nk\frac{\partial T}{\partial\boldsymbol{x}}+\frac{\partial p}{\partial\boldsymbol{x}}\right).\end{aligned}\right\}\qquad(42.6)$$

At constant pressure, the second term in this expression vanishes and the direct contribution of the thermal motion to the coefficient of thermal conductivity is $\frac{5}{6}nk^2T/(m\zeta)$. On the other hand, the contribution of the thermal motion to the coefficient of shearing viscosity is $\frac{1}{2}nkT/\zeta$, according to (42.5). In rare gases, the intermolecular forces make no appreciable contribution to either viscosity or thermal conduction, so the ratio of the coefficients of viscosity and thermal conduction should be $\frac{3m}{5k}$, if the theory is applicable. The value for this ratio derived from BOLTZMANN's equation in the theory of gases, and confirmed experimentally for the monatomic gases, is $\frac{4m}{15k}$, about one half of the predicted

[1] A discrepancy which arises when this condition is not satisfied may be attributed to the approximation involved in (42.1). See J. G. KIRKWOOD, F. P. BUFF and M. S. GREEN: J. Chem. Phys. **18**, 901 (1950).

[2] A slightly different result was published, due to the use of an incorrect mean value of $(v^2)^2$.

value. However, in liquids the contributions to the coefficients of viscosity and thermal conduction due to the thermal motion are small compared with those due to the intermolecular forces, and this particular discrepancy is unimportant.

To compute the coefficient of viscosity, it is necessary to determine n_2', or $g' = n_2'/n^2$. A method by which this can be done, again using the Eq. (36.10) but in a manner different from that described in Sect. 40, was developed by KIRKWOOD, BUFF and GREEN[1].

If the expression (42.2) for $\eta_2^{(1)}$ and a similar expression for $\eta_2^{(2)}$ are substituted into (36.10), the latter can be written

$$\left.\begin{aligned} \zeta g^0 (\boldsymbol{u}_2^{(2)} - \boldsymbol{u}^{(2)} - \boldsymbol{u}_2^{(1)} + \boldsymbol{u}^{(1)}) &= g(\eta_2^{(2)0} - \eta_2^{(1)0}) - \frac{2kT}{m}\frac{\partial g}{\partial \boldsymbol{r}} \\ &= -\frac{2kT}{m}\left(\frac{\partial g'}{\partial \boldsymbol{r}} - \frac{g'}{g^0}\frac{\partial g_0}{\partial \boldsymbol{r}}\right). \end{aligned}\right\} \tag{42.7}$$

On the other hand, the equation of continuity (35.17) for pairs of molecules can, with the help of (35.9), be written in the form

$$\frac{\partial g^0}{\partial t} + \frac{\partial}{\partial \boldsymbol{r}} \cdot \{g(\boldsymbol{u}_2^{(2)} - \boldsymbol{u}^{(2)} - \boldsymbol{u}_2^{(1)} + \boldsymbol{u}^{(1)})\} + \boldsymbol{r}\cdot\left(\frac{\partial}{\partial \boldsymbol{x}}\boldsymbol{u}\right)\cdot\frac{\partial g^0}{\partial \boldsymbol{r}} = 0.$$

Substituting from (42.7), one obtains, therefore

$$\frac{\partial}{\partial \boldsymbol{r}}\cdot\left(\frac{\partial g'}{\partial \boldsymbol{r}} - \frac{g'}{g^0}\frac{\partial g^0}{\partial \boldsymbol{r}}\right) = \frac{m\zeta}{2kT}\left\{\frac{\partial g^0}{\partial t} + \boldsymbol{r}\cdot\left(\frac{\partial}{\partial \boldsymbol{x}}\boldsymbol{u}\right)\cdot\frac{\partial g^0}{\partial \boldsymbol{r}}\right\}, \tag{42.8}$$

which is KIRKWOOD, BUFF and GREEN'S equation to determine g'. A difficulty arises over the choice of boundary conditions for large values of r. The simplest choice is that which allows the immediate integration of (42.8). When $\frac{\partial}{\partial \boldsymbol{x}}\cdot\boldsymbol{u} = 0$, $\frac{\partial g^0}{\partial t}$ vanishes and it is seen that

$$\frac{\partial g'}{\partial \boldsymbol{r}} - \frac{g'}{g^0}\frac{\partial g^0}{\partial \boldsymbol{r}} = \frac{m\zeta g^0}{2kT}\boldsymbol{r}\cdot\left(\frac{\partial_s \boldsymbol{u}}{\partial \boldsymbol{x}}\right) \tag{42.9}$$

is a first integral, and

$$g' = \frac{1}{4} m\beta\zeta g^0 \boldsymbol{r}\cdot\left(\frac{\partial_s \boldsymbol{u}}{\partial \boldsymbol{x}}\right)\cdot\boldsymbol{r} \tag{42.10}$$

a solution. A considerably more complicated solution was obtained by KIRKWOOD and his associates, due to a special choice of boundary conditions, but it does not appear to differ much from (42.10) in its quantitative aspects. The result obtained allows one to calculate

$$p_1' = -\left\{\frac{m\beta\zeta}{60}\int n_2^0(r)\,\phi'(r)\,r^3\,d^3r\right\}\frac{\partial_s \boldsymbol{u}}{\partial \boldsymbol{x}} \tag{42.11}$$

and infer the coefficient of viscosity in the form

$$\eta = \frac{1}{2} n kT/\zeta + \frac{m\zeta}{120kT}\int n_2^0\,\phi'(r)\,r^3\,d^3r. \tag{42.12}$$

It will be seen that this result does not differ to any important degree from (40.5), and if ζ could be calculated would determine the value of $c(r_1)$. KIRKWOOD, BUFF and GREEN used the relaxation frequency defined by (41.13) for ζ and were thus able to calculate a coefficient of shear viscosity in liquid argon at 89° K. They found a value of 1.27×10^{-3} poise, in rough agreement with the experimental value of 2.39×10^{-3} poise. Using another unpublished method of calculating ζ,

[1] J. G. KIRKWOOD, F. P. BUFF and M. S. GREEN: J. Chem. Phys. 17, 988 (1949).

ZWANZIG, KIRKWOOD, STRIPP and OPPENHEIM[1] later found a somewhat smaller value of the viscosity; they also calculated the coefficient of volume viscosity by a method which may, however, be in error due to neglect of the change of temperature which accompanies dilatation and affects the time derivative of g_0. The coefficient of bulk viscosity can be calculated by using BORN and GREEN's formula (36.9) in conjunction with (42.6); however, it is not yet possible to make any comparison with experiment.

A theory of thermal conductivity has been developed by ZWANZIG, KIRKWOOD, OPPENHEIM and ALDER[2], based on the equation (36.11). The approximations of (40.1) are used for $\langle \boldsymbol{v}_2^{(i)} \boldsymbol{v}_2^{(j)} \rangle$, and $\eta_2^{(1)}$, and $\eta_2^{(2)}$ are represented as in (42.2), so that (36.11) reduces to

$$\left.\begin{aligned} &\frac{kT}{m}\frac{\partial}{\partial \boldsymbol{x}}\left(\frac{n_2^0}{n^2}\right)+\frac{k(T^{(2)}-T^{(1)})}{m}\frac{\partial n_2^0}{\partial \boldsymbol{r}}\\ &\qquad = n_2^0\{(\eta_2^{(1)0}+\eta_2^{(2)0})-\zeta(\boldsymbol{u}_2^{(1)}+\boldsymbol{u}_2^{(2)}-\boldsymbol{u}^{(1)}-\boldsymbol{u}^{(2)})\}.\end{aligned}\right\} \tag{42.13}$$

Now, according to (42.3),

$$\eta_2^{(1)0}+\eta_2^{(2)0}=\boldsymbol{r}\cdot\frac{\partial}{\partial \boldsymbol{x}}\left\{\frac{kT}{m}\frac{\partial}{\partial \boldsymbol{r}}\log g^0(r,\boldsymbol{x})\right\} \tag{42.14}$$

so that, when $\boldsymbol{u}=0$

$$\zeta g^0(\boldsymbol{u}_2^{(1)}+\boldsymbol{u}_2^{(2)})=\frac{kT}{m}\left\{-\frac{\partial g^0}{\partial \boldsymbol{x}}+g^0\boldsymbol{r}\cdot\frac{\partial}{\partial \boldsymbol{x}}\left(\frac{1}{g^0}\frac{\partial g^0}{\partial \boldsymbol{r}}\right)\right\}. \tag{42.15}$$

This result was used by the authors mentioned to calculate the coefficient of thermal conduction. It cannot be regarded as completely satisfactory, since it does not fulfil the condition (31.9); however it may nevertheless be a good approximation. The coefficient of thermal conduction in KIRKWOOD's theory is

$$\left.\begin{aligned}\lambda = \frac{5nk^2T}{6m\zeta}+\frac{n^2kT}{12m\zeta}\int\Big[&\{r\phi'(r)-\phi(r)\}r\,g(r)\frac{d}{dr}\left\{\frac{1}{g(r)}\frac{\partial g(r)}{\partial T}\right\}+\\ &+\left\{\phi(r)-\frac{1}{3}r\phi'(r)\right\}\frac{\partial g(r)}{\partial T}\Big]d^3r.\end{aligned}\right\} \tag{42.16}$$

The numerical results depend sensitively on the form of $\phi(r)$ and $g(r)$ adopted, so that a precise comparison with experiment is difficult. However, the experimental value $\lambda=2.9\times10^{-4}$ cal/g sec °K obtained by UHLIR[3] for liquid argon at its normal boiling point is closely confirmed by substitution of the best values of $\phi(r)$ and $g(r)$ in (42.16).

The result (42.16) is not the analogue of (42.12) which one might expect. The contribution to the viscosity from the intermolecular forces is proportional to the friction constant, whereas it would appear from (42.16) that the corresponding contribution to the thermal conductivity is *inversely* proportional to the friction constant. The alternative Eq. (40.6), however, yields a different result, more closely analogous to (42.12), in which the leading contribution to the thermal conductivity is proportional to ζ. The results obtained by the two methods, however, are not necessarily incompatible; assuming they are approximately equivalent, one finds a condition which ought to be satisfied by the friction constant. When accurate determinations of the friction constant are available, it will be possible to test the consistency of KIRKWOOD's theory in this way.

[1] R. W. ZWANZIG, J. G. KIRKWOOD, K. F. STRIPP and I. OPPENHEIM: J. Chem. Phys. **21**, 2050 (1953).

[2] R. W. ZWANZIG, J. G. KIRKWOOD, I. OPPENHEIM and B. J. ALDER: J. Chem. Phys. **22**, 783 (1954).

[3] A. UHLIR: J. Chem. Phys. **20**, 463 (1952).

V. Structure of quantum liquids.

a) Liquid helium.

43. Phenomenology of He II. Up to this point the theory of liquid structure and its applications has been founded on the tacit assumption that classical mechanics can be used without introducing any significant error. The theoretical justification of this assumption will be considered in Sect. 46; pragmatically it may already be accepted on the ground that, disregarding the obvious mathematical difficulties, the theory appears to apply to most liquids fairly well. Liquid helium is a remarkable exception, and though its behaviour is not yet accounted for by any universally accepted theory, it is generally agreed that the reason for its anomalous behaviour is to be found in the failure of classical mechanics: it is a "quantum" liquid.

A full review of the experimental facts concerning liquid helium should be sought elsewhere[1]; they will be mentioned here only very briefly in so far as they are immediately relevant to the molecular theories. First, mention should be made of the thermodynamic discontinuity which occurs at the so-called λ-point (a temperature of 2.186° K at the pressure of the saturated vapour, decreasing slowly as the pressure is increased). The thermodynamic functions, such as the pressure, internal energy and isothermal compressibility, which depend on the radial distribution function, are not themselves discontinuous, but their temperature derivative shows a discontinuity. For example, the specific heat can be represented as a monotonically increasing function of temperature, of the type

$$c = \alpha T^3 + \beta T^{\frac{3}{2}} \exp(-\Delta/kT) \tag{43.1}$$

between absolute zero and the λ-point, where α and β are positive constants and Δ/k is about 9° K. But above the λ-point, the specific heat falls abruptly to a small fraction of value given by (43.1). The inference to be drawn from this is that $\frac{\partial n_2(r)}{\partial T}$ is likely to be discontinuous at the λ-point.

Helium remains liquid at all temperatures from the condensation point (4.2° K) down to absolute zero, except at pressures above 25 atmospheres, where it solidifies in a hexagonal close-packed structure. The fact that it does not solidify except under high pressure is ascribed to the rather large mean kinetic energy (zero-point energy) of the molecules expected on quantum-theoretical grounds to persist even at absolute zero. Also it should be remembered that the attractive van der Waals forces are weaker between helium molecules than between molecules of any other kind.

Above the λ-point, liquid helium is called He I and is not radically different in its behaviour from other liquids. But below the λ-point, it is called He II and has a number of bizarre properties which point to a peculiar molecular constitution. In narrow slits and capillaries, and in a film which is formed on all solid surfaces exposed to the saturated vapour, it exhibits a practically complete lack of viscosity. The liquid which flows through such narrow channels carries practically no entropy; it can be set in motion by either a temperature gradient or a pressure gradient between the ends. There is evidence that this "superfluid" also forms part of the liquid in bulk.

---

[1] See W. H. Keesom: Helium „Elsevier". Amsterdam 1942. — R. B. Dingle, K. R. Atkins and J. G. Daunt: Adv. Physics 1, No. 2 (1952). — Cf. also the article of K. Mendelssohn in Vol. XV of this Encyclopedia.

The properties of He II can be explained qualitatively and often also quantitatively by a phenomenological theory, due in essence to LONDON[1], TISZA[2] and LANDAU[3]. This theory supposes the liquid to be a mixture of the superfluid which passes without friction through narrow channels in the way described, and the "normal fluid", which is similar in its properties to ordinary fluids. The superfluid and the normal fluid can move with different velocities without mutual friction, provided their relative velocity does not exceed a certain critical figure. The normal fluid is present as well as the superfluid in narrow channels, but owing to its viscosity cannot move.

The phenomenological theory will now be developed in a radical form which enables the liquid to be regarded as a mixture of "superfluid atoms"—i.e., atoms in a "superfluid state", and "normal atoms". Let ϱ_s and ϱ_n denote the mass densities of the superfluids and normal fluid respectively, and $\boldsymbol{u}_s$ and $\boldsymbol{u}_n$ their velocities of flow. If Γ is the rate of creation of superfluid mass per unit volume, they will satisfy

$$\left.\begin{aligned} \frac{\partial \varrho_s}{\partial t} + \frac{\partial}{\partial \boldsymbol{x}} \cdot (\varrho_s \boldsymbol{u}_s) &= \Gamma, \\ \frac{\partial \varrho_n}{\partial t} + \frac{\partial}{\partial \boldsymbol{x}} \cdot (\varrho_n \boldsymbol{u}_n) &= -\Gamma. \end{aligned}\right\} \tag{43.2}$$

The superfluid and the normal fluid each contribute partial pressures p_s and p_n to the resultant hydrostatic pressure p^0; but the superfluid does not contribute to the viscosity so that the partial pressure tensors are

$$\left.\begin{aligned} \mathsf{p}_s &= p_s \boldsymbol{\delta}, \\ \mathsf{p}_n &= \left(p_n - \eta^0 \frac{\partial}{\partial \boldsymbol{x}} \cdot \boldsymbol{u}_n\right) \boldsymbol{\delta} - 2\eta \frac{\partial_s \boldsymbol{u}_n}{\partial \boldsymbol{x}}. \end{aligned}\right\} \tag{43.3}$$

The equations of conservation of momentum for the two components are

$$\left.\begin{aligned} \frac{\partial}{\partial t}(\varrho_s \boldsymbol{u}_s) + \frac{\partial}{\partial \boldsymbol{x}} \cdot (\varrho_s \boldsymbol{u}_s \boldsymbol{u}_s + \mathsf{p}_s) &= \Gamma \boldsymbol{u}_s, \\ \frac{\partial}{\partial t}(\varrho_n \boldsymbol{u}_n) + \frac{\partial}{\partial \boldsymbol{x}} \cdot (\varrho_n \boldsymbol{u}_n \boldsymbol{u}_n + \mathsf{p}_n) &= -\Gamma \boldsymbol{u}_s. \end{aligned}\right\} \tag{43.4}$$

With the help of (43.2), these equations of motion are reduced to

$$\left.\begin{aligned} \varrho_s \frac{d_s \boldsymbol{u}_s}{d t_s} + \frac{\partial}{\partial \boldsymbol{x}} \cdot \mathsf{p}_s &= 0, \\ \varrho_n \frac{d_n \boldsymbol{u}_n}{d t_n} + \frac{\partial}{\partial \boldsymbol{x}} \cdot \mathsf{p}_n &= \Gamma(\boldsymbol{u}_n - \boldsymbol{u}_s), \end{aligned}\right\} \tag{43.5}$$

where $\frac{d_s}{dt_s} = \frac{\partial}{\partial t} + \boldsymbol{u}_s \cdot \frac{\partial}{\partial \boldsymbol{x}}$ and $\frac{d_n}{dt_n} = \frac{\partial}{\partial t} + \boldsymbol{u}_n \cdot \frac{\partial}{\partial \boldsymbol{x}}$.

Only the normal fluid contributes to the thermal flux $\boldsymbol{q}$, so, of U_s and U_n are the internal energies per unit mass of the two components, the equations of conservation of energy read

$$\left.\begin{aligned} &\frac{\partial}{\partial t}\left\{\varrho_s\left(\frac{1}{2}\boldsymbol{u}_s^2 + U_s\right)\right\} + \frac{\partial}{\partial \boldsymbol{x}} \cdot \left\{\varrho_s \boldsymbol{u}_s\left(\frac{1}{2}\boldsymbol{u}_s^2 + U_s\right) + \boldsymbol{u}_s \cdot \mathsf{p}_s\right\} \\ &\qquad = \Gamma\left(\frac{1}{2}\boldsymbol{u}_s^2 + U_s + \frac{p_s}{\varrho_s}\right), \\ &\frac{\partial}{\partial t}\left\{\varrho_n\left(\frac{1}{2}\boldsymbol{u}_n^2 + U_n\right)\right\} + \frac{\partial}{\partial \boldsymbol{x}} \cdot \left\{\varrho_n \boldsymbol{u}_n\left(\frac{1}{2}\boldsymbol{u}_n^2 + U_n\right) + \boldsymbol{u}_n \cdot \mathsf{p}_n + \boldsymbol{q}\right\} \\ &\qquad = -\Gamma\left(\frac{1}{2}\boldsymbol{u}_s^2 + U_s + \frac{p_s}{\varrho_s}\right). \end{aligned}\right\} \tag{43.6}$$

[1] F. LONDON: Phys. Rev. **54**, 947 (1938).
[2] L. TISZA: C.R. Acad. Sci., Paris **207**, 1035, 1186 (1938). — J. Phys. Rad. **1**, 164, 350 (1940). — Phys. Rev. **72**, 838 (1947).
[3] L. LANDAU: J. Phys. USSR. **5**, 71 (1941); **8**, 110 (1944); **11**, 91 (1947).

Using (43.2), (43.3) and (43.5), one finds that (43.6) reduces to

$$\left.\begin{aligned} \varrho_s \frac{d_s U_s}{d t_s} - \frac{p_s}{\varrho_s}\frac{d_s \varrho_s}{d t_s} &= 0, \\ \varrho_n \frac{d_n U_n}{d t_n} - \frac{p_n}{\varrho_n}\frac{d_n \varrho_n}{d t_n} &= \varrho_n T \frac{d_n S_n}{d t_n}, \end{aligned}\right\} \tag{43.7}$$

where

$$\left.\begin{aligned} \varrho_n T \frac{d_n S_n}{d t_n} + \frac{\partial}{\partial \boldsymbol{x}}\cdot \boldsymbol{q} &= \eta\left(\frac{\partial}{\partial \boldsymbol{x}}\cdot \boldsymbol{u}_n\right)^2 + 2\eta\left(\frac{\partial_s \boldsymbol{u}_n}{\partial \boldsymbol{x}}\right)^2 + \\ &+ \Gamma\left(U_n + \frac{p_n}{\varrho_n} - U_s - \frac{p_s}{\varrho_s} - \frac{1}{2}(\boldsymbol{u}_n - \boldsymbol{u}_s)^2\right). \end{aligned}\right\} \tag{43.8}$$

The first of the results (43.7) implies that the superfluid carries no entropy; in order to obtain this result it was assumed in (43.6) that the superfluid is created with mean energy $U_s + p_s/\varrho_s$, as well as its macroscopic kinetic energy. If the masses of superfluid and normal fluid within the volume V of a fixed total mass are M_s and M_n respectively, so that the interval energies and entropy in the liquid are $U_s^V = M_s U_s$, $U_n^V = M_n U_n$ and $S^V = M_n S_n$, it follows from (43.7) that

$$\left.\begin{aligned} dU_s^V + p_s dV &= \mu\, dM_s, \\ dU_n^V + p_n dV &= \mu\, dM_n + T\, d S^V \end{aligned}\right\} \tag{43.9}$$

where μ is the chemical potential, given by

$$\left.\begin{aligned} M_s \mu &= U_s^V + p_s V, \\ M_n \mu &= U_n^V + p_n V - T S^V. \end{aligned}\right\} \tag{43.10}$$

The relations

$$\left.\begin{aligned} M_s\, d\mu &= V d p_s, \\ M_n\, d\mu &= V\, d p_n - S^V dT, \end{aligned}\right\} \tag{43.11}$$

follow immediately, and enable one to determine

$$\left.\begin{aligned} \frac{\partial p_s}{\partial \boldsymbol{x}} &= \varrho_s \frac{\partial \mu}{\partial \boldsymbol{x}} = \frac{\varrho_s}{\varrho}\left(\frac{\partial p^0}{\partial \boldsymbol{x}} - \varrho_n S_n \frac{\partial T}{\partial \boldsymbol{x}}\right), \\ \frac{\partial p_n}{\partial \boldsymbol{x}} &= \frac{\varrho_n}{\varrho}\left(\frac{\partial p^0}{\partial \boldsymbol{x}} + \varrho_s S_n \frac{\partial T}{\partial \boldsymbol{x}}\right) \end{aligned}\right\} \tag{43.12}$$

which are the hydrostatic pressure forces acting on the superfluid and normal fluid respectively, in the form adopted by LONDON and TISZA.

The requirements of the second law of thermodynamics, as it relates to irreversible processes, are often overlooked. The result (43.8) shows that when superfluid is created, it is necessary that $T S_n \geq \frac{1}{2}(\boldsymbol{u}_n - \boldsymbol{u}_s)^2$, and that it can be annihilated only if $T S_n \leq \frac{1}{2}(\boldsymbol{u}_n - \boldsymbol{u}_s)^2$.

The phenomenological theory is one which was devised originally with the sole object of fitting the experimental facts, and may well be approximate in its details. For example, it cannot be regarded as certain that the entropy carried by the superflow is rigorously zero, though it is small enough to have escaped experimental detection[1]. However, in spite of possible limitations the macroscopic theory provided a goal towards which a molecular theory may be directed, and it has in fact suggested many of the attempts to account for the properties of helium in terms of molecular structure, which will be discussed in the following section.

---

[1] See D. F. BREWER, D. O. EDWARDS and K. MENDELSSOHN: Proc. Phys. Soc. Lond. A **68**, 939 (1955).

Another important fact which must be taken into account in formulating a molecular theory is that superfluidity appears to be a characteristic only of the common isotope He^4 in mixtures of the two isotopes He^3 and He^4. As the proportion of He^3 is increased, the λ-point is depressed in a way which suggests that it may approach absolute zero in pure He^3. As He^4 atoms obey Bose statistics and He^3 atoms obey Fermi statistics, it can be concluded at least that Fermi statistics do not favour superfluidity. Thus account must be taken of quantum statistics as well as quantum mechanics in formulating a satisfactory molecular theory of liquid helium.

44. Molecular theories. A brief review will now be made of the several molecular theories of liquid helium which appear to have an important bearing on its structure. In spite of obvious differences in emphasis and point of view, the theories have much in common. Thus there are many different explanations of the λ-phenomenon, but they all reduce fundamentally to one of two explanations which are by no means inconsistent with one another.

The first explanation of the λ-phenomenon is that it results from a quantum condensation. The resemblance between the temperature dependence of the specific heat in liquid helium near the λ-point, and that in an ideal Bose-Einstein gas near the transition temperature, was first noticed by LONDON[1], who also showed that the resemblance becomes more pronounced if the interatomic forces are approximately taken into account. It is tempting to identify the superfluid with the condensed phase of a Bose gas below the transition temperature, modified by the atomic interactions. The changes produced by the interactions are profound, and it is not easy to reach even a qualitative understanding of their effect on the Bose condensation and the condensed phase. Nevertheless the λ-phenomenon is undoubtedly associated with a "condensation" of some type influenced by quantum statistics.

The second explanation of the λ-phenomenon is that it marks the transition to a partially ordered state as the temperature is lowered. The condensed phase of a Bose gas may be regarded as having an ordered character, so there is no real conflict between this and the point of view already described. As to the type of order, the liquid at absolute zero is pure superfluid, so the fully ordered state is the quantum-mechanical ground state. But since the quantum-mechanical problem of finding the proper states of an assembly of similar interacting particles cannot be solved explicitly, one can only guess at the structure of the liquid in this state. LONDON[2] examined the possibility that the structure of liquid helium near absolute zero might approximate to some type of crystalline structure. Using the known form of interaction energy $\phi(r)$ between two helium atoms, he calculated the potential energy of various lattice structures with a helium atom at each lattice point, and made an estimate of the energy of the zero-point vibrations. This enabled him to predict the internal energy of liquid helium, on the assumption that its structure was similar to that of one of the lattice structures he investigated. LONDON concluded, on the basis of his rough calculation, that the most stable structure would be the one based on the diamond lattice, which also had an energy compatible with the observed internal energy of liquid helium.

There is some reason to believe that even if the superfluid has not a true lattice structure, the atoms resemble those in a crystal at least in so far as they are vibrating about mean positions, which move with the macroscopic velocity $\boldsymbol{u}_s$ of the superfluid. The term in (43.1), proportional to T^3 and dominant near

[1] F. LONDON: Phys. Rev. **54**, 947 (1938). — J. Chem. Phys. **43**, 49 (1939).
[2] F. LONDON: Proc. Roy. Soc. Lond., Ser. A **153**, 576 (1936).

absolute zero, is what one would expect of a system of harmonic oscillators and could hardly be explained in any other way. The superfluid, then, is an assembly of atoms, which, in spite of their zero-point motion, do not diffuse relative to one another. If this is accepted, the next question concerns the nature of the excitations which constitute the normal component; to this various answers have been given.

FRÖHLICH[1] and JONES[2] have examined the possibility that the excitations consist of atoms moving into the interstitial positions in the lattice structure proposed by LONDON; LONDON[3], on the other hand adopted the more appealing but perhaps not contradictory view that the excited atoms are moving freely through the superfluid.

The excitations, of course, may be associated with groups of atoms in a macroscopic region rather than individual atoms; this is the opinion of LANDAU[4], who suggested the existence of two different kinds of excitations, which he called phonons and rotons. The phonons are simply energy quanta associated with the acoustic vibrations of the system of atoms, and are responsible for the contribution αT^3 to the specific heat. The rotons are excited at somewhat higher temperatures, and their increase is responsible for the eventual dissolution of the superfluid at the λ-point. LANDAU identified the rotons with quantized states of vortex motion in the liquid, but this identification is not generally accepted, and FEYNMAN[5], in particular, has maintained that they cannot be essentially different from the phonons.

FEYNMAN has attempted to determine the energy Δ in the formula (43.1), which he regards as a minimum of the energy $E(k)$ of an excitation, expressed as a function of its momentum k. His result is to a great extent independent of certain assumptions which he made concerning the form of the wave function of the excited state, and the properties of the radial distribution function; and can be understood qualitatively in the following way. An atom with a de Broglie wave-length either less than or greater than the distance between nearest atoms is described by a wave-function whose maxima overlap at least some atoms in its environment, and will therefore have a rather high energy of interaction. But if its de Broglie wave-length is the same as the distance between nearest atoms, the maxima of its wave-function can lie in interstitial positions and its interaction energy will be low. The energy Δ is therefore that corresponding to a mean de Broglie wave-length equal to the mean distance of nearest neighbours. Atoms with this energy may move freely through the superfluid, subject only to collisions with one another; they have, therefore, just the properties required of those entering into the constitution of the normal component. It should be emphasised, however, that FEYNMAN associated the excitations with groups of atoms rather than individual atoms.

Mention should be made of one or two attempts to calculate the radial distribution function in liquid helium directly. LONDON[6] has made a calculation which takes account correctly of the Bose-Einstein statistics, but treats the interatomic forces as small. The author[7] has also made an approximate calculation in which the effect of the environment of a pair of molecules is represented

[1] H. FRÖHLICH: Physica, Haag **4**, 639 (1937).
[2] H. JONES: Proc. Cambridge Phil. Soc. **34**, 253 (1938).
[3] F. LONDON: Proc. Roy. Soc. Lond., Ser. A **171**, 484 (1939).
[4] L. LANDAU: J. Phys. USSR **5**, 71 (1941).
[5] R. P. FEYNMAN: Phys. Rev. **94**, 262 (1954).
[6] F. LONDON: J. Chem. Phys. **11**, 203 (1943).
[7] H. S. GREEN: Proc. Roy. Soc. Lond., Ser. A **194**, 244 (1948).

as a correction to their interaction energy, and in the boundary conditions. The results, which are strictly applicable only near absolute zero, indicate that the usual peak in the radial distribution function, corresponding to the first co-ordination shell, may be practically non-existent in liquid helium at very low temperatures. The experimental evidence, which relates to somewhat higher temperatures, will be discussed in the following section.

45. Experimental evidence. The direct experimental evidence on the structure of liquid helium has been obtained by X-ray and neutron diffraction techniques. Due mainly to recent work by REEKIE and his associates[1], fairly reliable and detailed information is now available from X-ray scattering data down to about 1.27° K. This is supplemented by results obtained from neutron diffraction, which, though more difficult to interpret, will be considered first.

The experiments, carried out by HURST and HENSHAW[2], were made with an almost monochromatic beam of neutrons and with excellent precision. The recoil of the scattering helium atoms was not negligible, but the energy loss of the neutrons was measured for certain scattering angles and proved to be the same as for free atoms, within the experimental error. The angular distribution of the scattered neutrons was observed at various temperatures between 1.65 and 5.04° K. The observations were analysed using the relation (19.13) in a form appropriate for scattering without loss of energy, and this seems to the author the only reason to question the accuracy of the results. The results are not, in fact, in detailed agreement with those obtained by X-ray scattering experiments. However, their broad features are certainly correct.

The function $g(r)$ inferred from these experiments shows no striking change in the range of temperature explored, although the density increased by 50% between 5.04 and 1.65° K. The distance of the first co-ordination peak remained near 3.5 Å over most of the temperature range, rather more than the generally accepted position of the minimum of $\phi(r)$ would lead one to expect. The height of the first maximum of $g(r)$ was low, (even assuming that it is not composite) corresponding to an occupation number of only 7 or 8 atoms. There was no evidence of any marked change of structure below the λ-point, but since the temperature derivative of $g(r)$ was not obtained, this would not necessarily be apparent even if it had occurred in the manner expected.

The most detailed discussion of the results obtained by the X-ray scattering technique appears in the paper by GOLDSTEIN and REEKIE[1]. The radial distribution functions computed by these authors for 2.06 and 4.20° K are shown in Fig. 8 below. These curves supersede those published earlier by BEAUMONT and REEKIE[1], though they are based on the same experimental data. It was found that a straight-forward analysis of the diffraction data failed to give a function $n_2(r)$ tending to zero for small values of r and satisfying the relation (23.26). Such discrepancies could, presumably be attributed to the absence of data for very small scattering angles, and the fact that the Fourier transform $n_2^F(k)$ of $n_2(r) - n^2$ cannot be determined accurately for large values of k (see Sect. 19). The curves shown are therefore modified to secure approximate

[1] J. REEKIE, T. S. HUTCHISON and C. F. A. BEAUMONT: Proc. Phys. Soc. Lond. A **66**, 409 (1953). — J. REEKIE and T. S. HUTCHISON: Phys. Rev. **92**, 827 (1953). — C. F. A. BEAUMONG and J. REEKIE: Proc. Roy. Soc. Lond., A **228**, 363 (1955). — L. GOLDSTEIN and J. REEKIE: Phys. Rev. **98**, 857 (1955).

[2] D. G. HENSHAW and D. G. HURST: Phys. Rev. **91**, 1222 (1953). — D. G. HENSHAW, D. G. HURST and N. K. POPE: Phys. Rev. **92**, 1229 (1953). — D. G. HURST and D. G. HENSHAW: Phys. Rev. **100**, 994 (1955).

fulfilment of the relation (23.26), using the known values of the isothermal compressibility, and are also made to vanish for small values of r. While it is perhaps premature to say that the curves thus obtained are more accurate than those inferred from the neutron diffraction experiments, it is difficult to believe that they are seriously in error.

Though the curves for 2.06 and 4.20° K bear a general resemblance to one another, there are very important differences in the shape and position of the first maximum. At the higher temperature there is apparently a first-co-ordination shell centred at about 3 Å, which almost merges with a second shell centred near 3.8 Å. At the lower temperature the coalescence of the two shells is effectively complete, and one would say either that they are moving together or that the first is disappearing, to the aggrandizement of the second. Evidence exists also for a similar but less obvious duplication of the outer shells at the higher temperature, which disappears at the lower temperatures. There is an obvious need for confirmation of this remarkable phenomenon, of which the first superficial analyses gave no indication. It is hard to resist the conclusion that it is connected with the discontinuities which occur at the λ-point.

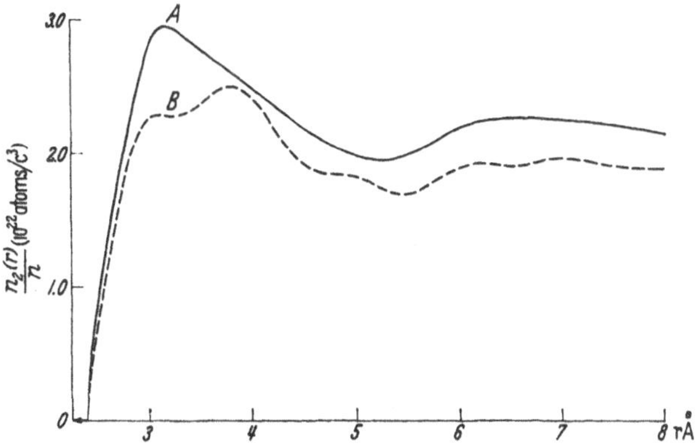

Fig. 8. Radial distribution functions in liquid helium. Curve A: at 2.06° K, 2.194×10^{22} atoms/cm³. Curve B: at 4.20° K; 1.892×10^{22} atoms/cm³.

Goldstein and Reekie have devoted much care to the accurate evaluation of $n_2(r)$ for large values of r, which is necessary to determine the integral $\int\{n_2(r)-n^2\}\,d^3r$ within reasonable limits. However, some quantities of physical interest, such as the mean potential energy per atom

$$\Phi=\int n_2(r)\,\phi(r)\,d^3r/n \tag{45.1}$$

depend most sensitively on the behaviour of $n_2(r)$ in the neighbourhood of the first co-ordination shell, especially where it first differs appreciably from zero. Goldstein and Reekie, using forms of $\phi(r)$ which are now obsolescent, found that the short-range repulsive interaction gave little or no contribution to Φ. In view of more recent determinations of $\phi(r)$, due to Hooton[1] and Mason and Rice[2], there is a need for this part of their work to be re-examined. Their principal conclusion, however, is likely to be unaffected: the abrupt change in the rate of increase of kinetic energy at the λ-point is partially compensated by an opposite change in the rate of increase of potential energy. It was already expected on theoretical grounds[3] that changes in the molecular kinetic energy are intimately related to changes in potential energy at very low temperatures, and opposite in sign. This appears to be confirmed by a study of the temperature dependence of Φ.

[1] D. J. Hooton: Phil. Mag. **46**, 701 (1955).
[2] E. A. Mason and W. E. Rice: J. Chem. Phys. **22**, 522 (1954).
[3] See F. London: J. Chem. Phys. **11**, 203 (1943); also p. 960 of H. S. Green: J. Chem. Phys. **19**, 955 (1951).

The scattering of X-radiation through small angles in liquid helium has been observed by Tweet[1]. It is evident from (19.9) and (19.10) that the intensity at small angles is an indirect measure of the isothermal compressibility. Tweet found that the isothermal compressibility determined in this way was in satisfactory agreement with that deduced from the experimentally determined velocity of sound and ratio of the specific heats. There was no obvious singularity of any kind at the λ-point in the curve obtained by either method. Actually one would expect a slight discontinuity in the temperature derivative of the isothermal compressibility at the λ-point, but as the isothermal compressibility depends only very weakly on the values of $n_2(r)$ in the first co-ordination shell, this would not be easy to detect experimentally at ordinary pressures. An earlier analysis of experimental data by W.H. and A.P. Keesom[2] showed a discontinuity in the isothermal compressibility at the λ-point which was quite large at high pressures, but insignificant at ordinary pressures; the accuracy of the analysis was not, however, sufficiently great to enable one to distinguish between a discontinuity of the isothermal compressibility and its temperature derivative.

The balance of the evidence, then, suggests that while there is no abrupt change of structure in passing below the λ-point, probably a change of structure abruptly commences there. At ordinary pressures this affects mainly the inner coordination shells, and might be due to the gradual coalescence of the inner shell with the second, the third with the fourth, and so on. This picture is in harmony with the trend of the molecular theories described in Sect. 44, especially those which suggest the gradual growth in the superfluid of a normal component physically separated from though interlacing the superfluid component, and attaining maturity at the λ-point.

b) Quantum theory of structure.

46. Quantum theory of liquids. One task of the quantum theory of liquid structure is to explain the anomalous properties of liquid helium and to determine quantum-mechanically the structure and structural properties of liquids at low temperatures. Another of equal importance is to justify the application of classical mechanics at higher temperatures and to ascertain at what point the quantum corrections first become important as the temperature is decreased.

The most convenient approach to the quantum mechanics of liquids is provided by the theory of the density matrix, introduced into quantum mechanics by Dirac[3] and elaborated by Husimi[4]. The application of this theory to liquids was worked out by Born and Green[5], and it is an adaption of their method which will be described here. An alternative and closely related approach is possible by way of the quantum theory of the phase space distribution function, which was initiated by Wigner[6] and developed in a general form by Moyal[7]. The practical application of this method to the molecular theory of fluids was pointed out by the author[8] and elaborated by Irving and Zwanzig[9].

---

[1] A. G. Tweet: Phys. Rev. **93**, 15 (1954).
[2] W. H. Keesom and A. P. Keesom: Physica, Haag **3**, 105 (1936).
[3] P. A. M. Dirac: The Principles of Quantum Mechanics. Oxford 1935.
[4] K. Husimi: Proc. Phys.-Math. Soc. Japan (3) **22**, 264 (1940).
[5] M. Born and H. S. Green: Proc. Roy. Soc. Lond., Ser. A **191**, 48 (1947).
[6] E. P. Wigner: Phys. Rev. **40**, 749 (1932).
[7] J. E. Moyal: Proc. Cambridge Phil. Soc. **45**, 99 (1949).
[8] H. S. Green: J. Chem. Phys. **19**, 955 (1951); **20**, 1274 (1952).
[9] J. H. Irving and R. W. Zwanzig: J. Chem. Phys. **19**, 1173 (1951).

A system of q similar particles is often described quantum-mechanically by means of a set of ortho-normal wave-functions Ψ_q^A satisfying the Schrödinger wave equation

$$i\hbar \frac{\partial \Psi_q^A}{\partial t} = H_q \Psi_q^A \tag{46.1}$$

where H_q is the Hamiltonian operator

$$H_q \equiv \left\{ -\frac{\hbar^2}{2m} \sum_{i=1}^{q} \frac{\partial^2}{\partial \boldsymbol{x}^{(i)2}} \right\} + \Phi_q(\boldsymbol{x}) \tag{46.2}$$

and Φ_q is the total potential energy, including the interaction energy of the particles with their average environment, expressed as a function of their positions $\boldsymbol{x}^{(1)}, \ldots, \boldsymbol{x}^{(q)}$. Each wave function Ψ_q^A describes the system in a pure state A; the condition of ortho-normality will be formulated so that, if B is any similar state,

$$\int \Psi_q^A(\boldsymbol{x}) \{\Psi_q^B(\boldsymbol{x})\}^* \, dx_q = \begin{cases} q! & \text{if } A \equiv B; \\ 0, & \text{otherwise.} \end{cases} \tag{46.3}$$

A liquid in a particular macroscopic condition may, under certain circumstances, be regarded as a member of an ensemble of systems in the same macroscopic condition, each in a pure state; if the state is one of q molecules, the appropriate density matrix is defined by

$$P_q(\boldsymbol{x}, \boldsymbol{x}') = \sum_A a_q^{AA} \Psi_q^A(\boldsymbol{x}) \{\Psi_q^A(\boldsymbol{x}')\}^*, \tag{46.4}$$

where a_q^{AA} is the probability that A is the pure state of such a system. The effect of the interaction of the liquid with its environment is generally to modify the density matrix, so that it has the form

$$P_q(\boldsymbol{x}, \boldsymbol{x}') = \sum_{A,B} a_q^{AB} \Psi_q^A(\boldsymbol{x}) \{\Psi_q^B(\boldsymbol{x}')\}^*; \tag{46.5}$$

but in any event it will be seen from (46.1) that $P_q(\boldsymbol{x}, \boldsymbol{x}')$ satisfies

$$i\hbar \frac{\partial P_q}{\partial t} = (H_q - H_q') P_q = [H_q, P_q] \tag{46.6}$$

where H_q is given by (46.2) and H_q' is a similar operator with the coordinates $\boldsymbol{x}^{(1)}, \ldots, \boldsymbol{x}^{(q)}$ changed to $\boldsymbol{x}^{(1)\prime}, \ldots, \boldsymbol{x}^{(q)\prime}$; the brackets represent the commutator of matrix notation.

The density matrix $P_q(\boldsymbol{x}, \boldsymbol{x}')$ is simply related to the quantum mechanical analogue of the velocity distribution function $F_q(\boldsymbol{x}, \boldsymbol{\xi})$ defined in Sect. 35; the relation is, in fact

$$F_q(\boldsymbol{x}, \boldsymbol{\xi}) = (m/h)^3 \int P_q(\boldsymbol{x} - \tfrac{1}{2}\boldsymbol{r}, \boldsymbol{x} + \tfrac{1}{2}\boldsymbol{r}) \exp(i m \boldsymbol{\xi} \cdot \boldsymbol{r}/\hbar) \, d^3r \tag{46.7}$$

where $h = 2\pi\hbar$ is Planck's constant. It is readily confirmed from this relation that

$$N_q(\boldsymbol{x}) = \int F_q(\boldsymbol{x}, \boldsymbol{\xi}) \, d\boldsymbol{\xi}_q = P_q(\boldsymbol{x}, \boldsymbol{x}) \tag{46.8}$$

is the distribution function relating to precisely q molecules in the finite region occupied by the liquid.

The density matrix for a group of q molecules in the liquid, when other molecules are also possibly present, will be denoted by ϱ_q. If

$$\chi^{(i)} = \iint d^3x^{(i)} \, d^3x^{(i)\prime} \, \delta(\boldsymbol{x}^{(i)} - \boldsymbol{x}^{(i)\prime}) \tag{46.9}$$

is an operator which replaces $x^{(i)\prime}$ by $x^{(i)}$, and then integrates with respect to $x^{(i)}$ (in the terminology of matrix theory, takes the trace, or spur),

$$\varrho_q = P_q + \chi^{(q+1)} P_{q+1} + \chi^{(q+1)} \chi^{(q+2)} P_{q+2}/2! + \cdots. \tag{46.10}$$

From (46.6) it is easily deduced that

$$i\hbar \frac{\partial \varrho_q}{\partial t} = [H_q, \varrho_q] + \sum_{i=1}^{q} \chi^{(q+1)} [\phi^{(iq+1)}, \varrho_{q+1}], \tag{46.11}$$

where

$$[\phi^{(iq+1)}, \varrho_{q+1}] = \{\phi(|x^{(i)} - x^{(q+1)}|) - \phi(|x^{(i)\prime} - x^{(q+1)\prime}|)\} \varrho_{q+1}(x, x').$$

The important property of the density matrix is that the mean value of any function $\varphi_q(\xi)$ of the velocities of q molecules, situated at the point $x^{(1)}, \ldots, x^{(q)}$ can be obtained by applying the operator $\varphi_q(\xi_{\mathrm{op}})$ to ϱ_q, where

$$\xi^{(i)}_{\mathrm{op}} = -\frac{i\hbar}{2m}\left(\frac{\partial}{\partial x^{(i)}} - \frac{\partial}{\partial x^{(i)\prime}}\right). \tag{46.12}$$

To be precise, the mean value $\langle \varphi_q \rangle$ of $\varphi_q(\xi)$ is given by

$$n_q \langle \varphi_q \rangle = \{\varphi_q(\xi_{\mathrm{op}})\, \varrho_q(x, x')\}_{x'=x}, \tag{46.13}$$

where

$$n_q = \varrho_q(x, x) = \{\varrho_q(x, x')\}_{x'=x}$$

and the suffix $x' = x$ means that each of the coordinates $x^{(i)\prime}$ is to be set equal to $x^{(i)}$ in the expression within the braces. Thus the mean velocity $u^{(i)}_q$ of a molecule within a group of q molecules with assigned positions is given by

$$n_q u^{(i)}_q = \{\xi^{(i)}_{\mathrm{op}}\, \varrho_q(x, x')\}_{x'=x};$$

so, on setting $x' = x$ in (46.11) one obtains after cancelling a factor $i\hbar$, the result

$$\frac{\partial n_q}{\partial t} + \sum_{i=1}^{p} \frac{\partial}{\partial x^{(i)}} \cdot (n_q u^{(i)}_q) = 0, \tag{46.14}$$

in the same form as in the classical theory.

In a similar way, by multiplying (46.11) by $\xi^{(j)}_{\mathrm{op}}$ before setting $x' = x$ one has

$$\left.\begin{aligned} &\frac{\partial}{\partial t}(n_q u^{(j)}_q) + \sum_{i=1}^{q} \frac{\partial}{\partial x^{(i)}} \cdot (n_q \langle \xi^{(i)} \xi^{(j)} \rangle) \\ &\quad = -\frac{n_q}{m} \frac{\partial \Phi_q}{\partial x^{(j)}} - \int \frac{n_{q+1}}{m} \frac{\partial \phi^{(jq+1)}}{\partial x^{(j)}}\, d^3x^{(q+1)}. \end{aligned}\right\} \tag{46.15}$$

Finally, by multiplying (46.11) by $(\xi^{(j)}_{\mathrm{op}} \xi^{(k)}_{\mathrm{op}})$ before setting $x' = x$, one has also

$$\left.\begin{aligned} &\frac{\partial}{\partial t}(n_q \langle \xi^{(j)} \xi^{(k)} \rangle) + \sum_{i=1}^{q} \frac{\partial}{\partial x^{(i)}} \cdot (n_q \langle \xi^{(i)} \xi^{(j)} \xi^{(k)} \rangle) \\ &\quad = -\frac{n_q u^{(j)}_q}{m} \frac{\partial \Phi_q}{\partial x^{(k)}} - \int \frac{n_{q+1} u^{(j)}_{q+1}}{m} \frac{\partial \phi^{(kq+1)}}{\partial x^{(k)}}\, d^3x^{(q+1)} - \\ &\qquad - \frac{\partial \Phi_q}{\partial x^{(j)}} \frac{n_q u^{(k)}_q}{m} - \int \frac{\partial \phi^{(jq+1)}}{\partial x^{(j)}} \frac{n_{q+1} u^{(k)}_{q+1}}{m}\, d^3x^{(q+1)}. \end{aligned}\right\} \tag{46.16}$$

These results do not involve PLANCK's constant explicitly, and are therefore the same in quantum and classical theory. As might have been anticipated, the equations of conservation of mass, momentum and energy and all their consequences are in fact unaltered. It does not of course follow that quantities like n_q, $\boldsymbol{u}_q^{(i)}$ and $\langle \xi_q^{(i)} \xi_q^{(j)} \rangle$ will have the same values in the quantum theory as in the classical theory, and indeed there are quantum corrections to the classical values which will be computed in Sects. 47 and 48 and which become increasingly important as the temperature is lowered.

47. Quantum statistical mechanics. In thermal and mechanical equilibrium, the probability of finding a system in a particular state can depend only on the constants of the motion of the system. If a liquid is prevented by containing walls from moving in any direction or rotating freely, these constants are reduced to the total energy and the number of molecules which it contains. The energies and the numbers of molecules of systems in weak interaction are additive, whilst their configurational probabilities are multiplicative. From these considerations it follows that, for a liquid in equilibrium, the coefficients a_q^{AA} appearing in (46.4) or (46.5) must be given by

$$a_q^{AA} = \text{const} \times (\vartheta z)^q \exp(-\beta E_q^A) \tag{47.1}$$

where E_q^A is the total energy of the state A, and z and β are in the first instance parameters which, however, can be identified with the thermodynamic activity and with $1/kT$. The factor ϑ^q is inserted for convenience, on the understanding that ϑ is a function of the temperature only. The constant multiplier in (47.1) is found to be $\exp(-\beta p V)$, where p is the thermodynamic pressure and V the volume occupied by the liquid. If $P_q(\boldsymbol{x}, \boldsymbol{x}')$ is known, the pressure and activity can be calculated by making use of the relations

$$\left.\begin{aligned}
\sum_q \int P_q(\boldsymbol{x}, \boldsymbol{x})\, d\boldsymbol{x}_q/q! &= \sum_q \sum_A a_q^{AA} \\
&= \exp(-\beta p V) \sum_q (\vartheta z)^q \sum_A \exp(-\beta E_q^A) = 1, \\
\sum_q q \int P_q(\boldsymbol{x}, \boldsymbol{x})\, d\boldsymbol{x}_q/q! &= \sum_q q \sum_A a_q^{AA} \\
&= \exp(-\beta p V) \sum_q q (\vartheta z)^q \sum_A \exp(-\beta E_q^A) = n V.
\end{aligned}\right\} \tag{47.2}$$

In principle, therefore, the thermodynamic problem is solved by determining $P_q(\boldsymbol{x}, \boldsymbol{x}')$, or alternatively the function

$$R_q(\boldsymbol{x}, \boldsymbol{x}') = \sum_A \exp(-\beta E_q^A)\, \Psi_q^A(\boldsymbol{x}) \{\Psi_q^A(\boldsymbol{x}')\}^*/q!. \tag{47.3}$$

The most convenient approach to the calculation of $R_q(\boldsymbol{x}, \boldsymbol{x}')$ is to make use of the equation

$$-\frac{\partial R_q}{\partial \beta} = H_q R_q \tag{47.4}$$

which it obviously satisfies, together with the "boundary condition" for $\beta = 0$. Only the boundary condition is affected by the type of statistics satisfied by the molecules. If it assumed that classical statistics are applicable, the summation over A embraces the complete ortho-normal set of solutions of (46.1), so that (47.3) reduces to

$$R_q(\boldsymbol{x}, \boldsymbol{x}') = \prod_{i=1}^{q} \delta(\boldsymbol{x}^{(i)} - \boldsymbol{x}^{(i)\prime}) \qquad \text{for } \beta = 0. \tag{47.5}$$

The functions corresponding to R_q for Bose-Einstein statistics or Fermi-Dirac statistics will henceforth be denoted by R_q^+ or R_q^- respectively. They are obtained by limiting the summation in (47.3) to states with wave functions which are unchanged or change sign respectively, when any two coordinates are interchanged. These functions can be obtained from R_q by using the relations

$$\left.\begin{aligned} R_q^+ &= \sum_P R_q(P\boldsymbol{x}, \boldsymbol{x}'), \\ R_q^- &= \sum_P (-1)^{S(P)} R_q(P\boldsymbol{x}, \boldsymbol{x}'), \end{aligned}\right\} \tag{47.6}$$

where $P\boldsymbol{x}$ represents a typical permutation of the coordinates $\boldsymbol{x}$, $\sum_P$ denotes summation over all $q!$ permutations, and $S(P)$ is even or odd according as P is an even or odd permutation. As the density matrices for quantum statistics are obtained so easily from the one for classical statistics, the remainder of this section will be devoted to the discussion of methods for calculating R_q.

The first method which will be considered is well adapted for calculating quantum-mechanical corrections at moderately low temperatures, where they first become important. One substitutes

$$\left.\begin{aligned} R_q &= e_q(\boldsymbol{x} - \boldsymbol{x}', \beta) \exp(-\beta V_q) K_q, \\ e_q(\boldsymbol{x} - \boldsymbol{x}', \beta) &= \left(\frac{m}{2\pi\beta\hbar^2}\right)^{3q/2} \exp\left\{\frac{-m\sum_{i=1}^{q}(\boldsymbol{x}^{(i)} - \boldsymbol{x}^{(i)\prime})^2}{2\beta\hbar^2}\right\}, \\ V_q &= \int_0^1 \Phi_q\{\boldsymbol{x}' + \lambda(\boldsymbol{x} - \boldsymbol{x}')\}\, d\lambda, \end{aligned}\right\} \tag{47.7}$$

whereupon (47.4) reduces to

$$\left.\begin{aligned} \frac{\partial K_q}{\partial\beta} + \frac{1}{\beta}\sum_{i=1}^{q}(\boldsymbol{x}^{(i)} - \boldsymbol{x}^{(i)\prime})\cdot\frac{\partial K_q}{\partial \boldsymbol{x}^{(i)}} &= L_q(\boldsymbol{x}, \beta). \\ L_q(\boldsymbol{x}, \beta) = \frac{\hbar^2}{2m}\sum_{i=1}^{q}\left(\frac{\partial^2 K_q}{\partial \boldsymbol{x}^{(i)2}} - 2\beta\frac{\partial V_q}{\partial \boldsymbol{x}^{(i)}}\cdot\frac{\partial K_q}{\partial \boldsymbol{x}^{(i)}}\right) &- W_q(\boldsymbol{x}, \beta) K_q, \\ W_q(\boldsymbol{x}, \beta) = \frac{\beta\hbar^2}{2m}\sum_{i=1}^{q}\left\{\beta\left(\frac{\partial V_q}{\partial \boldsymbol{x}^{(i)}}\right)^2 - \frac{\partial^2 V_q}{\partial \boldsymbol{x}^{(i)2}}\right\}&. \end{aligned}\right\} \tag{47.8}$$

The "boundary condition" (47.5) requires that $K_q = 1$ when $\beta = 0$. Integrating the first equation of (47.8), one has, therefore

$$\left.\begin{aligned} K_q &= 1 + \beta\int_0^1 L_q\{\boldsymbol{x}' + \lambda(\boldsymbol{x} - \boldsymbol{x}'), \lambda\beta\}\, d\lambda \\ &= 1 - \beta\int_0^1 W_q\{\boldsymbol{x}' + \lambda(\boldsymbol{x} - \boldsymbol{x}'), \lambda\beta\}\, d\lambda + 0(\hbar^4). \end{aligned}\right\} \tag{47.9}$$

When $\boldsymbol{x}' = \boldsymbol{x}$,

$$\frac{\partial V_q}{\partial \boldsymbol{x}^{(i)}} = \frac{1}{2}\frac{\partial \Phi_q}{\partial \boldsymbol{x}^{(i)}} \quad\text{and}\quad \frac{\partial^2 V_q}{\partial \boldsymbol{x}^{(i)2}} = \frac{1}{3}\frac{\partial^2 \Phi_q}{\partial \boldsymbol{x}^{(i)2}},$$

so that

$$K_q(\boldsymbol{x}, \boldsymbol{x}) = 1 + \frac{\beta\hbar^2}{12m}\sum_{i=1}^{q}\left\{\frac{\beta}{2}\left(\frac{\partial \Phi_q}{\partial \boldsymbol{x}^{(i)}}\right)^2 - \frac{\partial^2 \Phi_q}{\partial \boldsymbol{x}^{(i)2}}\right\} + 0(\hbar^4). \tag{47.10}$$

This result was first obtained by WIGNER[1]. The method has been used by the author[2] to compute various thermodynamical quantities in a form applicable

[1] E. P. WIGNER: Phys. Rev. 40, 749 (1932).
[2] H. S. GREEN: J. Chem. Phys. 19, 955 (1951).

to liquids. The following results may be mentioned:

$$\left.\begin{aligned} p &= p^c + \frac{\beta\hbar^2}{24m}\int\left(n\frac{\partial n_2^c}{\partial n} - n_2^c\right)\frac{\partial^2\phi}{\partial \boldsymbol{r}^2}\,d\boldsymbol{r} + O(\hbar^4),\\ \log z &= \log z^c + \frac{\beta\hbar^2}{24m}\int\frac{\partial n_2^c}{\partial n}\,\frac{\partial^2\phi}{\partial \boldsymbol{r}^2}\,d\boldsymbol{r} + O(\hbar^4),\\ \left\langle\frac{1}{2}m\,\xi^2\right\rangle &= \frac{3}{2\beta} + \frac{\beta\hbar^2}{24m}\int n_2^c\,\frac{\partial^2\phi}{\partial \boldsymbol{r}^2}\,d\boldsymbol{r} + O(\hbar^4), \end{aligned}\right\} \tag{47.11}$$

where the superfix c indicates the corresponding quantity calculated by classical mechanics. It will be seen from the last of these results that the thermodynamic temperature is not a precise measure of the mean kinetic energy of the molecules at low temperatures.

At very low temperatures where β is large, the expansion in powers of $\hbar^2$ obtained by the method just described fails to converge, and no entirely satisfactory alternative method is yet available. There are, however, several noteworthy methods of approximation which future developments may favour. With the help of (47.5), the Eq. (47.4) can be integrated to form an integral equation

$$R_q(\boldsymbol{x},\boldsymbol{x}';\beta) = e_q(\boldsymbol{x}-\boldsymbol{x}',\beta) - \iint\limits_0^\beta e_q(\boldsymbol{x}-\boldsymbol{x}_1,\beta-\beta_1)\,\Phi_q(\boldsymbol{x}_1)\,R_q(\boldsymbol{x}_1,\boldsymbol{x}';\beta_1)\,d\beta_1\,d\boldsymbol{x}_{q1} \tag{47.12}$$

which, in various forms, has provided the starting point of several investigations. The author[1] and GOLDBERGER and ADAMS[2] have examined "perturbation" developments of R_q in terms of the potential energy Φ_q by iteration of (47.12). FEYNMAN[3] has used an essentially similar approach, aided by some intuitive reasoning, in a first attempt to evaluate the partition function. The results obtained by such methods, though interesting and suggestive, cannot command much confidence in view of the rather drastic approximations necessary to overcome the twin difficulties of the strong intermolecular repulsion at short distances and the very low temperatures.

A somewhat different approach to the evaluation of R_q was adopted by BUTLER and FRIEDMAN[4], who observed that

$$\left.\begin{aligned} R_q(\boldsymbol{x},\boldsymbol{x}';\beta) = \int\overset{(n-1)}{\cdots}\int R_q(\boldsymbol{x},\boldsymbol{x}_1;\beta/n)\,R_q(\boldsymbol{x}_1,\boldsymbol{x}_2;\beta/n)\cdots\\ R_q(\boldsymbol{x}_{n-1},\boldsymbol{x}';\beta/n)\,d^3x_{q1}\cdots d^3x_{q\,n-1}, \end{aligned}\right\} \tag{47.13}$$

so that if R_q can be calculated for a moderately low temperature T, it can be derived for the much lower temperature T/n by performing integrations over $n-1$ sets of coordinates. The technical difficulties in performing the integrations are, however, very great.

FEYNMAN, in his second attempt to evaluate the partition function[5], has found a method, which, sufficiently refined, may prove the most convenient of those described here for the calculation of the thermodynamic functions. In essence it aims at the direct determination of the lowest energy levels of a system of molecules, which can then be used to calculate the sum $\sum\limits_A \exp(-\beta E_A)$.

The two pressures. The calculations so far reviewed provide no clear indication of a basis for the phenomenological theory of He II described in Sect. 43. To see how this matter should be approached, one may examine the pheno-

[1] See previous reference, also H. S. GREEN: J. Chem. Phys. **20**, 1274 (1952).
[2] M. L. GOLDBERGER and E. N. ADAMS: J. Chem. Phys. **20**, 240 (1952).
[3] R. P. FEYNMAN: Phys. Rev. **90**, 1116; **91**, 1291 (1953).
[4] S. T. BUTLER and M. H. FRIEDMAN: Phys. Rev. **98**, 287 (1955).
[5] R. P. FEYNMAN: Phys. Rev. **94**, 262 (1954).

menological equations (43.4), which, when added together, give

$$\frac{\partial}{\partial t}(\varrho\,\boldsymbol{u}) + \frac{\partial}{\partial \boldsymbol{x}}\cdot\left\{\varrho\,\boldsymbol{u}\boldsymbol{u} + \frac{\varrho_s\varrho_n}{\varrho}(\boldsymbol{u}_s-\boldsymbol{u}_n)(\boldsymbol{u}_s-\boldsymbol{u}_n) + \mathsf{p}\right\} = 0 \tag{47.14}$$

where $\varrho\boldsymbol{u} = \varrho_s\boldsymbol{u}_s + \varrho_n\boldsymbol{u}_n$, and $\mathsf{p} = \mathsf{p}_s + \mathsf{p}_n$ is the thermodynamic pressure tensor. One sees from this that the mean pressure which tends to move the liquid is not the thermodynamic pressure p, but

$$p_1 = p + \tfrac{1}{3}(\varrho_s\varrho_n/\varrho)(\boldsymbol{u}_s-\boldsymbol{u}_n)^2. \tag{47.15}$$

The difference $\pi = p_1 - p$ may be regarded as an auxiliary thermodynamic variable. One task of the molecular theory is then to account for a difference between the thermodynamic pressure, and what might be called the kinetic pressure p_1.

The author[1] has published a calculation of the thermodynamic pressure which led to the result

$$p = \frac{1}{3}m\langle\xi^2\rangle - \frac{1}{6}\int n_2\,\boldsymbol{r}\cdot\frac{\partial\phi}{\partial\boldsymbol{r}}\,d\boldsymbol{r} - \pi \tag{47.16}$$

where π appeared formally as a series in $\hbar^2$. This result was challenged by several authors[2] who furnished proofs that π must be zero, using, however, the implicit assumptions (not adopted by the author) that $\boldsymbol{u}=0$ at the liquid boundary, and that the liquid is in a pure state. It can be verified that these proofs lead to an opposite result, if the form (46.5) is adopted for the density matrix instead of (46.4). To establish the phenomenon of superfluidity, it seems necessary to assume that the density matrix, has non-diagonal terms connecting the state of lowest energy with excited states. These do not, of course, affect the value of the partition function.

BUTLER and BLATT[3] have recently given an interesting argument leading to the conclusion that the state of thermal and mechanical equilibrium of a liquid cannot be superfluid, based on the fact that the positions and motions of molecules separated by a large distance are uncorrelated in a liquid. If this is accepted, it would seem to follow that thermomechanical equilibrium is never attained in He II, except at absolute zero.

48. The influence of quantum statistics. The method of taking account of quantum statistics has already been indicated; in effect, it means using not the density matrix R_q which satisfied the condition (47.5), but one or other of the density matrices R_q^+ and R_q^- defined by (47.6). The statistical mechanics of liquids satisfying Bose or Fermi statistics depends on the evaluation of the integrals

$$I_q^{\pm} = \int R_q^{\pm}(\boldsymbol{x},\boldsymbol{x})\,dx_q = \Sigma_P \int (\pm 1)^{S(P)} R_q(P\boldsymbol{x},\boldsymbol{x})\,dx_q. \tag{48.1}$$

The first of Eqs. (47.2) can be written

$$\sum_{q=0}^{\infty} (\vartheta z)^q I_q^{\pm}/q! = \exp(\beta p V) \tag{48.2}$$

in terms of these integrals.

Now, a typical permutation of coordinates can be decomposed into a sequence of cyclic permutations, in which, for example, k_1 coordinates remain unchanged k_2 pairs of coordinates are interchanged, k_3 sets of 3 coordinates suffer cyclic permutation, etc. Such a permutation will be denoted by $P_{k_1 k_2 \ldots k_q}$, or more briefly by $P_{\boldsymbol{k}}$. Since the number of distinct permutations of this type which can be applied to a group of q coordinates is

$$\frac{q!}{k_1!\,k_2!\ldots k_q!}\;\frac{1}{2^{k_2}3^{k_3}\ldots q^{k_q}}$$

[1] H. S. GREEN: Proc. Roy. Soc. Lond., Ser. A **194**, 80 (1948).
[2] See P. J. PRICE: Phil. Mag. **41**, 948 (1950).
[3] S. T. BUTLER and J. M. BLATT: Phys. Rev. **100**, 495 (1955).

where of course $k_1+2k_2+\cdots+qk_q=q$, (48.1) can be written

$$I_q^{\pm}=\sum_k\int\frac{R_q(P_k\boldsymbol{x},\boldsymbol{x})\,d\,x_q}{k_1!\ldots k_q!\,(\pm 2)^{k_2}3^{k_3}(\pm 4)^{k_4}\ldots} \tag{48.3}$$

and (48.2) becomes

$$\sum_k\left(\frac{\vartheta z}{1}\right)^{k_1}\left(\frac{\vartheta z}{\pm 2}\right)^{k_2}\left(\frac{\vartheta z}{3}\right)^{k_3}\cdots\int\frac{R_{q(k)}(P_k\boldsymbol{x},\boldsymbol{x})}{k_1!\ldots k_{q(k)}!}\,d\boldsymbol{x}_{q(k)}. \tag{48.4}$$

Further progress depends on a knowledge of the structure of R_q. At sufficiently high temperatures, this is given by the conjunction of (47.7) and (47.10). Now if r_0 is the distance of nearest approach of two molecules, $\beta\ll m r_0^2/\hbar^2$ in the temperature range where terms in $O(\hbar^4)$ can be neglected. Also, if P_k is any but the identical permutation with $k_2=k_3=\cdots=0$, $e_{q(k)}(P_k\boldsymbol{x}-\boldsymbol{x},\beta)$ is very small in this range unless at least one pair of molecules are separated by a distance less than r_0, in which case $\exp\{-\beta V_q(P_k\boldsymbol{x},\boldsymbol{x})\}$ is extremely small. It follows that the effect of quantum statistics is perfectly negligible in the temperature range where the development of R_q in powers of $\hbar^2$ is quickly convergent. What happens at lower temperatures is unfortunately still a matter for speculation, though it is clear that the statistics become progressively more important, and probably play a dominant role at the temperatures of He II.

Feynman[1] has published some interesting speculations on the effect of the Bose statistics on the structure of He II. He supposes that there is a large number of molecular configurations with almost the same energy. These configurations can be transformed continuously into one another by the movement of a group of atoms which form, for example, a ring, together round the perimeter of the ring, until each molecule occupies the position of one of its neighbours. The wave function Ψ_q^A then returns to its original value, after changing sign in the intermediate configurations. From these considerations Feynman deduces an approximate relation between the ground state and the lowest excited states. This relation is probably not very accurately satisfied, but it does serve to illustrate the general effect of the statistics on the wave function; and, as remarked in Sect. 44, it is not essential to Feynman's conclusion that the de Broglie wavelength of the excited atoms must be of the same order as the intermolecular distances.

There is some independent evidence that quantum statistics play a very minor role until the temperatures of He II are reached. De Boer and Lunbeck[2] have remarked that the equation of state of all fluids whose molecular interaction energy can be represented by Lennard-Jones' approximation (21.26) can be expressed as relation between the reduced pressure $p^*=p a^3/\varepsilon$, the reduced temperature $T^*=kT/\varepsilon$, the reduced molecular density $n a^3$, and a parameter $\Lambda^*=\frac{h}{a\,(m\varepsilon)^{\frac{1}{2}}}$ depending on Planck's constant h—provided that the statistics are ignored. This relation they determined empirically and were thus able to predict the critical temperature (3.34° K), the critical pressure and the vapour pressure of fluid He^3, with remarkable accuracy. The success of their predictions may be regarded as evidence that the Fermi statistics of He^3 atoms have little effect on the macroscopic properties of the fluid down to 3° K or beyond. The differences in the behaviour of He^3 and He^4 below the λ-point, however, show that the statistics are of great importance near absolute zero.

---

[1] R. P. Feynman: Phys. Rev. **91**, 1301 (1953); **94**, 262 (1953).
[2] J. de Boer and R. J. Lunbeck: Physica, Haag **14**, 318, 510 (1948).

Molecular Theory of Surface Tension in Liquids.

By

Syu Ono and Sohei Kondo.

With 32 Figures.

1. Introduction. This article is devoted to the description of surface tension and physical adsorption in pure liquids and solutions from the molecular standpoint. It is based upon the methods of thermodynamics and statistical mechanics.

When a system in equilibrium is composed of two or more phases, the interface region between any two phases has a small but perceptible contribution to the mechanical and thermodynamic behavior of the system. Among the numerous phenomena associated with the interface the most important is that of surface tension. Since the days of Young[1] the interface between two fluids, say, a liquid and its vapor, has been considered from the mechanical standpoint as if it were a uniformly stretched membrane of zero thickness. Surface tension has been defined from a macroscopic standpoint as the tractional force, γ, acting across any unit length of line on this fictitious membrane.

When two fluids, in mutual mechanical equilibrium, are separated by a spherical interface of radius a, the pressure of the fluid inside, p^α, differs from that of the fluid outside, p^β. If the interface is assumed to be of zero thickness, the condition for mechanical equilibrium provides a simple relation between p^α and p^β:

$$p^\alpha - p^\beta = \frac{2\gamma}{a}, \tag{1.1}$$

which is well known as the Kelvin relation[2].

The above mentioned method for treating surface tension is simple, intuitive and useful but it is of an approximate nature from the molecular view point, for the structure of the fluid undergoes not a discontinuous but a progressive modification across the actual interface, although its thickness estimated from the ellipticity of reflected light is extremely small. Based upon such a concept, van der Waals and his school [*10*] proposed to express the surface tension as the integral, taken over this interface zone, of the differences between two pressures, normal and tangential to the interface, assuming the applicability of the laws of macroscopic hydrodynamics to this exceedingly small portion of nonhomogeneous matter.

In this article we treat the interface zone, following Gibbs [*1*], by relating its properties to a mathematical surface of separation called the dividing surface, to avoid the necessity of assigning some thickness to it. We can then strictly define surface tension with reference to the dividing surface. In addition, we can define any other property of the interface zone as a superficial quantity of the dividing surface with a definite value once a definite position of the dividing surface is chosen. When the quantities in the interface zone are so treated, their thermodynamic equations take forms analogous to those in bulk phases.

[1] T. Young: Phil. Trans. Roy. Soc. Lond. **95**, 65 (1805).

[2] W. Thomson (Lord Kelvin): Proc. Roy. Soc. Lond. **9**, 255 (1858). — Phil. Mag. **17**, 61 (1859).

We first describe the thermodynamic theory of surface tension and adsorption, by the method of the dividing surface of GIBBS. The use of a dividing surface or its equivalent is indispensable for the treatment of a curved interface, as otherwise the concepts of the area and curvature of the interface, cannot be precisely defined.

In the case of a plane interface, however, the concept of the dividing surface is not necessary and a valid alternative exposition has been proposed by GUGGENHEIM [3], [4] in treating the interface zone as a separate entity of some definite thickness bounded by two mathematical planes. We make, however, little mention of this method, since it seems to be of only minor importance in connection with the statistical treatment of an interface.

To avoid any ambiguity, the treatment of a spherical interface given in this article is based not on the original method of GIBBS but on the method modified by HILL [8] and KONDO [9]. This method, however, is not applicable to non-spherical interfaces, which will not be dealt with in this article.

Although all the relations for a plane interface can be deduced from the corresponding ones for a spherical interface by putting the curvature equal to zero, the planar and the spherical cases are considered separately because of the practical importance and easy physical visualization of a plane interface.

The thermodynamic method for defining surface tension based upon the second law is convenient but restricted only to systems in thermodynamic equilibrium. As the phenomenon of surface tension is not confined only to systems in equilibrium, the hydrostatic approach to the theory of surface tension, which is applicable even to systems out of equilibrium, will also be described.

A brief description is given of the irreversible thermodynamics of surface phenomena. Quasi-thermodynamic theory is developed as a transitional one from the macroscopic theory to the molecular.

On the other hand, several empirical formulae have been proposed for the surface phenomena. We deal with some of them in connection with the principle of corresponding states. The principle of corresponding states applied to surface tension of simple molecules is found to be very accurate. Its extreme usefulness becomes apparent when the principle, after being correctly modified to take into account quantum effects, is applied to surface tension of lighter substances. The statistical-mechanical development of the classical principle of corresponding states is also presented later.

It is statistical mechanics that provides the general description of macroscopic phenomena from the molecular point of view. The thermodynamic quantities of a system in thermodynamic equilibrium can, in principle, be given in terms of statistical thermodynamics which is based on the relation between the Helmholtz free energy and the partition function. We shall obtain the Helmholtz free energy as a function of the area of the interface, from which exact expressions are derived for relating surface tension and other superficial quantities to the intermolecular potentials and molecular distribution functions. We also elucidate the grand partition function method because it is especially convenient for the treatment of dilute solutions.

In the case of surface tension, the same expression can be derived by the method of statistical hydrodynamics on the basis of the direct relation between the pressure tensor and distribution functions. This method has the advantage of being applicable to surface tension in systems out of equilibrium since it is not based on the condition for equilibrium.

At the present, in statistical mechanics the pair distribution function in the interface region is unknown. We are obliged to calculate the surface tension on the basis of some approximations to the distribution functions. The theoretical results are compared with experimental data.

Another characteristic feature of the interface is physical adsorption. The amount adsorbed is defined as the superficial density of molecules with reference to the dividing surface. It is treated from the general statistical mechanical considerations.

The classical statistical expression for surface tension is generalized in terms of quantum statistics.

Another method for calculating the surface tension from the intermolecular potential is that based on some models of liquids or solutions. In this article the free volume model and the hole model are used to calculate surface tension and the results are compared with the experimental data for pure liquids, regular solutions and athermal polymer solutions. But these models are not applicable so the case of a spherical surface.

A. Thermodynamics and quasithermodynamics.

I. Thermodynamics.

a) Thermodynamic quantities of interface layer.

2. Dividing surface. Let us consider a system made up of two phases meeting at a thin transition zone, which may be plane or curved. Following Gibbs [*1*] we shall imagine a mathematical surface, to be called the *dividing surface*, constructed so as to give a precise separation of the two bulk phases. Throughout this article the two phases shall be referred to as α and β, respectively, and any quantity referred to either phase is indicated by superscript α or β and subscript i will be used to refer to molecular species i.

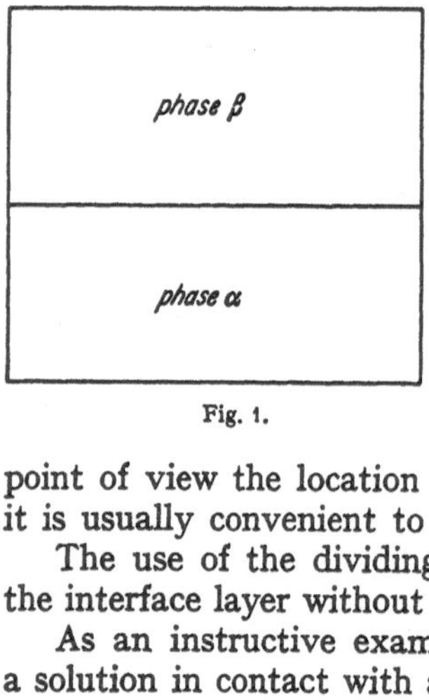

Fig. 1.

In general the dividing surface should be chosen to be normal to the density gradient in the transition zone and hence, in the case of a plane interface, it becomes plane. From the purely mathematical point of view the location of the dividing surface is rather arbitrary although it is usually convenient to choose it within or very near to the transition zone.

The use of the dividing surface has the advantage of enabling us to treat the interface layer without reference to its thickness.

As an instructive example, we shall consider a multicomponent system of a solution in contact with air, in which certain components have a tendency to become more abundant in the interface zone than in either phase. At first sight it may seem feasible to define the concentration or amount adsorbed of any component in the interface just as in an ordinary homogeneous bulk phase, but actually the amount adsorbed depends on the choice of the two boundaries which are necessary to distinguish the interface zone as a separate entity from the adjacent two bulk phases [*3*], [*4*]. To reduce the arbitrariness in the thickness between the two boundaries, we may adopt, instead of two, only one mathematical surface, that is, the dividing surface. Its usefulness will be apparent in this article.

Having made some particular choice for the dividing surface, we can divide the whole volume V occupied by the system precisely into two volumes V^α and V^β. If the dividing surface is chosen to lie in the interface zone as usual,

V^α contains the bulk phase α together with a small amount of matter in the transition zone and V^β contains the bulk phase β together with the rest of that in the transition zone.

Following GIBBS [1] we shall imagine a hypothetical system composed of two bulk phases α and β which are both strictly homogeneous right up to the above dividing surface and hence have volumes V^α and V^β, respectively. Let N_i^α and N_i^β be the numbers of molecules of species i in the above hypothetical bulk phases α and β, respectively. Namely, N_i^α is equal to the number of molecules of species i per unit volume in the phase α, n_i^α, multiplied by V^α and N_i^β is equal to the corresponding $n_i^\beta V^\beta$. It should now be noticed that in general the total number N_i of molecules of species i contained in the actual system is not necessarily equal to that of the above-mentioned hypothetical system, $N_i^\alpha + N_i^\beta$, but can be expressed in the form

$$N_i = N_i^\alpha + N_i^\beta + N_i^s, \tag{2.1}$$

where N_i^s may be considered as the contribution due to the presence of the interface in the actual system.

In the case of a single component system we may choose the dividing surface so as to make this contribution of the interface vanish. This particular dividing surface plays an important role in the theory of surface tension and is called the *equimolecular dividing surface.*

It should be noted and emphasized that N_i^s is not always to be called the number of molecules of species i adsorbed at the interface, because the value of N_i^s is not independent of the location of the artificial dividing surface. The number N_i^s or any other similar quantity will be called a superficial quantity and indicated by superscript s throughout this article.

3. Contributions of interface layer to thermodynamic quantities. In much the same way as in the case of the number of molecules, any extensive thermodynamic property of the entire system may be considered, in reference to an arbitrary dividing surface, as the sum of three contributions: that due to the phase α, that due to the phase β and that due to the interface. For example the Helmholtz free energy of the entire system may be written as

$$F = F^\alpha + F^\beta + F^s. \tag{3.1}$$

Here F^α is the free energy of the bulk phase α which remains strictly homogeneous right up to the dividing surface and F^β is defined similarly for the bulk phase β. Thus $F^\alpha + F^\beta$ is the free energy of the hypothetical system in which intensive physical properties discontinuously change from those of the bulk phase α to those of bulk phase β at the dividing surface, and F^s is the contribution of the interface layer to the actual free energy. Since the total free energy of the two-phase system with an interface is a well-defined quantity, (3.1) may be regarded as defining the superficial free energy F^s.

Similarly, the superficial internal energy U^s and the superficial entropy S^s is defined respectively by

$$U = U^\alpha + U^\beta + U^s, \tag{3.2}$$

$$S = S^\alpha + S^\beta + S^s. \tag{3.3}$$

Making use of the relation $F = U - TS$ and the above definitions, we obtain

$$F^s = U^s - TS^s, \tag{3.4}$$

the absolute temperature T being assumed to be constant throughout the system.

b) Plane interface.

4. Definition of surface tension. Let a two-phase system having a plane interface of area A be contained in a rectangular parallelepiped vessel one of whose edges is normal to the interface (see Fig. 1).

Let us now imagine that the area A of the interface is increased by the amount dA through a reversible and isothermal displacement of the side walls of the vessel. If the system is in hydrostatic equilibrium with pressure p, the work done by the system in the above process may be divided into two parts: the work, pdV, to cause the increase dV in the volume and the excess work to increase the area of the interface by dA, which we shall denote by $-\gamma dA$. Then the work done by the system in this process is $pdV - \gamma dA$. Let us now push the top and bottom walls of the vessel in such a manner as to return the system to its original volume. The work done by the system in this process is $-pdV$. At the end of these processes the system will be found to be in the same volume, pressure, composition and temperature as it was initially. The only change is that the area of the interface has been increased by dA. This implies that the work done on the system to cause the increase dA in the area of the interface at constant pressure, volume and temperature is γdA. Hence the quantity γ is the work done on the system to increase unit area of the interface and is called the *surface tension*. This surface tension is equivalent to that briefly discussed in the introduction as the force acting across unit length of an arbitrary line at the interface.

Thus the elementary work done by the system, dW, when its volume and area have changed by dV and dA, respectively, is to be expressed as

$$dW = p\,dV - \gamma\,dA. \tag{4.1}$$

This may be regarded as defining the surface tension γ. It may be seen from (4.1) that in the case of a plane interface the surface tension is independent of the location of the dividing surface for the change of the latter has nothing to do with the area of interface in this case.

If the value of the surface tension γ defined in the above manner were negative, the work done by the system to increase the area of the interface would be positive and consequently the interface would increase its area spontaneously. This implies that the plane interface is stable only if γ is positive. Then the value of the surface tension defined here cannot be negative, just as the pressure in a fluid cannot be negative in the customary definition.

The internal energy, U, of the two-phase system also changes with increasing area of the interface. If the amount of heat received by the system during an infinitesimal increase dA is denoted by dQ, the first law of thermodynamics is written in the form

$$dQ = dU + p\,dV - \gamma\,dA. \tag{4.2}$$

It should be noted that (4.2) is valid for a non-equilibrium system as well so long as the system is in mechanical equilibrium with constant pressure p.

5. Surface free energy. If the system is in thermodynamic equilibrium, the heat dQ absorbed in the course of a reversible transformation becomes, according to the second law, equal to the increase in the entropy of the system dS multiplied by the temperature T. Combining with the first law, (4.2), we obtain

$$dU = T\,dS - p\,dV + \gamma\,dA. \tag{5.1}$$

This important relation can be written in a more convenient form with use of the Helmholtz free energy

$$dF = -p\,dV - S\,dT + \gamma\,dA. \tag{5.2}$$

In an open system, which can exchange materials as well as energy with the surroundings, the numbers of molecules may change and hence the total differential of the Helmholtz free energy F given by (5.2) should have the additional terms due to such changes, as follows:

$$dF = -p\,dV - S\,dT + \gamma\,dA + \sum_{i=1}^{\varkappa} \mu_i\,dN_i, \tag{5.3}$$

where $\varkappa$ is the number of species; μ_i is the chemical potential per molecule of the species i and is to be constant throughout the system if the system is in chemical equilibrium. The Gibbs fundamental equation (5.3) for the two-phase system may be regarded as an equation defining the surface tension γ in an open system.

If the free energy F is known as a function of the volume, the temperature, the area of the interface and the numbers of molecules, γ may readily be evaluated through the relation

$$\gamma = \left(\frac{\partial F}{\partial A}\right)_{T,V,\boldsymbol{N}}, \tag{5.4}$$

where the bold-face notation $\boldsymbol{N}$ represents a set of the numbers $N_1, N_2, \ldots, N_\varkappa$.

Let us now increase the volume of the two-phase system illustrated in Fig. 1, in which for the present we restrict ourselves to a thermodynamic equilibrium case, keeping its temperature, pressure, composition and the height of the vessel unaltered. Then the quantities, F, V, $\boldsymbol{N}$, A, change in the same proportion. Thus F is a homogeneous function of first degree in V, $\boldsymbol{N}$ and A, and, therefore, by EULER's theorem we can obtain from (5.3)

$$F = \sum_{i=1}^{\varkappa} \mu_i N_i - p\,V + \gamma\,A. \tag{5.5}$$

For the bulk phases α and β we have

$$F^\alpha = \sum_{i=1}^{\varkappa} \mu_i N_i^\alpha - p\,V^\alpha, \tag{5.6}$$

$$F^\beta = \sum_{i=1}^{\varkappa} \mu_i N_i^\beta - p\,V^\beta. \tag{5.7}$$

Subtracting (5.6) and (5.7) from (5.5) and making use of (2.1), (3.1) and the additivity of the volumes, we obtain

$$F^s = \sum_{i=1}^{\varkappa} \mu_i N_i^s + \gamma\,A. \tag{5.8}$$

In contrast with the surface tension γ, the superficial quantities such as F^s, U^s and N_i^s depend on the location of the dividing surface. If we choose the dividing surface so as to make the sum $\sum_{i=1}^{\varkappa} \mu_i N_i^s$ vanish, (5.8) reduces to

$$F^s = \gamma\,A. \tag{5.9}$$

This particular dividing surface becomes identical with the equimolecular dividing surface defined in Sect. 2 in the case of a single component system. It is seen from (5.8) that the surface tension becomes equal to the superficial density of the Helmholtz free energy only for this special choice of the dividing surface[1]. It should be noted that the surface tension is neither the internal energy nor the potential energy per unit area of the surface.

[1] R.H. FOWLER: Proc. Roy. Soc. Lond., Ser. A **159**, 229 (1937). — Physica, Haag **5**, 39 (1938).

To treat an open system it is convenient to introduce the grand potential defined by[1,2] [26],

$$\Omega = F - \sum_{i=1}^{\varkappa} \mu_i N_i. \tag{5.10}$$

Then we obtain from (5.5)

$$\Omega = \gamma A - p V, \tag{5.11}$$

and from (5.3)

$$d\Omega = - S\, dT - p\, dV + \gamma\, dA - \sum_{i=1}^{\varkappa} N_i\, d\mu_i. \tag{5.12}$$

Using the bold-face notation $\boldsymbol{\mu} = \mu_1, \mu_2, \ldots, \mu_\varkappa$, from (5.12) we immediately obtain[3]

$$S = - \left(\frac{\partial \Omega}{\partial T}\right)_{V, \boldsymbol{\mu}, A}, \tag{5.13}$$

$$p = - \left(\frac{\partial \Omega}{\partial V}\right)_{T, \boldsymbol{\mu}, A}, \tag{5.14}$$

$$N_i = - \left(\frac{\partial \Omega}{\partial \mu_i}\right)_{V, T, A, \mu_j}. \tag{5.15}$$

Then from (5.10) and (5.13) we have the relation

$$\sum_{i=1}^{\varkappa} \mu_i N_i - U = T^2 \frac{\partial}{\partial T}\left(\frac{\Omega}{T}\right). \tag{5.16}$$

6. Thermodynamic relations for plane interface. The Gibbs fundamental equations for the bulk phases, α and β, are, respectively

$$dF^\alpha = - p\, dV^\alpha - S^\alpha\, dT + \sum_{i=1}^{\varkappa} \mu_i\, dN_i^\alpha, \tag{6.1}$$

$$dF^\beta = - p\, dV^\beta - S^\beta\, dT + \sum_{i=1}^{\varkappa} \mu_i\, dN_i^\beta. \tag{6.2}$$

Subtracting (6.1) and (6.2) from (5.3), and making use of (2.1), (3.1) and (3.3), we obtain

$$dF^s = - S^s\, dT + \gamma\, dA + \sum_{i=1}^{\varkappa} \mu_i\, dN_i^s, \tag{6.3}$$

which may be regarded as the Gibbs fundamental equation for the interface layer.

It is thus seen that superficial thermodynamic quantities are connected with each other by a relation similar to that for an ordinary bulk phase, and hence they may be treated as if they belong to a third phase.

Differentiating (5.8) and using (6.3), we obtain the Gibbs adsorption equation in its general form for a plane interface layer

$$A\, d\gamma + S^s\, dT + \sum_{i=1}^{\varkappa} N_i^s\, d\mu_i = 0. \tag{6.4}$$

Since N_i^s, S^s, and U^s are all proportional to the area, A, of the interface, it is convenient to introduce the following notation to express the superficial

[1] H.A. Kramers: Proc. Kon. Nederl. Akad. Wetensch., Amsterd. **41**, 10 (1938).

[2] S. Ono: Busseiron Kenkyu No. 23, 10 (1950) [in Japanese].

[3] Throughout this article we use the subscript μ_j to indicate that all of the variables $\mu_1, \ldots, \mu_\varkappa$ except μ_i are kept constant.

densities of these quantities:

$$\Gamma_i = \frac{N_i^s}{A}, \tag{6.5}$$

$$\sigma = \frac{S^s}{A}, \tag{6.6}$$

$$v = \frac{U^s}{A}. \tag{6.7}$$

The superficial density of the entropy, σ, and that of the internal energy, v, are often called the *surface entropy* and the *surface energy*, respectively. Then Γ_i will be called the superficial number density of molecules of species i. Then (6.4) is written as

$$d\gamma + \sigma\, dT + \sum_{i=1}^{\varkappa} \Gamma_i\, d\mu_i = 0, \tag{6.8}$$

and (5.8) as

$$\gamma = v - \sigma T - \sum_{i=1}^{\varkappa} \Gamma_i \mu_i, \tag{6.9}$$

where (3.4) has been used.

For simplicity, we shall now restrict ourselves to the case where every species is present appreciably in both phases. A more general case will be discussed later in Sect. 11. Let us define the mole fraction of species i by

$$x_i = N_i \Big/ \sum_{i=1}^{\varkappa} N_i. \tag{6.10}$$

Then we can take T and $x_2^\alpha, \ldots, x_\varkappa^\alpha$, i.e., the mole fractions in phase α, as the complete set of $\varkappa$ independent variables, since according to Gibbs' phase rule a system consisting of two phases and composed of $\varkappa$ components has $\varkappa$ degrees of freedom. Thus we may write

$$\left(\frac{\partial \mu_i}{\partial T}\right)_{x^\alpha} = - s_i^\alpha + v_i^\alpha \left(\frac{\partial p}{\partial T}\right)_{x^\alpha}, \tag{6.11}$$

where s_i^α and v_i^α are respectively the partial entropy and volume per molecule of species i in phase α, defined by[1]

$$s_i^\alpha = \left(\frac{\partial S^\alpha}{\partial N_i^\alpha}\right)_{T,p,N_j^\alpha}; \qquad v_i^\alpha = \left(\frac{\partial V^\alpha}{\partial N_i^\alpha}\right)_{T,p,N_j^\alpha}. \tag{6.12}$$

Using (6.8) and (6.11) we obtain

$$\left(\frac{\partial \gamma}{\partial T}\right)_{x^\alpha} = -\sigma + \sum_{i=1}^{\varkappa} \Gamma_i \left(s_i^\alpha - v_i^\alpha \left(\frac{\partial p}{\partial T}\right)_{x^\alpha}\right). \tag{6.13}$$

From (6.9) and (6.13) we have

$$v = \gamma - T\left(\frac{\partial \gamma}{\partial T}\right)_{x^\alpha} + \sum_{i=1}^{\varkappa} \Gamma_i \left(\mu_i + T s_i^\alpha - T v_i^\alpha \left(\frac{\partial p}{\partial T}\right)_{x^\alpha}\right). \tag{6.14}$$

We may choose the dividing surface in such a manner that the sum of $v_i^\alpha \Gamma_i$ vanishes, denoting the superficial number density of species i referred to this particular dividing surface by $\Gamma_i^{(v)}$:

$$\sum_{i=1}^{\varkappa} v_i^\alpha \Gamma_i^{(v)} = 0. \tag{6.15}$$

[1] See footnote 3, p. 140.

For this special choice of the dividing surface (6.14) reduces to

$$v = \gamma - T\left(\frac{\partial \gamma}{\partial T}\right)_{x^\alpha} + \sum_{i=1}^{\varkappa} h_i^\alpha \Gamma_i^{(v)}, \tag{6.16}$$

where h_i^α is the partial enthalpy per molecule of species i in the phase α.

It is of special interest to consider the case of pure liquids. If the equimolecular dividing surface is chosen, (6.13) reduces to

$$\frac{d\gamma}{dT} = -\sigma, \tag{6.17}$$

and (6.14) to the Gibbs-Helmholtz equation for a plane interface

$$v = \gamma - T\frac{d\gamma}{dT}, \tag{6.18}$$

or in an alternative form:

$$v = -T^2 \frac{d}{dT}\left(\frac{\gamma}{T}\right). \tag{6.19}$$

It should be emphasized that although Γ_i, v and σ depend on the choice of the dividing surface, the general forms of the thermodynamic equations thus far obtained, excluding special ones such as (5.9), (6.16), are invariant with respect to the shift of the dividing surface, as will be shown in the next section.

7. Adsorption at plane interface. Let us suppose that the phase α is a slightly volatile solution having the vapor phase β. We shall choose the dividing surface so as to make the superficial number density of the solvent vanish. The number densities of the solutes in the interface may in general be different from those in the interior of either bulk phase.

If some of the solutes are only slightly soluble, their concentrations near the surface will be much greater than those in the interior of either bulk phase, in other words, they will be adsorbed at the interface. Then it is obvious that Γ_i for any such solute is a positive and appreciable quantity irrespective of the location of the dividing surface and hence may in general be considered as a measure of adsorption of molecules of species i. But it should be noted that for a highly volatile or soluble substance, Γ_i depends sensitively on the choice of the artificial dividing surface and that it sometimes even changes its sign.

Let us now discuss adsorption at a plane interface between two phases in thermodynamic equilibrium, on the basis of the relation (6.8).

At constant temperature, (6.8) reduces to

$$-d\gamma = \sum_{i=1}^{\varkappa} \Gamma_i \, d\mu_i. \tag{7.1}$$

This equation relates the surface tension to the adsorption and plays a fundamental role in the thermodynamics of surface layers.

Since Γ_i depends upon the choice of the dividing surface, it is possible and convenient to make a special choice for the dividing surface in such a manner as to make the superficial number density of a chosen species, say 1, which shall be called the reference species, vanish. We shall denote the superficial number density of species i defined in this way by $\Gamma_i^{(1)}$, indicating that species 1 is the reference species. Then we have instead of (7.1)

$$-d\gamma = \sum_{i=2}^{\varkappa} \Gamma_i^{(1)} \, d\mu_i, \tag{7.2}$$

which is the relation well-known as the *Gibbs adsorption formula* [*1*].

Provided the density of the vapor phase is so low that the vapor of each species obeys the perfect gas law, the chemical potential may be given by

$$\mu_i = \varphi_i + kT \log p_i, \tag{7.3}$$

where p_i is the partial vapor pressure of species i and φ_i is a function of temperature alone. Then we may write as an alternative to (7.2) the following:

$$-d\gamma = kT \sum_{i=2}^{\varkappa} \Gamma_i^{(1)} d \log p_i. \tag{7.4}$$

This equation implies that $\Gamma_i^{(1)}$ is to be positive or negative according as the observed value of γ decreases or increases when we increase the concentration of species i in the interior of the liquid phase keeping the other species at their given concentrations.

We shall now show that the right-hand side of (7.1) remains unaffected by the choice of the dividing surface [2]. Let n_i^α and n_i^β be the number densities of species i in the interiors of the bulk phases α and β, respectively. Let us consider two arbitrary dividing surfaces, A and A', and distinguish the superficial number density Γ_i referred to A from Γ_i' referred to A'. Then it is obvious from the definition of the dividing surface that

$$\Gamma_i' = \Gamma_i - (n_i^\alpha - n_i^\beta)\, \Delta', \tag{7.5}$$

where the distance Δ' between A and A' is positive for A' located nearer to phase β than A.

On the other hand the familiar Gibbs-Duhem equations at constant temperature for the bulk phases α and β, which can readily be derived from (5.6) and (5.7) combined respectively with (6.1) and (6.2), are as follows

$$-dp + \sum_{i=1}^{\varkappa} n_i^\alpha\, d\mu_i = 0, \tag{7.6}$$

$$-dp + \sum_{i=1}^{\varkappa} n_i^\beta\, d\mu_i = 0. \tag{7.7}$$

Since dp and $d\mu_i$ in (7.6) should be equal to those in (7.7), respectively, when the phases α and β remain in equilibrium, we have

$$\sum_{i=1}^{\varkappa} (n_i^\alpha - n_i^\beta)\, d\mu_i = 0. \tag{7.8}$$

With use of (7.5) and (7.8), we may rewrite (7.1) as

$$-d\gamma = \sum_{i=1}^{\varkappa} \Gamma_i'\, d\mu_i. \tag{7.9}$$

Thus the relation (7.1) has been proved to remain invariant.

In much the same way, we can prove the invariance of other relations, (6.3), (6.4), (6.9), (6.14) and the like, with respect to displacement of the dividing surface.

Eliminating $d\mu_1$ from (7.1) with use of (7.8), we have for an arbitrary dividing surface

$$-d\gamma = \sum_{i=2}^{\varkappa} \left(\Gamma_i - \Gamma_1 \frac{n_i^\alpha - n_i^\beta}{n_1^\alpha - n_1^\beta}\right) d\mu_i, \tag{7.10}$$

which is also invariant with respect to displacement of the dividing surface.

Let us introduce

$$\Gamma_{i1} = \Gamma_i - \Gamma_1 \frac{n_i^\alpha - n_i^\beta}{n_1^\alpha - n_1^\beta}. \tag{7.11}$$

Then it is evident from (7.5) that Γ_{i1} is unaffected by displacement of the dividing surface. This invariant quantity is called the *relative adsorption* [13]. Gibbs denoted this quantity by $\Gamma_i^{(1)}$.

It is possible to treat the interface layer as a separate entity of non-zero thickness bounded by two such planes as $K'H'$ and $K''H''$ in Fig. 2, parallel to the dividing surface KH, following Guggenheim [3], [4]. If the remaining portions of the system are completely homogeneous so that the spatial variation of the thermodynamic quantities is restricted to the surface phase between $K'H'$ and $K''H''$ of thickness Δ, the number of molecules of species i per unit area in this surface phase is given by

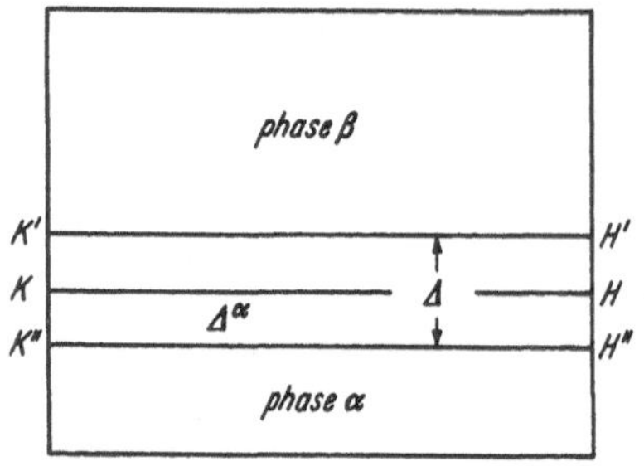

Fig. 2.

$$\tilde{\Gamma}_i = \Gamma_i + \Delta^\alpha (n_i^\alpha - n_i^\beta) + \Delta n_i^\beta, \tag{7.12}$$

where Δ^α is the distance between KH and $K''H''$. Then we have

$$\left.\begin{aligned} \tilde{\Gamma}_{i1} &= \tilde{\Gamma}_i - \tilde{\Gamma}_1 \frac{n_i^\alpha - n_i^\beta}{n_1^\alpha - n_1^\beta} \\ &= \Gamma_i - \Gamma_1 \frac{n_i^\alpha - n_i^\beta}{n_1^\alpha - n_1^\beta} + \Delta \frac{n_1^\alpha n_i^\beta - n_1^\beta n_i^\alpha}{n_1^\alpha - n_1^\beta}. \end{aligned}\right\} \tag{7.13}$$

If every species has such low vapor density that we may neglect n_i^β comparing with n_i^α, (7.13) reduces to

$$\tilde{\Gamma}_{i1} \approx \tilde{\Gamma}_i - \tilde{\Gamma}_1 \frac{n_i^\alpha}{n_1^\alpha} \approx \Gamma_{i1}. \tag{7.14}$$

It is now obvious that, although $\tilde{\Gamma}_1$ and $\tilde{\Gamma}_i$ are not precisely determined unless the exact locations of the two boundary surfaces are given, practically $\tilde{\Gamma}_{i1}$ is almost invariant with respect to those locations and equal to the relative adsorption defined by (7.11), insofar as the above approximation is valid. If the vapors obey the perfect gas law and consequently the approximate relations (7.3) and (7.14) are valid, (7.10) reduces to,

$$-d\gamma = kT \sum_{i=2}^{\varkappa} \left(\tilde{\Gamma}_i - \tilde{\Gamma}_1 \frac{n_i^\alpha}{n_1^\alpha}\right) d \log p_i. \tag{7.15}$$

For a two-component system, (7.14) and (7.15) are written as

$$\Gamma_{21} \approx \tilde{\Gamma}_{21} \approx \tilde{\Gamma}_2 - \tilde{\Gamma}_1 \frac{x}{1-x}, \tag{7.16}$$

$$-d\gamma = kT \left(\tilde{\Gamma}_2 - \tilde{\Gamma}_1 \frac{x}{1-x}\right) d \log p_2, \tag{7.17}$$

where $x = n_2^\alpha/(n_1^\alpha + n_2^\alpha)$.

Although the intuitive quantity $\tilde{\Gamma}_i$ is more convenient than the abstract quantity Γ_i referred to the imaginary dividing surface, this method seems to be inadequate when applied to the case of a curved interface.

For some purposes a dividing surface which makes the total sum of the superficial number densities vanish may be useful. We shall denote any superficial number density referred to this dividing surface by $\Gamma_i^{(N)}$:

$$\sum_{i=1}^{\varkappa} \Gamma_i^{(N)} = 0. \tag{7.18}$$

For a system composed of two components, (7.18) reduces to

$$\Gamma_1^{(N)} = - \Gamma_2^{(N)}. \tag{7.19}$$

We have already utilized the convention $\Gamma_i^{(v)}$ defined by (6.15). This is another example of the choice of the dividing surface.

c) Spherical interface.

8. Definition of surface tension. In this Part A I c we shall derive the thermodynamic relations for a spherical interface. One of the most familiar examples of a spherical interface is a drop of liquid with its surrounding vapor.

Let us consider a closed system consisting of two phases separated by a spherical interface layer, i.e., a transition zone with constant curvature. For convenience we shall limit our treatment to a portion of the system contained in a conical vessel such as that shown in Fig. 3, where ω and r denote, respectively, the solid angle of the cone and the radial coordinate with the origin O at the apex of the cone. The vessel consists of that part of the cone between $r = R^\alpha$ and $r = R^\beta$ where $R^\alpha < R^\beta$. We assume that the system is composed of $\varkappa$ independent components under no external fields and that the center of curvature of the spherical interface layer lies at O. Then any intensive property of the system can depend only on r. We shall refer to the so-called interior and exterior phases as α and β, respectively.

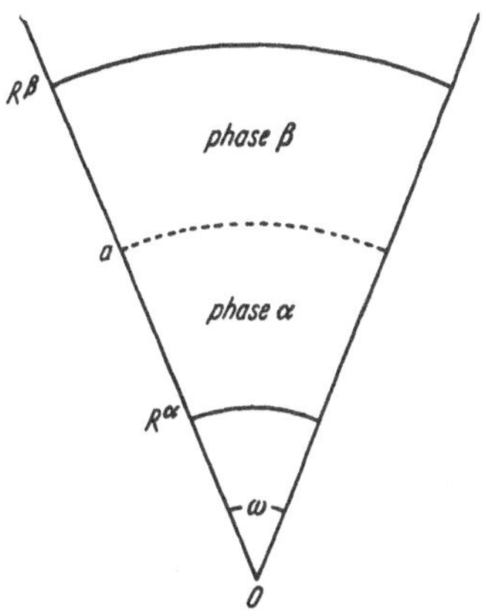

Fig. 3.

Let us choose the sphere $r = a$ as the dividing surface. Then it divides the total volume V into two volumes V^α and V^β given respectively by

$$V^\alpha = \tfrac{1}{3}\,\omega\,\{a^3 - (R^\alpha)^3\}, \tag{8.1}$$

$$V^\beta = \tfrac{1}{3}\,\omega\,\{(R^\beta)^3 - a^3\}. \tag{8.2}$$

The area of the dividing surface is

$$A = \omega\, a^2. \tag{8.3}$$

In contrast with the case of a plane interface, the pressure in the interior of phase α, p^α, cannot be equal to the pressure in the interior of the phase β, p^β, as long as the system is in mechanical equilibrium. In the present section the temperature need not be uniform. The only restriction is that p^α and p^β are uniform in the interiors of the bulk phases α and β, repsectively. It should, however, be noted that the following method cannot be applied to the case of so small a spherical droplet that the homogeneous bulk properties may not be attained even at the center of the phase α.

Let us imagine that the solid angle is increased by an amount $d\omega$ through a reversible and isothermal displacement of the side wall of the vessel keeping all other variables constant. Then the work done by the system in this process is proportional to $d\omega$ and may be denoted by $-\eta\, d\omega$.

In a way similar to the planar case this work may be considered as consisting of two parts: the work $p^\alpha dV^\alpha + p^\beta dV^\beta$, to cause a change in the volumes of the two homogeneous bulk phases; and the excess work which we shall denote by $-\gamma\, dA$. Then we can write

$$dW = -\eta\, d\omega = p^\alpha dV^\alpha + p^\beta dV^\beta - \gamma\, dA. \tag{8.4}$$

From a purely mathematical view point we may say that $-\gamma\, dA$ is merely a correction term to make up the total work dW which is a well-defined quantity including the complicated effects arising from the presence of the interface layer besides the contributions due to homogeneous parts in phases α and β. The coefficient γ may, however, be taken as the work done on the system associated with a unit increase in the area of the dividing surface at the given curvature. This implies that we may regard the actual system as if it were composed of two homogeneous phases α and β meeting at a simple spherical membrane of zero thickness with radius a, having a tension γ uniform in all directions. Then it may be reasonable to take γ defined by (8.4) as the surface tension of the spherical interface. With use of (8.1) to (8.3), we obtain[1] from (8.4)

$$\gamma = \frac{1}{3}\, a\,(p^\alpha - p^\beta) + \frac{K}{a^2}, \tag{8.5}$$

where

$$K = \eta + \tfrac{1}{3}(R^\beta)^3 p^\beta - \tfrac{1}{3}(R^\alpha)^3 p^\alpha. \tag{8.6}$$

The quantity K does not depend on the choice of the dividing surface. Then, according to (8.5) the surface tension defined by (8.4) depends on the choice of the dividing surface in contrast with the planar case.

With use of (8.1) to (8.3) and (8.6) we can rewrite (8.5) in an alternative form

$$\eta\,\omega = \gamma A - p^\alpha V^\alpha - p^\beta V^\beta. \tag{8.7}$$

Differentiating (8.5) with respect to a, we obtain[1]

$$\gamma + \frac{1}{2}\, a\left[\frac{\partial\gamma}{\partial a}\right] = \frac{1}{2}\, a\,(p^\alpha - p^\beta). \tag{8.8}$$

Here $[\partial\gamma/\partial a]\, da$ is the change in the value of the surface tension accompanying the mathematical displacement of the dividing surface by the amount da, keeping all the physical quantities inside the system and the external conditions unaltered[2]. That is, this change in the value of the surface tension is merely due to the arbitrariness in the choice of the dividing surface and is not to be confused with a change in γ accompanying increase in the radius of the physical interface. The derivative $[\partial\gamma/\partial a]$ plays an important role in the thermodynamics of a spherical interface, as will be seen later.

Since the work done by the system associated with changes dR^α and dR^β is $-p^\alpha\omega(R^\alpha)^2\, dR^\alpha + p^\beta\omega(R^\beta)^2 dR^\beta$, the general expression for the elementary work dW is, instead of (8.4), to be written in the form

$$dW = -\eta\, d\omega + p^\beta\omega(R^\beta)^2\, dR^\beta - p^\alpha\omega(R^\alpha)^2\, dR^\alpha. \tag{8.9}$$

---

[1] S. KONDO [9].

[2] The notation $[\partial\gamma/\partial a]$ is to be distinguished from the derivative for the actual radius dependence, which will be expressed by $\partial\gamma/\partial a$.

Every term on the right-hand side of (8.9) is independent of the choice of the dividing surface. We have from (8.1) to (8.3),

$$dV^\alpha = \frac{d\omega}{3}\{a^3 - (R^\alpha)^3\} + \omega a^2\, da - \omega (R^\alpha)^2\, dR^\alpha, \tag{8.10}$$

$$dV^\beta = \frac{d\omega}{3}\{(R^\beta)^3 - a^3\} + \omega (R^\beta)^2\, dR^\beta - \omega a^2\, da, \tag{8.11}$$

$$dA = a^2\, d\omega + 2\omega a\, da. \tag{8.12}$$

Then solving the above equations for dR^α, dR^β and $d\omega$ in terms of dV^α, dV^β, dA and da, substituting the results into (8.9), and using (8.1) to (8.3) and (8.7), we obtain

$$dW = p^\alpha dV^\alpha + p^\beta dV^\beta - \gamma\, dA - \left\{(p^\alpha - p^\beta) - \frac{2\gamma}{a}\right\} A\, da. \tag{8.13}$$

The last term on the right-hand side of this equation plays an important role in the thermodynamics of a spherical interface.

Utilizing (8.8) we can rewrite (8.13) in a more transparent form

$$dW = p^\alpha dV^\alpha + p^\beta dV^\beta - \gamma\, dA - \left[\frac{\partial \gamma}{\partial a}\right] A\, da. \tag{8.14}$$

It is easily shown that (8.14) is unaffected by an infinitesimal variation of a non-physical nature:

$$dV^\alpha = -dV^\beta = A\, da, \qquad dA = 2A\frac{da}{a}, \tag{8.15}$$

arising from an artificial mathematical displacement of the location of the dividing surface.

Let us next consider the changes not only in ω, R^α, R^β but also in other state variables. Even in such a general case (8.14) remains valid. Thus if we denote by dQ the heat received by the system and by dU the change in the internal energy of the system in the course of this process, we have the expression for the first law of thermodynamics as follows

$$dU = dQ - p^\alpha dV^\alpha - p^\beta dV^\beta + \gamma\, dA + \left\{(p^\alpha - p^\beta) - \frac{2\gamma}{a}\right\} A\, da. \tag{8.16}$$

It is to be remarked that this equation is valid even for systems out of equilibrium insofar as they have well-defined p^α and p^β.

9. Surface of tension. As already mentioned in Sect. 1 the condition for mechanical equilibrium of a system composed of two phases α and β separated by a fictitious membrane with the radius a, having a uniform tension γ, leads to the familiar Kelvin relation given by (1.1).

On the other hand we have the relation (8.8) which is applicable to any system with a spherical interface. We shall now choose the dividing surface so as to make $[\partial\gamma/\partial a]$ vanish, and hereafter distinguish the quantities referred to this particular dividing surface by the subscript s. Then the radius of this particular dividing surface, a_s, is defined by [9]

$$\left[\frac{\partial \gamma}{\partial a}\right]_{a=a_s} = 0. \tag{9.1}$$

This is one of the most important dividing surfaces, because (8.8) assumes the simple form of the Kelvin relation (1.1) for and only for this particular dividing surface:

$$p^\alpha - p^\beta = \frac{2\gamma_s}{a_s}. \tag{9.2}$$

Following Gibbs [1] this particular dividing surface is called the *surface of tension.* Agreement in form between (9.2) and the Kelvin relation (1.1) implies that the mechanical effect of an actual spherical interface of complicated structure can be replaced by that of a simple flexible membrane streched at the location of the surface of tension, having zero thickness, and tension γ_s uniform in all directions [1], [11].

Substitution of (8.5) into (9.2) leads to

$$a_s = \left(\frac{6K}{p^\alpha - p^\beta}\right)^{\frac{1}{3}}. \tag{9.3}$$

Using the above relation we can eliminate K from (8.5) and obtain [9]

$$\gamma = \frac{a_s^2 \gamma_s}{3a^2} + \frac{2\gamma_s a}{3a_s}. \tag{9.4}$$

Since values of the surface tension are always positive, it will readily be seen from (9.4) that γ has its minimum value γ_s at $a = a_s$. Hence we may conclude that so long as a_s is large compared with the thickness of the interface layer, all the γ's referred to those dividing surfaces located within or near to the spherical interface layer, within which the surface of tension is expected to lie (see Sects. 38, 39), are almost equal to the minimum value γ_s [9][1]. This means that from the macroscopic standpoint we may treat the surface tension as being practically independent of the location of the dividing surface so long as the latter is placed within the interface layer (except for droplets or bubbles so small that their radii are comparable with the thickness of the interface layer). Thus the physical surface tension of a spherical interface may be well defined by such a γ.

Finally it must be emphasized that the relations stated in this section are valid even for systems out of equilibrium.

10. Fundamental thermodynamic equations for spherical interface [8], [9]. The heat dQ received by the system in the course of a reversible transformation is equal to $T\,dS$ where dS is the increase in the entropy. Then (8.16) reduces to

$$dU = T\,dS - p^\alpha dV^\alpha - p^\beta dV^\beta + \gamma\,dA + \left\{(p^\alpha - p^\beta) - \frac{2\gamma}{a}\right\} A\,da. \tag{10.1}$$

Thus the change in the Helmholtz free energy is given from its definition as follows

$$dF = -p^\alpha dV^\alpha - p^\beta dV^\beta - S\,dT + \gamma\,dA + \left\{(p^\alpha - p^\beta) - \frac{2\gamma}{a}\right\} A\,da. \tag{10.2}$$

It should now be emphasized that in contrast with (8.16) the above relations are valid only for systems in thermodynamic equilibrium. The condition for chemical equilibrium is $\mu_i^\alpha = \mu_i^\beta = \mu_i$ $(i = 1, 2, 3, \ldots, \varkappa)$, $\varkappa$ being the number of independent components.

In the case of open systems the change in the Helmholtz free energy is, instead of (10.2), to be given[2] by

$$\left.\begin{aligned} dF = &-p^\alpha dV^\alpha - p^\beta dV^\beta - S\,dT + \gamma\,dA + \\ &+ \left\{(p^\alpha - p^\beta) - \frac{2\gamma}{a}\right\} A\,da + \sum_{i=1}^{\varkappa} \mu_i\,dN_i. \end{aligned}\right\} \tag{10.3}$$

---

[1] From (9.4) it is easily shown that $\gamma = \gamma_s \left\{1 + \frac{(a - a_s)^2}{a_s^2} + O\left(\frac{a - a_s}{a_s}\right)^3\right\}$.

[2] T.L. Hill [8].

With use of (8.8) we can rewrite (10.3) as [9]

$$dF = -p^\alpha dV^\alpha - p^\beta dV^\beta - S\,dT + \gamma\,dA + \left[\frac{\partial\gamma}{\partial a}\right] A\,da + \sum_{i=1}^{\varkappa} \mu_i\,dN_i. \qquad (10.4)$$

Likewise we obtain the fundamental equation for the change in the internal energy

$$dU = T\,dS - p^\alpha dV^\alpha - p^\beta dV^\beta + \gamma\,dA + \left[\frac{\partial\gamma}{\partial a}\right] A\,da + \sum_{i=1}^{\varkappa} \mu_i\,dN_i. \qquad (10.5)$$

Substituting (8.10) to (8.12) into (10.3) and making use of (8.6) and (8.8) we obtain [8],

$$dF = \eta\,d\omega + p^\alpha \omega (R^\alpha)^2\,dR^\alpha - p^\beta \omega (R^\beta)^2\,dR^\beta - S\,dT + \sum_{i=1}^{\varkappa} \mu_i\,dN_i, \qquad (10.6)$$

from which we immediately find [8]

$$\eta = \left(\frac{\partial F}{\partial \omega}\right)_{R^\alpha, R^\beta, T, \boldsymbol{N}}. \qquad (10.7)$$

This is the definition of the quantity η based on the second law.

Since F is a homogeneous function of first degree in $N_1, N_2, \ldots, N_\varkappa$ and ω, by using EULER's theorem we obtain from (10.6)

$$F = \sum_{i=1}^{\varkappa} \mu_i N_i + \eta\omega. \qquad (10.8)$$

Substitution of (8.7) into (10.8) leads to

$$F = \sum_{i=1}^{\varkappa} \mu_i N_i - p^\alpha V^\alpha - p^\beta V^\beta + \gamma A. \qquad (10.9)$$

This may be regarded as an equation defining the surface tension γ for a system in thermodynamic equilibrium.

From (10.9) we obtain

$$U = \sum_{i=1}^{\varkappa} \mu_i N_i + TS - p^\alpha V^\alpha - p^\beta V^\beta + \gamma A. \qquad (10.10)$$

11. Thermodynamic relations between superficial quantities of spherical interface. In a similar fashion to the case of a plane interface the following equations hold for the bulk phases α and β in chemical equilibrium:

$$F^\alpha = -p^\alpha V^\alpha + \sum_{i=1}^{\varkappa} \mu_i N_i^\alpha, \qquad (11.1)$$

$$F^\beta = -p^\beta V^\beta + \sum_{i=1}^{\varkappa} \mu_i N_i^\beta. \qquad (11.2)$$

Subtracting the above equations from (10.9) we obtain, with use of (2.1) and (3.1), the following relation for the superficial Helmholtz energy F^s:

$$F^s = \sum_{i=1}^{\varkappa} \mu_i N_i^s + \gamma A. \qquad (11.3)$$

If the dividing surface is chosen so as to make $\sum_{i=1}^{\varkappa} \mu_i N_i$ equal to zero, (11.3) reduces to

$$F^s = \gamma A. \qquad (11.4)$$

This implies that the surface tension of a spherical interface becomes equal to the superficial Helmholtz free energy per unit area of the dividing surface if selected so as to make $\sum_{i=1}^{\varkappa} \mu_i N_i$ vanish. It should, however, be remembered that γ referred to any other dividing surface cannot be equal to the superficial free energy.

The changes in F^α and F^β are respectively given by

$$dF^\alpha = -p^\alpha dV^\alpha - S^\alpha dT + \sum_{i=1}^{\varkappa} \mu_i dN_i^\alpha, \tag{11.5}$$

$$dF^\beta = -p^\beta dV^\beta - S^\beta dT + \sum_{i=1}^{\varkappa} \mu_i dN_i^\beta. \tag{11.6}$$

Subtracting both the above equations from (10.4) we obtain

$$dF^s = -S^s dT + \gamma\, dA + \sum_{i=1}^{\varkappa} \mu_i dN_i^s + \left[\frac{\partial \gamma}{\partial a}\right] A\, da. \tag{11.7}$$

Differentiating (11.3) and combining with (11.7) we obtain the Gibbs-Duhem relation for a spherical interface

$$A\, d\gamma + S^s dT + \sum_{i=1}^{\varkappa} N_i^s d\mu_i = \left[\frac{\partial \gamma}{\partial a}\right] A\, da, \tag{11.8}$$

or in terms of superficial quantities defined by (6.5) and (6.6) [7], [9]

$$d\gamma + \sigma\, dT + \sum_{i=1}^{\varkappa} \Gamma_i d\mu_i = \left[\frac{\partial \gamma}{\partial a}\right] da. \tag{11.9}$$

This is a generalization of the Gibbs adsorption equation; (11.9) differs from the adsorption equation (6.8) for a plane interface only in the appearance of the term $\left[\frac{\partial \gamma}{\partial a}\right] da$.

Following KOENIG [6] we now divide the species into three classes to be called Classes 1, 2 and 3, respectively, and defined as follows. Class 1: those species appreciably present in both phases; Class 2: those appreciably present only in the phase α; Class 3: those appreciably present only in the phase β. We shall denote the number of species in Class 1 by $\varkappa'$, in Class 2 by $\varkappa''$, in Class 3 by $\varkappa'''$, and the number of species appreciably present in α by $\varkappa^\alpha$ and in β by $\varkappa^\beta$. Then we have the relations

$$\varkappa = \varkappa' + \varkappa'' + \varkappa''', \qquad \varkappa^\alpha = \varkappa' + \varkappa'', \qquad \varkappa^\beta = \varkappa' + \varkappa'''. \tag{11.10}$$

As an example let us consider an aqueous solution of non-volatile salts in contact with air. Then water belongs to Class 1 and the salts to Class 2.

If the interface of our system were constrained to remain plane ($a \to \infty$), in accordance with the usual phase rule the system would have $\varkappa$ degrees of freedom. Removal of this constraint evidently increases the number of degrees of freedom by one. Thus to describe completely each of the intensive variables of the system, $\varkappa + 1$ independent intensive variables are necessary and sufficient [6].

Let us now denote the mole fraction of component i by x_i. Then we have the following restrictions:

$$\left.\begin{aligned} \sum_{i=1}^{\varkappa^\alpha} x_i^\alpha &= \sum_{i=1}^{\varkappa'} x_i'^\alpha + \sum_{i=1}^{\varkappa''} x_i''^\alpha = 1, \\ \sum_{i=1}^{\varkappa^\beta} x_i^\beta &= \sum_{i=1}^{\varkappa'} x_i'^\beta + \sum_{i=1}^{\varkappa'''} x_i'''^\beta = 1. \end{aligned}\right\} \tag{11.11}$$

There can be many combinations of $\varkappa - 1$ independent variables selected out of $(x_1'^{\alpha}, \ldots, x_{\varkappa'}'^{\alpha}, x_1'^{\beta}, \ldots, x_{\varkappa'}'^{\beta}, x_1''^{\alpha}, \ldots, x_{\varkappa''}''^{\alpha}, x_1'''^{\beta}, \ldots, x_{\varkappa'''}'''^{\beta})$.

Thus we may select $(a, T, x_2^{\alpha}, \ldots, x_{\varkappa^{\alpha}}^{\alpha}, x_1'''^{\beta}, \ldots, x_{\varkappa'''}'''^{\beta})$, to be called Set A, as a complete set of independent variables. The radius of dividing surface can be taken as an independent variable, if the dividing surface condition is fixed. In addition we shall need three auxiliary sets of independent variables: Set I $(p^{\alpha}, T, x_2^{\alpha}, \ldots, x_{\varkappa^{\alpha}}^{\alpha}, x_1''', \ldots, x_{\varkappa'''}''')$; Set II $(p^{\alpha}, T, x_2^{\alpha}, \ldots, x_{\varkappa^{\alpha}}^{\alpha})$ and Set III $(p^{\beta}, T, x_2^{\beta}, \ldots, x_{\varkappa^{\beta}}^{\beta})$. Partial derivatives corresponding to Set A will be written without subscript and partial derivatives corresponding to Sets I to III will be distinguished by the subscripts I to III, respectively.

Then we have [see (6.12)]

$$d\mu_i^{\alpha} = -s_i^{\alpha}\,dT + v_i^{\alpha}\,dp^{\alpha} + \sum_{j=2}^{\varkappa^{\alpha}} \left(\frac{\partial \mu_i^{\alpha}}{\partial x_j^{\alpha}}\right)_{\mathrm{II}} dx_j^{\alpha} \qquad (i = 1, 2, \ldots, \varkappa^{\alpha}), \tag{11.12}$$

$$d\mu_i'''^{\beta} = -s_i'''^{\beta}\,dT + v_i'''^{\beta}\,dp^{\beta} + \sum_{j=2}^{\varkappa^{\beta}} \left(\frac{\partial \mu_i'''^{\beta}}{\partial x_j^{\beta}}\right)_{\mathrm{III}} dx_j^{\beta} \qquad (i = 1, \ldots, \varkappa'''). \tag{11.13}$$

Thus at the given composition in Set A we obtain [*6*],

$$d\mu_i^{\alpha} = -s_i^{\alpha}\,dT + v_i^{\alpha}\,dp^{\alpha}, \tag{11.14}$$

$$d\mu_i'''^{\beta} = -s_i'''^{\beta}\,dT + \left\{v_i'''^{\beta}\left(\frac{\partial p^{\beta}}{\partial p^{\alpha}}\right)_{\mathrm{I}} + w_i'''^{\beta}\right\} dp^{\alpha}, \tag{11.15}$$

where

$$w_i'''^{\beta} = \sum_{j=2}^{\varkappa'} \left(\frac{\partial \mu_i'''^{\beta}}{\partial x_j'^{\beta}}\right)_{\mathrm{III}} \left(\frac{\partial x_j'^{\beta}}{\partial p^{\alpha}}\right)_{\mathrm{I}}. \tag{11.16}$$

Hence we can rewrite (11.9) in a useful form (see [*6*])

$$d\gamma = (\Sigma_{s^{\alpha}} - \sigma)\,dT - \Sigma_{v^{\alpha}}\,dp^{\alpha} + \left[\frac{\partial \gamma}{\partial a}\right] da \quad (\text{const } x_2^{\alpha}, \ldots, x_{\varkappa'''}'''), \tag{11.17}$$

where

$$\Sigma_{s^{\alpha}} = \sum_{i=1}^{\varkappa'} \Gamma_i'\, s_i'^{\alpha} + \sum_{i=1}^{\varkappa''} \Gamma_i''\, s_i''^{\alpha} + \sum_{i=1}^{\varkappa'''} \Gamma_i'''\, s_i'''^{\beta}, \tag{11.18}$$

$$\Sigma_{v^{\alpha}} = \sum_{i=1}^{\varkappa'} \Gamma_i'\, v_i'^{\alpha} + \sum_{i=1}^{\varkappa''} \Gamma_i''\, v_i''^{\alpha} + \sum_{i=1}^{\varkappa'''} \Gamma_i''' \left\{v_i'''^{\beta}\left(\frac{\partial p^{\beta}}{\partial p^{\alpha}}\right)_{\mathrm{I}} + w_i'''^{\beta}\right\}. \tag{11.19}$$

From (11.17) we can obtain the expression for the temperature dependence of surface tension as follows

$$\left.\begin{aligned} \frac{\partial \gamma}{\partial T} = -\sigma &+ \sum_{i=1}^{\varkappa'} \Gamma_i' \left(s_i'^{\alpha} - v_i'^{\alpha}\left(\frac{\partial p^{\alpha}}{\partial T}\right)\right) + \sum_{i=1}^{\varkappa''} \Gamma_i'' \left(s_i''^{\alpha} - v_i''^{\alpha}\left(\frac{\partial p^{\alpha}}{\partial T}\right)\right) + \\ &+ \sum_{i=1}^{\varkappa'''} \Gamma_i''' \left\{s_i'''^{\beta} - \left(v_i'''^{\beta}\left(\frac{\partial p^{\beta}}{\partial p^{\alpha}}\right)_{\mathrm{I}} + w_i'''^{\beta}\right)\left(\frac{\partial p^{\alpha}}{\partial T}\right)\right\}. \end{aligned}\right\} \tag{11.20}$$

On the other hand from (3.4) and (11.3) we readily obtain

$$U^s = \sum_{i=1}^{\varkappa} \mu_i N_i^s + T\,S^s + \gamma A, \tag{11.21}$$

or, in terms of superficial densities defined by (6.5) to (6.7),

$$v = \sum_{i=1}^{\varkappa} \mu_i \Gamma_i + \sigma T + \gamma. \tag{11.22}$$

Making use of (11.20) we can eliminate σ from (11.22) with the result

$$\left.\begin{aligned} v = \gamma - T\left(\frac{\partial\gamma}{\partial T}\right) + \sum_{i=1}^{\varkappa} \Gamma_i' \left[\mu_i'^{\alpha} + T\left\{s_i'^{\alpha} - v_i'^{\alpha}\left(\frac{\partial p^{\alpha}}{\partial T}\right)\right\}\right] + \\ + \sum_{i=1}^{\varkappa''} \Gamma_i'' \left[\mu_i''^{\alpha} + T\left\{s_i''^{\alpha} - v_i''^{\alpha}\left(\frac{\partial p^{\alpha}}{\partial T}\right)\right\}\right] + \\ + \sum_{i=1}^{\varkappa'''} \Gamma_i''' \left[\mu_i'''^{\beta} + T\left\{s_i'''^{\beta} - \left(v_i'''^{\beta}\left(\frac{\partial p^{\beta}}{\partial p^{\alpha}}\right)_{\mathrm{I}} + w_i'''^{\beta}\right)\left(\frac{\partial p^{\alpha}}{\partial T}\right)\right\}\right]. \end{aligned}\right\} \quad (11.23)$$

If we assume $\varkappa'' = \varkappa''' = 0$ as in Sect. 6, (11.20) and (11.23) become identical in form with (6.13) and (6.14), respectively.

We shall choose the special dividing surface which makes $\Sigma_{v\alpha}$, given by (11.19), vanish and denote Γ_i, γ, σ and v referred to this dividing surface by $\Gamma_i^{(v)}$, γ_v, σ_v and v_v, respectively. Then the condition for this dividing surface is

$$\Sigma_{v\alpha} = \sum_{i=1}^{\varkappa'} \Gamma_i^{(v)'} v_i'^{\alpha} + \sum_{i=1}^{\varkappa''} \Gamma_i^{(v)''} v_i''^{\alpha} + \sum_{i=1}^{\varkappa'''} \Gamma_i^{(v)'''} \left(v_i'''^{\beta}\left(\frac{\partial p^{\beta}}{\partial p^{\alpha}}\right)_{\mathrm{I}} + w_i'''^{\beta}\right) = 0, \quad (11.24)$$

which is a generalization of the convention (6.15). Then (11.20) and (11.23) reduce respectively to

$$\frac{\partial\gamma_v}{\partial T} = -\sigma_v + \sum_{i=1}^{\varkappa'} \Gamma_i^{(v)'} s_i'^{\alpha} + \sum_{i=1}^{\varkappa''} \Gamma_i^{(v)''} s_i''^{\alpha} + \sum_{i=1}^{\varkappa'''} \Gamma_i^{(v)'''} s_i'''^{\beta}, \quad (11.25)$$

and

$$v_v = \gamma_v - T\left(\frac{\partial\gamma_v}{\partial T}\right) + \sum_{i=1}^{\varkappa'} \Gamma_i^{(v)'} h_i'^{\alpha} + \sum_{i=1}^{\varkappa''} \Gamma_i^{(v)''} h_i''^{\alpha} + \sum_{i=1}^{\varkappa'''} \Gamma_i^{(v)'''} h_i'''^{\beta}, \quad (11.26)$$

where h_i is the partial enthalpy per molecule of species i.

In the case of pure liquids, if we choose the equimolecular dividing surface, 11.20) and (11.23) reduce respectively to

$$\frac{\partial\gamma_v}{\partial T} = -\sigma_v, \quad (11.27)$$

and

$$v_v = -T^2 \frac{\partial}{\partial T}\left(\frac{\gamma_v}{T}\right), \quad (11.28)$$

which are identical in form with (6.17) and (6.19), respectively.

12. Curvature dependence of surface tension. Let us consider the special dividing surface defined by the condition (11.24). Then we can easily obtain from (11.17) [9]

$$\frac{\partial\gamma_v}{\partial a_v} = \left[\frac{\partial\gamma}{\partial a}\right]_{a=a_v}. \quad (12.1)$$

It should now be noted and emphasized that $\frac{\partial\gamma_v}{\partial a_v}$ means the ratio of the increase in the surface tension γ_v to the increase in the actual radius a_v of a drop or bubble. On the other hand, as introduced in Sect. 8 (see footnote 2 on p. 146), $[\partial\gamma/\partial a]$ denotes the ratio of the change in the surface tension to the imaginary change in the choice of the radius of dividing surface. Thus (12.1) implies that for and only for the dividing surface condition $\Sigma_{v\alpha} = 0$ given by (11.24), the rate of change in surface tension due to the actual change in the radius of the drop or bubble (which is associated with some change in p^{α} and p^{β}) happens to agree with that due to the artificial displacement of the chosen dividing surface (which does not involve any change in the physical situation of the system).

On the other hand, differentiation of (9.4) leads to

$$d\gamma = \frac{1}{3}\left(\frac{a_s^2}{a^2} + \frac{2a}{a_s}\right) d\gamma_s + \frac{2}{3}\gamma_s\left(\frac{a_s}{a^2} - \frac{a}{a_s^2}\right) d a_s + \frac{2}{3}\gamma_s\left(\frac{1}{a_s} - \frac{a_s^2}{a^3}\right) d a, \tag{12.2}$$

from which we immediately obtain

$$\left[\frac{\partial \gamma}{\partial a}\right]_{a=a_v} = \frac{2}{3}\,\frac{\gamma_s(a_v^3 - a_s^3)}{a_v^3\, a_s}, \tag{12.3}$$

and

$$\frac{\partial \gamma_v}{\partial a_v} = \frac{1}{3}\,\frac{(a_s^3 + 2a_v^3)}{a_v^2\, a_s}\,\frac{\partial \gamma_s}{\partial a_v} + \frac{2}{3}\gamma_s\,\frac{(a_s^3 - a_v^3)}{a_v^2\, a_s^2}\,\frac{\partial a_s}{\partial a_v} + \frac{2}{3}\gamma_s\,\frac{(a_v^3 - a_s^3)}{a_v^3\, a_s}. \tag{12.4}$$

In accordance with (12.1) we equate (12.3) to (12.4), obtaining

$$\frac{a_s}{\gamma_s}\cdot\frac{\partial \gamma_s}{\partial a_s} = \frac{2(a_v^3 - a_s^3)}{a_s^3 + 2a_v^3}. \tag{12.5}$$

Substituting (12.3) into (12.1) and remembering (9.4) we can readily obtain

$$a_v\,\frac{\partial \gamma_v}{\partial a_v} = \frac{2}{3}\,\frac{\gamma_s(a_v^3 - a_s^3)}{a_v^2\, a_s} = \frac{2\gamma_v(a_v^3 - a_s^3)}{a_s^3 + 2a_v^3}. \tag{12.6}$$

Finally combining (12.3) and (12.5) with (12.6) we obtain [9] the Gibbs-Tolman-Koenig-Buff equation [1], [5], [6], [7],

$$\left[\frac{\partial \log \gamma}{\partial \log a}\right]_{a=a_v} = \frac{\partial \log \gamma_v}{\partial \log a_v} = \frac{\partial \log \gamma_s}{\partial \log a_s} = \frac{2\left(\frac{\delta}{a_s}\right)\left[1 + \frac{\delta}{a_s} + \frac{1}{3}\left(\frac{\delta}{a_s}\right)^2\right]}{1 + 2\frac{\delta}{a_s}\left[1 + \frac{\delta}{a_s} + \frac{1}{3}\left(\frac{\delta}{a_s}\right)^2\right]}, \tag{12.7}$$

where

$$\delta = a_v - a_s. \tag{12.8}$$

As stated in Sect. 11, once the dividing surface condition is fixed the radius of the dividing surface can be taken as an independent state variable. Thus a_s or a_v used in the above equations should be regarded as state variable [6] and we shall often use the conventional symbol R in place of a_s when it is used as a measure of the actual radius of a droplet or bubble. Then integration of (12.7) readily leads to the desired expression for the curvature dependence of the surface tension as follows [5],

$$\log\frac{\gamma_s(R)}{\gamma_\infty} = \int_\infty^R \frac{\frac{2\delta}{R^2}\left\{1 + \frac{\delta}{R} + \frac{1}{3}\left(\frac{\delta}{R}\right)^2\right\}}{1 + \frac{2\delta}{R}\left\{1 + \frac{\delta}{R} + \frac{1}{3}\left(\frac{\delta}{R}\right)^2\right\}}\, dR, \tag{12.9}$$

where γ_∞ is the surface tension for the plane interface.

It would be plausible to assume that both the surface of tension and the special dividing surface given by (11.24) lie within or very near to the interface layer and consequently δ is, at the largest, of the order of magnitude of the thickness Δ of the interface layer. If we restrict our treatment to a drop or bubble with radius R large compared with Δ, we may regard δ as being approximately constant and equal to δ_∞, the value of δ for the plane interface. Then we can easily carry out the integration (12.9), obtaining

$$\gamma_s(R) = \gamma_\infty\left(1 - \frac{2\delta_\infty}{R} + \cdots\right), \tag{12.10}$$

and similarly from (12.7)

$$\gamma_v(R_v) = \gamma_\infty\left(1 - \frac{2\delta_\infty}{R_v} + \cdots\right), \tag{12.11}$$

where R_v has been used in place of a_v. As will easily be seen from (11.24), R_v corresponds to the radius of the equimolecular dividing surface for the case of pure liquids.

It is evident from (12.8) and (12.7) or (12.10) that the surface tension decreases or increases with a decrease in the radius of the drop or bubble according as a_v is greater or smaller than a_s, by a fraction of the order of magnitude of $2\delta_\infty/R$ in ratio.

Although there is no experimental information about the sign and the value of $\delta = a_v - a_s$, approximate calculations for liquid argon, based upon statistical mechanics, which will be elucidated later in Sects. 38, 39, suggest that δ is to be of the order of magnitude of 3 Å and is positive for the case of liquid drops. Thus taking δ as 3 Å we can expect that the surface tension of a spherical interface of radius 100 Å decreases by 6% for a drop and increase for a bubble as compared with the surface tension of a plane interface. Table 1 gives the values of γ/γ_∞ computed from (12.9) by assuming δ as constant [5].

Table 1. *Change in surface tension with radius of curvature of the surface of tension, R* (TOLMAN [5]).

δ/R	γ_s/γ_∞	δ/R	γ_s/γ_∞	δ/R	γ_s/γ_∞
0.00	1.00	0.20	0.70	0.70	0.36
0.01	0.98	0.30	0.60	0.80	0.33
0.02	0.96	0.40	0.52	0.90	0.30
0.05	0.91	0.50	0.46	1.00	0.28
0.10	0.83	0.60	0.41		

13. The vapor pressure and work of formation of drop and bubble. Let us consider a liquid phase in equilibrium with its vapor phase. If the two phases are separated by a plane interface, the pressures p^α of the phase α and p^β of the phase β are both equal to the ordinary vapor pressure, which will be denoted by p^∞. In the case of a spherical interface, however, according to the Kelvin relation (1.1) the two pressures p^α and p^β cannot have the same value and consequently the vapor pressure should depend on the curvature of the interface.

We shall now inquire into the above-mentioned curvature dependence of the vapor pressure. For simplicity, we restrict our treatment to the case where all of the $\varkappa$ species are appreciably present in both phases. For a fixed temperature and composition in phase α, we obtain from (11.12)

$$d\mu_i^\alpha = v_i^\alpha \, dp^\alpha . \tag{13.1}$$

If the interior phase α is liquid and the exterior phase β is a vapor phase which may be regarded as a perfect gas, the chemical potential of species i is given by

$$\mu_i^\beta = \varphi_i + kT \log p_i^\beta , \tag{13.2}$$

where p_i^β is the partial pressure of species i and φ_i depends only on the temperature.

Since $\mu_i^\alpha = \mu_i^\beta$ for the system in equilibrium, we obtain from (13.1) and (13.2)

$$v_i^\alpha \, dp^\alpha = kT \frac{dp_i^\beta}{p_i^\beta} . \tag{13.3}$$

Assuming incompressibility of the liquid, we can integrate both sides of (13.3) with respect to p^α to obtain

$$v_i^\alpha (p^\alpha - p^\infty) = kT \log \frac{p_i^\beta}{p_i^\infty} , \tag{13.4}$$

where p_i^∞ is the partial pressure of species i for the case of a plane interface.

If we eliminate p^α from (13.4) by using (8.8), we have the relation

$$\frac{2\gamma}{a} + \left[\frac{\partial \gamma}{\partial a}\right] = \frac{kT}{v_i^\alpha} \log \frac{p_i^\beta}{p_i^\infty} - (p^\beta - p^\infty). \tag{13.5}$$

Since $1/v^\beta$ is sufficiently small compared with $1/v_i^\alpha$ at temperatures appreciably below the critical temperature, we may neglect $(kT/v^\beta) \log (p^\beta/p^\infty)$ compared with $(kT/v_i^\alpha) \log (p_i^\beta/p_i^\infty)$. Then, with use of the relation $v^\beta = kT/p^\beta$, we may rewrite (13.5) in the form

$$\log \frac{p_i^\beta}{p_i^\infty} = \frac{v_i^\alpha}{kT} \left\{ \frac{2\gamma}{a} + \left[\frac{\partial \gamma}{\partial a}\right] \right\}, \tag{13.6}$$

which is identical with GUGGENHEIM'S formula[1] apart from the small correction $[\partial\gamma/\partial a]$. For a pure liquid this formula reduces to the generalized Gibbs-Thomson formula,

$$\log \frac{p^\beta}{p^\infty} = \frac{v^\alpha}{kT} \left\{ \frac{2\gamma}{a} + \left[\frac{\partial \gamma}{\partial a}\right] \right\}, \tag{13.7}$$

which relates the radius of the liquid drop to its saturated vapor pressure in the case of a one-component system. Taking R and R_v as the radii of the surface of tension and of the dividing surface with the dividing surface condition (11.24), respectively, and making use of (9.1) and (12.1) we can rewrite (13.7) as follows [7],

$$\log \frac{p^\beta}{p^\infty} = \frac{2v^\alpha \gamma_s}{kTR} = \frac{v^\alpha}{kT} \left(\frac{2\gamma_v}{R_v} + \frac{\partial \gamma_v}{\partial R_v} \right) \quad \text{(drop)}. \tag{13.8}$$

In a similar manner we obtain the generalized Gibbs-Thomson formula for the case of bubbles:

$$\log \frac{p^\alpha}{p^\infty} = -\frac{2v^\beta \gamma_s}{kTR} = -\frac{v^\beta}{kT} \left(\frac{2\gamma_v}{R_v} + \frac{\partial \gamma_v}{\partial R_v} \right) \quad \text{(bubble)}, \tag{13.9}$$

where the interior phase α has been assumed to be made up of a perfect gas.

Next, we consider the work of formation of a drop from a large amount of its vapor at a given temperature. According to (10.10) the internal energy of the two-phase system is given by

$$U = \sum_{i=1}^{\varkappa} \mu_i N_i + TS - p^\alpha V^\alpha - p^\beta V^\beta + \gamma A. \tag{13.10}$$

We suppose that initially the system under consideration is composed of a uniform vapor phase having the same total number of molecules for each species, the same volume and entropy as the entire final system. Then the internal energy U^0 of this original system is given by

$$U^0 = \sum_{i=1}^{\varkappa} \mu_i^0 N_i + T^0 S - p^0 V, \tag{13.11}$$

where V is assumed to be equal to $V^\alpha + V^\beta$. In general the chemical potential μ_i^0, the temperature T^0 and the pressure p^0 may differ from the corresponding μ_i, T and p^β of the bulk vapor phase β having the number of molecules N_i^β, the entropy S^β and the volume V^β in the final system.

[1] E. A. GUGGENHEIM: Modern Thermodynamics by the Methods of J. W. GIBBS. London: Methuen 1933.

If V^α is sufficiently small compared with V, we may neglect square terms in $(N_j^\beta - N_j)$, $(S^\beta - S)$ and $(V^\beta - V)$, obtaining

$$\mu_i = \mu_i^0 + \sum_{j=1}^{\varkappa} \left(\frac{\partial \mu_i}{\partial N_j}\right)^0_{V,S,N_k} (N_j^\beta - N_j) + \left(\frac{\partial \mu_i}{\partial S}\right)^0_{V,\boldsymbol{N}} (S^\beta - S) + \left(\frac{\partial \mu_i}{\partial V}\right)^0_{S,\boldsymbol{N}} (V^\beta - V), \quad (13.12)$$

$$T = T^0 + \sum_{j=1}^{\varkappa} \left(\frac{\partial T}{\partial N_j}\right)^0_{V,S,N_k} (N_j^\beta - N_j) + \left(\frac{\partial T}{\partial S}\right)^0_{V,\boldsymbol{N}} (S^\beta - S) + \left(\frac{\partial T}{\partial V}\right)^0_{S,\boldsymbol{N}} (V^\beta - V), \quad (13.13)$$

$$p^\beta = p^0 + \sum_{j=1}^{\varkappa} \left(\frac{\partial p}{\partial N_j}\right)^0_{V,S,N_k} (N_j^\beta - N_j) + \left(\frac{\partial p}{\partial S}\right)^0_{V,\boldsymbol{N}} (S^\beta - S) + \left(\frac{\partial p}{\partial V}\right)^0_{S,\boldsymbol{N}} (V^\beta - V), \quad (13.14)$$

where the superscript 0 indicates the values of derivatives taken in the initial system[1].

On the other hand we have the thermodynamic relations

$$\left(\frac{\partial \mu_i}{\partial N_j}\right)_{V,S,N_k} = \left(\frac{\partial \mu_j}{\partial N_i}\right)_{V,S,N_k}, \quad (13.15)$$

$$\left(\frac{\partial T}{\partial N_j}\right)_{V,S,N_k} = \left(\frac{\partial \mu_j}{\partial S}\right)_{V,\boldsymbol{N}}, \quad (13.16)$$

$$\left(\frac{\partial p}{\partial N_j}\right)_{V,S,N_k} = -\left(\frac{\partial \mu_j}{\partial V}\right)_{S,\boldsymbol{N}}, \quad (13.17)$$

$$\left(\frac{\partial T}{\partial V}\right)_{S,\boldsymbol{N}} = -\left(\frac{\partial p}{\partial S}\right)_{V,\boldsymbol{N}}. \quad (13.18)$$

Since μ_i, T and p of a bulk phase are homogeneous functions of zeroth degree in V, N_i and S, making use of (13.15) to (13.18) we obtain

$$\left.\begin{aligned} &\sum_{i=1}^{\varkappa} \left(\frac{\partial \mu_i}{\partial N_j}\right)_{V,S,N_k} N_i + \left(\frac{\partial T}{\partial N_j}\right)_{V,S,N_k} S - \left(\frac{\partial p}{\partial N_j}\right)_{V,S,N_k} V \\ &\quad = \sum_{i=1}^{\varkappa} \left(\frac{\partial \mu_j}{\partial N_i}\right)_{V,S,N_k} N_i + \left(\frac{\partial \mu_j}{\partial S}\right)_{V,\boldsymbol{N}} S + \left(\frac{\partial \mu_j}{\partial V}\right)_{S,\boldsymbol{N}} V = 0, \end{aligned}\right\} \quad (13.19)$$

$$\left.\begin{aligned} &\sum_{i=1}^{\varkappa} \left(\frac{\partial \mu_i}{\partial S}\right)_{V,\boldsymbol{N}} N_i + \left(\frac{\partial T}{\partial S}\right)_{V,\boldsymbol{N}} S - \left(\frac{\partial p}{\partial S}\right)_{V,\boldsymbol{N}} V \\ &\quad = \sum_{i=1}^{\varkappa} \left(\frac{\partial T}{\partial N_i}\right)_{V,S,N_k} N_i + \left(\frac{\partial T}{\partial S}\right)_{V,\boldsymbol{N}} S + \left(\frac{\partial T}{\partial V}\right)_{S,\boldsymbol{N}} V = 0, \end{aligned}\right\} \quad (13.20)$$

$$\left.\begin{aligned} &\sum_{i=1}^{\varkappa} \left(\frac{\partial \mu_i}{\partial V}\right)_{S,\boldsymbol{N}} N_i + \left(\frac{\partial T}{\partial V}\right)_{S,\boldsymbol{N}} S - \left(\frac{\partial p}{\partial V}\right)_{S,\boldsymbol{N}} V \\ &\quad = -\sum_{i=1}^{\varkappa} \left(\frac{\partial p}{\partial N_i}\right)_{V,S,N_k} N_i - \left(\frac{\partial p}{\partial S}\right)_{V,\boldsymbol{N}} S - \left(\frac{\partial p}{\partial V}\right)_{S,\boldsymbol{N}} V = 0. \end{aligned}\right\} \quad (13.21)$$

If we neglect the square terms, with use of the above relations, (13.11) can be written as

$$U^0 = \sum_{i=1}^{\varkappa} \mu_i N_i + T S - p^\beta V. \quad (13.22)$$

It is noteworthy that the differences $N_i \mu_i - N_i \mu_i^0$, $TS - T^0 S$ and $p^0 V - p^\beta V$ themselves may not necessarily be negligible.

---

[1] See footnote 3, p. 140.

Comparing (13.10) with (13.22) we readily obtain the increase in the internal energy of the closed system at constant volume and entropy due to the formation of the drop as follows

$$W = U - U^0 = \gamma A - (p^\alpha - p^\beta) V^\alpha. \tag{13.23}$$

It can also be proved that this work of formation is equal to the increase in the Helmholtz free energy when the temperature instead of the entropy is kept constant.

In the case of spherical drops, the work of formation of a drop, from (13.23), is expressed as

$$W = 4\pi a^2 \gamma - \frac{4\pi a^3}{3} (p^\alpha - p^\beta). \tag{13.24}$$

With use of (8.5) or (8.8) the above equation can be rewritten as

$$W = 4\pi K, \tag{13.25}$$

or

$$W = \frac{4\pi a^2 \gamma}{3} \left(1 - \left[\frac{\partial \log \gamma}{\partial \log a}\right]\right). \tag{13.26}$$

From (13.25) we find that the constant K introduced in (8.6) is equal to the work of drop formation divided by 4π.

If we choose the surface of tension, whose radius is denoted by R, as the dividing surface, (9.1) reduces (13.26) to the simple form:

$$W = \frac{4\pi R^2 \gamma_s}{3} = \frac{A \gamma_s}{3}. \tag{13.27}$$

The work of drop formation plays an important role in the theory of nucleation in supersaturated vapors, the details of which are beyond the scope of the present article. It should, however, be noted that the number of molecules contained in the drop of pure liquid is given by $4\pi R_v^3/3 v^\alpha$ instead of $4\pi R^3/3 v^\alpha$, R_v being the radius of the equimolecular dividing surface. Making use of (12.1) we can rewrite (13.26) as [7]

$$W = \frac{4\pi R_v^2 \gamma_v}{3} \left[1 - \frac{\partial \log \gamma_v}{\partial \log R_v}\right], \tag{13.28}$$

where $\partial \log \gamma_v / \partial \log R_v$ has been given by (12.7).

II. Hydrostatic approach.

14. Mechanical definition of surface tension of plane interface. As mentioned before the physical interface is not a geometrical surface of zero thickness but a transition layer of finite thickness. Since the density varies appreciably with position in the transition zone along the direction normal to the interface, within this zone the pressure tensor defined in the hydrostatic sense may change with position, although it becomes an isotropic and constant hydrostatic pressure in the interior of either phase.

Let us define a rectangular coordinate system (x, y, z) with the z axis normal to the plane interface layer directed from phase α to phase β and with the (x, y) plane in the interface layer.

In any homogeneous bulk phase the pressure is isotropic, i.e., the force acting across any unit area is normal to it and equal in all directions. Hence in the interior of either homogeneous phase α or β the pressure tensor $\mathbf{p}$ reduces to the hydrostatic pressure p multiplied by the unit tensor $\mathbf{1}$;

$$\mathbf{p} = p\mathbf{1}; \quad p_{xx} = p_{yy} = p_{zz} = p, \quad p_{xy} = p_{yz} = p_{zx} = 0. \tag{14.1}$$

Within the interface layer the force acting across unit area is not the same in all directions. It will, however, easily be seen from symmetry requirements that for the planar case, even in an interface layer p_{xy}, p_{yz}, p_{zx} must vanish and that p_{xx} must be equal to p_{yy}, neither being dependent on x nor y. Then it will be convenient to introduce the symbol $p_T(z)$ and $p_N(z)$, to which we shall refer respectively as the tangential and the normal component of the pressure tensor at z:

$$\mathbf{p} = p_T(z)\,(\mathbf{e}_x\,\mathbf{e}_x + \mathbf{e}_y\,\mathbf{e}_y) + p_N(z)\,\mathbf{e}_z\,\mathbf{e}_z; \qquad p_N(z) = p, \tag{14.2}$$

where $\mathbf{e}_x$, $\mathbf{e}_y$ and $\mathbf{e}_z$ are unit vectors directed along the coordinate axes. The tangential component p_T may change in a complicated manner depending on z but the normal component p_N is required from the condition of hydrostatic equilibrium to be equal to p even in the transition zone.

Let us take a strip of unit width in the y direction extending from $-l/2$ to $l/2$ in the z direction. It is evident that the actual total stress $\Delta\Sigma_x$ acting in the x direction across this strip may be given by

$$\Delta\Sigma_x = -\int_{-l/2}^{l/2} p_T(z)\,dz. \tag{14.3}$$

If there were no interfacial contribution, the total stress acting across this strip in the x direction would have the value $-pl$. Then the excess of the stress due to the interface is expressed by [*24*]

$$\gamma = -\int_{-l/2}^{l/2} p_T(z)\,dz + pl = \int_{-l/2}^{l/2} \{p - p_T(z)\}\,dz. \tag{14.4}$$

This equation means that the actual system with a plane interface can be considered as if it were composed of two homogeneous phases separated by a plane membrane of zero thickness with tension γ given by (14.4). Thus it is seen that (14.4) is an equation defining the surface tension and we shall refer to (14.4) as the *mechanical definition.*

It is also evident that the surface tension defined by (14.4) is independent of l provided l is of such a macroscopic length that $p_T(z)$ at $z=l/2$ and $z=-l/2$ reduces to the hydrostatic pressure p. For this reason we may take l as infinity and rewrite (14.4) in the conventional form

$$\gamma = \int_{-\infty}^{\infty} \{p - p_T(z)\}\,dz. \tag{14.5}$$

If the tangential pressure p_T is known as a function of z, we can calculate the position of the particular membrane which is mechanically equivalent to the actual interface layer with respect to both the resultant force and moment exerted on the strip. The position of this mechanically equivalent membrane is given by [*8*]

$$z_s = \frac{1}{\gamma}\int_{-\infty}^{\infty} \{p - p_T(z)\}\,z\,dz. \tag{14.6}$$

As will be shown in Sect. 16, the surface of tension discussed in Sect. 9 turns out in the limit of zero curvature of the interface to be the dividing surface located at z_s. Thus if we choose the surface of tension as the (x, y) plane, the condition for the surface of tension, (14.6), reduces to

$$\int_{-\infty}^{\infty} \{p - p_T(z)\}\,z\,dz = 0. \tag{14.7}$$

Using the fact that p_N is independent of z and equal to p, we may obtain from (14.5) the Bakker equation [10],

$$\gamma = \int_{-\infty}^{\infty} (p_N - p_T)\, dz. \tag{14.8}$$

It should, however, be noted that, though the concept of tangential pressure is useful and has played an important role in the theory of surface tension since the early work of VAN DER WAALS and his school [10], in ordinary hydrodynamics physical quantities are usually assumed to remain constant over the range of intermolecular forces and therefore $p_T(z)$ used so far cannot be exactly identified with the pressure in the hydrodynamic sense.

15. Mechanical definition of surface tensions of spherical interface. Let us consider that portion of a two-phase system with a spherical interface enclosed in a container of the same shape as that in Sect. 8 (see Fig. 3), i.e. a spherical cone.

We define a rectangular coordinate system (x, y, z) with origin at the apex of the cone O and with the axis of the cone taken as the z-axis, as illustrated in Fig. 4.

As in the planar case, the symmetry of the system requires that the pressure tensor $\mathbf{p}$ be expressed in the form

$$\mathbf{p} = p_T(r)\,(\boldsymbol{e}_\vartheta \boldsymbol{e}_\vartheta + \boldsymbol{e}_\varphi \boldsymbol{e}_\varphi) + p_N(r)\, \boldsymbol{e}_r \boldsymbol{e}_r, \tag{15.1}$$

where $\boldsymbol{e}_r$, $\boldsymbol{e}_\vartheta$, $\boldsymbol{e}_\varphi$ are orthogonal unit vectors corresponding to r, ϑ, φ, respectively, and $p_T(r)$ and $p_N(r)$ are called the tangential and normal pressure, both being independent of ϑ and φ.

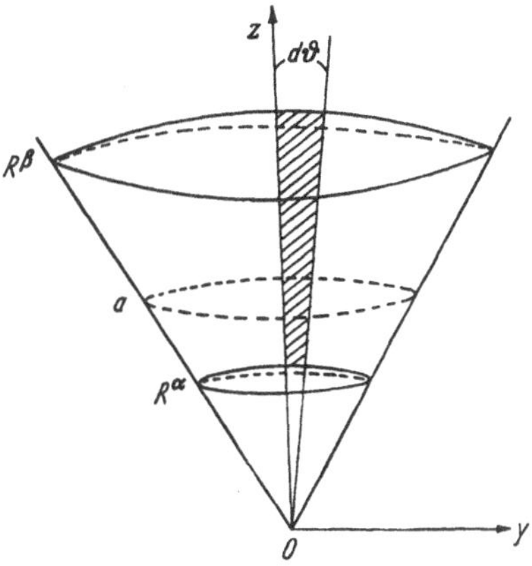

Fig. 4. The hatched portion represents the sectorial strip.

Both $p_T(r)$ and $p_N(r)$ reduce to p^α and p^β, respectively, in the interiors of the bulk phases α and β. In the spherical case $p_N(r)$ is no longer a constant but a function of r, which can be determined from the equation for hydrostatic equilibrium.

In the absence of external fields, the equation for hydrostatic equilibrium is

$$\nabla \cdot \mathbf{p} = 0. \tag{15.2}$$

From (15.1) and (15.2), we have the condition of hydrostatic equilibrium in the following equivalent forms [26]:

$$\frac{d p_N}{d r} = \frac{2(p_T - p_N)}{r}, \qquad \frac{d(r^3 p_N)}{d r} = r^2 (p_N + 2 p_T), \qquad \frac{d(r^2 p_N)}{d(r^2)} = p_T. \tag{15.3}$$

Integrating the last form of (15.3) with respect to r from R^α to R^β, we obtain the integral form for hydrostatic equilibrium

$$(R^\beta)^2 p^\beta - (R^\alpha)^2 p^\alpha = 2 \int_{R^\alpha}^{R^\beta} p_T\, r\, dr. \tag{15.4}$$

We take a sectorial strip of sector angle $d\vartheta$, radially extending from $r = R^\alpha$ to $r = R^\beta$ in the (y, z) plane as illustrated in Fig. 4.

To obtain the mechanical definition of the surface tension of a spherical interface in a manner similar to the planar case, let us imagine a hypothetical system composed of the bulk phases α and β, each being homogeneous right up to a spherical membrane with uniform tension γ_s and radius a_s. Then we may specify γ_s and a_s by the conditions that the hypothetical system is mechanically equivalent to the actual system with respect to both the resultant force and moment exerted on the sectorial strip. We may adopt the γ_s thus specified as the mechanical definition of the surface tension of a spherical interface.

The resultant stress, $d\Sigma_x$, acting in the x direction across the sectorial strip in the actual system is given by

$$d\Sigma_x = -d\vartheta \int_{R^\alpha}^{R^\beta} p_T(r)\, r\, dr. \tag{15.5}$$

On the other hand, in the hypothetical system, the stress, $d\Sigma^{\alpha\beta}$, acting in the x direction across the sectorial strip may be expressed in the form[1]

$$d\Sigma^{\alpha\beta} = -d\vartheta \int_{R^\alpha}^{R^\beta} p^{\alpha\beta} r\, dr + \gamma_s a_s d\vartheta, \tag{15.6}$$

$$p^{\alpha\beta} = p^\alpha[1 - A(r-a)] + p^\beta A(r-a); \quad A(r-a) = \begin{cases} 1, & r \geqq a \\ 0, & r < a. \end{cases} \tag{15.7}$$

Equating (15.5) to (15.6) we obtain the expression for the mechanical definition of surface tension of a spherical interface:

$$\gamma_s = \frac{1}{a_s} \int_{R^\alpha}^{R^\beta} (p^{\alpha\beta} - p_T)\, r\, dr. \tag{15.8}$$

The value of a_s is given by the condition for equivalence with respect to the resultant moment:

$$\int_{R^\alpha}^{R^\beta} (p^{\alpha\beta} - p_T)\, r\, (r - a_s)\, dr = 0. \tag{15.9}$$

If the actual interface were a membrane of zero thickness uniformly stretched with tension γ_s, the Kelvin equation (1.1) would be exactly satisfied as a consequence of mechanical equilibrium. Since the spherical membrane located at the position satisfying (15.9) with tension γ_s given by (15.8) is equivalent in the resultant mechanical effects to that of the actual interface, it will easily be understood that the Kelvin equation should be automatically satisfied by a_s and γ_s insofar as the actual system is in mechanical equilibrium.

This fact implies that the dividing surface satisfying the condition (15.9) is identical with the surface of tension which was defined in Sect. 9 as the dividing surface to exactly satisfy the Kelvin equation. In other words, (15.9) is the condition to determine the radius of the surface of tension, a_s. Anticipating these results we have used the notation γ_s and a_s.

Utilizing the fact that p_T reduces to p^α and p^β in the interiors of phases α and β, respectively, we may rewrite (15.9) in the form

$$\int_{-\infty}^{\infty} (p^{\alpha\beta} - p_T)\left(1 + \frac{1}{a_s}\xi\right) \xi\, d\xi = 0, \tag{15.10}$$

where

$$\xi = r - a_s. \tag{15.11}$$

In the limit of zero curvature (15.10) reduces to (14.7) to determine the surface of tension in the planar case.

[1] For $p^{\alpha\beta}$ appearing in (15.6), (15.8), (15.9), (15.10), (15.12) and (15.13) we set $a = a_s$, but generally a in (15.7) is the radius of an arbitrary dividing surface.

Similarly (15.8) is expressed in the form

$$\gamma_s = \int_{-\infty}^{\infty} (p^{\alpha\beta} - p_T)\left(1 + \frac{1}{a_s}\xi\right) d\xi. \tag{15.12}$$

With use of (15.10), we have an alternative expression for (15.12)

$$\gamma_s = \int_{-\infty}^{\infty} (p^{\alpha\beta} - p_T)\left(1 + \frac{1}{a_s}\xi\right)^2 d\xi. \tag{15.13}$$

Starting from the hydrostatic equilibrium condition, we now give a direct proof that the Kelvin relation is exactly satisfied by the dividing surface given by (15.10). For this purpose we rewrite the integral form of the hydrostatic equilibrium condition, (15.4) [*26*]:

$$p^\alpha - p^\beta = \frac{2}{a^2}\int_{R^\alpha}^{R^\beta} (p^{\alpha\beta} - p_T)\, r\, dr = \frac{2}{a}\int_{-\infty}^{\infty} (p^{\alpha\beta} - p_T)\left(1 + \frac{1}{a}\xi\right) d\xi, \tag{15.14}$$

$$\xi = r - a, \tag{15.15}$$

where a is the radius of an arbitrary dividing surface[1], and where we have used the relation

$$\int_{R^\alpha}^{R^\beta} p^{\alpha\beta}\, r\, dr = \tfrac{1}{2}\left[(R^\beta)^2 p^\beta - a^2 p^\beta + a^2 p^\alpha - (R^\alpha)^2 p^\alpha\right], \tag{15.16}$$

which can easily be obtained from the definition (15.7).

Then combination of (15.12) and (15.14) with $a = a_s$ at once leads to the Kelvin relation. Although the condition (15.10) has not been explicitly utilized in the above derivation of the Kelvin equation, it must be noted that γ_s given by (15.12) is identified with the surface tension only when it is referred to the surface of tension subject to the condition (15.10).

In the same manner as in the derivation of (15.14), we obtain from the second equation of (15.3)

$$p^\alpha - p^\beta = \frac{1}{a}\left[2\int_{-\infty}^{\infty} (p^{\alpha\beta} - p_T)\left(1 + \frac{1}{a}\xi\right)^2 d\xi + \int_{-\infty}^{\infty} (p^{\alpha\beta} - p_N)\left(1 + \frac{1}{a}\xi\right)^2 d\xi\right]. \tag{15.17}$$

Putting $a = a_s$ in (15.14) and (15.17), and combining the results with (15.10), we can eliminate $p^{\alpha\beta}$ from (15.13), obtaining the form of the Bakker equation:

$$\gamma_s = \int_{-\infty}^{\infty} (p_N - p_T)\left(1 + \frac{1}{a}\xi\right)^2 d\xi. \tag{15.18}$$

The exact derivation of (15.18) was first achieved by Buff [*26*].

16. Mechanical definition of surface tension through the work done by the system. The surface tension of a spherical interface has been uniquely defined in the last section from a purely mechanical standpoint. This method of defining the surface tension is, in contrast with the thermodynamic method, restricted to the surface of tension.

Let us again consider the two-phase system with a spherical interface that was used in Sect. 15. Then the pressure tensor is given by (15.1) and hence the force acting across a surface element of the conical wall of the vessel (see Fig. 4),

---

[1] See footnote 1, p. 160.

which lies between r and $r+dr$ and between φ and $\varphi+d\varphi$, may be expressed as $p_T(r)\, r \sin\vartheta\, dr\, d\varphi$. We now increase the solid angle of the cone by an infinitesimal amount $d\omega$. Then the total work done by the system across the conical wall is given by

$$dW = d\omega \int_{R^\alpha}^{R^\beta} p_T(r)\, r^2\, dr. \tag{16.1}$$

Combining with (8.4) we obtain

$$d\omega \int_{R^\alpha}^{R^\beta} p_T(r)\, r^2\, dr = p^\alpha dV^\alpha + p^\beta dV^\beta - \gamma\, dA, \tag{16.2}$$

where now $dA = a^2\, d\omega$, a being the radius of an arbitrary dividing surface. According to (8.1) to (8.3), the increases in V^α and V^β accompanying $d\omega$ are given respectively by

$$dV^\alpha = d\omega \int_{R^\alpha}^{a} r^2\, dr, \tag{16.3}$$

$$dV^\beta = d\omega \int_{a}^{R^\beta} r^2\, dr. \tag{16.4}$$

Substitution of (16.3) and (16.4) into (16.2) leads to

$$\gamma = \frac{1}{a^2} \int_{R^\alpha}^{R^\beta} (p^{\alpha\beta} - p_T)\, r^2\, dr, \tag{16.5}$$

where $p^{\alpha\beta}$ is defined by (15.7). The γ defined by (16.5) may be regarded as another mechanical definition of surface tension referred to an arbitrary dividing surface.

Differentiation of γ in (16.5) with respect to a gives [*9*]

$$\left[\frac{\partial \gamma}{\partial a}\right] = -\frac{2}{a^3} \int_{R^\alpha}^{R^\beta} (p^{\alpha\beta} - p_T)\, r^2\, dr + (p^\alpha - p^\beta), \tag{16.6}$$

where the term $(p^\alpha - p^\beta)$ results from the differentiation of $p^{\alpha\beta}$. Now we find that (16.6) is identical with the generalized Kelvin relation (8.8) and hence (16.5) is the desired expression conforming to the thermodynamic definition of surface tension.

If we use the variable $\xi = r - a$, we can rewrite (16.5) in an alternative form

$$\gamma = \int_{-\infty}^{\infty} (p^{\alpha\beta} - p_T) \left(1 + \frac{1}{a}\xi\right)^2 d\xi. \tag{16.7}$$

We then find that the condition for hydrostatic equilibrium, (15.14), reduces (16.6) to the Tolman-Buff equation [*11*], [*26*],

$$\left[\frac{\partial \gamma}{\partial a}\right] = -\frac{2}{a^2} \int_{-\infty}^{\infty} (p^{\alpha\beta} - p_T) \left(1 + \frac{1}{a}\xi\right) \xi\, d\xi. \tag{16.8}$$

Comparing (16.6) with another form of hydrostatic condition (15.17) we obtain

$$\left[\frac{\partial \gamma}{\partial a}\right] = \frac{1}{a} \int_{-\infty}^{\infty} (p^{\alpha\beta} - p_N) \left(1 + \frac{1}{a}\xi\right)^2 d\xi, \tag{16.9}$$

which was first derived by Buff [*26*].

It is seen from the Tolman-Buff equation (16.8) that the condition for determining the radius of the surface of tension, $[\partial\gamma/\partial a]=0$ (see Sect. 9), is identified with (15.9) or (15.10) derived from purely mechanical considerations.

It is obvious from (15.13) and (16.7)that for $a=a_s$ the γ defined here is identical with γ defined in Sect. 15.

III. Quasithermodynamics.

17. Introductory remarks. In the previous sections we have assumed that the pressure may be defined at every point as a function of that precise point even in the microscopic interface zone. This procedure is advantageous in providing insight into the microscopic structure of the interface. The extension of such a detailed treatment to all other thermodynamic quantities provides the fundamental basis for the quasithermodynamic theory to be developed in the following.

In general, the local formulation of thermodynamics is possible, as it is in the case of the thermodynamics of irreversible processes. But there is a prerequisite for such a local formulation of thermodynamics, namely, that all the spatial variations in physical conditions within any volume element of molecular dimensions be entirely negligible. However, since the interface layer is several molecules thick, as will be shown in Chaps. B and C, such spatial variation is only to be anticipated. Consequently, thermodynamic functions used in their local forms cannot be expected to be so well defined in the interface as in ordinary bulk phases. The theory of interface layers based on the local formulation of thermodynamics is called the quasithermodynamics of interface layers.

Thus the following quasithermodynamic theory, which is somewhat different from the original development of TOLMAN [*11*], is not strictly exact but is, nevertheless, useful in providing physical insight into the structure of interface layers without use of the exact but more complicated equations based on statistical mechanics. That is, quasithermodynamics serves as a bridge between macroscopic thermodynamics and molecular theory based on statistical mechanics.

18. Fundamental postulate of the quasithermodynamics for plane interface. For simplicity let us confine ourselves to the case of a one-component system. We shall assume that our system composed of two phases in equilibrium with a plane interface is confined, for convenience, to a rectangular parallelepiped vessel (see also Fig. 8, p. 181). Let us define a rectangular coordinate system (x, y, z) with the dividing surface in the interface layer as the (x, y) plane, and z axis normal to the interface directed from the phase α to the phase β. Let l_1, l_2, l_3, denote the edge lengths of the vessel directed along the coordinate axes, respectively, and let the lower and upper boundaries of the vessel be located at $z=-l^\alpha$ and $z=l^\beta$, respectively. Hence the volume of the vessel, V, is equal to $l_1 l_2 l_3$ and the area of the interface A equal to $l_1 l_2$.

In the case of ordinary macroscopic thermodynamics all intensive quantities are uniquely determined as functions of the density and temperature. Our fundamental postulate of quasithermodynamics is that any thermodynamic intensive quantity is uniquely determined at each point as a function of the temperature and the number density of molecules at that precise point[1].

Let $n(z)$, $v(z)$ and $f(z)$ be the number of molecules per unit volume, the volume per molecule and the Helmholtz free energy per molecule, respectively. From

[1] It seems to be an improvement over this postulate to take account of a contribution (from the derivatives of local density [J. W. CAHN and J. E. HILLIARD, J. Chem. Phys. **28**, 258 1958)]. — Seealso R. C. TOLMAN [*11*] and Sect. 34.

the definitions we have

$$v(z) = \frac{1}{n(z)}, \tag{18.1}$$

$$N = l_1 l_2 \int_{-l^\alpha}^{l^\beta} n(z)\, dz, \tag{18.2}$$

$$F = l_1 l_2 \int_{-l^\alpha}^{l^\beta} n(z)\, f(z)\, dz. \tag{18.3}$$

Furthermore the pressure tensor $\mathbf{p}$ is assumed to be expressed in the form (14.2) in terms of the tangential and normal pressure $p_T(z)$ and $p_N(z)$.

Let us imagine that the area A is increased by an arbitrary amount δA through a reversible and isothermal displacement of the side wall of the vessel. Then the elementary work done by the portion of the system located between z and $z + dz$ is

$$\delta w = p_T(z)\, \delta A\, dz, \tag{18.4}$$

provided that there are no external fields. This portion may not, in general, be a closed system because a certain number of molecules may flow into or out of the region due to the variation in the distribution of molecules accompanying the expansion process δA.

In order to proceed, however, it is convenient to consider a virtual variation so that the above infinitesimal system is a closed one. That is,

$$A\, \delta n(z) + n(z)\, \delta A = 0. \tag{18.5}$$

We shall often use the symbol δ indicating the change in the thermodynamic state of a system when we want to distinguish it from d indicating the change in position as used in dz.

Then, the work done (18.4) is to be equal to the decrease in the Helmholtz free energy of that portion, i.e., $-A\, n(z)\, \delta f(z)\, dz$, and hence we have

$$A\, n(z)\, \delta f(z) = -p_T(z)\, \delta A. \tag{18.6}$$

We obtain from the above two equations

$$\frac{\partial f(z)}{\partial n(z)} = \frac{p_T(z)}{(n(z))^2}. \tag{18.7}$$

Using $v(z)$ given by (18.1), instead of $n(z)$, we can rewrite (18.7) in an alternative form

$$\frac{\partial f(z)}{\partial v(z)} = -p_T(z). \tag{18.8}$$

19. Condition for equilibrium of plane interface. Due to the postulate we may write

$$f(z, T) = f(n(z, T); T). \tag{19.1}$$

This implies that at a given temperature the Helmholtz free energy of the system as a whole, F, is a functional of $n(z)$. Then the equilibrium form of $n(z)$ is to be determined by the condition that F is minimized consistent with the given temperature and volume.

The variation in F at constant T, V and N, derived from (18.3), can be written in the following form with the help of (18.2) and (18.7):

$$\delta F = l_1 l_2 \int_{-l^\alpha}^{l^\beta} \left(f(z) + \frac{p_T(z)}{n(z)} \right) \delta n(z)\, dz, \tag{19.2}$$

where the variation $\delta n(z)$ is subject to the assumed restrictions $\delta N = 0$ and $\delta V = 0$:

$$l_1 l_2 \int_{-l^\alpha}^{l^\beta} \delta n(z)\, dz = 0. \tag{19.3}$$

Using LAGRANGE's undetermined multiplier μ, we readily obtain from (19.2) and (19.3) the quasithermodynamic expression for equilibrium (see [9], [11]),

$$f(z) + p_T(z) \frac{1}{n(z)} = \mu. \tag{19.4}$$

It is evident that μ is the chemical potential common to the bulk phases α and β, for in the interiors of those bulk phases $p_T(z)$ reduces to the ordinary hydrostatic pressure p and hence the left-hand side of (19.4) to the chemical potential. Eq. (19.4) is one of the most fundamental equations for quasithermodynamics.

Differentiation of both sides of (18.3) with respect to T at constant V and N yields

$$-S = l_1 l_2 \int_{-l^\alpha}^{l^\beta} \left[n(z) \frac{\partial f(z)}{\partial T} + \left(n(z) \frac{\partial f(z)}{\partial n(z)} + f(z) \right) \frac{\partial n(z)}{\partial T} \right] dz, \tag{19.5}$$

where (19.1) has been utilized and the variation in $n(z)$ is subject to (19.3). Substitution of (18.7), combined with (19.4), into (19.5) and the use of (19.3) lead to the more transparent form

$$S = l_1 l_2 \int_{-l^\alpha}^{l^\beta} n(z)\, s(z)\, dz, \tag{19.6}$$

where

$$s(z) = -\frac{\partial f(z)}{\partial T}. \tag{19.7}$$

It is evident from (19.6) that $s(z)$ defined by (19.7) is the entropy per molecule at z. Hence the internal energy per molecule at z, $u(z)$, is given by

$$u(z) = f(z) + T s(z). \tag{19.8}$$

The internal energy $u(z)$ satisfies a relation similar to (19.6).

20. Quasithermodynamic relations for plane interface. We are now ready to derive the quasithermodynamic fundamental equations. From (19.1), (18.7) and (19.7) we have

$$\delta f = -s\, \delta T - p_T\, \delta\left(\frac{1}{n}\right). \tag{20.1}$$

This may also be written with use of (19.8) in the following form, (see [9])

$$\delta u = T\, \delta s - p_T\, \delta\left(\frac{1}{n}\right). \tag{20.2}$$

Differentiating (19.4) and combining the result with (20.1) we obtain [9]

$$-\delta p_T + n\, s\, \delta T + n\, \delta\mu = 0, \tag{20.3}$$

which corresponds to the Gibbs-Duhem equation.

Integrating both sides of (19.4), multiplied by $n(z)$, with respect to z and making use of (18.2) and (18.3), we obtain

$$-l_1 l_2 \int_{-l^\alpha}^{l^\beta} p_T\, dz = F - \mu N. \tag{20.4}$$

Combining with (5.5) we at once obtain the expression for the surface tension

$$\gamma = \int_{-\infty}^{\infty} (p - p_T)\, dz, \tag{20.5}$$

identical with (14.5) derived from the mechanical definition of the surface tension.

This implies that we can derive (14.5) from the quasithermodynamic equation for equilibrium, (19.4). This result is an empirical justification for our fundamental postulate of quasithermodynamics, but that is all.

We shall rewrite the thermodynamic relations obtained in Sect. 6 in quasithermodynamic form. For this purpose we introduce the following notation, in analogy with (15.7),

$$x^{\alpha\beta} = \big(1 - A(z)\big)\, x^{\alpha} + A(z)\, x^{\beta}, \tag{20.6}$$

$$\left.\begin{aligned} A(z) &= 1 \quad z \geqq 0, \\ &= 0 \quad z < 0, \end{aligned}\right\} \tag{20.7}$$

where x stands for an arbitrary thermodynamic quantity such as number density n.

It will be obvious from the definitions (2.1), (3.2) and (3.3) that for the superficial entropy $\sigma = S^s/A$, superficial energy $v = U^s/A$, and superficial number density $\Gamma = N^s/A$ we have respectively

$$\sigma = \int_{-\infty}^{\infty} (n\, s - n^{\alpha\beta} s^{\alpha\beta})\, dz, \tag{20.8}$$

$$v = \int_{-\infty}^{\infty} (n\, u - n^{\alpha\beta} u^{\alpha\beta})\, dz, \tag{20.9}$$

and

$$\Gamma = \int_{-\infty}^{\infty} (n - n^{\alpha\beta})\, dz, \tag{20.10}$$

where the (x, y) plane is assumed as the dividing surface to which the superficial quantities are referred.

On the other hand, the ordinary Gibbs-Duhem equations for phases α and β may be written as

$$-\delta p + n^{\alpha\beta} s^{\alpha\beta}\, \delta T + n^{\alpha\beta}\, \delta\mu = 0. \tag{20.11}$$

Subtracting (20.3) and integrating the result with respect to z, we obtain

$$\int_{-\infty}^{\infty} \delta(p - p_T)\, dz + \delta T \int_{-\infty}^{\infty} (n\, s - n^{\alpha\beta} s^{\alpha\beta})\, dz + \delta\mu \int_{-\infty}^{\infty} (n - n^{\alpha\beta})\, dz = 0. \tag{20.12}$$

With use of (20.5) and (20.8) to (20.10) we can at once show that (20.12) is identical with the Gibbs adsorption equation (6.8). This is another empirical justification of our fundamental postulate of quasithermodynamics.

21. Quasithermodynamic relations for spherical interface. Let us consider the same system as that considered in Sects. 8 and 15, and the same notation will be used here (see Figs. 3 and 4). In the spherical case thermodynamic intensive quantities are to be considered as functions of r alone. Then, corresponding to (18.1) to (18.3) we have

$$v(r) = \frac{1}{n(r)}, \tag{21.1}$$

$$N = \omega \int_{R^{\alpha}}^{R^{\beta}} n(r)\, r^2\, dr, \tag{21.2}$$

$$F = \omega \int_{R^{\alpha}}^{R^{\beta}} n(r)\, f(r)\, r^2\, dr. \tag{21.3}$$

The pressure tensor $\mathbf{p}$ assumes the form (15.1), i.e., both the tangential pressure p_T and the normal pressure p_N depend only on r.

Just as in the planar case we can derive the quasithermodynamic equation for the equilibrium of a spherical interface,

$$f(r) + p_T(r)\,\frac{1}{n(r)} = \mu. \tag{21.4}$$

This result is identical in form with (19.4) for a plane interface.

If we integrate both sides of (21.4), multiplied by $\omega r^2 n$, with respect to r, from R^α to R^β, we obtain with the help of (21.2) and (21.3),

$$-\omega \int_{R^\alpha}^{R^\beta} p_T(r)\, r^2\, dr = F - \mu N. \tag{21.5}$$

Substituting the thermodynamic relation (10.9) combined with (8.1) to (8.3) into (21.5) and rewriting the result by making use of the notation $p^{\alpha\beta}$ given by (15.7), we readily obtain the following expression as the surface tension γ referred to an arbitrary dividing surface of radius a:

$$\gamma = \frac{1}{a^2} \int_{R^\alpha}^{R^\beta} (p^{\alpha\beta} - p_T)\, r^2\, dr, \tag{21.6}$$

or in the conventional form,

$$\gamma = \int_{-\infty}^{\infty} (p^{\alpha\beta} - p_T) \left(1 + \frac{1}{a}\,\xi\right)^2 d\xi, \tag{21.7}$$

which is identical with the expression (16.7) derived from the mechanical definition of surface tension.

For the quasithermodynamic forms of the superficial entropy σ, superficial energy v and superficial number density Γ we have, (see [11])

$$\sigma = \int_{-\infty}^{\infty} (n\, s - n^{\alpha\beta} s^{\alpha\beta}) \left(1 + \frac{1}{a}\,\xi\right)^2 d\xi, \tag{21.8}$$

$$v = \int_{-\infty}^{\infty} (n\, u - n^{\alpha\beta} u^{\alpha\beta}) \left(1 + \frac{1}{a}\,\xi\right)^2 d\xi, \tag{21.9}$$

$$\Gamma = \int_{-\infty}^{\infty} (n - n^{\alpha\beta}) \left(1 + \frac{1}{a}\,\xi\right)^2 d\xi, \tag{21.10}$$

where the notation (15.15) and (20.6) have been used.

Since the Gibbs-Duhem equation expressed in the form of (20.11) is valid for both the bulk phases α and β, whether the interface is plane or spherical, we obtain[1] from (21.4)

$$\left.\begin{aligned} &\int_{-\infty}^{\infty} (\delta(p^{\alpha\beta} - p_T)) \left(1 + \frac{1}{a}\,\xi\right)^2 d\xi + \delta T \int_{-\infty}^{\infty} (n\, s - n^{\alpha\beta} s^{\alpha\beta}) \left(1 + \frac{1}{a}\,\xi\right)^2 d\xi + \\ &\qquad + \delta\mu \int_{-\infty}^{\infty} (n - n^{\alpha\beta}) \left(1 + \frac{1}{a}\,\xi\right)^2 d\xi = 0. \end{aligned}\right\} \tag{21.11}$$

[1] A similar relation to (20.1) has been used.

On the other hand, total differentiation of (21.7) yields

$$\delta\gamma = \int_{-\infty}^{\infty} \left(\delta(p^{\alpha\beta} - p_T)\right)\left(1 + \frac{1}{a}\xi\right)^2 d\xi + \left[\frac{\partial\gamma}{\partial a}\right] da. \tag{21.12}$$

Substitution of (21.12), (21.8) and (21.10) into (21.11) leads at once to [9]

$$\delta\gamma + \sigma\,\delta T + \Gamma\,\delta\mu = \left[\frac{\partial\gamma}{\partial a}\right] da, \tag{21.13}$$

which is just the generalized Gibbs adsorption equation (11.9) derived on the basis of ordinary thermodynamics.

IV. Application of thermodynamics of irreversible processes.

22. Interface between two phases in non-equilibrium [12], [13]. *α) Affinity of adsorption.* Let us now consider a system with a plane interface separating two phases α and β which are in mechanical and thermal equilibrium but not in chemical equilibrium. That is, temperature T and pressure p are uniform throughout the system but the chemical potential μ_i^α in phase α is not necessarily equal to the chemical potential μ_i^β in phase β. In the transition zone, the chemical potential μ_i may vary continuously with position along the normal to the interface.

We apply thermodynamics of irreversible processes[1,2] to this non-equilibrium system. Then, the change in entropy, dS, is split into two parts: $d_e S$, the entropy flow due to interaction with the exterior, and $d_i S$, the contribution due to irreversible processes inside the system. Thus, we have

$$dS = d_e S + d_i S. \tag{22.1}$$

Furthermore, if the system is closed, we have

$$d_e S = \frac{dQ}{T}, \tag{22.2}$$

where dQ is the heat received by the system. The internal production of entropy $d_i S$ can never be negative according to the second law of thermodynamics.

For simplicity, we shall not consider other chemical reactions than adsorption on the interface and phase transition between bulk phases α and β.

At first, we assume that the system is closed. Then, N_i^α, N_i^β and N_i^s, which are now variables dependent on time, are expressed, with use of initial values N_{i0}^α, N_{i0}^β and N_{i0}^s, in the forms

$$N_i^\alpha - N_{i0}^\alpha = -\xi_i^\alpha; \quad N_i^\beta - N_{i0}^\beta = -\xi_i^\beta; \quad N_i^s - N_{i0}^s = \xi_i^\alpha + \xi_i^\beta, \tag{22.3}$$

where ξ_i^α and ξ_i^β are called the degrees of advancement[2] of adsorption reaction in phases α and β, respectively. The above equations mean that the migration of molecules from phase α to phase β is a linear combination of two adsorption reactions: adsorption from phase α and that from phase β.

The irreversible production of entropy during an infinitesimal spontaneous transformation of ξ_i^α's and ξ_i^β's, is expressed in the form

$$T\,d_i S = \sum_{i=1}^{\varkappa} K_i^\alpha\, d\xi_i^\alpha + \sum_{i=1}^{\varkappa} K_i^\beta\, d\xi_i^\beta, \tag{22.4}$$

[1] I. PRIGOGINE: Étude thermodynamique des phénomènes irréversibles. Liège: Editions Desoer 1947.

[2] T. DE DONDER: L'Affinité. Paris: Gauthier-Villars 1927.

where K_i^α and K_i^β are the affinities of adsorption at the interface of molecules of species i from phases α and β, respectively. If temperature T, volume V and area A of the interface are kept constant, the increase in the internal energy due to the above infinitesimal transformation, can be expressed in the form

$$dU = -\sum_{i=1}^{\varkappa} r_i^\alpha \, d\xi_i^\alpha - \sum_{i=1}^{\varkappa} r_i^\beta \, d\xi_i^\beta, \tag{22.5}$$

which is equal to the amount of heat received by the system in the above process, according to the first law. This means that r_i^α is the heat of adsorption from phase α, i.e. the heat released from the system when one molecule is adsorbed at the interface from bulk phase α. With use of (22.2), (22.4) and (22.5), we can rewrite express (22.1) in the form

$$T\,dS = \sum_{i=1}^{\varkappa} (K_i^\alpha - r_i^\alpha)\, d\xi_i^\alpha + \sum_{i=1}^{\varkappa} (K_i^\beta - r_i^\beta)\, d\xi_i^\beta. \tag{22.6}$$

From (22.5) and (22.6) we obtain the expression for the increase in the Helmholtz free energy,

$$dF = -\sum_{i=1}^{\varkappa} K_i^\alpha \, d\xi_i^\alpha - \sum_{i=1}^{\varkappa} K_i^\beta \, d\xi_i^\beta. \tag{22.7}$$

In the thermodynamics of irreversible processes it is generally assumed that the entropy of the system is given as a function of macroscopic variables T, A, V, $\boldsymbol{N}$, $\boldsymbol{\xi}^\alpha$, $\boldsymbol{\xi}^\beta$[1]. Then we immediately obtain from (22.6)

$$\left.\begin{aligned} K_i^\alpha &= r_i^\alpha + T\left(\frac{\partial S}{\partial \xi_i^\alpha}\right)_{T,V,A,\boldsymbol{N},\xi_j^\alpha,\boldsymbol{\xi}^\beta}, \\ K_i^\beta &= r_i^\beta + T\left(\frac{\partial S}{\partial \xi_i^\beta}\right)_{T,V,A,\boldsymbol{N},\xi_j^\beta,\boldsymbol{\xi}^\alpha}. \end{aligned}\right\} \tag{22.8}$$

These equations enable us, in principle, to calculate affinites of adsorption, K_i^α and K_i^β, from heat of adsorption and entropy change due to adsorption.

Thus far we have dealt only with closed systems. For the theory of adsorption, however, it is more convenient to consider open systems. It is assumed, as in the case of equilibrium, that the Helmholtz free energy is expressible as a function of T, V^α, V^β, A, $\boldsymbol{N}^\alpha$, $\boldsymbol{N}^\beta$ and $\boldsymbol{N}^s$. Then we have

$$dF = -S\,dT - p\,dV^\alpha - p\,dV^\beta + \gamma\,dA + \sum_{i=1}^{\varkappa} (\bar\mu_i^\alpha\, dN_i^\alpha + \bar\mu_i^\beta\, dN_i^\beta + \mu_i^s\, dN_i^s), \tag{22.9}$$

where $\bar\mu_i^\alpha$ and $\bar\mu_i^\beta$ are called the *complete chemical potentials*, to be distinguished from the ordinary ones and defined by

$$\bar\mu_i^\alpha = (\partial F/\partial N_i^\alpha)_{T,V,A,N_j^\alpha,\boldsymbol{N}^\beta,\boldsymbol{N}^s}, \tag{22.10}$$

$$\bar\mu_i^\beta = (\partial F/\partial N_i^\beta)_{T,V,A,N_j^\beta,\boldsymbol{N}^\alpha,\boldsymbol{N}^s}, \tag{22.11}$$

$$\mu_i^s = (\partial F/\partial N_i^s)_{T,V,A,N_j^s,\boldsymbol{N}^\alpha,\boldsymbol{N}^\beta}. \tag{22.12}$$

Recalling the discussion on superficial quantities given in Sect. 3 we assume

$$\left.\begin{aligned} F^s/A = f^s &= f^s(T, \Gamma_1, \ldots, \Gamma_\varkappa, n_i^\alpha, \ldots, n_\varkappa^\alpha, n_i^\beta, \ldots, n_\varkappa^\beta), \\ n_i^\alpha &= N_i^\alpha/V^\alpha; \quad n_i^\beta = N_i^\beta/V^\beta. \end{aligned}\right\} \tag{22.13}$$

[1] The bold-face symbol $\boldsymbol{\xi}$ stands for the set of $\varkappa$ variables $\xi_1, \ldots, \xi_\varkappa$; see also footnote 3, p. 140.

Then by utilizing (3.1) and (22.13), we can readily rewrite (22.10) to (22.12) in the alternative forms,

$$\bar{\mu}_i^\alpha = \mu_i^\alpha + \frac{A}{V^\alpha}\,\varepsilon_i^\alpha, \tag{22.14}$$

$$\bar{\mu}_i^\beta = \mu_i^\beta + \frac{A}{V^\beta}\,\varepsilon_i^\beta, \tag{22.15}$$

$$\mu_i^s = (\partial f^s/\partial \Gamma_i)_{T,\Gamma_j,\boldsymbol{n}^\alpha,\boldsymbol{n}^\beta}, \tag{22.16}$$

where

$$\mu_i^\alpha = (\partial F^\alpha/\partial N_i^\alpha)_{T,V^\alpha,N_j^\alpha}, \tag{22.17}$$

$$\mu_i^\beta = (\partial F^\beta/\partial N_i^\beta)_{T,V^\beta,N^\beta}, \tag{22.18}$$

$$\varepsilon_i^\alpha = (\partial f^s/\partial n_i^\alpha)_{T,\boldsymbol{\Gamma},n_j^\alpha,\boldsymbol{n}^\beta}, \tag{22.19}$$

$$\varepsilon_i^\beta = (\partial f^s/\partial n_i^\beta)_{T,\boldsymbol{\Gamma},n_j^\beta,\boldsymbol{n}^\alpha}. \tag{22.20}$$

The symbols μ_i^α and μ_i^β denote ordinary chemical potentials, which are in the present case not necessarily equal to each other. The new quantities ε_i^α and ε_i^β are called, following Defay [12], the *lateral chemical potentials*, which are responsible for the difference between $\bar{\mu}_i^\alpha$ and μ_i^α, and that between $\bar{\mu}_i^\beta$ and μ_i^β, respectively.

Thus, with use of (22.14) and (22.15), we can rewrite (22.9) in the form

$$\left.\begin{aligned} dF = -\,S\,dT - p\,dV^\alpha - p\,dV^\beta + \gamma\,dA + \sum_{i=1}^{\varkappa}\left(\mu_i^\alpha + \frac{A}{V^\alpha}\,\varepsilon_i^\alpha\right) dN_i^\alpha + {}&\\ + \sum_{i=1}^{\varkappa}\left(\mu_i^\beta + \frac{A}{V^\beta}\,\varepsilon_i^\beta\right) dN_i^\beta + \sum_{i=1}^{\varkappa}\mu_i^s\,dN_i^s.&\end{aligned}\right\} \tag{22.21}$$

β) Gibbs adsorption formula generalized to non-equilibrium [12], [13]. We now consider the Helmholtz free energy change in the simple adsorption process given by (22.3). Comparing (22.7) with (22.21) which is now subject to (22.3), we find

$$K_i^\alpha = \mu_i^\alpha - \mu_i^s + \frac{A}{V^\alpha}\,\varepsilon_i^\alpha, \tag{22.22}$$

$$K_i^\beta = \mu_i^\beta - \mu_i^s + \frac{A}{V^\beta}\,\varepsilon_i^\beta. \tag{22.23}$$

With use of (22.10) to (22.16), we obtain from (22.22) and (22.23)

$$K_i^\alpha = \left(\frac{\partial F}{\partial N_i^\alpha}\right)_{T,V,A,N_j^\alpha,\boldsymbol{N}^\beta,\boldsymbol{N}^s} - \left(\frac{\partial F}{\partial N_i^s}\right)_{T,V,A,N_j^s,\boldsymbol{N}^\alpha,\boldsymbol{N}^\beta}, \tag{22.24}$$

$$K_i^\beta = \left(\frac{\partial F}{\partial N_i^\beta}\right)_{T,V,A,N_j^\beta,\boldsymbol{N}^\alpha,\boldsymbol{N}^s} - \left(\frac{\partial F}{\partial N_i^s}\right)_{T,V,A,N_j^s,\boldsymbol{N}^\alpha,\boldsymbol{N}^\beta}. \tag{22.25}$$

The Gibbs fundamental equations for the bulk phases α and β are

$$dF^\alpha = -\,S^\alpha\,dT - p\,dV^\alpha + \sum_{i=1}^{\varkappa}\mu_i^\alpha\,dN_i^\alpha, \tag{22.26}$$

$$dF^\beta = -\,S^\beta\,dT - p\,dV^\beta + \sum_{i=1}^{\varkappa}\mu_i^\beta\,dN_i^\beta. \tag{22.27}$$

Subtracting (22.26) and (22.27) from (22.21) and recalling (3.1) and (3.3) we obtain

$$dF^s = -\,S^s\,dT + \gamma\,dA + \sum_{i=1}^{\varkappa}\frac{A}{V^\alpha}\,\varepsilon_i^\alpha\,dN_i^\alpha + \sum_{i=1}^{\varkappa}\frac{A}{V^\beta}\,\varepsilon_i^\beta\,dN_i^\beta + \sum_{i=1}^{\varkappa}\mu_i^s\,dN_i^s. \tag{22.28}$$

Since F^s is independent of V^α and V^β as easily seen from (22.28), we obtain from (22.13)

$$\sum_{i=1}^{\varkappa} n_i^\alpha \left(\frac{\partial f^s}{\partial n_i^\alpha}\right)_{T,\Gamma,n_j^\alpha,n^\beta} = 0, \tag{22.29}$$

$$\sum_{i=1}^{\varkappa} n_i^\beta \left(\frac{\partial f^s}{\partial n_i^\beta}\right)_{T,\Gamma,n_j^\beta,n^\alpha} = 0. \tag{22.30}$$

Combining with (22.19) and (22.20), respectively, we obtain the important relations

$$\sum_{i=1}^{\varkappa} n_i^\alpha \varepsilon_i^\alpha = 0, \tag{22.31}$$

$$\sum_{i=1}^{\varkappa} n_i^\beta \varepsilon_i^\beta = 0. \tag{22.32}$$

Since F is a homogeneous function of first degree in V^α, V^β, A, $\boldsymbol{N}^\alpha$, $\boldsymbol{N}^\beta$, $\boldsymbol{N}^s$, using EULER's theorem we obtain from (22.21)

$$\left.\begin{aligned} F = \sum_{i=1}^{\varkappa}\left(\mu_i^\alpha + \frac{A}{V^\alpha}\varepsilon_i^\alpha\right)N_i^\alpha + \sum_{i=1}^{\varkappa}\left(\mu_i^\beta + \frac{A}{V^\beta}\varepsilon_i^\beta\right)N_i^\beta + \sum_{i=1}^{\varkappa}\mu_i^\beta N_i^s - \\ - p V^\alpha - p V^\beta + \gamma A. \end{aligned}\right\} \tag{22.33}$$

Subtracting the corresponding integral formulae for bulk phases α and β, $F^\alpha = \sum_{i=1}^{\varkappa} \mu_i^\alpha N_i^\alpha - p V^\alpha$ and $F^\beta = \sum_{i=1}^{\varkappa} \mu_i^\beta N_i^\beta - p V^\beta$, from (22.33) and using (22.31) and (22.32), we have

$$F^s = \sum_{i=1}^{\varkappa} \mu_i^s N_i^s + \gamma A, \tag{22.34}$$

or in terms of superficial densities

$$\gamma = f^s - \sum_{i=1}^{\varkappa} \mu_i^s \Gamma_i. \tag{22.35}$$

This is identical in form with (5.8) for surface tension of an equilibrium system.

Differentiating (22.34), equating the result to (22.28) and using (22.31) and (22.32) we obtain the desired equation [12], [13]

$$-d\gamma = \sigma\, dT + \sum_{i=1}^{\varkappa} \Gamma_i\, d\mu_i^s + \sum_{i=1}^{\varkappa} n_i^\alpha\, d\varepsilon_i^\alpha + \sum_{i=1}^{\varkappa} n_i^\beta\, d\varepsilon_i^\beta, \tag{22.36}$$

σ being the surface entropy given by (6.6). This is a generalization of the Gibbs equation (6.8). Eq. (22.36) differs from (6.8) by the presence of terms due to lateral chemical potentials. These terms vanish for the case of thermodynamic equilibrium and hence (22.36) reduces to (6.8).

It should be noted that μ_i^s, ε_i^α and ε_i^β are not independent of the location of the dividing surface and (22.36) and any other equations derived from (22.4) are valid only when referred to the special dividing surface, which satisfies both of the relations (22.31) and (22.32). This implies that the relation (22.4) itself is not valid except for this special dividing surface. The condition for invariance has been discussed by DEFAY [14] in some detail.

However, it is also possible from a purely logical point of view that the dividing surface which satisfies (22.31) may generally differ from that which satisfies (22.32), since no relation between ε_i^α and ε_i^β has been furnished from the present theory. This situation suggests that the validity of the macroscopic irreversible thermodynamics is restricted to the special cases when applied to the surface layer in which the thermodynamic quantities change appreciable even within the volume of molecular dimensions, and that the further progress in this direction seems to be difficult without considering the detailed changes within the surface layer explicitly as done in the quasithermodynamic theory[1].

V. Empirical equations for temperature dependence of surface tension.

23. The principle of corresponding states and empirical formulae for surface tension. α) *The principle of corresponding states.* To describe the p-V-T behavior of gases and liquids, various semi-quantitative equations of state have been proposed, of which the most familiar is that of VAN DER WAALS:

$$\left(p + \frac{N_0^2 a}{V^2}\right)(V - N_0 b) = N_0 kT \tag{23.1}$$

where N_0 is the Avogadro number, V the molar volume and a and b VAN DER WAALS' constants per molecule.

The critical point is usually determined as the point for which $(\partial p/\partial V)_T$ and $(\partial^2 p/\partial V^2)_T$ vanish. The values of p, V, T at this point are denoted by p_c, V_c, T_c, respectively. For temperatures lower than the critical, the isotherms of (23.1) show the van der Waals loop. In this region horizontal lines drawn to connect the volumes at the same pressure and the same Gibbs free energy on the two branches of the curve for which $(\partial p/\partial V)_T < 0$ are considered as corresponding to the coexistence of liquid and vapor phase.

We introduce here the reduced variables

$$p_r = \frac{p}{p_c}; \quad V_r = \frac{V}{V_c}; \quad T_r = \frac{T}{T_c}. \tag{23.2}$$

Use of those variables enables us to eliminate a and b from (23.1). This means that for all substances obeying the van der Waals equation one of p_r, V_r and T_r is a universal function of the remaining two.

Although the validity of the equation deduced from (23.1) is only semi-quantitative, it has been empirically justified for some groups of similar substances[2] that one of the three reduced variables is a universal function of the remaining two. This is the principle of corresponding states. According to this principle, any dimensionless quantity such as the compressibility factor pV/NkT is a universal function of p_r and T_r.

For surface tension, we chose the dimensionless quantity $\gamma_r = \gamma v_c^{\frac{2}{3}} (kT_c)^{-1}$ to which we refer as the *reduced surface tension*, v_c being the critical volume per molecule. Then we would expect that γ_r would be a universal function of T_r for substances obeying the principle of corresponding states:

$$\gamma = \frac{kT_c}{v_c^{\frac{2}{3}}} G\left(\frac{T}{T_c}\right), \tag{23.3}$$

where G is a universal function. The value of γ_r is plotted against T/T_c in Fig. 5 for neon, argon, nitrogen and oxygen. As seen from Fig. 5, the principle of

[1] It has been shown by S. KONDO [Besseiron Kenkyu **2**, 926 (1956) (in Japanese)] that a more general treatment of non-equilibrium surface tension is possible without using the concept of lateral chemical potentials.

[2] K. S. PITZER: J. Chem. Phys. **7**, 583 (1939). — E. A. GUGGENHEIM [*15*].

corresponding states applies well to surface tension for A, N_2, O_2. The appreciable deviations for neon will be discussed in Sect. 23γ.

It is also possible to employ the constants of intermolecular force to define the reduced variables. For chemically saturated molecules the potential energy of the system is assumed to be the sum of terms each representing the interaction energy of a pair of molecules. The intermolecular potential energy $\phi(r)$ of a pair of spherical or nearly spherical molecules depends only on the distance r between these two molecules.

Furthermore, it is reasonable to assume that the intermolecular potential can be represented in the form $\phi(r) = \varepsilon f(r/D)$, where f is a universal function[1] and ε and D are two scale factors characteristic of molecular species. As an approximate form of this function we often use the Lennard-Jones potential:

$$\phi(r) = 4\varepsilon\left\{\left(\frac{D}{r}\right)^{12} - \left(\frac{D}{r}\right)^{6}\right\}. \tag{23.4}$$

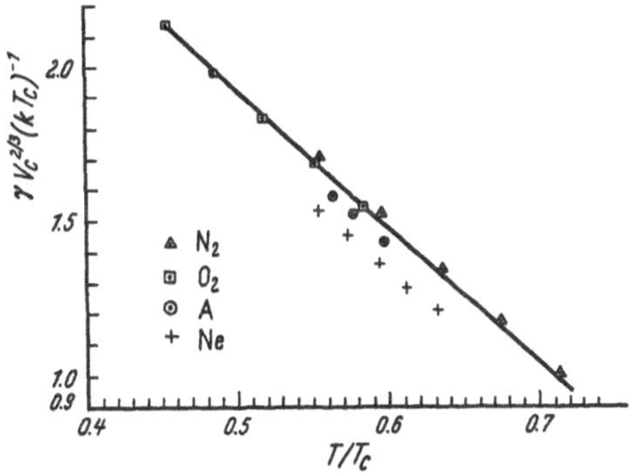

Fig. 5. Relation between surface tension and temperature in reduced units.

In terms of this potential function many properties of gases and liquids have been calculated, based on the method of statistical mechanics. The values of ε and D determined from the experimental data on second virial coefficients are given in Table 2; D is the distance for which $\phi(r) = 0$ and $-\varepsilon$ is the minimum value of $\phi(r)$. Then, corresponding to (23.2), we can conveniently define another set of three reduced variables

$$p^* = \frac{p D^3}{\varepsilon}; \quad v^* = \frac{V}{N D^3}; \quad T^* = \frac{kT}{\varepsilon}, \tag{23.5}$$

where N is the number of molecules in volume V. Similarly, the reduced surface tension can be expressed by $\gamma^* = \gamma D^2/\varepsilon$. Then corresponding to (23.3) we have

$$\gamma^* = \frac{\gamma D^2}{\varepsilon} = \gamma^*(T^*). \tag{23.6}$$

That is, γ^* is a universal function of T^*. If γ^* is plotted against T^* for A, N_2, O_2, we obtain a curve similar to that shown in Fig. 5 (see also Fig. 6).

Table 2. *Force constants for the Lennard-Jones potential.*

Gas	ε/k (°K)	D (Å)	Gas	ε/k (°K)	D (Å)
He (*Qu*)	10.22	2.556	Ne	34.9	2.78
He (*Cl*)	6.03	2.63	A	119.8	3.405
H_2 (*Qu*)	37.00	2.93	Kr	171	3.60
H_2 (Cl)	29.2	2.87	Xe	221	4.10
D_2 (*Qu*)	37.00	2.93	N_2	95.9	3.71
D_2 (*Cl*)	31.1	2.87	O_2	118	3.46

There are two sets of force constants for the lighter elements. The values labeled *Qu* were determined from quantum mechanical formulae and those labelled *Cl* were determined from classical ones, which are not strictly applicable at very low temperatures. The former are the true force constants, whereas the latter are only effective ones. For accurate calculations, and high-temperature extrapolations the *Qu* parameters must be used.

[1] It can be proved by statistical mechanics that the principle of the corresponding state actually follows from the form of the intermolecular potential $\phi(r) = \varepsilon f(r/D)$; cf. Sect. 27.

β) Empirical equations for surface tension. In general, observed values of surface tension of liquids decrease with increasing temperature. One of the earliest and best known empirical equations for the temperature dependence of surface tension is that of EÖTVÖS[1]:

$$\gamma \propto \frac{1}{(V^\alpha)^{\frac{2}{3}}}(T_0 - T), \tag{23.7}$$

where V^α is the molar volume of a liquid and T_0 is a temperature nearly equal to T_c. However, this formula was found to be inexact and various modifications of it have been proposed, of which the most satisfactory is that of KATAYAMA[2]:

$$\gamma \propto y^{\frac{2}{3}}\left(1 - \frac{T}{T_c}\right), \tag{23.8}$$

$$y V_c = \frac{\varrho^\alpha - \varrho^\beta}{\varrho_c}, \tag{23.9}$$

where V_c, ϱ^α, ϱ^β, and ϱ_c are the critical molar volume, orthobaric densities of the liquid and the vapor, and the critical density, respectively. This is a striking improvement over the Eötvös formula as was shown by KATAYAMA for various organic compounds.

On the other hand, GUGGENHEIM [*15*], [*4*] showed that the orthobaric densities ϱ^α and ϱ^β of simple molecules satisfy with high accuracy the following empirical formula

$$\frac{\varrho^\alpha - \varrho^\beta}{\varrho_c} = \frac{7}{3}\left(1 - \frac{T}{T_c}\right)^{\frac{1}{3}}. \tag{23.10}$$

Combining (23.10) with (23.9), we obtain from (23.8) the Katayama-Guggenheim formula [*15*], [*4*]

$$\gamma = \gamma_0 \left(1 - \frac{T}{T_c}\right)^{\frac{11}{9}}. \tag{23.11}$$

As shown in Table 3, the calculated values are in very good agreement with experiment.

As seen from (23.3), $\gamma_0 V_c^{\frac{2}{3}} T_c^{-1}$ should assume a common value for substances obeying the principle of corresponding states. The data of Table 3 show that this is in fact the case.

MACLEOD[3] proposed an empirical equation for surface tension

$$\gamma = \text{const}\,(\varrho^\alpha - \varrho^\beta)^4. \tag{23.12}$$

This was rewritten by SUGDEN[4] in the form

$$P = \frac{M \gamma^{\frac{1}{4}}}{\varrho^\alpha - \varrho^\beta}, \tag{23.13}$$

where M is the molecular weight. The constant P is called the "parachor". The comparative study of parachors for various liquids has proved that P is a constant (very nearly) for any given substance, independent of T over a wide range and that the parachor of a molecule is approximately equal to the sum of the parachors of its component atoms. FOWLER [*27*] proposed a tentative theory to derive MCLEOD's equation from general statistical considerations, and obtained a statistical mechanical expression for the parachor P.

---

[1] R. EÖTVÖS: Wied. Ann. Physik **27**, 448 (1886).
[2] M. KATAYAMA: Sci. Rep. Tôhoku Imp. Univ. **4**, 373 (1916).
[3] D.B. MACLEOD: Trans. Faraday Soc. **19**, 38 (1923).
[4] S. SUGDEN: The Parachor and Valency. London: Routledge 1930.

Table 3. *A comparison of the Katayama-Guggenheim formula* $\gamma = \gamma_0 \left(1 - \frac{T_c}{T}\right)^{\frac{11}{9}}$ *with experimental values for simple molecules* (GUGGENHEIM, [*15*]).

Ne			A			N_2			O_2		
$T_c = 44.8°$ K $\gamma_0 = 15.1$ dyne/cm			$T_c = 150.7°$ K $\gamma_0 = 36.31$ dyne/cm			$T_c = 126.0°$ K $\gamma_0 = 28.4$ dyne/cm			$T_c = 154.3°$ K $\gamma_0 = 38.4$ dyne/cm		
T °K	γ (dyne/cm)		T °K	γ (dyne/cm)		T °K	γ (dyne/cm)		T °K	γ (dyne/cm)	
	calc.	obs.		calc.	obs.		calc.	obs.		calc.	obs.
24.8	5.64	5.61	85.0	13.16	13.19	70.0	10.54	10.53	70.0	18.34	18.35
25.7	5.33	5.33	87.0	12.67	12.68	95.0	9.40	9.39	75.0	17.02	17.0
26.6	5.02	4.99	90.0	11.95	11.91	80.0	8.29	8.27	80.0	15.72	15.73
27.4	4.75	4.69				85.0	7.20	7.20	85.0	14.44	14.5
28.3	4.45	4.44				90.0	6.14	6.16	90.0	13.17	13.23
$V_c = 41.7$ cm³/mole $\gamma_0 V_c^{\frac{2}{3}} T_c^{-1}$ $= 4.05$ erg deg.$^{-1}$ mole$^{-\frac{2}{3}}$			$V_c = 75.3$ cm³/mole $\gamma_0 V_c^{\frac{2}{3}} T_c^{-1}$ $= 4.3$ erg deg.$^{-1}$ mole$^{-\frac{2}{3}}$			$V_c = 90.2$ cm³/mole $\gamma_0 V_c^{\frac{2}{3}} T_c^{-1}$ $= 4.5$ erg deg.$^{-1}$ mole$^{-\frac{2}{3}}$			$V_c = 74.5$ cm³/mole $\gamma_0 V_c^{\frac{2}{3}} T_c^{-1}$ $= 4.4$ erg deg.$^{-1}$ mole$^{-\frac{2}{3}}$		

But, the Katayama-Guggenheim equation (23.11) combined with (23.10) leads to [*15*], [*4*]

$$\gamma = \text{const}\,(\varrho^{\alpha} - \varrho^{\beta})^{\frac{11}{3}}. \tag{23.14}$$

Actually, for simple molecules this equation is more accurate than McLEOD's equation.

γ) Quantum corrections to the principle of corresponding states. As seen in Fig. 5, deviations from the principle of corresponding states are appreciable for neon and such deviations become more important for D_2, H_2 and He. Since these substances exist in the liquid state at very low temperatures where quantum mechanical effects become important, these deviations are expected to be due to quantum effects.

It was first suggested by BYK[1] that the classical principle of corresponding states should be modified so as to include PLANCK's constant h (which is responsible for quantum effects) in dimensionless form. If we use molecular parameters D and ε, we have a dimensionless combination

$$\Lambda^* = \frac{h}{D\sqrt{m\,\varepsilon}}, \tag{23.15}$$

which was first introduced by DE BOER and his coworkers[2-4]. Since $h/\sqrt{m\varepsilon}$ is the de Broglie wavelength (corresponding to a system of reduced mass $\mu = m/2$ and energy ε), Λ^* expresses the importance of diffraction effects due to the wave nature of molecules. Then we may expect the quantum deviations from the classical principle of corresponding states to depend on the magnitude of parameter Λ^*.

Then, by analogy with the classical relation (23.6), the reduced surface tension may be expressed in the form

$$\gamma^* = \gamma^*(T^*, \Lambda^*). \tag{23.16}$$

[1] A. BYK: Ann. d. Phys. **66**, 157 (1921); **69**, 161 (1922).
[2] J. DE BOER: Physica, Haag **14**, 139 (1948).
[3] J. DE BOER and B. S. BLAISSE: Physica, Haag **14**, 149 (1948).
[4] J. DE BOER and R. J. LUNBECK: Physica, Haag **14**, 520 (1948).

This means that the reduced surface tension is a function of the reduced temperature and depends on the quantum mechanical parameter Λ^*.

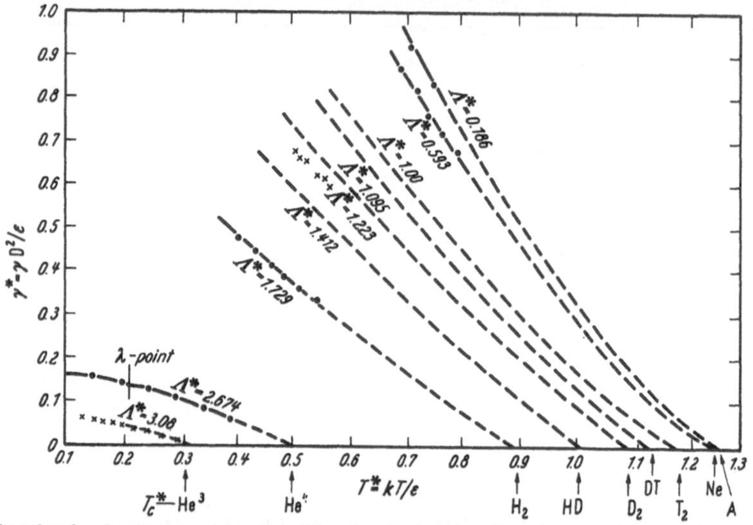

Fig. 6. The reduced surface tension as a function of the reduced temperature T^* and the quantum mechanical parameter Λ^*. The arrows above the atomic and molecular symbols are pointing to the reduced critical temperatures of the various substances (after J. DE BOER and R. B. BIRD: Quantum Theory and Equation of State. In: Molecular Theory of Gases and Liquids by J. O. HIRSCHFELDER, C. F. CURTISS and R. B. BIRD. New York: John Wiley & Sons 1954 [16]). [Experimental values for D_2 and for He^3 (×) are added for comparison with theoretical values (- - - -) predicted by de BOER and BIRD.]

Values of γ^* are plotted in Fig. 6 as a function of T^* and Λ^*. We see that the quantum deviations are greater for those substances with greater values of Λ^*.

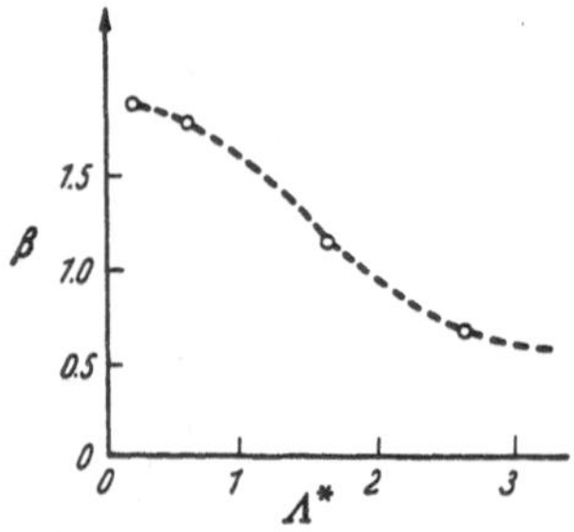

Fig. 7. $\beta(\Lambda^*)$ as a function of Λ^* (after DE BOER and BIRD, see Fig. 6).

As an explicit form for the universal function $\gamma^*(T^*, \Lambda^*)$ in (23.16), DE BOER and BIRD [16] assumed

$$\gamma^* = \beta(\Lambda^*)\,(T_c^* - T^*)^{\frac{11}{9}}, \qquad (23.17)$$

which corresponds to the Katayama-Guggenheim classical equation (23.11). By fitting the surface tension data of H_2, He^4, Ne and A with (23.17), the values of β can be plotted as a function of Λ^* as shown in Fig. 7. By extrapolation, DE BOER and BIRD [16] found $\beta = 0.56$ for He^3. The results predicted from this quantum mechanical principle of corresponding states are compared in Table 4 with experimental values[1] of the surface tension of He^3. The agreement is good. The values of surface tension of hydrogen isotopes HD, HT, DT and T_2 have not yet been measured.

Table 4. *Surface tension of He^3.*

T(°K)	1.75	2.00	2.25	2.50	2.75	3.00
γ_{extrapol} (dyne/cm)	0.123	0.101	0.078	0.057	0.037	0.019
γ_{exper} (dyne/cm)	0.102	0.086	0.070	0.054	0.039	0.022

[1] Numerical figures of γ_{exper} are estimated from K. N. ZINOVEVA's experimental curve [J. Exp. Theor. Phys. USSR. **29**, 899 (1955)]. See also B. N. ESELSON and N. G. BEREZNIAK: Dokl. Akad. Nauk USSR. **99**, 365 (1954). — K. N. ZINOVEVA: J. Exp. Theor. Phys. USSR. **28**, 125 (1954). — D. R. LOVEJOY [43]. — K. R. ATKINS: Liquid Helium, pp. 233—234. London: Cambridge University Press 1959.

δ) Limitations and modifications. As seen in Fig. 6, the surface tension of He^4 exhibits quite clear deviations from (23.17) below the λ-point and tends to be independent of T as T approaches absolute zero. This fact is expected from the third law of thermodynamics, for the surface entropy also approaches zero as $T \to 0$ and consequently $d\gamma/dT$ vanishes at absolute zero.

We have so far limited the principle of corresponding states to simple molecules. This principle can be extended to polyatomic and polar molecules by introducing additional parameters[1-4] [*51*]. Following MEISSNER and SEFERIAN's[1] and RIEDEL's[2] modifications of the principle of corresponding states, BROCK and BIRD[5] have proposed the modified Katayama-Guggenheim formulae

$$\gamma = \left(-0.951 + 0.432 \frac{1}{Z_c}\right)\left(1 - \frac{T}{T_c}\right)^{\frac{11}{9}}; \quad Z_c = p_c V_c/RT_c \tag{23.18}$$

and

$$\gamma = (-0.281 + 0.133\, Y_c)\left(1 - \frac{T}{T_c}\right)^{\frac{11}{9}}; \qquad Y_c = (d \log p\, V/d \log T). \tag{23.19}$$

Both are sufficiently accurate for simple inorganic substances and for a wide variety of organic compounds.

B. Statistical mechanics.

It is statistical mechanics that provides the general description of macroscopic phenomena from the molecular point of view. In this article the exact theories of classical and quantum statistical mechanics will be used to describe surface tension in terms of intermolecular potentials and molecular distribution functions. The quantum statistical treatment of surface tension will, however be limited to the brief description in Sect. 43. Theoretical values based on some approximations in classical statistics will be compared with experimental data.

I. Statistical thermodynamic method[6].

a) Canonical ensemble.

24. Partition function and surface tension. In statistical thermodynamics, which deals exclusively with matter in equilibrium, the Helmholtz free energy F of a closed system is usually expressed in terms of the partition function Z by the relation

$$F = -kT \log Z. \tag{24.1}$$

Let us consider a closed system, containing $N_1, \ldots, N_\varkappa$ molecules of the species $1, \ldots, \varkappa$. We shall denote the position vector of the center-of-mass of the s-th molecule of species i by $\boldsymbol{r}_i^{(s)}$, the set of the position vector $\boldsymbol{r}_i^{(1)}, \ldots, \boldsymbol{r}_i^{(N_i)}$ of the N_i molecules by $\{\boldsymbol{r}_i\}$, and the product of the volume elements of these N_i molecules

[1] H.P. MEISSNER and R. SEFERIAN: Chem. Engng. Progr. **47**, 579 (1951).
[2] L. RIEDEL: Chem. Ing. Tech. **26**, 83, 259 (1954).
[3] J.S. ROWLINSON: Trans. Faraday Soc. **50**, 647 (1954).
[4] I. PRIGOGINE, A. BELLEMANS and C. NAAR-COLIN: J. Chem. Phys. **26**, 710, 751 (1957).
[5] J.R. BROCK and R.B. BIRD: Amer. Inst. Chem. Engng. J. **1**, 174 (1955).
[6] For a general description on statistical thermodynamics, see, for example, T.L. HILL: Statistical Mechanics, New York: McGraw-Hill 1956.

$d\boldsymbol{r}_i^{(1)} \dots d\boldsymbol{r}_i^{(N_i)}$ by $d\{\boldsymbol{r}_i\}$. Then the classical partition function Z is given by

$$Z(\boldsymbol{N}, V, T) = Z(N_1, \dots, N_\varkappa, V, T) = \prod_{i=1}^{\varkappa} (\lambda_i^{-3} j_i(T))^{N_i} Q(N_1, \dots, N_\nu, V, T) \tag{24.2}$$

$$\left.\begin{aligned} Q(\boldsymbol{N}, V, T) &= Q(N_1, \dots, N_\varkappa, V, T) \\ &= \frac{1}{\prod_{i=1}^{\varkappa} N_i!} \int \cdots \int \exp\left[-\frac{\Phi(\{\boldsymbol{r}_1\}, \dots, \{\boldsymbol{r}_\varkappa\})}{kT}\right] d\{\boldsymbol{r}_1\} \dots d\{\boldsymbol{r}_\varkappa\}, \end{aligned}\right\} \tag{24.3}$$

$$\lambda_i^{-1} = \frac{(2\pi m_i kT)^{\frac{1}{2}}}{h}, \tag{24.4}$$

in which $\Phi(\{\boldsymbol{r}_1\}, \dots, \{\boldsymbol{r}_\varkappa\})$ is the potential energy function and h Planck's constant, m_i the mass of a molecule of species i, Q is called the configuration integral and $j_i(T)$ is the rotational and vibrational contributions of a molecule of species i to the partition function. Throughout this article we assume that the vibrational and rotational degrees of freedom are all separable from the translational degrees although the potential energy of a system does, in general, depend on the rotational coordinates as well as on the translational coordinates of molecules. This assumption seems to be rather drastic for the statistical thermodynamics of interfaces because the rotational condition of motion depends rather sensibly on the state of the aggregation of matter. Hence, the theory to be developed below is exact only when we treat monoatomic substances for which $j_i(T) = 1$ and only approximate in treating diatomic or higher polyatomic substances.

Furthermore, we often need to use the probability density function $P^{(\boldsymbol{N})}(\{\boldsymbol{r}_1\}, \dots, \{\boldsymbol{r}_\varkappa\}; \{\boldsymbol{p}_1\}, \dots, \{\boldsymbol{p}_\varkappa\})$ defined such that the probability of finding the system in the state $d\{\boldsymbol{r}_1\}, \dots, d\{\boldsymbol{r}_\varkappa\}, d\{\boldsymbol{p}_1\}, \dots, d\{\boldsymbol{p}_\varkappa\}$ around $\{\boldsymbol{r}_1\}, \dots, \{\boldsymbol{r}_\varkappa\}, \{\boldsymbol{p}_1\}, \dots, \{\boldsymbol{p}_\varkappa\}$ is $P^{(\boldsymbol{N})}(\{\boldsymbol{r}_1\}, \dots, \{\boldsymbol{r}_\varkappa\}; \{\boldsymbol{p}_1\}, \dots, \{\boldsymbol{p}_\varkappa\})\, d\{\boldsymbol{r}_1\} \dots d\{\boldsymbol{r}_\varkappa\}\, d\{\boldsymbol{p}_1\} \dots d\{\boldsymbol{p}_\varkappa\}$, where $\{\boldsymbol{p}_i\}$ denotes the set of momenta $\boldsymbol{p}_i^{(1)}, \dots, \boldsymbol{p}_i^{(N_i)}$, $\boldsymbol{p}_i^{(s)}$ being the momentum of the s-th molecule of species i, and $d\{\boldsymbol{p}_i\} = d\boldsymbol{p}_i^{(1)} \dots d\boldsymbol{p}_i^{(N_i)}$.

In the case of a closed system in equilibrium contact with a reservoir at temperature T, the probability density function $P^{(\boldsymbol{N})}$ is given by

$$\left.\begin{aligned} &P^{(\boldsymbol{N})}(\{\boldsymbol{r}_1\}, \dots, \{\boldsymbol{r}_\varkappa\}; \{\boldsymbol{p}_1\}, \dots, \{\boldsymbol{p}_\varkappa\}) \\ &\quad = Z^{-1}\left[\prod_{i=1}^{\varkappa} \frac{(j_i(T)\, h^{-3})^{N_i}}{N_i!}\right] \exp\left[-\frac{H(\{\boldsymbol{r}_1\}, \dots, \{\boldsymbol{r}_\varkappa\}; \{\boldsymbol{p}_1\}, \dots, \{\boldsymbol{p}_\varkappa\})}{kT}\right], \end{aligned}\right\} \tag{24.5}$$

where H is the Hamiltonian of the system

$$H = \sum_{i=1}^{\varkappa} \frac{1}{2m_i} \sum_{s=1}^{N_i} (\boldsymbol{p}_i^{(s)})^2 + \Phi(\{\boldsymbol{r}_1\}, \dots, \{\boldsymbol{r}_\varkappa\}). \tag{24.6}$$

An ensemble of systems whose probability distribution is given by (24.5) is called a canonical ensemble.

Once the partition function Z is calculated as a function of the external parameter V, using (24.2) and (24.3), it is, in principle, possible to derive all the thermodynamic properties of the homogeneous system from Z. The fundamental relations relevant for this purpose may by summarized as follows:

$$U = kT^2 \left(\frac{\partial \log Z}{\partial T}\right)_{V, \boldsymbol{N}}, \tag{24.7}$$

$$p = kT \left(\frac{\partial \log Z}{\partial V}\right)_{T, \boldsymbol{N}}, \tag{24.8}$$

$$S = k \log Z + kT \left(\frac{\partial \log Z}{\partial T}\right)_{V, \boldsymbol{N}}. \tag{24.9}$$

In the case of a two-phase system, if the area of the interface, A, could be regarded as an external parameter, we would obtain from (5.4) and (24.1), the statistical-mechanical expression for the surface tension:

$$\gamma = \left(\frac{\partial F}{\partial A}\right)_{V,T,N} = -kT\left(\frac{\partial \log Z}{\partial A}\right)_{V,T,N}. \tag{24.10}$$

The area of the interface may, however, not always be regarded as an external parameter because the thermodynamic quantities of the system depend, not only on the area, but also on the shape and the orientation of the interface. Then we have to define the area of the interface as an external parameter in each case, as will be seen from the following examples.

Let us consider a two-phase system contained in a vessel of rectangular parallelepiped shape. If the system is placed in a uniform gravitational field in such a way that the top and bottom of the container are horizontal, then the area of the interface can be considered from a macroscopic standpoint as being determined uniquely by the area of either horizontal wall. Then we may identify the area of the interface with the area of either horizontal wall, which is certainly an external parameter. In other words, we have only to calculate the partition function as a function of the area of the above mentioned wall to obtain the equation for the surface tension from the statistical-mechanical expression (24.10).

It is to be noted that since the contribution of gravitation to the superficial properties is practically negligible on account of the very small thickness of the interface layer, we may, in general, treat a plane interface without explicit reference to such an auxiliary field maintaining the plane interface.

When we discuss the surface tension of a spherical interface an auxiliary field of a central force is assumed (as will be used in Sect. 28) instead of a uniform gravitational field.

25. Molecular distribution functions. To proceed further in the statistical-mechanical theory of interfaces, we need to define distribution functions of molecules. These functions play an important role in the molecular interpretation of thermodynamic properties of macroscopic systems. The singlet distribution function $n_i(\boldsymbol{r})$ specifies the average number, $n_i(\boldsymbol{r})\,d\boldsymbol{r}$, of molecules of species i in the volume element $d\boldsymbol{r}$ at $\boldsymbol{r}$. The pair distribution function $n_{ij}(\boldsymbol{r},\boldsymbol{r}')$ is defined so that the average number of molecular pairs of species i and j, one member of which, molecule i, is found in the volume element $d\boldsymbol{r}$ at $\boldsymbol{r}$ and at the same time the other molecule j, is found in the volume $d\boldsymbol{r}'$ at $\boldsymbol{r}'$, is equal to $n_{ij}\,d\boldsymbol{r}\,d\boldsymbol{r}'$[1].

It is readily seen from (24.5) that in the canonical ensemble n_i and n_{ij} are given by

$$n_i(\boldsymbol{r}^{(1)}) = \frac{N_i}{(\prod_j N_j!)\,Q}\int\cdots\int \delta(\boldsymbol{r}^{(1)} - \boldsymbol{r}_i^{(s)}) \exp\left(-\frac{\Phi}{kT}\right) d\{\boldsymbol{r}_1\}\cdots d\{\boldsymbol{r}_\varkappa\}, \tag{25.1}$$

$$\left.\begin{aligned} n_{ij}(\boldsymbol{r}^{(1)},\boldsymbol{r}^{(2)}) = \frac{N_i(N_j-\delta_{ij})}{(\prod_k N_k!)\,Q}\int\cdots\int \delta(\boldsymbol{r}^{(1)}-\boldsymbol{r}_i^{(s)})\,\delta(\boldsymbol{r}^{(2)}-\boldsymbol{r}_j^{(t)})\times \\ \times\exp\left(-\frac{\Phi}{kT}\right) d\{\boldsymbol{r}_1\}\cdots d\{\boldsymbol{r}_\varkappa\}, \end{aligned}\right\} \tag{25.2}$$

where Q is the configuration integral defined as (24.3), and $\delta(\boldsymbol{r}-\boldsymbol{r}')$ the Dirac δ-function[2].

[1] It follows from this definition that the pair distribution functions satisfy the relation $n_{ij}(\boldsymbol{r}^{(1)},\boldsymbol{r}^{(2)}) = n_{ji}(\boldsymbol{r}^{(2)},\boldsymbol{r}^{(1)})$ but in general $n_{ij}(\boldsymbol{r}^{(1)},\boldsymbol{r}^{(2)}) \neq n_{ij}(\boldsymbol{r}^{(2)},\boldsymbol{r}^{(1)})$.

[2] The Dirac δ-function $\delta(\boldsymbol{r}-\boldsymbol{r}')$ has the property that it is zero everywhere except when $|\boldsymbol{r}-\boldsymbol{r}'| = 0$, and at this point it is so large that the integral of $\delta(\boldsymbol{r}-\boldsymbol{r}')$ over all space is unity (see, for example, L. I. SCHIFF: Quantum Mechanics, New York: McGraw-Hill 1949).

In the interior of a homogeneous bulk fluid phase, $n_i(\boldsymbol{r}^{(1)})$ is independent of $\boldsymbol{r}^{(1)}$ and equal to the average number density of molecules of species i, N_i/V; and $n_{ij}(\boldsymbol{r}^{(1)}, \boldsymbol{r}^{(2)})$ depends only on the distance $|\boldsymbol{r}^{(1)}-\boldsymbol{r}^{(2)}|$. But within or near the interface layer, both functions may appreciably change depending on the position. The complexity encountered in the molecular theory of surface tension arises from the fact that the physical properties of the interface layer depend sensibly on the spatial variation of the distribution functions.

In the present article we assume that the total potential energy function of the system is given by

$$\Phi = \tfrac{1}{2} \sum_{i,j=1}^{\varkappa} \sum_{s=1}^{N_i} \sum_{t=1}^{N_j} \phi_{ij}(|\boldsymbol{r}_i^{(s)} - \boldsymbol{r}_j^{(t)}|), \tag{25.3}$$

in which $\phi_{ij}(r)$ denotes the intermolecular potential of two molecules of species i and j at distance r from one another. This condition may be satisfied for most species of spherical or nearly spherical molecules which are chemically saturated.

Then the internal energy of the system can be expressed in the form

$$U = \tfrac{3}{2} \sum_{i=1}^{\varkappa} N_i kT + \tfrac{1}{2} \sum_{i,j=1}^{\varkappa} \int \cdots \int n_{ij}(\boldsymbol{r}^{(1)}, \boldsymbol{r}^{(2)})\, \phi_{ij}(r^{(12)})\, d\boldsymbol{r}^{(1)}\, d\boldsymbol{r}^{(2)}, \tag{25.4}$$

where $\boldsymbol{r}^{(12)} = \boldsymbol{r}^{(2)} - \boldsymbol{r}^{(1)}$. The first term on the right-hand side represents the kinetic energy of the system and the second the potential energy.

If the potential energy function is assumed to take the form of (25.3), on differentiating (25.1) with respect to $\boldsymbol{r}^{(1)}$ we obtain

$$\left.\begin{aligned} \nabla^{(1)} n_i(\boldsymbol{r}^{(1)}) = \frac{1}{(\prod_j N_j!)\, Q} \sum_{j=1}^{\varkappa} \sum_{s=1}^{N_i} \sum_{t=1}^{N_j} \int \cdots \int \left(-\frac{1}{kT}\right) \nabla_i^{(s)} \phi_{ij}(r^{(st)})\, e^{-\Phi/kT} \times \\ \times \delta(\boldsymbol{r}_i^{(s)} - \boldsymbol{r}^{(1)})\, d\{\boldsymbol{r}_1\} \ldots d\{\boldsymbol{r}_\varkappa\}, \end{aligned}\right\} \tag{25.5}$$

i.e., with use of (25.2)

$$\nabla^{(1)} n_i(\boldsymbol{r}^{(1)}) = \frac{1}{kT} \sum_{j=1}^{\varkappa} \int \frac{\boldsymbol{r}^{(12)}}{r^{(12)}} \phi'_{ij}(r^{(12)})\, n_{ij}(\boldsymbol{r}^{(1)}, \boldsymbol{r}^{(2)})\, d\boldsymbol{r}^{(2)}, \tag{25.6}$$

where the term corresponding to $s=t$ and $i=j$, should be omitted from the summation. Eq. (25.6) is a generalization of the Born-Green-Yvon equation[1,2] for the singlet distribution function to a multicomponent system[3] and is one of the basic equations of the statistical mechanics of surface tension.

By exactly the same manipulation as that used in the derivation of (25.6), differentiation of (25.2) with respect to $\boldsymbol{r}^{(1)}$, gives the Born-Green-Yvon equation of the pair distribution function, generalized to multicomponent systems[3,4]

$$\left.\begin{aligned} \nabla^{(1)} n_{ij}(\boldsymbol{r}^{(1)}, \boldsymbol{r}^{(2)}) + \frac{1}{kT} n_{ij}(\boldsymbol{r}^{(1)}, \boldsymbol{r}^{(2)})\, \nabla^{(1)} \phi_{ij}(r^{(12)}) \\ = \frac{1}{kT} \sum_{k=1}^{\varkappa} \int n_{ijk}(\boldsymbol{r}^{(1)}, \boldsymbol{r}^{(2)}, \boldsymbol{r}^{(3)})\, \nabla^{(3)} \phi_{ik}(r^{(13)})\, d\boldsymbol{r}^{(3)}, \end{aligned}\right\} \tag{25.7}$$

---

[1] M. BORN and H. S. GREEN: Proc. Roy. Soc. Lond., Ser. A **188**, 10 (1946). — H. S. GREEN [*25*].

[2] J. YVON: La théorie statistique des fluides et l'equation d'état. Paris: Herman & Cie. 1935.

[3] S. ONO: Progr. Theoret. Phys. **5**, 822 (1950).

[4] G. FOURNET: J. Phys. Radium **12**, 592 (1951).

where $n_{ijk}(\boldsymbol{r}^{(1)}, \boldsymbol{r}^{(2)}, \boldsymbol{r}^{(3)})$ is a triplet distribution function defined by

$$\left.\begin{aligned} n_{ijk}(\boldsymbol{r}^{(1)}, \boldsymbol{r}^{(2)}, \boldsymbol{r}^{(3)}) &= \frac{N_i(N_j-\delta_{ij})(N_k-\delta_{ik}-\delta_{jk})}{(\prod_l N_l!)\,Q} \times \\ &\times \int\cdots\int \delta(\boldsymbol{r}^{(1)}-\boldsymbol{r}_i^{(s)})\,\delta(\boldsymbol{r}^{(2)}-\boldsymbol{r}_j^{(t)})\,\delta(\boldsymbol{r}^{(3)}-\boldsymbol{r}_k^{(u)}) \exp\left(-\frac{\Phi}{kT}\right) d\{\boldsymbol{r}_1\}\cdots d\{\boldsymbol{r}_k\}. \end{aligned}\right\} \quad (25.8)$$

Since the set of equations for n_{ij}, (25.7), involves triplet distribution functions, we have to introduce some approximation to the triplet functions in terms of pair distribution functions in order to reduce (25.7) to a closed set of equations for n_{ij}. In the case of a homogeneous system, the following approximation, which was first introduced by KIRKWOOD[1] and is usually called the superposition approximation, is often used:

$$n_{ijk}(\boldsymbol{r}^{(1)}, \boldsymbol{r}^{(2)}, \boldsymbol{r}^{3)}) = \frac{n_{ij}(\boldsymbol{r}^{(1)}, \boldsymbol{r}^{(2)})\, n_{jk}(\boldsymbol{r}^{(2)}, \boldsymbol{r}^{(3)})\, n_{ki}(\boldsymbol{r}^{(3)}, \boldsymbol{r}^{(1)})}{n_i(\boldsymbol{r}^{(1)})\, n_j(\boldsymbol{r}^{(2)})\, n_k(\boldsymbol{r}^{(3)})}. \quad (25.9)$$

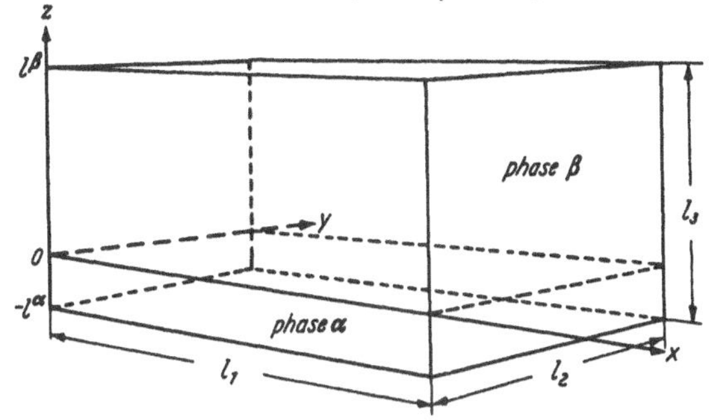

Fig. 8.

Substitution of (25.9) into (25.7) leads to

$$\left.\begin{aligned} &\nabla^{(1)} \log n_{ij}(\boldsymbol{r}^{(1)}, \boldsymbol{r}^{(2)}) + \frac{1}{kT} \nabla^{(1)} \phi_{ij}(r^{(12)}) \\ &= \frac{1}{kT} \sum_k \int \frac{n_{jk}(\boldsymbol{r}^{(2)}, \boldsymbol{r}^{(3)})\, n_{ki}(\boldsymbol{r}^{(3)}, \boldsymbol{r}^{(1)})}{n_i(\boldsymbol{r}^{(1)})\, n_j(\boldsymbol{r}^{(2)})\, n_k(\boldsymbol{r}^{(3)})} \frac{\boldsymbol{r}^{(13)}}{r^{(13)}} \phi'_{ik}(r^{(13)})\, d\boldsymbol{r}^{(3)}. \end{aligned}\right\} \quad (25.10)$$

In the case of a one-component and homogeneous system, (25.10) has been solved numerically for rigid sphere molecules[2] and molecules with a modified Lennard-Jones potential[3]. In principle we can use (25.10) for a two-phase system with an interface as well, but it seems difficult to obtain a numerical solution for such a system. It is indeed the crux of the problem in the interface layer to know the pair distribution function.

In a homogeneous fluid, the pair distribution function is often expressed in the form

$$n_{ij}(r) = n_i n_j g_{ij}(r), \quad (25.11)$$

where n_i and n_j are independent of r and $g_{ij}(r)$ is called the correlation function.

26. Surface tension of plane interface in multicomponent systems. Let us consider a system contained in a rectangular parallelepiped vessel which has edge-lengths l_1, l_2, l_3 in the directions of rectangular coordinate axes x, y, z, respectively, with the z axis taken in the vertical direction opposite to the gravitational force. The notation is the same as given in Sect. 18 (see also Fig. 8). Then we have

$$V = l_1 l_2 l_3, \quad (26.1)$$

$$A = l_1 l_2. \quad (26.2)$$

[1] J. G. KIRKWOOD: J. Chem. Phys. 3, 300 (1935).
[2] J. G. KIRKWOOD, E. K. MAUN and B. J. ALDER: J. Chem. Phys. 18, 1040 (1950).
[3] J. G. KIRKWOOD, V. A. LEWINSON and B. J. ALDER: J. Chem. Phys. 20, 929 (1952).

If the Helmholtz free energy of the two-phase system depends on A and V, we obtain the following relation:

$$\left(\frac{\partial F}{\partial l_1}\right)_{l_2, l_3, T, N} = l_2 \left(\frac{\partial F}{\partial A}\right)_{V, T, N} + l_2 l_3 \left(\frac{\partial F}{\partial V}\right)_{A, T, N}. \tag{26.3}$$

Since $(\partial F/\partial A)_{V,T,N}$ and $-(\partial F/\partial V)_{A,T,N}$ are equal to the surface tension γ and the pressure p, respectively, we obtain

$$\gamma = p\, l_3 + \frac{1}{l_2} \left(\frac{\partial F}{\partial l_1}\right)_{l_2, l_3, T, N}. \tag{26.4}$$

Alternatively, making use of the relation $(\partial F/\partial V)_{A,T,N} = \frac{1}{l_1 l_2} \left(\frac{\partial F}{\partial l_3}\right)_{l_1, l_2, T, N}$ we can rewrite (26.3) in the form

$$\gamma = \frac{1}{l_1 l_2} \left[l_1 \left(\frac{\partial F}{\partial l_1}\right) - l_3 \left(\frac{\partial F}{\partial l_3}\right) \right]. \tag{26.5}$$

The pressure of the system expressed in terms of the molecular distribution functions cannot be obtained by the straightforward differentiation of $kT \log Z$ with respect to V, but is derived by using either the virial theorem or a special device introduced by BOGOLIUBOV and GREEN[1]. The expression of surface tension, in terms of the distribution functions, was obtained for the case of a pure liquid by using a generalization of the Bogoliubov-Green device [*17*], [*18*], [*19*]. To utilize this device, (26.4) or (26.5) is more convenient than (24.10) as will be seen below.

To differentiate the partition function, we introduce the reduced variables[2]:

$$x_i^{(s)} = l_1 \xi_i^{(s)}; \quad y_i^{(s)} = l_2 \eta_i^{(s)}; \quad z_i^{(s)} + l^\alpha = l_3 \zeta_i^{(s)};$$
$$0 \leqq \xi_i^{(s)}, \eta_i^{(s)}, \zeta_i^{(s)} \leqq 1, \quad \begin{pmatrix} s = 1, 2, \ldots, N_i \\ i = 1, 2, \ldots, \varkappa \end{pmatrix}. \tag{26.6}$$

Then, the configuration integral (24.3) may be written in the form

$$Q_N = \frac{\prod_i (l_1 l_2 l_3)^{N_i}}{\prod_i N_i!} \int_0^1 \cdots \int_0^1 \exp\left[-\frac{\Phi^*}{kT}\right] \prod_{i=1}^{\varkappa} d\xi_i^{(1)} \cdots d\zeta_i^{(N_i)}, \tag{26.7}$$

where Φ^* is the potential energy Φ expressed as a function of the reduced variables given by (26.6). We also have

$$\frac{\partial r_{ij}^{(st)}}{\partial l_1} = \frac{\partial}{\partial l_1} \left[l_1^2 (\xi_i^{(s)} - \xi_j^{(t)})^2 + l_2^2 (\eta_i^{(s)} - \eta_j^{(t)})^2 + l_3^2 (\zeta_i^{(s)} - \zeta_j^{(t)})^2 \right]^{\frac{1}{2}}, \tag{26.8}$$

and hence [see (25.3)]

$$l_1 \frac{\partial \Phi^*}{\partial l_1} = \frac{1}{2} \sum_{i,j} \sum_{s,t} \frac{(x_i^{(s)} - x_j^{(t)})^2}{r_{ij}^{(st)}} \phi'_{ij}(r_{ij}^{(st)}). \tag{26.9}$$

Using the above relations we can readily obtain from (24.1), (24.2) and (25.2)

$$\left. \begin{aligned} l_1 \left(\frac{\partial F}{\partial l_1}\right)_{l_2, l_3, T, N} = &-\sum_{i=1}^{\varkappa} N_i kT + \\ &+ \frac{1}{2} \sum_{i,j} \iint n_{ij}(\boldsymbol{r}^{(1)}, \boldsymbol{r}^{(2)}) \, \phi'_{ij}(r^{(12)}) \frac{(x^{(12)})^2}{r^{(12)}} \, d\boldsymbol{r}^{(1)} d\boldsymbol{r}^{(2)}, \end{aligned} \right\} \tag{26.10}$$

---

[1] N. N. BOGOLIUBOV: Problems of a Dynamical Theory in Statistical Physics (in Russian), Chap. 1. Moscow: Gostekhizdat 1946. — H. S. GREEN: Proc. Roy. Soc. Lond., Ser. A **189**, 103 (1947).

[2] See Fig 8.

where $x^{(12)} = x^{(2)} - x^{(1)}$ is the x component of the relative position vector $\boldsymbol{r}^{(12)} = \boldsymbol{r}^{(2)} - \boldsymbol{r}^{(1)}$ measured from the point $\boldsymbol{r}^{(1)}$.

If edge effects due to the walls of the container are neglected, $n_i(\boldsymbol{r}^{(1)})$ depends only on $z^{(1)}$ so that it may be denoted by $n_i(z^{(1)})$ and $n_{ij}(\boldsymbol{r}^{(1)}, \boldsymbol{r}^{(2)})$ depends only on the coordinates $z^{(1)}$ and $\boldsymbol{r}^{(12)}$ so that it may be denoted by $n_{ij}(z^{(1)}; \boldsymbol{r}^{(12)})$. Then (26.10) reduces to

$$\left.\begin{aligned} l_1 \left(\frac{\partial F}{\partial l_1}\right)_{l_2, l_3, T, N} &= - l_1 l_2 \sum_i \int_{-l^\alpha}^{l^\beta} n_i(z^{(1)})\, kT\, dz^{(1)} + \\ &+ \frac{l_1 l_2}{2} \sum_{i,j} \int_{-l^\alpha}^{l^\beta} dz^{(1)} \int n_{ij}(z^{(1)}; \boldsymbol{r}^{(12)})\, \phi'_{ij}(r^{(12)})\, \frac{(x^{(12)})^2}{r^{(12)}}\, d\boldsymbol{r}^{(12)}. \end{aligned}\right\} \tag{26.11}$$

Likewise we obtain

$$\left.\begin{aligned} l_3 \left(\frac{\partial F}{\partial l_3}\right)_{l_1, l_2, T, N} &= - l_1 l_2 \sum_i \int_{-l^\alpha}^{l^\beta} n_i(z^{(1)})\, kT\, dz^{(1)} + \\ &+ \frac{l_1 l_2}{2} \sum_{i,j} \int_{-l^\alpha}^{l^\beta} dz^{(1)} \int n_{ij}(z^{(1)}; \boldsymbol{r}^{(12)})\, \phi'_{ij}(r^{(12)})\, \frac{(z^{(12)})^2}{r^{(12)}}\, d\boldsymbol{r}^{(12)}. \end{aligned}\right\} \tag{26.12}$$

Substitution of (26.11) and (26.12) into (26.5) leads at once to the desired expression for the surface tension of a plane interface[1] [*17*], [*18*], [*19*]

$$\gamma = \frac{1}{2} \sum_{i,j} \int_{-\infty}^{\infty} dz^{(1)} \int n_{ij}(z^{(1)}; \boldsymbol{r}^{(12)})\, \phi'_{ij}(r^{(12)})\, \frac{(x^{(12)})^2 - (z^{(12)})^2}{r^{(12)}}\, d\boldsymbol{r}^{(12)}. \tag{26.13}$$

Alternatively, from (26.4) and (26.11) we obtain

$$\gamma = \int_{-l^\alpha}^{l^\beta} p\, dz^{(1)} - \int_{-l^\alpha}^{l^\beta} \left[\sum_i n_i kT - \frac{1}{2} \sum_{i,j} \int n_{ij} \phi'_{ii} \frac{(x^{(12)})^2}{(r^{(12)})}\, d\boldsymbol{r}^{(12)}\right] dz^{(1)}. \tag{26.14}$$

If the entire system is homogeneous and has no interfacial contribution, $l_1(\partial F/\partial l_1)_{l_2, l_3, T, N}$ in (26.11) should be equal to $V(\partial F/\partial V) = -pV$ and hence we have

$$p = \sum_i n_i kT - \frac{1}{2} \sum_{i,j} \int n_{ij}(r^{(12)})\, \frac{(x^{(12)})^2}{r^{(12)}}\, \phi'_{ij}(r^{(12)})\, d\boldsymbol{r}^{(12)}. \tag{26.15}$$

Thus in the interiors of phases α and β we have, respectively,

$$p = \sum_i n_i^\alpha kT - \frac{1}{2} \sum_{i,j} \int n_{ij}^\alpha \frac{(x^{(12)})^2}{r^{(12)}}\, \phi'_{ij}(r^{(12)})\, d\boldsymbol{r}^{(12)}, \tag{26.16}$$

$$p = \sum_i n_i^\beta kT - \frac{1}{2} \sum_{i,j} \int n_{ij}^\beta \frac{(x^{(12)})^2}{r^{(12)}}\, \phi'_{ij}(r^{(12)})\, d\boldsymbol{r}^{(12)}. \tag{26.17}$$

Substituting the above two equations into (26.14) we obtain

$$\gamma = kT \sum_i \int_{-\infty}^{\infty} (n_i^{\alpha\beta} - n_i)\, dz^{(1)} - \frac{1}{2} \sum_{i,j} \int_{-\infty}^{\infty} dz^{(1)} \int (n_{ij}^{\alpha\beta} - n_{ij})\, \phi'_{ij} \frac{(x^{(12)})^2}{r^{(12)}}\, d\boldsymbol{r}^{(12)}, \tag{26.18}$$

[1] We have utilized the fact that we may replace l^α and l^β by ∞ if they are sufficiently large compared with the depth of the interface zone.

where $n_i^{\alpha\beta}$ and $n_{ij}^{\alpha\beta}$ are defined as

$$n_i^{\alpha\beta} = n_i^{\alpha}[1 - A(z^{(1)})] + n_i^{\beta} A(z^{(1)}), \tag{26.19}$$

$$n_{ij}^{\alpha\beta} = n_{ij}^{\alpha}[1 - A(z^{(1)})] + n_{ij}^{\beta} A(z^{(1)}), \tag{26.20}$$

$A(z)$ being the unit step function given by (20.7).

The superficial number density of molecules of species i is given by

$$\Gamma_i = \int_{-\infty}^{\infty} (n_i(z^{(1)}) - n_i^{\alpha\beta})\, dz^{(1)}, \tag{26.21}$$

and we may also introduce a similar quantity, the superficial pair density of species i and j,

$$\Gamma_{ij}(\boldsymbol{r}^{(12)}) = \int_{-\infty}^{\infty} (n_{ij}(z^{(1)}; \boldsymbol{r}^{(12)}) - n_{ij}^{\alpha\beta})\, dz^{(1)}. \tag{26.22}$$

Then (26.18) can be written in the form

$$\gamma = -kT \sum_i \Gamma_i + \frac{1}{2} \sum_{i,j} \int_{-\infty}^{\infty} \Gamma_{ij}(\boldsymbol{r}^{(12)})\, \phi'_{ij}(r^{(12)})\, \frac{(x^{(12)})^2}{r^{(12)}}\, d\boldsymbol{r}^{(12)}. \tag{26.23}$$

Eqs. (26.13), (26.14) and (26.23) are equivalent rigorous expressions for the surface tension of a plane interface. They are completely free from any other assumption than (25.3), which is responsible for the fact that the surface tension can be expressed simply by the singlet and pair distribution functions without reference to distribution functions of higher order.

It will be seen from (26.13) that we can readily calculate the numerical value of the surface tension if we know the pair distribution function in the interface zone. But, as briefly mentioned in the previous section, at the present stage of statistical mechanics, solution of the approximate integral equation for the pair distribution function in the interface region has not yet been obtained even for the case of a one-component system. We must employ some other more drastic approximations for calculating the value of the surface tension by the distribution function method. In Sects. 38 to 40 we shall summarize some numerical calculations based on such approximations.

In a homogeneous fluid, because of its isotropy, we have the equations which can be obtained by replacing $x^{(12)}$ in (26.15) by $y^{(12)}$ or $z^{(12)}$. Then adding together the three equations corresponding to $x^{(12)}, y^{(12)}, z^{(12)}$, we obtain

$$\left.\begin{aligned} p &= \sum_i n_i kT - \frac{1}{6} \sum_{i,j} \int n_{ij}(r^{(12)})\, r^{(12)}\, \phi'_{ij}(r^{(12)})\, d\boldsymbol{r}^{(12)} \\ &= \sum_i n_i kT - \frac{2\pi}{3} \sum_{i,j} n_i n_j \int_0^{\infty} g_{ij}(r^{12})\, \phi'_{ij}(r)\, r^3\, dr, \end{aligned}\right\} \tag{26.24}$$

which is usually used as the equation of state.

27. Surface tension of plane interface between pure liquid and its vapor. *α) General expressions* [24]. For the case of a single component we shall rewrite the equations obtained in the previous section into convenient forms for numerical calculation. For this purpose we denote the singlet distribution function by $n(\boldsymbol{r})$ and the pair distribution function by $n_2(\boldsymbol{r}, \boldsymbol{r}')$. Then the expression for surface tension (26.13) reduces to

$$\gamma = \frac{1}{2} \int_{-\infty}^{\infty} dz^{(1)} \int n_2(z^{(1)}; \boldsymbol{r}^{(12)})\, \phi'(r^{(12)})\, \frac{(x^{(12)})^2 - (z^{(12)})^2}{r^{(12)}}\, d\boldsymbol{r}^{(12)}, \tag{27.1}$$

and (26.23) to

$$\gamma = -\Gamma kT + \frac{1}{2}\int \Gamma_2(\boldsymbol{r}^{(12)})\, \phi'(r^{(12)})\, \frac{(x^{(12)})^2}{r^{(12)}}\, d\boldsymbol{r}^{(12)}, \tag{27.2}$$

where Γ and Γ_2 are respectively defined by [instead of (26.21) and (26.22)]

$$\Gamma = \int_{-\infty}^{\infty} (n - n^{\alpha\beta})\, dz^{(1)}, \tag{27.3}$$

and

$$\Gamma_2(\boldsymbol{r}^{(12)}) = \int_{-\infty}^{\infty} [n_2(z^{(1)}; \boldsymbol{r}^{(12)}) - n_2^{\alpha\beta}]\, dz^{(1)}. \tag{27.4}$$

The generalized Born-Green-Yvon equation for the singlet distribution function (25.6) is reduced to the following form for a one-component system:

$$\frac{dn(z^{(1)})}{dz^{(1)}} = \frac{1}{kT}\int \frac{z^{(12)}}{r^{(12)}}\, \phi'(r^{(12)})\, n_2(z^{(1)}; \boldsymbol{r}^{(12)})\, d\boldsymbol{r}^{(12)}, \tag{27.5}$$

where the singlet distribution function is assumed to depend only on $z^{(1)}$.

We shall define the ν-th moments of $[n(z) - n^{\alpha\beta}]$ and $[n_2(z; \boldsymbol{r}^{(12)}) - n_2^{\alpha\beta}]$ as follows:

$$[\Gamma]_\nu = \int_{-\infty}^{\infty} z^\nu\, [n(z) - n^{\alpha\beta}]\, dz, \tag{27.6}$$

$$[\Gamma_2]_\nu = \int_{-\infty}^{\infty} z^\nu\, [n_2(z; \boldsymbol{r}^{(12)}) - n_2^{\alpha\beta}]\, dz. \tag{27.7}$$

Integration of (27.5) with respect to $z^{(1)}$ from $-\infty$ to ∞ yields

$$n^\beta - n^\alpha = \frac{1}{kT}\int \frac{z^{(12)}}{r^{(12)}}\, \phi'(r^{(12)})\, \Gamma_2(\boldsymbol{r}^{(12)})\, d\boldsymbol{r}^{(12)}, \tag{27.8}$$

where we have used the following relations

$$\lim_{z\to-\infty} n(z) = n^\alpha, \qquad \lim_{z\to\infty} n(z) = n^\beta, \tag{27.9}$$

$$\int \frac{z^{(12)}}{r^{(12)}}\, \phi'(r^{(12)})\, n_2^{\alpha\beta}\, d\boldsymbol{r}^{(12)} = 0. \tag{27.10}$$

Likewise, for the ν-th moment we obtain

$$[\Gamma]_\nu = -\frac{1}{kT}\,\frac{1}{\nu+1}\int \frac{z^{(12)}}{r^{(12)}}\, \phi'(r^{(12)})\, [\Gamma_2]_{\nu+1}\, d\boldsymbol{r}^{(12)}. \tag{27.11}$$

Since $[\Gamma]_0$ is equal to the superficial number density Γ given by (27.3), substitution of (27.11) for $\nu=0$ into (27.2) leads to

$$\gamma = \int \frac{1}{r^{(12)}}\, \phi'(r^{(12)}) \left[z^{(12)}\, [\Gamma_2]_1 + \frac{(x^{(12)})^2}{2}\, [\Gamma_2]_0\right] d\boldsymbol{r}^{(12)}, \tag{27.12}$$

which is an alternative expression for the surface tension, originally obtained by Kirkwood and Buff [*24*].

Let us now consider the statistical expression for the superficial internal energy which is some times called the surface energy. From (3.2) and (25.4) we can easily find

$$U^s = \frac{3}{2} N^s kT + \frac{A}{2}\iint [n_2(z^{(1)}; \boldsymbol{r}^{(12)}) - n_2^{\alpha\beta}]\, \phi(r^{(12)})\, d\boldsymbol{r}^{(12)}\, dz^{(1)}, \tag{27.13}$$

the area A arising from the integration with respect to $x^{(1)}$ and $y^{(1)}$. Using (6.5), (6.7) and (27.4), we can rewrite (27.13) in the form of a superficial density

$$v = \tfrac{3}{2}\Gamma kT + \tfrac{1}{2}\int \Gamma_2(\boldsymbol{r}^{(12)})\,\phi(r^{(12)})\,d\boldsymbol{r}^{(12)}. \tag{27.14}$$

If we choose the equimolecular dividing surface as the (x, y) plane, (27.2) reduces to

$$\gamma = \frac{1}{2}\int \Gamma_2(\boldsymbol{r}^{(12)})\,\phi'(r^{(12)})\,\frac{(x^{(12)})^2}{r^{(12)}}\,d\boldsymbol{r}^{(12)}, \tag{27.15}$$

and (27.14) to

$$v = \tfrac{1}{2}\int \Gamma_2(\boldsymbol{r}^{(12)})\,\phi(r^{(12)})\,d\boldsymbol{r}^{(12)}. \tag{27.16}$$

β) Principle of corresponding states. Although the expression (27.1) gives the surface tension of a pure fluid in terms of the intermolecular potential and of the pair distribution function, it is extremely difficult to calculate the numerical value of surface tension by means of this equation without the use of rather crude approximation to the distribution function. However, we can rather easily obtain the statistical interpretation of the principle of corresponding states for such molecules the intermolecular potential of which can be represented by some universal function, f, together with two scale factors D and ε characteristic of molecular species:

$$\phi(r) = \varepsilon f(r/D). \tag{27.17}$$

The Lennard-Jones potential given by (23.4) is a familiar example of this two constant form to represent an approximate universal function. As in Sect. 23 we define the reduced variables

$$v^* = \frac{V}{N D^3}; \qquad T^* = \frac{kT}{\varepsilon}; \qquad \gamma^* = \frac{\gamma D^2}{\varepsilon}. \tag{27.18}$$

If we use (24.3) and (25.2), the pair distribution function in a pure fluid can then be expressed as

$$n_2(\boldsymbol{r}, \boldsymbol{r}') = \frac{N(N-1)\iint \delta(\boldsymbol{r}-\boldsymbol{r}^{(1)})\,\delta(\boldsymbol{r}'-\boldsymbol{r}^{(2)})\exp[-\Phi/kT]\prod_{s=1}^{N} d\boldsymbol{r}^{(s)}}{\iint \exp[-\Phi/kT]\prod_{s=1}^{N} d\boldsymbol{r}^{(s)}}, \tag{27.19}$$

where Φ is the potential energy of the system.

To express the pair distribution function in a reduced form, we introduce the variables

$$\boldsymbol{\vartheta}^{(s)} = \frac{\boldsymbol{r}^{(s)}}{D}, \qquad \boldsymbol{\vartheta}^{(st)} = \frac{\boldsymbol{r}^{(st)}}{D}, \qquad \vartheta^{(st)} = \frac{r^{(st)}}{D}. \tag{27.20}$$

For the sake of simplicity, we consider a system contained in a cubic vessel with the edge length l. Then, (27.19) can be rewritten in the reduced form[1]

$$D^6 n_2(D\boldsymbol{\vartheta}, D\boldsymbol{\vartheta}') = \frac{N(N-1)\int_0^{l/D}\cdots\int_0^{l/D}\delta(\boldsymbol{\vartheta}-\boldsymbol{\vartheta}^{(1)})\,\delta(\boldsymbol{\vartheta}'-\boldsymbol{\vartheta}^{(2)})\exp[-\Phi/kT]\prod_{s=1}^{N} d\boldsymbol{\vartheta}^{(s)}}{\int_0^{l/D}\cdots\int_0^{l/D}\exp[-\Phi/kT]\prod_{s=1}^{N} d\boldsymbol{\vartheta}^{(s)}}, \tag{27.21}$$

$$\Phi = \frac{\varepsilon}{2}\sum_{s,t=1}^{N} f(\vartheta^{(st)}). \tag{27.22}$$

It can be easily seen that the right-hand side of (27.21) depends only on l/D and ε/kT. Then $D^6 n_2(\boldsymbol{r}, \boldsymbol{r}')$ is a universal function of the reduced volume v^*

[1] Strictly speaking, as stated in Sect. 24, it is necessary to consider an auxiliary external field like a gravitational field to maintain the plane interface. It is, however, not necessary to consider explicitly the external field, because the results are independent of the magnitude of the auxiliary field insofar as this field is of the order of magnitude of the usual gravitational field.

and the reduced temperature T^* defined by (27.18), insofar as the number of molecules N is kept constant.

Using (27.22) and the reduced variables $\boldsymbol{\vartheta}^{(s)}$ and $\boldsymbol{\vartheta}^{(st)}$ given by (27.20), we can rewrite (27.1) in the following form:

$$\gamma^* = \frac{1}{2} \int_{-\infty}^{\infty} d\vartheta_z^{(1)} \int D^6 n_2(D\vartheta_z^{(1)}; D\boldsymbol{\vartheta}^{(12)})\, f'(\vartheta^{(12)})\, \frac{(\vartheta_x^{(12)})^2 - (\vartheta_z^{(12)})^2}{\vartheta^{(12)}}\, d\boldsymbol{\vartheta}^{(12)}, \tag{27.23}$$

where γ^* is the reduced surface tension given by (27.18). Since $D^6 n_2(D\vartheta_z^{(1)}; D\boldsymbol{\vartheta}^{(12)})$ has been proved to be the universal function of T^*, v^*, $\vartheta_z^{(1)}$ and $\boldsymbol{\vartheta}^{(12)}$, γ^* must be a universal function of T^* and v^* only.

If the value of N/V is between those of number densities of the liquid and vapor under orthobaric pressure, the density of the bulk phase on either side of the interface does not depend on the value of N/V but on temperature only. Thus, for a given temperature, the pair distribution function $n_2(z^{(1)}, \boldsymbol{r}^{(12)})$ has a specified value in the interface region as well as in the interior of the both bulk phases. Thus the value of γ^* given by (27.23) is independent of the number of molecules and of the volume and shape of the vessel. That is, γ^* is a universal function of T^* only. This conforms with the statement of the principle of corresponding states given by (23.6).

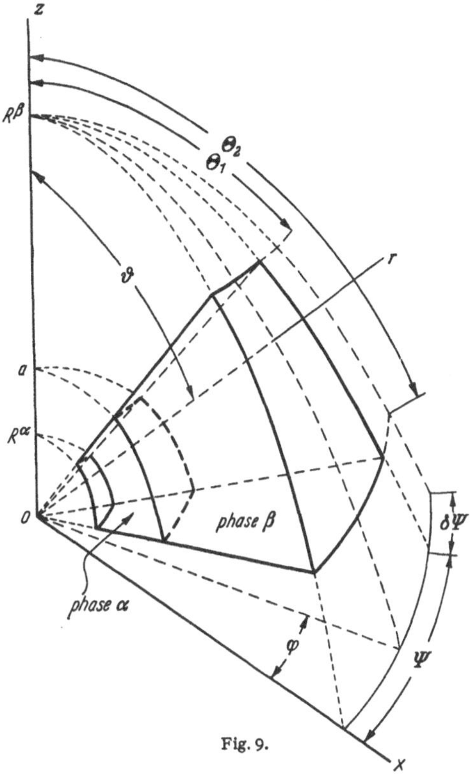

Fig. 9.

28. Surface tension of spherical interface. To avoid unnecessary complication we shall restrict ourselves to a one-component system. The generalization to multicomponent is not very difficult[1].

For thermodynamic treatments of a spherical interface it is convenient to consider a portion of a sphere. Although the same portion of the sphere as considered in Sect. 8 can be used [*26*], we shall consider, for simplicity, the section of a spherical shell given by (see also Fig. 9)

$$\left.\begin{aligned} R^\alpha \leqq r \leqq R^\beta, \\ \Theta_1 \leqq \vartheta \leqq \Theta_2, \\ 0 \leqq \varphi \leqq \Psi, \end{aligned}\right\} \tag{28.1}$$

where r, ϑ and φ are polar coordinates, the origin of which is located at the center of curvature of the spherical interface, with the corresponding orthogonal unit vectors $\boldsymbol{e}_r$, $\boldsymbol{e}_\vartheta$ and $\boldsymbol{e}_\varphi$ [*20*]. A weak field of central force with center at the origin

[1] Most of the equations which are derived in the present section may be generalized to the multicomponent case simply by a summation over the component species, as may be surmised from the results obtained in the preceding sections for the plane interface. We shall show an alternative method of derivation for the multicomponent case in Sect. 37.

of the above coordinates would suffice to produce such an interface, insofar as edge effects of the container are negligible.

Let us choose the spherical surface $r=a$ as the dividing surface and consider an infinitesimal variation $\delta\Psi$ only, keeping a, R^α, R^β, Θ_1, Θ_2, T and N constant. Then taking account of the accompanied changes in V^α, V^β and A; and constancy of a, and making use of (10.4) we obtain

$$\frac{\partial F}{\partial \Psi} = -p^\alpha \frac{\partial V^\alpha}{\partial \Psi} - p^\beta \frac{\partial V^\beta}{\partial \Psi} + \gamma \frac{\partial A}{\partial \Psi}. \tag{28.2}$$

In the present case we have

$$\left.\begin{aligned} V^\alpha &= \{a^3-(R^\alpha)^3\}(\cos\Theta_1-\cos\Theta_2)\frac{\Psi}{3},\\ V^\beta &= \{(R^\beta)^3-a^3\}(\cos\Theta_1-\cos\Theta_2)\frac{\Psi}{3},\\ A &= a^2(\cos\Theta_1-\cos\Theta_2)\,\Psi. \end{aligned}\right\} \tag{28.3}$$

Substituting into (28.2) we obtain [see (15.7)]

$$\gamma = \frac{1}{a^2}\left[\int\limits_{R^\alpha}^{R^\beta} p^{\alpha\beta} r^2\, dr + \frac{1}{\cos\Theta_1-\cos\Theta_2}\frac{\partial F}{\partial\Psi}\right]. \tag{28.4}$$

We can carry out the differentiation $\partial F/\partial\Psi$ just in the same way as in the planar case by using the reduced variables defined as follows:

$$\varphi^{(s)} = \Psi\tilde{\omega}^{(s)};\quad 0 \leqq \tilde{\omega}^{(s)} \leqq 1 \quad (s=1,\ldots,N). \tag{28.5}$$

Then the volume element of the s-th molecule and the distance between the s-th and t-th molecules are respectively given by

$$d\boldsymbol{r}^{(s)} = (r^{(s)})^2\,\Psi \sin\vartheta^{(s)}\, dr^{(s)}\, d\vartheta^{(s)}\, d\tilde{\omega}^{(s)}, \tag{28.6}$$

and

$$\left.\begin{aligned} r^{(st)} = \big[&(r^{(s)})^2+(r^{(t)})^2-2r^{(s)}r^{(t)}\times\\ &\times\{\sin\vartheta^{(s)}\sin\vartheta^{(t)}\cos(\Psi(\tilde{\omega}^{(t)}-\tilde{\omega}^{(s)}))+\cos\vartheta^{(s)}\cos\vartheta^{(t)}\}\big]^{\frac{1}{2}}. \end{aligned}\right\} \tag{28.7}$$

Hence the configuration integral (24.3) can be rewritten in the form

$$Q = \frac{\Psi_N}{N!}\int\cdots\int \exp\left[-\frac{\Phi^*}{kT}\right]\prod_{s=1}^{N}(r^{(s)})^2\sin\vartheta^{(s)}\,dr^{(s)}\,d\vartheta^{(s)}\,d\tilde{\omega}^{(s)}, \tag{28.8}$$

where the ranges of integration with respect to $r^{(s)}$, $\vartheta^{(s)}$ and $\tilde{\omega}^{(s)}$ extend from R^α to R^β, Θ_1 to Θ_2 and 0 to 1, respectively and Φ^* the potential energy as a function of reduced variables.

If Φ has the form given by (25.3), we obtain from (24.1), (28.7) and (28.8)

$$\left.\begin{aligned} \Psi\frac{\partial F}{\partial\Psi} = &-NkT+\frac{1}{2}\iint n_2(\boldsymbol{r}^{(1)},\boldsymbol{r}^{(2)})\,\phi'(r^{(12)})\times\\ &\times\frac{r^{(1)}r^{(2)}(\varphi^{(2)}-\varphi^{(1)})\sin(\varphi^{(2)}-\varphi^{(1)})}{r^{(12)}}\sin\vartheta^{(1)}\sin\vartheta^{(2)}\,d\boldsymbol{r}^{(1)}d\boldsymbol{r}^{(2)}. \end{aligned}\right\} \tag{28.9}$$

If edge effects are neglected, in accordance with the symmetry of the system we can write

$$n(\boldsymbol{r}^{(1)}) = n(r^{(1)});\qquad n_2(\boldsymbol{r}^{(1)},\boldsymbol{r}^{(2)}) = n_2(r^{(1)};\ \boldsymbol{r}^{(12)}). \tag{28.10}$$

The component of $\boldsymbol{r}^{(12)}$ in the direction of $\boldsymbol{e}_\varphi$ takes the following forms corresponding to $\boldsymbol{e}_\varphi^{(1)}$ at $\boldsymbol{r}^{(1)}$ and $\boldsymbol{e}_\varphi^{(2)}$ at $\boldsymbol{r}^{(2)}$:

$$\boldsymbol{r}^{(12)}\cdot\boldsymbol{e}_\varphi^{(1)} = r^{(2)}\sin\vartheta^{(2)}\sin(\varphi^{(2)}-\varphi^{(1)});\quad \boldsymbol{r}^{(12)}\cdot\boldsymbol{e}_\varphi^{(2)} = r^{(1)}\sin\vartheta^{(1)}\sin(\varphi^{(2)}-\varphi^{(1)}). \tag{28.11}$$

Then we have

$$\left.\begin{aligned} r^{(1)} r^{(2)} (\varphi^{(2)}-\varphi^{(1)}) \sin(\varphi^{(2)}-\varphi^{(1)}) \sin\vartheta^{(1)} \sin\vartheta^{(2)} = (\boldsymbol{r}^{(12)}\cdot \boldsymbol{e}_\varphi^{(1)})^2 \frac{\varphi^{(2)}-\varphi^{(1)}}{\sin(\varphi^{(2)}-\varphi^{(1)})} - \\ - (\boldsymbol{r}^{(12)}\cdot \boldsymbol{e}_\varphi^{(1)}) [\boldsymbol{r}^{(12)}\cdot (\boldsymbol{e}_\varphi^{(1)} - \boldsymbol{e}_\varphi^{(2)})] \frac{\varphi^{(2)}-\varphi^{(1)}}{\sin(\varphi^{(2)}-\varphi^{(1)})}\,. \end{aligned}\right\} \quad (28.12)$$

Substituting (28.10) and (28.12) into the second term on the right-hand side of (28.9) we obtain

$$\left.\begin{aligned} &\frac{1}{2}\iint n_2(\boldsymbol{r}^{(1)},\boldsymbol{r}^{(2)})\,\phi'(r^{(12)}) \frac{r^{(1)} r^{(2)} (\varphi^{(2)}-\varphi^{(1)}) \sin(\varphi^{(2)}-\varphi^{(1)}) \sin\vartheta^{(1)}\sin\vartheta^{(2)}}{r^{(12)}}\, d\boldsymbol{r}^{(1)} d\boldsymbol{r}^{(2)} \\ &= \frac{1}{2}\iint n_2(r^{(1)};\boldsymbol{r}^{(12)})\,\phi'(r^{(12)}) \frac{(\boldsymbol{r}^{(12)}\cdot\boldsymbol{e}_\varphi^{(1)})^2}{r^{(12)}} \frac{(\varphi^{(2)}-\varphi^{(1)})}{\sin(\varphi^{(2)}-\varphi^{(1)})}\, d\boldsymbol{r}^{(1)} d\boldsymbol{r}^{(2)} - \\ &- \frac{1}{4}\iint n_2(r^{(1)};\boldsymbol{r}^{(12)})\,\phi'(r^{(12)}) \frac{[\boldsymbol{r}^{(12)}\cdot(\boldsymbol{e}_\varphi^{(2)}-\boldsymbol{e}_\varphi^{(1)})]^2}{r^{(12)}} \frac{(\varphi^{(2)}-\varphi^{(1)})}{\sin(\varphi^{(2)}-\varphi^{(1)})}\, d\boldsymbol{r}^{(1)} d\boldsymbol{r}^{(2)}, \end{aligned}\right\} \quad (28.13)$$

where the second term is due to the following relation

$$\left.\begin{aligned} &\int n_2(\boldsymbol{r}^{(1)},\boldsymbol{r}^{(2)})\,\phi'(r^{(12)}) \frac{(\boldsymbol{r}^{(12)}\cdot\boldsymbol{e}_\varphi^{(1)})[\boldsymbol{r}^{(12)}\cdot(\boldsymbol{e}_\varphi^{(1)}-\boldsymbol{e}_\varphi^{(2)})](\varphi^{(2)}-\varphi^{(1)})}{r^{(12)}\sin(\varphi^{(2)}-\varphi^{(1)})}\, d\boldsymbol{r}^{(1)} d\boldsymbol{r}^{(2)} \\ &= \int n_2(\boldsymbol{r}^{(1)},\boldsymbol{r}^{(2)})\,\phi'(r^{(12)}) \frac{(\boldsymbol{r}^{(21)}\cdot\boldsymbol{e}_\varphi^{(2)})[\boldsymbol{r}^{(21)}\cdot(\boldsymbol{e}_\varphi^{(2)}-\boldsymbol{e}_\varphi^{(1)})](\varphi^{(1)}-\varphi^{(2)})}{r^{(12)}\sin(\varphi^{(1)}-\varphi^{(2)})}\, d\boldsymbol{r}^{(1)} d\boldsymbol{r}^{(2)} \\ &= \frac{1}{2}\int n_2(\boldsymbol{r}^{(1)},\boldsymbol{r}^{(2)})\,\phi'(r^{(12)}) \frac{[\boldsymbol{r}^{(12)}\cdot(\boldsymbol{e}_\varphi^{(1)}-\boldsymbol{e}_\varphi^{(2)})]^2(\varphi^{(2)}-\varphi^{(1)})}{r^{(12)}\sin(\varphi^{(2)}-\varphi^{(1)})}\, d\boldsymbol{r}^{(1)} d\boldsymbol{r}^{(2)}. \end{aligned}\right\} \quad (28.14)$$

Let us take τ as a length of the order of magnitude of the range of intermolecular force $\phi'(r^{(12)})$. Then the integrands on the right-hand side of (28.13) vanish for $r^{(12)} \gg \tau$. Since $|\varphi^{(2)}-\varphi^{(1)}|$ is at largest of the order of $r^{(12)}/r^{(1)}$, we have $(\varphi^{(2)}-\varphi^{(1)})/\sin(\varphi^{(2)}-\varphi^{(1)}) = 1 + 0(\tau^2/(r^{(1)})^2)$ for $r^{(12)} \lesssim \tau$. Similarly $|\boldsymbol{e}_\varphi^{(1)}-\boldsymbol{e}_\varphi^{(2)}|$ is of the order of $\tau/r^{(1)}$. Then the integrand on the right-hand side of (28.9) may be replaced by $n_2(r^{(1)};\boldsymbol{r}^{(12)})\,\phi'(r^{(12)})\,(\boldsymbol{r}^{(12)}\cdot\boldsymbol{e}_\varphi^{(1)})^2/r^{(12)}$ if we neglect quantities of the order of $(\tau/r^{(1)})^2$.

If the radius of the interface is sufficiently large compared with τ, we can choose R^α, R^β, Θ_1 and Θ_2 so that we may make in (28.9) the above replacement for the whole domain of the integration, obtaining [*20*]

$$\Psi\frac{\partial F}{\partial \Psi} = -NkT + \frac{A}{2a^2}\int_{R^\alpha}^{R^\beta} (r^{(1)})^2\, dr^{(1)} \int n_2(r^{(1)};\boldsymbol{r}^{(12)})\,\phi'(r^{(12)}) \frac{(\boldsymbol{r}^{(12)}\cdot\boldsymbol{e}_\varphi^{(1)})^2}{r^{(12)}}\, d\boldsymbol{r}^{(12)}, \quad (28.15)$$

which is valid up to the first order in τ/R.

If the homogeneous bulk phase α alone is enclosed in V^α, there is no interfacial contribution and hence by analogy with (28.4) we have

$$\frac{A}{a^2}\int_{R^\alpha}^{a} p^\alpha (r^{(1)})^2\, dr^{(1)} = -\Psi\frac{\partial F^\alpha}{\partial \Psi}. \quad (28.16)$$

From (28.15), it is then easily seen that

$$\left.\begin{aligned} &\frac{A}{a^2}\int_{R^\alpha}^{a} p^\alpha (r^{(1)})^2\, dr^{(1)} \\ &\quad = N^\alpha kT - \frac{A}{2a^2}\int_{R^\alpha}^{a} (r^{(1)})^2\, dr^{(1)} \int n_2^\alpha(r^{(12)})\,\phi'(r^{(12)}) \frac{(\boldsymbol{r}^{(12)}\cdot\boldsymbol{e}_\varphi^{(1)})^2}{r^{(12)}}\, d\boldsymbol{r}^{(12)}, \end{aligned}\right\} \quad (28.17)$$

and similarly

$$\left.\begin{aligned}&\frac{A}{a^2}\int_a^{R^\beta} p^\beta (r^{(1)})^2\, d r^{(1)}\\ &\quad = N^\beta kT - \frac{A}{2a^2}\int_a^{R^\beta} (r^{(1)})^2\, d r^{(1)} \int n_2^\beta (r^{(12)})\, \phi'(r^{(12)})\, \frac{(\boldsymbol{r}^{(12)}\cdot \boldsymbol{e}_\varphi^{(1)})^2}{r^{(12)}}\, d\boldsymbol{r}^{(12)}.\end{aligned}\right\} \tag{28.18}$$

The above two equations can be put together in a compact form

$$\int_{R^\alpha}^{R^\beta} p^{\alpha\beta} (r^{(1)})^2\, d r^{(1)} = \int_{R^\alpha}^{R^\beta}\left[n^{\alpha\beta} kT - \frac{1}{2}\int n_2^{\alpha\beta}\, \phi'(r^{(12)})\, \frac{(\boldsymbol{r}^{(12)}\cdot \boldsymbol{e}_\varphi^{(1)})^2}{r^{(12)}}\, d\boldsymbol{r}^{(12)}\right] (r^{(1)})^2\, d r^{(1)}, \tag{28.19}$$

where we have utilized the notation similar to (26.19) and (26.20) as follows

$$n^{\alpha\beta} = n^\alpha\{1 - A(r^{(1)} - a)\} + n^\beta A(r^{(1)} - a), \tag{28.20}$$

$$n_2^{\alpha\beta} = n_2^\alpha\{1 - A(r^{(1)} - a)\} + n_2^\beta A(r^{(1)} - a). \tag{28.21}$$

Substitution of (28.15) and (28.19) into (28.4) yields [*20*],

$$\left.\begin{aligned}\gamma = &-\frac{kT}{a^2}\int_{R^\alpha}^{R^\beta} (n - n^{\alpha\beta})\, (r^{(1)})^2\, d r^{(1)} +\\ &+ \frac{1}{2a^2}\int_{R^\alpha}^{R^\beta} (r^{(1)})^2\, d r^{(1)} \int (n_2 - n_2^{\alpha\beta})\, \phi'(r^{(12)})\, \frac{(\boldsymbol{r}^{(12)}\cdot \boldsymbol{e}_\varphi^{(1)})^2}{r^{(12)}}\, d\boldsymbol{r}^{(12)}.\end{aligned}\right\} \tag{28.22}$$

If we use the variable $\xi^{(1)} = r^{(1)} - a$ similar to (15.15), the above equation is written in the form

$$\gamma = -\Gamma kT + \frac{1}{2}\int \Gamma_2\, \phi'(r^{(12)})\, \frac{(\boldsymbol{r}^{(12)}\cdot \boldsymbol{e}_\varphi^{(1)})^2}{r^{(12)}}\, d\boldsymbol{r}^{(12)}, \tag{28.23}$$

in which

$$\Gamma = \int_{-\infty}^{\infty} (n - n^{\alpha\beta})\left(1 + \frac{\xi^{(1)}}{a}\right)^2 d\xi^{(1)}, \tag{28.24}$$

$$\Gamma_2 = \int_{-\infty}^{\infty} (n_2 - n_2^{\alpha\beta})\left(1 + \frac{\xi^{(1)}}{a}\right)^2 d\xi^{(1)}. \tag{28.25}$$

It is evident that in the limit $a \to \infty$, (28.23) reduces to (27.2), for (28.24) and (28.25) in the same limit reduce to (27.3) and (27.4), respectively.

The expression for the surface energy can be deduced in the same manner as in the case of a plane interface and an equation similar to (27.14) is obtained:

$$v = \tfrac{3}{2}\Gamma kT + \tfrac{1}{2}\int \Gamma_2\, \phi(r^{(12)})\, d\boldsymbol{r}^{(12)}, \tag{28.26}$$

where Γ and Γ_2 are given by (28.24) and (28.25), respectively.

b) Grand canonical ensemble[1].

29. Grand partition function and surface tension. We have hitherto considered only closed systems whose probability function is given by (24.5). To relate the thermodynamic properties of an open system, which can exchange matter as

[1] As for the generality and details of the grand canonical ensemble, see, for example, D. TER HAAR: Elements of Statistical Mechanics. New York: Rinehart 1954. — T.L. HILL: Statistical Mechanics. New York: McGraw-Hill 1956.

well as energy with its surroundings, to the intermolecular potentials by the method of statistical thermodynamics, we use the grand partition function.

For an open system with chemical potentials $\mu_1, \ldots, \mu_\varkappa$, temperature T, and volume V, the grand partition function is defined by

$$\Xi(\boldsymbol{\mu}, T, V) = \Xi(\mu_1, \ldots, \mu_\varkappa, T, V) = \sum_{N_1=0}^{\infty} \cdots \sum_{N_\varkappa=0}^{\infty} Z(\boldsymbol{N}, T, V) \prod_{i=1}^{\varkappa} e^{\frac{N_i \mu_i}{kT}}, \tag{29.1}$$

where Z is the partition function defined by (24.2). The probability that an open system has $N_1, \ldots, N_\varkappa$ molecules of the corresponding species, is given by

$$P(\boldsymbol{N}, T, V) = P(N_1, \ldots, N_\varkappa, T, V) = \Xi^{-1} Z(\boldsymbol{N}, T, V) \prod_{i=1}^{\varkappa} e^{\frac{N_i \mu_i}{kT}}, \tag{29.2}$$

and the probability density function in classical statistics, corresponding to (24.5) for a closed system, is

$$\left.\begin{aligned} &P_{\boldsymbol{N}}(\{\boldsymbol{r}_1\}_{N_1}, \ldots, \{\boldsymbol{r}_\varkappa\}_{N_\varkappa}; \{\boldsymbol{p}_1\}_{N_1}, \ldots, \{\boldsymbol{p}_\varkappa\}_{N_\varkappa}) \\ &= \Xi^{-1} \left[\prod_{i=1}^{\varkappa} \frac{e^{\frac{N_i \mu_i}{kT}}}{N_i!} \left\{ \frac{j_i(T)}{h^3} \right\}^{N_i} \right] \exp\left[-\frac{1}{kT} H_{\boldsymbol{N}}(\{\boldsymbol{r}_1\}_{N_1}, \ldots, \{\boldsymbol{r}_\varkappa\}_{N_\varkappa}; \{\boldsymbol{p}_1\}_{N_1}, \ldots, \{\boldsymbol{p}_\varkappa\}_{N_\varkappa}) \right], \end{aligned}\right\} \tag{29.3}$$

where $H_{\boldsymbol{N}}$ is the Hamiltonian of a system composed of N_1 molecules of species 1, N_2 molecules of species 2 and so forth:

$$H_{\boldsymbol{N}} = \sum_{i=1}^{\varkappa} \frac{1}{2m_i} \sum_{s=1}^{N_i} (\boldsymbol{p}_i^{(s)})^2 + \Phi_{\boldsymbol{N}}(\{\boldsymbol{r}_1\}_{N_1}, \ldots, \{\boldsymbol{r}_\varkappa\}_{N_\varkappa}), \tag{29.4}$$

$\Phi_{\boldsymbol{N}}$ being the potential energy function. The symbol $\{\boldsymbol{r}_i\}_{N_i}$ denotes the set of coordinate vectors $\boldsymbol{r}_i^{(s)}$ of N_i molecules of species i [1]. Integration of (29.3) over the whole phase-space leads to (29.2). An ensemble of systems, whose probability distribution is given by (29.2) and (29.3), is called a grand canonical ensemble.

The average value of the dynamical variables $X_{\boldsymbol{N}}(\{\boldsymbol{r}_1\}_{N_1}, \ldots, \{\boldsymbol{r}_\varkappa\}_{N_\varkappa}; \{\boldsymbol{p}_1\}_{N_1}, \ldots, \{\boldsymbol{p}_\varkappa\}_{N_\varkappa})$ over a grand canonical ensemble is given by

$$\overline{X} = \sum_{N_1=0}^{\infty} \cdots \sum_{N_\varkappa=0}^{\infty} \int \cdots \int P_{\boldsymbol{N}} X_{\boldsymbol{N}} \, d\{\boldsymbol{p}_1\}_{N_1} \ldots d\{\boldsymbol{p}_\varkappa\}_{N_\varkappa} \, d\{\boldsymbol{r}_1\}_{N_1} \ldots d\{\boldsymbol{r}_\varkappa\}_{N_\varkappa}. \tag{29.5}$$

As is the case of the canonical ensemble, the internal energy U is given by the average of the Hamiltonian:

$$U = \overline{H}, \tag{29.6}$$

and pressure p by

$$p = -\overline{\frac{\partial H}{\partial V}}. \tag{29.7}$$

The macroscopic value of N_i, which is to be regarded as the mole number multiplied by the Avogadro number, is given as the average of N_i:

$$N_{i\,(\text{macroscopic})} = \overline{N}_i = \sum_{N_1=0}^{\infty} \cdots \sum_{N_\varkappa=0}^{\infty} N_i P(\boldsymbol{N}, T, V). \tag{29.8}$$

[1] The suffix N_i is indispensable to specify the number of i molecules because the number N_i is variable for the case of an open system in contrast with the case of a closed system.

With use of (29.2), (29.3) and (29.5), we can rewrite (29.6) to (29.8), respectively, in the following forms[1]:

$$U = \sum_{i=1}^{\varkappa} \mu_i \left(\frac{\partial\, kT \log \Xi}{\partial \mu_i}\right)_{T, V, \mu_j} + kT^2 \left(\frac{\partial \log \Xi}{\partial T}\right)_{V, \mu}, \tag{29.9}$$

$$p = \left(\frac{\partial\, kT \log \Xi}{\partial V}\right)_{T, \mu}, \tag{29.10}$$

$$\overline{N}_i = \left(\frac{\partial\, kT \log \Xi}{\partial \mu_i}\right)_{T, V, \mu_j}. \tag{29.11}$$

Comparing (29.9) to (29.11) with (5.14) to (5.16), we obtain

$$\Omega = -kT \log \Xi, \tag{29.12}$$

where the integration constant has been determined in the limit of infinite dilution. Eq. (29.12) is the fundamental equation for the grand canonical ensemble. It relates the thermodynamic quantities to the intermolecular forces and corresponds to (24.1) for the canonical ensemble.

Substitution of (5.11) into (29.12) leads to the desired equation.

$$\gamma A = -kT \log \Xi + pV, \tag{29.13}$$

which reduces to the customary relation

$$pV = kT \log \Xi$$

for a homogeneous system. With use of (29.10) we can rewrite (29.13) in an alternative form

$$\gamma A = kT V^2 \left\{\frac{\partial}{\partial V}\left(\frac{\log \Xi}{V}\right)\right\}_{A, T, \mu}. \tag{29.14}$$

The above relations are based on the assumption that the area of the interface A is one of the external parameters of the system. But this assumption is, strictly speaking, not always valid, for the coexistence of two phases requires the presence of a small perturbation, such as a gravitational field as stated in Sect. 24, with no exception in contrast with the case of closed systems. We shall discuss this problem in detail in the next section.

30. Two-phase system in grand canonical ensemble. The previous section has concerned with only the formal discussion of the statistical-mechanical theory of surface tension based on the grand canonical ensemble. In the grand canonical ensemble, a system is characterized by the set of chemical potentials instead of the set of numbers of molecules. The chemical potential of each species is usually assumed to be uniform throughout the system even for the two-phase case. As seen below, however, the probability of finding the system in the state of coexistence of two phases, is almost zero.

Now we shall examine more carefully the condition for the applicability of the grand partion function method to the theory of surface tension. To make clear the essential point of the problem we shall treat a one-component system. Then according to (29.2) the probability that the system has N molecules is

$$P(N, V, T) = \Xi^{-1} Z(N, T, V)\, e^{N\mu/kT}. \tag{30.1}$$

If we express the grand potential Ω given by (5.10) as a function of N, T, and V, though Ω is customarily given in terms of μ, T and V, we can rewrite

[1] See footnote 3, p. 140.

(30.1) in the form

$$P(N, V, T) = \Xi^{-1} e^{-\Omega(N, V, T)/kT}. \tag{30.2}$$

It is customarily believed that the probability $P(N, T, V)$ takes, at given V and T, only one sharp maximum for a certain value of N, say $\overline{N}$, except when μ assumes a certain value μ^* corresponding to the condition for the vapor-liquid equilibrium. The number $\overline{N}$ is a definite function of V, T and μ, and corresponds to the macroscopic number of molecules in the system.

Let N' be the number of molecules in the system when the whole volume is filled by a uniform vapor phase with chemical potential μ^*, and N'' the number when filled by a uniform liquid phase with the same chemical potential. Then pV is independent of the value of N between N' and N''. If the interfacial contribution is completely ignored, the grand potential Ω given by (5.11) is $-pV$, and hence from (30.2), we obtain a $P(N, V, T)$ such as is shown schematically in Fig. 10.

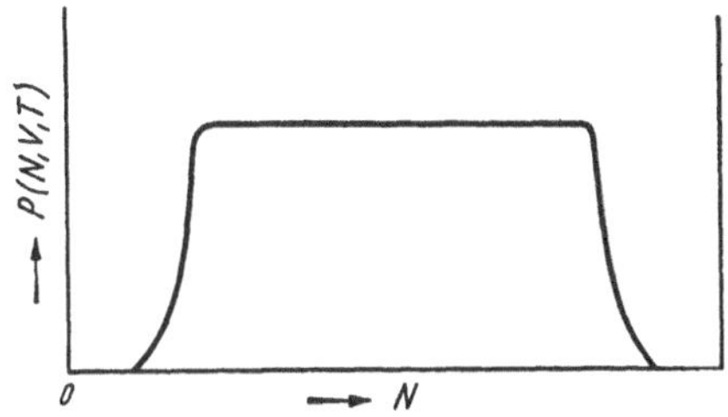

Fig. 10. $P(N, V, T)$ for $\gamma = 0$.

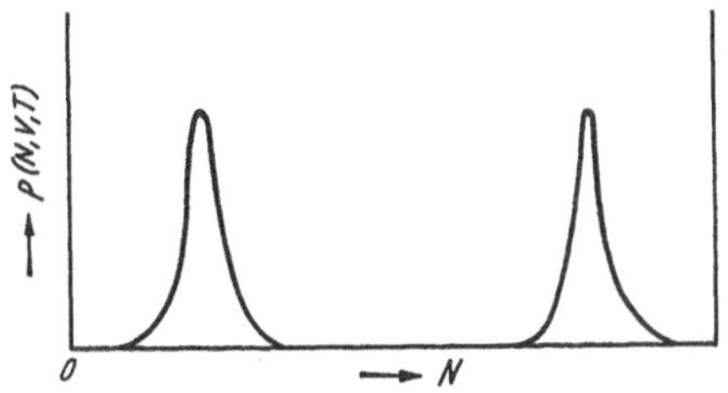

Fig. 11. $P(N, V, T)$ for $\gamma > 0$.

In an actual system composed of two phases, we must, however, take into account the interfacial contribution and therefore, have to write $\Omega = \gamma A - pV$ according to (5.11). Then we have a $P(N, V, T)$ for positive γ such as is illustrated schematically in Fig. 11. This curve has two sharp maxima at N' and N''.

If γ were negative, $\Omega = \gamma A - pV$ decreases with A so that according to (30.2), even for the case of $\mu = \mu^*$, $P(N, V, T)$ would take only one maximum at an N which maximizes $|\gamma| A$, i.e., submicroscopic drops or bubbles would be distributed throughout the volume V approximately uniformly, corresponding to the visible segregation of material into vapor and liquid below the critical temperature[1]. Thus we may conclude that the measurable surface tension is positive below the critical temperature and that it could only be negative above the critical temperature or between two miscible liquids[2].

Furthermore, $P(N, V, T)$ of the type schematically shown in Fig. 11, has actually been found by Katsura[3] from an exact calculation for a finite system consisting of a small number of molecules. We consider Katsura's result as a confirmation of the existence of a positive surface tension. His result leads to a loop in the $p - V$ diagram which vanishes in the limit of infinite N, as proved by Hill[4], and is thus not of the van der Waals type.

The behavior of $P(N, V, T)$ illustrated in Fig. 11 shows that the probability of finding the system at the state of coexistence of two phases is overwhelmingly

[1] O.K. Rice: J. Chem. Phys. **15**, 314 (1947). — Critical Phenomena. In: Thermodynamics and Physics of Matter (ed. F.D. Rossini). Princeton: Princeton University Press 1955.

[2] N.K. Adam: Physics and Chemistry of Surfaces. London: Oxford University Press 1941.

[3] S. Katsura: J. Chem. Phys. **22**, 1277 (1954). — Progr. Theoret. Phys. **11**, 476 (1954).

[4] T.L. Hill: J. Phys. Chem. **57**, 324 (1953). — J. Chem. Phys. **23**, 812 (1955). — Statistical Mechanics, pp. 413—423. New York: McGraw-Hill 1956.

small. Hence in the case of the grand canonical ensemble, it is especially indispensable to employ an auxiliary field such as a gravitational field for the case of a plane interface used in Sect. 24, whenever we deal with the interface.

The gravitational potential energy of a molecule with mass m at z is mgz, where the vertical z axis is directed upward and g denotes the gravitational acceleration. When the molecule is displaced from $z^{(1)}$ to $z^{(2)}$, the change in the Helmholtz free energy of the system is

$$\delta F = \left(\mu(z^{(2)}) + mgz^{(2)}\right) - \left(\mu(z^{(1)}) + mgz^{(1)}\right), \tag{30.3}$$

where $\mu(z)$ is the value of the chemical potential at z.

From the condition for thermodynamic equilibrium, $\delta F = 0$, we have

$$\mu(z) + mgz = \mu(0). \tag{30.4}$$

If the value of the chemical potential at $z=0$, $\mu(0)$, is equal to μ^*, a value enabling the two phases to be together at equilibrium, then, according to (30.4), $\mu(z)$ assumes a value larger than μ^* when $z<0$, and hence the portion of the fluid below the $z=0$ plane must be occupied by a denser phase, since chemical potential is a non-decreasing function of fluid density, whereas the upper portion is occupied by a less dense phase.

This corresponds to a grand canonical ensemble of systems connected exactly at $z=0$ with a material reservoir whose chemical potential is exactly equal to μ^*.

As has just been shown, a certain auxiliary field is indispensable for the substance to exist in two phases, but may be neglected from a thermodynamic point of view, since such a weak field that its contribution to the interfacial properties may be ignored is sufficient to allow the coexistence of two phases.

31. Molecular distribution functions and series expression of grand partition function. In classical statistical mechanics, the grand partition function $\Xi(\mu, T, V)$ given by (29.1) may be rewritten, with use of (24.2), in the following form:

$$\Xi(\zeta, T, V) = \sum_{N_1=0}^{\infty} \cdots \sum_{N_\varkappa=0}^{\infty} Q(\boldsymbol{N}, T, V)\, \zeta_1^{N_1} \ldots, \zeta_\varkappa^{N_\varkappa}, \tag{31.1}$$

$$\zeta_i = \lambda_i^{-3} j_i\, e^{\mu_i/kT}, \tag{31.2}$$

where λ_i is given by (24.4) and j_i is the rotational and vibrational partition function of a molecule of species i. We shall refer to ζ_i as the *activity* throughout this article[1].

For a mixture of gases sufficiently dilute to be regarded as being perfect, the chemical potential μ_i reduces to [see (7.3)]

$$\mu_i^0 = -kT \log(\lambda_i^{-3} j_i) + kT \log n_i. \tag{31.3}$$

Hence we get from (31.2)

$$\lim_{n\to 0} \left(\frac{\zeta_i}{n_i}\right) = 1. \tag{31.4}$$

The molecular distribution functions in a grand canonical ensemble depend on the set of activities ζ. Thus we may express the singlet and pair distribution

[1] This parameter has been referred to by MAYER and MAYER as the fugacity. The conventional symbol for activity is z_i but in this article ζ_i is used in order to be distinguished from the z coordinate of the rectangular coordinate system.

functions by $n_i(\zeta; \boldsymbol{r}^{(1)})$ and $n_{ij}(\zeta; \boldsymbol{r}^{(1)}, \boldsymbol{r}^{(2)})$, respectively. Using (29.3) and (29.4), we have

$$\left.\begin{aligned} n_i(\zeta; \boldsymbol{r}^{(1)}) &= \Xi^{-1} \sum_{N_1=0}^{\infty} \cdots \sum_{N_\varkappa=0}^{\infty} \frac{N_i \prod_{j=1}^{\varkappa} \zeta_j^{N_j}}{\prod_{j=1}^{\varkappa} N_j!} \times \\ &\times \int \cdots \int \exp[-\Phi/kT]\, \delta(\boldsymbol{r}^{(1)} - \boldsymbol{r}_i^{(s)}) \prod_{i=1}^{\varkappa} d\{\boldsymbol{r}_i\}_{N_i}, \end{aligned}\right\} \quad (31.5)$$

$$\left.\begin{aligned} n_{ij}(\zeta; \boldsymbol{r}^{(1)}, \boldsymbol{r}^{(2)}) &= \Xi^{-1} \sum_{N_1=1}^{\infty} \cdots \sum_{N_\varkappa=0}^{\infty} \frac{N_i (N_j - \delta_{ij}) \prod_{l=1}^{\varkappa} \zeta_l^{N_l}}{\prod_{l=1}^{\varkappa} N_l!} \times \\ &\times \int \cdots \int \exp[-\Phi/kT]\, \delta(\boldsymbol{r}^{(1)} - \boldsymbol{r}_i^{(s)})\, \delta(\boldsymbol{r}^{(2)} - \boldsymbol{r}_j^{(t)}) \prod_{i=1}^{\varkappa} d\{\boldsymbol{r}_i\}_{N_i}, \end{aligned}\right\} \quad (31.6)$$

where δ_{ij} is KRONECKER'S delta.

Furthermore we may define molecular distribution functions of higher order by[1]

$$\left.\begin{aligned} n_{ij\ldots p}(\zeta; \boldsymbol{r}^{(1)}, \ldots, \boldsymbol{r}^{(s)}) &= \Xi^{-1} \sum_{N_1=s_1}^{\infty} \cdots \sum_{N_\varkappa=s_\varkappa}^{\infty} \frac{\prod_{i=1}^{\varkappa} \zeta_i^{N_i}}{\prod_{i=1}^{\varkappa} (N_i - s_i)!} \times \\ &\times \int \cdots \int \exp[-\Phi/kT] \prod_{i=1}^{\varkappa} d\{\boldsymbol{r}_i\}_{N_i - s_i}, \end{aligned}\right\} \quad (31.7)$$

where s_i is the number of molecules of species i appearing in the set of $i, j, \ldots, p$ and $\sum_{i=1}^{\varkappa} s_i = s$. Differentiating (31.1) with respect to $(\partial/\partial\zeta_1)^{s_1} \ldots (\partial/\partial\zeta_\varkappa)^{s_\varkappa}$ we obtain

$$\zeta^{s_1} \ldots \zeta_\varkappa^{s_\varkappa} \frac{\partial^s \Xi}{\partial \zeta_1^{s_1} \ldots \partial \zeta_\varkappa^{s_\varkappa}} = \sum_{N_1=s_1}^{\infty} \cdots \sum_{N_\varkappa=s_\varkappa}^{\infty} \left\{ \prod_{i=1}^{\varkappa} \frac{N_i!\, \zeta_i^{N_i}}{(N_i - s_i)!} \right\} Q(\boldsymbol{N}, V, T) \quad (31.8)$$

Combining with (24.3) and (31.7), we obtain

$$\zeta_1^{s_1} \ldots \zeta_\varkappa^{s_\varkappa} \frac{\partial^s \Xi}{\partial \zeta_1^{s_1} \ldots \partial \zeta_\varkappa^{s_\varkappa}} = \Xi \int \cdots \int n_{ij\ldots p}(\zeta; \boldsymbol{r}^{(1)}, \ldots, \boldsymbol{r}^{(s)})\, d\boldsymbol{r}^{(1)} \ldots d\boldsymbol{r}^{(s)}. \quad (31.9)$$

Consequently TAYLOR'S expansion of the function $\Xi(\zeta)$ about the point ζ^*, is given by[2]

$$\left.\begin{aligned} \Xi(\zeta) &= \Xi(\zeta^*) \sum_{s_1=0}^{\infty} \cdots \sum_{s_\varkappa=0}^{\infty} \left\{ \prod_{i=1}^{\varkappa} \frac{1}{s_i!} \left(\frac{\zeta_i - \zeta_i^*}{\zeta_i^*} \right)^{s_i} \right\} \times \\ &\times \int \cdots \int n_{ij\ldots p}(\zeta^*; \boldsymbol{r}^{(1)}, \ldots, \boldsymbol{r}^{(s)})\, d\boldsymbol{r}^{(1)} \ldots d\boldsymbol{r}^{(s)}, \end{aligned}\right\} \quad (31.10)$$

where T and V are omitted, for simplicity, from the arguments of the grand partition functions and the distribution functions. This expansion of the grand partition function, which was first derived by MCMILLAN and MAYER[3], plays an important role in the theory of solutions.

---

[1] From the standpoint of consistency, it might be desirable to use such a notation as $n(\zeta; \{\boldsymbol{r}_i\}_{s_i}, \ldots, \{\boldsymbol{r}_\varkappa\}_{s_\varkappa})$. But the above one is used, for convenience.

[2] S. ONO: Progr. Theoret. Phys. **6**, 447 (1951). — See also T.L. HILL: Statistical Mechanics, Chap. 6. New York: McGraw-Hill 1956.

[3] W.G. MCMILLAN and J.E. MAYER: J. Chem. Phys. **13**, 276 (1945).

As with (31.8) and (31.9), we obtain from (31.5)

$$\left.\begin{aligned} &\zeta_1^{s_1}\ldots\zeta_i^{s_i-1}\ldots\zeta_\varkappa^{s_\varkappa}\frac{\partial^{s-1}\{\zeta_i^{-1}\Xi\, n_i(\zeta,\boldsymbol{r}^{(1)})\}}{\partial\zeta_1^{s_1}\ldots\partial\zeta_i^{s_i-1}\ldots\partial\zeta_\varkappa^{s_\varkappa}}\\ &\quad=\frac{1}{\zeta_i}\,\Xi\int\cdots\int n_{ij\ldots p}(\zeta;\boldsymbol{r}^{(1)},\boldsymbol{r}^{(2)},\ldots,\boldsymbol{r}^{(s)})\,d\boldsymbol{r}^{(2)}\ldots d\boldsymbol{r}^{(s)}.\end{aligned}\right\}\tag{31.11}$$

Then the TAYLOR's expansion of $\Xi(\zeta)\, n_i(\zeta;\boldsymbol{r}^{(1)})$ is

$$\left.\begin{aligned} \Xi(\zeta)\, n_i(\zeta;\boldsymbol{r}^{(1)}) &= \Xi(\zeta^*)\sum_{s_1=0}^{\infty}\cdots\sum_{s_\varkappa=0}^{\infty}\left[\left(\frac{s_i\zeta_i}{\zeta_i-\zeta_i^*}\right)\prod_{j=1}^{\varkappa}\frac{(\zeta_j-\zeta_j^*)^{s_j}}{s_j!\,\zeta_j^{*\,s_j}}\right]\times\\ &\quad\times\int\cdots\int n_{ij\ldots p}(\zeta^*;\boldsymbol{r}^{(1)},\boldsymbol{r}^{(2)},\ldots,\boldsymbol{r}^{(s)})\,d\boldsymbol{r}^{(2)}\ldots d\boldsymbol{r}^{(s)}.\end{aligned}\right\}\tag{31.12}$$

Likewise we obtain for the pair distribution function

$$\left.\begin{aligned} \Xi(\zeta)\, n_{ij}(\zeta;\boldsymbol{r}^{(1)},\boldsymbol{r}^{(2)}) &= \Xi(\zeta^*)\sum_{s_1=0}^{\infty}\cdots\sum_{s_\varkappa=0}^{\infty}\left[\frac{s_i(s_j-\delta_{ij})\,\zeta_i\zeta_j}{(\zeta_i-\zeta_i^*)\,(\zeta_j-\zeta_j^*)}\prod_{l=1}^{\varkappa}\frac{(\zeta_l-\zeta_l^*)^{s_l}}{s_l!\,\zeta_l^{*\,s_l}}\right]\times\\ &\quad\times\int\cdots\int n_{ijk\ldots p}(\zeta^*;\boldsymbol{r}^{(1)},\boldsymbol{r}^{(2)},\boldsymbol{r}^{(3)},\ldots,\boldsymbol{r}^{(s)})\,d\boldsymbol{r}^{(3)}\ldots d\boldsymbol{r}^{(s)}.\end{aligned}\right\}\tag{31.13}$$

One of the most useful applications of the above theory is obtained when we consider a solution of some solutes in a solvent, say species 1. Then we may define the standard state of the above solution by the conditions that all the activities of the solutes are zero and that the activity of solvent ζ_1^0 is such that the pure liquid has some convenient pressure p^0. The activity set of this standard state will be denoted by ζ^0. Now suppose we have a membrane that is permeable to the solvent but not to the solutes. Let the above solution with activity, say ζ, meet at this membrane in equilibrium with a solution containing the solvent species alone. Then the pressure difference across the membrane is, by definition, equal to the osmotic pressure π:

$$\pi = p - p^0,\tag{31.14}$$

where p corresponds to ζ. For this case the solvent has the same activity ζ_1^0 in the two states ζ and ζ^0 because of the permeability of the membrane to the solvent. With this particular choice of ζ and ζ^* equal to ζ^0, the terms contributing to the summation in (31.10), (31.12) and (31.13), are restricted to those which contain solute species alone. These conditions were first introduced by MCMILLAN and MAYER[1], and are called the *osmotic conditions*.

Let us now introduce the potentials of average forces, which play an important role especially in the theory of solutions[2], defined by

$$W_{ij\ldots p} = -\,kT\log\frac{n_{ij\ldots p}(\zeta;\boldsymbol{r}^{(1)},\ldots,\boldsymbol{r}^{(s)})}{n_i^\alpha(\zeta)\, n_j^\alpha(\zeta)\ldots n_p^\alpha(\zeta)},\tag{31.15}$$

where superscript α refers to phase α. For the present we shall restrict ourselves to the potentials of average forces on solute molecules, that is, the subscripts, $i, j, \ldots, p$, denote only the solute species.

We shall now, as in the foregoing, refer to the liquid phase of the solution as the phase α and its saturated vapor phase as the phase β. It is apparent from (31.15) that $W_{ij\ldots p}$ reduces exactly to its value in the homogeneous bulk phase α

[1] W. G. MCMILLAN and J. E. MAYER: J. Chem. Phys. 13, 276 (1945).

[2] In general, the potentials of average forces are similarly defined for a canonical ensemble. They play important roles even in the case of a pure liquid but we shall use them only for the grand canonical ensemble, in this article.

if all the position vectors $\boldsymbol{r}^{(1)}, \ldots, \boldsymbol{r}^{(s)}$ are in the interior of the phase α, whereas $W_{ij\ldots p}$ has its value in the bulk phase β, apart from an additive constant, if all of $\boldsymbol{r}^{(1)}, \ldots, \boldsymbol{r}^{(s)}$ are in the interior of the phase β.

The superposition approximation defined by (25.9) for a triplet distribution function is generalized as follows:

$$\frac{n_{ij\ldots p}(\boldsymbol{r}^{(1)}, \ldots, \boldsymbol{r}^{(s)})}{n_i(\boldsymbol{r}^{(1)})\, n_j(\boldsymbol{r}^{(2)}) \ldots n_p(\boldsymbol{r}^{(s)})} = \frac{n_{ij}(\boldsymbol{r}^{(1)}, \boldsymbol{r}^{(2)})}{n_i(\boldsymbol{r}^{(1)})\, n_j(\boldsymbol{r}^{(2)})} \frac{n_{ik}(\boldsymbol{r}^{(1)}, \boldsymbol{r}^{(3)})}{n_i(\boldsymbol{r}^{(1)})\, n_k(\boldsymbol{r}^{(3)})} \cdots \frac{n_{op}(\boldsymbol{r}^{(s-1)}, \boldsymbol{r}^{(s)})}{n_o(\boldsymbol{r}^{(s-1)})\, n_p(\boldsymbol{r}^{(s)})}, \tag{31.16}$$

which states that the probability of finding a set of s molecules of species $i, j, \ldots, p$, at $\boldsymbol{r}^{(1)}, \ldots, \boldsymbol{r}^{(s)}$, respectively, is approximately expressible as a product of independent pair probabilities. It is evident that this approximation can be rewritten in terms of the potentials of average forces (31.15) as follows:

$$\left.\begin{aligned} &W_{ij\ldots p}(\boldsymbol{r}^{(1)}, \ldots, \boldsymbol{r}^{(s)}) - w_i(\boldsymbol{r}^{(1)}) - w_j(\boldsymbol{r}^{(2)}) - \cdots - w_p(\boldsymbol{r}^{(s)}) \\ &= \left(W_{ij}(\boldsymbol{r}^{(1)}, \boldsymbol{r}^{(2)}) - w_i(\boldsymbol{r}^{(1)}) - w_j(\boldsymbol{r}^{(2)})\right) + \left(W_{ik}(\boldsymbol{r}^{(1)}, \boldsymbol{r}^{(3)}) - w_i(\boldsymbol{r}^{(1)}) - w_k(\boldsymbol{r}^{(3)})\right) + \\ &\qquad + \cdots + \left(W_{op}(\boldsymbol{r}^{(s-1)}, \boldsymbol{r}^{(s)}) - w_o(\boldsymbol{r}^{(s-1)}) - w_p(\boldsymbol{r}^{(s)})\right), \end{aligned}\right\} \tag{31.17}$$

where we use the symbol w_i instead of W_i for convenience.

On the other hand, the potential of average force $W_{ij}(\boldsymbol{r}^{(1)}, \boldsymbol{r}^{(2)})$ can be expressed in the form:

$$W_{ij}(\boldsymbol{r}^{(1)}, \boldsymbol{r}^{(2)}) = w_i(\boldsymbol{r}^{(1)}) + w_j(\boldsymbol{r}^{(2)}) + w_{ij}(\boldsymbol{r}^{(1)}, \boldsymbol{r}^{(2)}), \tag{31.18}$$

where we shall refer to $w_{ij}(\boldsymbol{r}^{(1)}, \boldsymbol{r}^{(2)})$ as the component potential of a pair of molecules, which is regarded as the potential of average intermolecular force acting between a molecule of species i at $\boldsymbol{r}^{(1)}$ and another molecule of species j at $\boldsymbol{r}^{(2)}$.

Inserting (31.18) into (31.17) we obtain an alternative expression for the superposition approximation:

$$\left.\begin{aligned} W_{ij\ldots p}(\boldsymbol{r}^{(1)}, \ldots, \boldsymbol{r}^{(s)}) &= w_i(\boldsymbol{r}^{(1)}) + \cdots + w_p(\boldsymbol{r}^{(s)}) + \\ &+ w_{ij}(\boldsymbol{r}^{(1)}, \boldsymbol{r}^{(2)}) + w_{ik}(\boldsymbol{r}^{(1)}, \boldsymbol{r}^{(3)}) + \cdots + w_{op}(\boldsymbol{r}^{(s-1)}, \boldsymbol{r}^{(s)}). \end{aligned}\right\} \tag{31.19}$$

Since in the limit of an infinitely dilute gas, $w_{ij}(\boldsymbol{r}^{(1)}, \boldsymbol{r}^{(2)})$ becomes equal to the intermolecular potential $\phi_{ij}(|\boldsymbol{r}^{(2)} - \boldsymbol{r}^{(1)}|)$ itself, and $w_i(\boldsymbol{r}^{(1)}) \to 0$, the approximation (31.19) in that limit is generally accepted as being as accurate as (25.3). It is, however, a much more serious approximation in the potential of average force.

As a special case of (31.15), the number denstiy $n_i(\boldsymbol{r}^{(1)})$ is given by

$$n_i(\boldsymbol{r}^{(1)}) = n_i^{\alpha}\, \mathrm{e}^{-w_i(\boldsymbol{r}^{(1)})/kT}. \tag{31.20}$$

Differentiating (31.12) with respect to $\boldsymbol{r}^{(1)}$ and using (31.13) and (31.19), the superposition approximation to $W^*_{ij\ldots p}$, we easily obtain the Born-Green-Yvon equation for the singlet distribution function

$$\left.\begin{aligned} \nabla^{(1)} n_i(\boldsymbol{r}^{(1)}) = &- \frac{1}{kT} n_i(\boldsymbol{r}^{(1)})\, \nabla^{(1)} w_i^*(\boldsymbol{r}^{(1)}) - \\ &- \frac{1}{kT} \sum_j \frac{\zeta_j - \zeta_j^*}{\zeta_j} \int n_{ij}(\boldsymbol{r}^{(1)}, \boldsymbol{r}^{(2)})\, \nabla^{(1)} w_{ij}^*(\boldsymbol{r}^{(1)}, \boldsymbol{r}^{(2)})\, d\boldsymbol{r}^{(2)}. \end{aligned}\right\} \tag{31.21}$$

32. Surface tension difference between solution and pure solvent. As mentioned in the previous section, the pressure difference between the solution and the pure solvent under osmotic conditions, is equal to the osmotic pressure. In the present section we shall consider the surface tension difference under the same conditions.

For this purpose, we shall consider a grand canonical ensemble of multicomponent systems with the activity set $\boldsymbol{\zeta}$ under a weak gravitational field. As in the previous section we denote the pressure by p and the surface tension, which we assume to exist at this state, by γ and the corresponding quantities at an arbitrary reference state $\boldsymbol{\zeta}^*$ by p^* and γ^*, respectively. Then we obtain from (29.13)

$$\exp\left[\frac{(p-p^*)\,V-(\gamma-\gamma^*)\,A}{kT}\right]=\frac{\Xi}{\Xi^*}\,. \tag{32.1}$$

Furthermore, we assume that the system is contained in a rectangular parallelepiped vessel which has edge lengths l_1, l_2, l_3, in the directions of rectangular coordinate axes, respectively, as shown in Fig. 8. The z axis is directed upwards in the direction opposite to the gravitational force. Then the volume and the area are given by $V=l_1\,l_2\,l_3$ and $A=l_1\,l_2$ respectively.

From (31.10) and (32.1) we obtain

$$\left.\begin{aligned}(p-p^*)\,V-(\gamma-\gamma^*)\,A &= kT\log\Bigg[\sum_{s_1=0}^{\infty}\cdots\sum_{s_\varkappa=0}^{\infty}\left\{\prod_{i=1}^{\varkappa}\frac{1}{s_i!}\left(\frac{\zeta_i-\zeta_i^*}{\zeta_i^*}\right)^{s_i}\right\}\times\\ &\quad\times\int\cdots\int n_{ij\ldots p}(\boldsymbol{\zeta}^*;\boldsymbol{r}^{(1)},\ldots,\boldsymbol{r}^{(s)})\,d\boldsymbol{r}^{(1)}\ldots d\boldsymbol{r}^{(s)}\Bigg].\end{aligned}\right\} \tag{32.2}$$

As in the case of osmotic conditions, we shall hereafter assume that the solvent species, say species 1 has the activity ζ_1 equal to ζ_1^*. Then (32.2) reduces to

$$\left.\begin{aligned}(p-p^*)\,V-(\gamma-\gamma^*)\,A &= kT\log\Bigg[\sum_{s_2=0}^{\infty}\cdots\sum_{s_\varkappa=0}^{\infty}\left\{\prod_{i=2}^{\varkappa}\frac{1}{s_i!}\left(\frac{\zeta_i-\zeta_i^*}{\zeta_i^*}\right)^{s_i}\right\}\times\\ &\quad\times\int\cdots\int n^*_{ij\ldots p}(\boldsymbol{r}^{(1)},\ldots,\boldsymbol{r}^{(s)})\,d\boldsymbol{r}^{(1)}\ldots d\boldsymbol{r}^{(s)}\Bigg],\end{aligned}\right\} \tag{32.3}$$

where $n^*_{ij\ldots p}(\boldsymbol{r}^{(1)},\ldots,\boldsymbol{r}^{(s)})$ stands for $n_{ij\ldots p}(\boldsymbol{\zeta}^*,\boldsymbol{r}^{(1)},\ldots,\boldsymbol{r}^{(s)})$, none of the subscripts $i, j, \ldots, p$ assume the value 1 corresponding to the solvent species, and hence for the present case, it is to be understood that $s=\sum_{i=2}^{\varkappa}s_i$, instead of $s=\sum_{i=1}^{\varkappa}s_i$.

Substituting $W_{ij\ldots p}$ from (31.15) into (32.3) we have

$$\left.\begin{aligned}(p-p^*)\,V-(\gamma-\gamma^*)\,A &= kT\log\Bigg[\sum_{s_2=0}^{\infty}\cdots\sum_{s_\varkappa=0}^{\infty}\prod_{i=2}^{\varkappa}\left\{\frac{(\zeta_i-\zeta_i^*)^{s_i}}{s_i!}\left(\frac{n_i^{*\alpha}}{\zeta_i^*}\right)^{s_i}\right\}\times\\ &\quad\times\int\cdots\int e^{-W^*_{ij\ldots p}(\boldsymbol{r}^{(1)},\ldots,\boldsymbol{r}^{(s)})/kT}\,d\boldsymbol{r}^{(1)}\ldots d\boldsymbol{r}^{(s)}\Bigg],\end{aligned}\right\} \tag{32.4}$$

in which the functions with asterisk refer to the reference state $\boldsymbol{\zeta}^*$. The above equation is the basic equation for the surface tension of an osmotic solution.

If we differentiate (32.4) with respect to l_1 using the relation $V=l_1\,l_2\,l_3$ and $A=l_1\,l_2$, there results the following equation:

$$\left.\begin{aligned}(p-p^*)\,l_2\,l_3-(\gamma-\gamma^*)\,l_2 &= kT\,\frac{\Xi^*}{\Xi}\sum_{s_2=0}^{\infty}\cdots\sum_{s_\varkappa=0}^{\infty}\left\{\prod_{i=2}^{\varkappa}\frac{(\zeta_i-\zeta_i^*)^{s_i}}{s_i!}\left(\frac{n_i^{*\alpha}}{\zeta_i^*}\right)^{s_i}\right\}\times\\ &\quad\times\frac{\partial}{\partial l_1}\int\cdots\int e^{-W^*_{ij\ldots p}(\boldsymbol{r}^{(1)},\ldots,\boldsymbol{r}^{(s)})/kT}\,d\boldsymbol{r}^{(1)}\ldots d\boldsymbol{r}^{(s)}.\end{aligned}\right\} \tag{32.5}$$

If edge effects are neglected, $w_i(\boldsymbol{r}^{(1)})$ is a function of $z^{(1)}$ alone and may be denoted by $w_i(z^{(1)})$. Also, $w_{ij}(\boldsymbol{r}^{(1)},\boldsymbol{r}^{(2)})$ is a function of $\boldsymbol{r}^{(12)}=\boldsymbol{r}^{(2)}-\boldsymbol{r}^{(1)}$ and $z^{(1)}$

or $z^{(2)}$. For convenience, we shall, however, restrict ourselves to the case in which w_{ij} is of the form $w_{ij}(r^{(12)}, r_{im}^{(12)})$, where (see also Fig. 17)

$$r_{im}^{(12)} = [(x^{(2)} - x^{(1)})^2 + (y^{(2)} - y^{(1)})^2 + (z^{(1)} + z^{(2)})]^{\frac{1}{2}}. \tag{32.6}$$

Then we may rewrite the superposition approximation, (31.19), as follows

$$\left.\begin{aligned} W_{ij\ldots p}(\boldsymbol{r}^{(1)}, \ldots, \boldsymbol{r}^{(s)}) &= w_i(z^{(1)}) + \cdots + w_p(z^{(s)}) + \\ &+ w_{ij}(r^{(12)}, r_{im}^{(12)}) + w_{ik}(r^{(13)}, r_{im}^{(13)}) + \cdots + w_{op}(r^{(s-1,s)}, r_{im}^{(s-1,s)}) . \end{aligned}\right\} \tag{32.7}$$

As in Sect. 26 employing the Bogoliubov-Green technique based on the change of variables (26.6), we obtain from (32.7)

$$\left.\begin{aligned} &\sum_{s_2=0}^{\infty}\cdots\sum_{s_\varkappa=0}^{\infty}\left\{\prod_{i=2}^{\varkappa}\frac{(\zeta_i-\zeta_i^*)^{s_i}}{s_i!}\left(\frac{n_i^{*\alpha}}{\zeta_i^*}\right)^{s_i}\right\}\frac{\partial}{\partial l_1}\int\cdots\int \mathrm{e}^{-W_{ij\ldots p}^*(\boldsymbol{r}^{(1)},\ldots,\boldsymbol{r}^{(s)})/kT}\,d\boldsymbol{r}^{(1)}\ldots d\boldsymbol{r}^{(s)} \\ &= \frac{1}{l_1}\sum_{i=2}^{\varkappa}\sum_{s_2=0}^{\infty}\cdots\sum_{s_\varkappa=0}^{\infty} s_i\left\{\prod_{j=2}^{\varkappa}\frac{(\zeta_j-\zeta_j^*)^{s_j}}{s_j!}\left(\frac{n_j^{*\alpha}}{\zeta_j^*}\right)^{s_j}\right\}\int\cdots\int \mathrm{e}^{-W_{ij\ldots p}^*/kT}\,d\boldsymbol{r}^{(1)}\ldots d\boldsymbol{r}^{(s)} - \\ &- \frac{1}{2l_1 kT}\sum_{i=2}^{\varkappa}\sum_{j=2}^{\varkappa}\sum_{s_2=0}^{\infty}\cdots\sum_{s_\varkappa=0}^{\infty}\left[s_i(s_j-\delta_{ij})\left\{\prod_{l=2}^{\varkappa}\frac{(\zeta_l-\zeta_l^*)^{s_l}}{s_l!}\left(\frac{n_l^{*\alpha}}{\zeta_l^*}\right)^{s_l}\right\}\times\right. \\ &\left.\times\int\cdots\int\left\{\frac{(x^{(2)}-x^{(1)})^2}{r^{(12)}}\frac{\partial w_{ij}^*}{\partial r^{(12)}} + \frac{(x^{(2)}-x^{(1)})^2}{r_{im}^{(12)}}\frac{\partial w_{ij}^*}{\partial r_{im}^{(12)}}\right\}\mathrm{e}^{-W_{ij\ldots p}^*/kT}\,d\boldsymbol{r}^{(1)}\ldots d\boldsymbol{r}^{(s)}\right]. \end{aligned}\right\} \tag{32.8}$$

Using (31.15), (31.12) and (31.13), we can rewrite (32.8) in the following form:

$$\left.\begin{aligned} &\sum_{s_2=0}^{\infty}\cdots\sum_{s_\varkappa=0}^{\infty}\prod_{i=2}^{\varkappa}\left\{\frac{(\zeta_i-\zeta_i^*)^{s_i}}{s_i!}\left(\frac{n_i^{*\alpha}}{\zeta_i^*}\right)^{s_i}\right\}\frac{\partial}{\partial l_1}\int\cdots\int \mathrm{e}^{-W_{ij\ldots p}^*(\boldsymbol{r}^{(1)},\ldots,\boldsymbol{r}^{(s)})/kT}\,d\boldsymbol{r}^{(1)}\ldots d\boldsymbol{r}^{(s)} \\ &= \frac{1}{l_1}\frac{\Xi}{\Xi^*}\sum_{i=2}^{\varkappa}\frac{\zeta_i-\zeta_i^*}{\zeta_i}\int n_i(z^{(1)})\,d\boldsymbol{r}^{(1)} - \frac{1}{2l_1 kT}\frac{\Xi}{\Xi^*}\sum_{i,j=2}^{\varkappa}\left(\frac{\zeta_i-\zeta_i^*}{\zeta_i}\right)\left(\frac{\zeta_j-\zeta_j^*}{\zeta_j}\right)\times \\ &\times\int\cdots\int\left[\frac{(x^{(2)}-x^{(1)})^2}{r^{(12)}}\frac{\partial w_{ij}^*}{\partial r^{(12)}} + \frac{(x^{(2)}-x^{(1)})^2}{r_{im}^{(12)}}\frac{\partial w_{ij}^*}{\partial r_{im}^{(12)}}\right] n_{ij}(z^{(1)}; \boldsymbol{r}^{(12)})\,d\boldsymbol{r}^{(1)}\,d\boldsymbol{r}^{(2)}. \end{aligned}\right\} \tag{32.9}$$

Substituting (32.9) into (32.5) and multiplying the result by l_1, we obtain, with use of $V = l_1\, l_2\, l_3$ and $A = l_1\, l_2$,

$$\left.\begin{aligned} &(p-p^*)\,V - (\gamma-\gamma^*)\,A = A\sum_{i=2}^{\varkappa}\frac{\zeta_i-\zeta_i^*}{\zeta_i}\,kT\int_{-l^\alpha}^{l^\beta} n_i(z^{(1)})\,dz^{(1)} - \\ &- \frac{A}{2}\sum_{i,j=2}^{\varkappa}\frac{(\zeta_i-\zeta_i^*)(\zeta_j-\zeta_j^*)}{\zeta_i\zeta_j}\int_{-l^\alpha}^{l^\beta} dz^{(1)}\int\left[\frac{(x^{(12)})^2}{r^{(12)}}\frac{\partial w_{ij}^*}{\partial r^{(12)}} + \frac{(x^{(12)})^2}{r_{im}^{(12)}}\frac{\partial w_{ij}^*}{\partial r_{im}^{(12)}}\right]\times \\ &\times n_{ij}(z^{(1)}, \boldsymbol{r}^{(12)})\,d\boldsymbol{r}^{(12)}, \end{aligned}\right\} \tag{32.10}$$

where l^α and l^β have the same meaning as in Sect. 26 (see Fig. 8).

Since $l_3 = l^\alpha + l^\beta$, (32.10) can be rewritten as

$$\left.\begin{aligned} \gamma-\gamma^* &= \int_{-l^\alpha}^{l^\beta}(p-p^*)\,dz^{(1)} - \sum_{i=2}^{\varkappa}\int_{-l^\alpha}^{l^\beta} dz^{(1)}\left[\frac{\zeta_i-\zeta_i^*}{\zeta_i}\,n_i(z^{(1)})\,kT - \right. \\ &- \frac{1}{2}\sum_{j=2}^{\varkappa}\frac{(\zeta_i-\zeta_i^*)(\zeta_j-\zeta_j^*)}{\zeta_i\zeta_j}\int\left\{\frac{(x^{(12)})^2}{r^{(12)}}\frac{\partial w_{ij}^*}{\partial r^{(12)}} + \frac{(x^{(12)})^2}{r_{im}^{(12)}}\frac{\partial w_{ij}^*}{\partial r_{im}^{(12)}}\right\}\times \\ &\left.\times\, n_{ij}(z^{(1)}; \boldsymbol{r}^{(12)})\,d\boldsymbol{r}^{(12)}\right]. \end{aligned}\right\} \tag{32.11}$$

If we differentiate (32.4) with respect to l_3, instead of l_1, we obtain in exactly the same manner as in the derivation of (32.10),

$$\left.\begin{aligned}(p-p^*)\,V &= A\sum_{i=2}^{\varkappa}\frac{\zeta_i-\zeta_i^*}{\zeta_i}\,kT\int_{-l^\alpha}^{l^\beta} n_i(z^{(1)})\,dz^{(1)} - A\sum_{i=2}^{\varkappa}\frac{\zeta_i-\zeta_i^*}{\zeta_i}\int_{-l^\alpha}^{l^\beta} z^{(1)}\times\\ &\times\frac{dw_i^*}{dz^{(1)}}\,n_i(z^{(1)})\,dz^{(1)} - \frac{A}{2}\sum_{i,j=2}^{\varkappa}\frac{(\zeta_i-\zeta_i^*)(\zeta_j-\zeta_j^*)}{\zeta_i\zeta_j}\int_{-l^\alpha}^{l^\beta} dz^{(1)}\times\\ &\times\int\left[\frac{(z^{(12)})^2}{r^{(12)}}\,\frac{\partial w_{ij}^*}{\partial r^{(12)}}+\frac{(z^{(1)}+z^{(2)})^2}{r_{im}^{(12)}}\,\frac{\partial w_{ij}^*}{\partial r_{im}^{(12)}}\right]n_{ij}(z^{(1)};\boldsymbol{r}^{(12)})\,d\boldsymbol{r}^{(12)}.\end{aligned}\right\}\quad(32.12)$$

Substracting (32.10) from (32.12), we obtain the following expression for the surface tension[1]:

$$\left.\begin{aligned}\gamma-\gamma^* &= -\sum_{i=2}^{\varkappa}\frac{\zeta_i-\zeta_i^*}{\zeta_i}\int_{-\infty}^{\infty} z^{(1)}\frac{dw_i^*}{dz^{(1)}}\,n_i(z^{(1)})\,dz^{(1)}+\frac{1}{2}\sum_{i,j=2}^{\varkappa}\frac{(\zeta_i-\zeta_i^*)(\zeta_j-\zeta_j^*)}{\zeta_i\zeta_j}\times\\ &\times\int_{-\infty}^{\infty} dz^{(1)}\int\left[\frac{(x^{(12)})^2-(z^{(12)})^2}{r^{(12)}}\,\frac{\partial w_{ij}^*}{\partial r^{(12)}}+\frac{(x^{(12)})^2-(z^{(1)}+z^{(2)})^2}{r_{im}^{(12)}}\,\frac{\partial w_{ij}^*}{\partial r_{im}^{(12)}}\right]\times\\ &\times n_{ij}(z^{(1)};\boldsymbol{r}^{(12)})\,d\boldsymbol{r}^{(12)}.\end{aligned}\right\}\quad(32.13)$$

By making use of this equation, we can calculate the difference of surface tension between the two states ζ and ζ^* under osmotic conditions, from the singlet and pair distribution functions and the potentials of average forces, w_i^* and w_{ij}^*.

For the homogeneous bulk phases α and β, (32.10) reduces respectively to

$$\left.\begin{aligned}(p-p^*)\,V^\alpha &= A\sum_{i=2}^{\varkappa}\frac{\zeta_i-\zeta_i^*}{\zeta_i}\,kT\int_{-l^\alpha}^{0} n_i^\alpha(z^{(1)})\,dz^{(1)}-\\ &-\frac{A}{2}\sum_{i,j=2}^{\varkappa}\frac{(\zeta_i-\zeta_i^*)(\zeta_j-\zeta_j^*)}{\zeta_i\zeta_j}\int_{-l^\alpha}^{0} dz^{(1)}\int\frac{(x^{(12)})^2}{r^{(12)}}\,\frac{\partial w_{ij}^{*\alpha}}{\partial r^{(12)}}\,n_{ij}^\alpha(r^{(12)})\,d\boldsymbol{r}^{(12)},\end{aligned}\right\}\quad(32.14)$$

$$\left.\begin{aligned}(p-p^*)\,V^\beta &= A\sum_{i=2}^{\varkappa}\frac{\zeta_i-\zeta_i^*}{\zeta_i}\,kT\int_{0}^{l^\beta} n_i^\beta(z^{(1)})\,dz^{(1)}-\\ &-\frac{A}{2}\sum_{i,j=2}^{\varkappa}\frac{(\zeta_i-\zeta_i^*)(\zeta_j-\zeta_j^*)}{\zeta_i\zeta_j}\int_{0}^{l^\beta} dz^{(1)}\int\frac{(x^{(12)})^2}{r^{12}}\,\frac{\partial w_{ij}^{*\beta}}{\partial r^{(12)}}\,n_{ij}^\beta(r^{(12)})\,d\boldsymbol{r}^{(12)},\end{aligned}\right\}\quad(32.15)$$

where $w_{ij}^{*\alpha}$ and $w_{ij}^{*\beta}$ are expressible as functions of $r^{(12)}$ alone.

Subtracting (32.14) and (32.15) from (32.10), recalling the additivity of V^α and V^β and utilizing the notation $n_{ij}^{\alpha\beta}$ defined by (26.20) and the similarly defined notation $w_{ij}^{*\alpha\beta}$, we obtain

$$\left.\begin{aligned}\gamma-\gamma^* &= -kT\sum_{i=2}^{\varkappa}\Gamma_i\,\frac{\zeta_i-\zeta_i^*}{\zeta_i}+\frac{1}{2}\sum_{i,j=2}^{\varkappa}\frac{(\zeta_i-\zeta_i^*)(\zeta_j-\zeta_j^*)}{\zeta_i\zeta_j}\int_{-\infty}^{\infty} dz^{(1)}\times\\ &\times\int\left[\frac{(x^{(12)})^2}{r^{(12)}}\left(\frac{\partial w_{ij}^*}{\partial r^{(12)}}\,n_{ij}-\frac{\partial w_{ij}^{*\alpha\beta}}{\partial r^{(12)}}\,n_{ij}^{\alpha\beta}\right)+\frac{(x^{(12)})^2}{r_{im}^{(12)}}\,\frac{\partial w_{ij}^*}{\partial r_{im}^{(12)}}\,n_{ij}\right]d\boldsymbol{r}^{(12)},\end{aligned}\right\}\quad(32.16)$$

where the relation (26.21) has been used.

[1] See the footnote 1 on p. 183.

As will be seen in the following, the expressions for surface tension, (32.11), (32.13) and (32.16) are, respectively, the generalizations of (26.14), (26.13) and (26.23).

Let us take the reference state $\boldsymbol{\zeta}^*$ as the standard state $\boldsymbol{\zeta}^0$, that is, $\zeta_1^* = \zeta_1^0$, $\zeta_2^* = \cdots \zeta_\varkappa^* = 0$. Then (32.11), (32.13) and (32.16) reduce, respectively, to

$$\left.\begin{aligned} \gamma - \gamma^0 = & \int_{-l^\alpha}^{l^\beta} \pi\, d z^{(1)} - \sum_{i=2}^{\varkappa} \int_{-l^\alpha}^{l^\beta} d z^{(1)} \Big[n_i(z^{(1)})\, kT - \\ & - \frac{1}{2} \sum_{j=2}^{\varkappa} \int \Big\{ \frac{(x^{(12)})^2}{r^{(12)}} \frac{\partial w_{ij}^0}{\partial r^{(12)}} + \frac{(x^{(12)})^2}{r_{im}^{(12)}} \frac{\partial w_{ij}^0}{\partial r_{im}^{(12)}} \Big\} n_{ij}(z^{(1)}, \boldsymbol{r}^{(12)})\, d\boldsymbol{r}^{(12)} \Big], \end{aligned}\right\} \tag{32.17}$$

$$\left.\begin{aligned} \gamma - \gamma^0 = & - \sum_{i=2}^{\varkappa} \int_{-\infty}^{\infty} d z^{(1)} \Big[z^{(1)} \frac{d w_i^0}{d z^{(1)}}\, n_i(z^{(1)}) - \\ & - \frac{1}{2} \sum_{j=2}^{\varkappa} \int \Big\{ \Big(\frac{(x^{(12)})^2}{r^{(12)}} \frac{\partial w_{ij}^0}{\partial r^{(12)}} + \frac{(x^{(12)})^2}{r_{im}^{(12)}} \frac{\partial w_{ij}^0}{\partial r_{im}^{(12)}} \Big) - \\ & - \Big(\frac{(z^{(12)})^2}{r^{(12)}} \frac{\partial w_{ij}^0}{\partial r^{(12)}} + \frac{(z^{(1)} + z^{(2)})^2}{r_{im}^{(12)}} \frac{\partial w_{ij}^0}{\partial r_{im}^{(12)}} \Big) \Big\} n_{ij}(z^{(1)}; \boldsymbol{r}^{(12)})\, d\boldsymbol{r}^{(12)} \Big], \end{aligned}\right\} \tag{32.18}$$

$$\left.\begin{aligned} \gamma - \gamma^0 = & - \sum_{i=2}^{\varkappa} \Gamma_i kT + \frac{1}{2} \sum_{i,j=2}^{\varkappa} \int_{-\infty}^{\infty} d z^{(1)} \times \\ & \times \int \Big\{ \frac{(x^{(12)})^2}{r^{(12)}} \Big(\frac{\partial w_{ij}^0}{\partial r^{(12)}} n_{ij} - \frac{\partial w^{0\alpha\beta}}{\partial r^{(12)}} n_{ij}^{\alpha\beta} \Big) + \frac{(x^{(12)})^2}{r_{im}^{(12)}} \frac{\partial w_{ij}^0}{\partial r_{im}^{(12)}} n_{ij} \Big\} d\boldsymbol{r}^{(12)}, \end{aligned}\right\} \tag{32.19}$$

where $p - p^*$ has been replaced by the osmotic pressure π [see (31.14)], γ_0 is the surface tension of the pure solvent and w_{ij}^0, $w_{ij}^{0\alpha\beta}$ are the potentials of average forces between the two referred solute molecules in the infinitely dilute solution, that is, in the pure solvent. The above three equations, based on the superposition approximation (32.7), are basic for the surface tension of osmotic solutions. These equations were originally derived by BUFF and STILLINGER [*23*] by an alternative method based on the Kirkwood-Buff theory of solutions[1].

For the further special case when the asterisk refers to the infinitely dilute gas, the potential of average force w_{ij} reduces to the intermolecular potential and w_i vanishes, and consequently (32.11), (32.13) and, (32.16) reduce, respectively, to (26.14), (26.13) and (26.23) as p^* and γ^* vanish.

33. Application to surface layer of rigorous power series expansion in activities. The theory given in the preceding section is based upon the superposition approximation. In the present section we shall present a rigorous but rather formal theory of the surface tension of solutions.

For simplicity let us introduce the notation

$$\omega_i = (\zeta_i - \zeta_i^*) \frac{n_i^{*\alpha}}{\zeta_i^*}. \tag{33.1}$$

Then the rigorous fundamental relation for an osmotic solution, (32.3), is rewritten as

$$\left.\begin{aligned} & \frac{(p - p^*)\, V - (\gamma - \gamma^*)\, A}{kT} \\ & = \log \Big[\sum_{s_2=0}^{\infty} \cdots \sum_{s_\varkappa=0}^{\infty} \Big\{ \prod_{i=2}^{\varkappa} \frac{(\omega_i)^{s_i}}{s_i!\, (n_i^{*\alpha})^{s_i}} \Big\} \int \cdots \int n_{ij\ldots p}^*(\boldsymbol{r}^{(1)}, \ldots, \boldsymbol{r}^{(s)})\, d\boldsymbol{r}^{(1)} \ldots d\boldsymbol{r}^{(s)} \Big], \end{aligned}\right\} \tag{33.2}$$

[1] J. G. KIRKWOOD and F. P. BUFF: J. Chem. Phys. **19**, 774 (1951).

where we consider the same system as used in connection with (32.2). Expanding the right-hand side of (33.2) in powers of ω_i, we can write

$$\frac{(p-p^*)V-(\gamma-\gamma^*)A}{kT}=\sum_{m_2=0}^{\infty}\cdots\sum_{m_\varkappa=0}^{\infty}V\,b^*(m_2,\ldots,m_\varkappa)\prod_{i=2}^{\varkappa}\omega_i^{m_i},\tag{33.3}$$

where $b^*(m_2,\ldots,m_\varkappa)$ is a generalized cluster integral[1,2] referred to ζ^* and $m_2!\ldots m_\varkappa!\,V\,b^*(m_2,\ldots,m_\varkappa)$ corresponds to the semi-invariant of a distribution in multi-dimensions in the theory of probability[3].

On the other hand, we have the similar relations for the homogeneous bulk phases α and β, whose volumes are V^α and V^β, respectively,

$$\frac{(p-p^*)V^\alpha}{kT}=\sum_{m_2=0}^{\infty}\cdots\sum_{m_\varkappa=0}^{\infty}V^\alpha b^{*\alpha}(m_2,\ldots,m_\varkappa)\prod_{i=2}^{\varkappa}\omega_i^{m_i},\tag{33.4}$$

$$\frac{(p-p^*)V^\beta}{kT}=\sum_{m_2=0}^{\infty}\cdots\sum_{m_\varkappa=0}^{\infty}V^\beta b^{*\beta}(m_2,\ldots,m_\varkappa)\prod_{i=2}^{\varkappa}\omega_i^{m_i}.\tag{33.5}$$

Let us now introduce the notation $\Lambda^*(m_2,\ldots,m_\varkappa)$:

$$\left.\begin{aligned}A\,\Lambda^*(m_2,\ldots,m_\varkappa)\prod_{i=2}^{\varkappa}\left(\frac{\Gamma_i^*}{n_i^{*\alpha}}\right)^{m_i}&=V\,b^*(m_2,\ldots,m_\varkappa)-\\&-V^\alpha b^{*\alpha}(m_2,\ldots,m_\varkappa)-V^\beta b^{*\beta}(m_2,\ldots,m_\varkappa),\end{aligned}\right\}\tag{33.6}$$

where Γ_i^* is the superficial number density of molecules of species i in the reference state ζ^*. We shall refer to $\Lambda^*(m_2,\ldots,m_\varkappa)$ as the superficial cluster integral referred to ζ^*.

Then from (33.1), (33.3) to (33.5) and the additivity of volumes V^α and V^β, we obtain the following expression for the surface tension difference[4],

$$\gamma-\gamma^*=-kT\sum_{m_2=0}^{\infty}\cdots\sum_{m_\varkappa=0}^{\infty}\Lambda^*(m_2,\ldots,m_\varkappa)\prod_{i=2}^{\varkappa}\left\{(\zeta_i-\zeta_i^*)\frac{\Gamma_i^*}{\zeta_i^*}\right\}^{m_i}.\tag{33.7}$$

We shall show that $\Lambda^*(m_2,\ldots,m_\varkappa)$ is independent of the area A. First, to understand the meaning of $\Lambda^*(m_2,\ldots,m_\varkappa)$, we consider small sets of $m_2,\ldots,m_\varkappa$. For one of the smallest sets, $m_i=1$, and $m_j=0$ for all $i\neq j$, we obtain from (33.2) and (33.3)

$$V\,b^*(0,\ldots,1,\ldots,0)=\int\frac{n_i^*(\boldsymbol{r}^{(1)})}{n_i^{*\alpha}}\,d\boldsymbol{r}^{(1)}.\tag{33.8}$$

By definition we have

$$V^\alpha b^{*\alpha}(0,\ldots,1,\ldots,0)=V^\alpha,\tag{33.9}$$

$$V^\beta b^{*\beta}(0,\ldots,1,\ldots,0)=V^\beta\frac{n_i^{*\beta}}{n_i^{*\alpha}}.\tag{33.10}$$

Substituting (33.8) to (33.10) into (33.6), we obtain

$$A\,\Lambda^*(0,\ldots,1,\ldots,0)=\frac{1}{\Gamma_i^*}\int\left(n_i^*(\boldsymbol{r}^{(1)})-n_i^{*\alpha\beta}\right)d\boldsymbol{r}^{(1)},\tag{33.11}$$

---

[1] S. ONO: J. Chem. Phys. **19**, 504 (1951). — Progr. Theoret. Phys. **8**, 1 (1952).

[2] J.E. KILPATRICK: J. Chem. Phys. **21**, 274 (1953).

[3] H. CRAMER: Mathematical Methods of Statistics, pp. 185—187. Princeton, N.J.: Princeton University Press 1951.

[4] N. SAITO [*21*]. His result is, however, not exactly equivalent to (33.7), since in his treatment the two states are assumed to be under the same pressure instead of the osmotic pressure conditions.

where $n_i^{\alpha\beta}$ is defined by (26.19). From (26.21) and (33.11), we find

$$\Lambda^*(0, \ldots, 1, \ldots, 0) = 1, \tag{33.12}$$

which is evidently independent of A.

For the set $m_i = 1$, $m_j = 1$ and $m_k = 0$ for $k \neq i$ and $k \neq j$, the cluster integral $b^*(0, \ldots, 1, \ldots, 1, \ldots, 0)$ is expressed by (33.2) and (33.3) as

$$\left.\begin{aligned} V\, b^*(0, \ldots, 1, \ldots, 1, \ldots, 0) &= \frac{1}{(1+\delta_{ij})\, n_i^{*\alpha} n_j^{*\alpha}} \times \\ &\times \iint_V \left(n_{ij}^*(\boldsymbol{r}^{(1)}, \boldsymbol{r}^{(2)}) - n_i^*(\boldsymbol{r}^{(1)})\, n_j^*(\boldsymbol{r}^{(2)})\right) d\boldsymbol{r}^{(1)}\, d\boldsymbol{r}^{(2)}, \end{aligned}\right\} \tag{33.13}$$

and likewise

$$\left.\begin{aligned} V^\alpha\, b^{*\alpha}(0, \ldots, 1, \ldots, 1, \ldots, 0) &= \frac{1}{(1+\delta_{ij})\, n_i^{*\alpha} n_j^{*\alpha}} \times \\ &\times \iint_{V^\alpha} \left(n_{ij}^{*\alpha}(\boldsymbol{r}^{(12)}) - n_i^{*\alpha} n_j^{*\alpha}\right) d\boldsymbol{r}^{(1)}\, d\boldsymbol{r}^{(2)}, \end{aligned}\right\} \tag{33.14}$$

$$\left.\begin{aligned} V^\beta\, b^{*\beta}(0, \ldots, 1, \ldots, 1, \ldots, 0) &= \frac{1}{(1+\delta_{ij})\, n_i^{*\alpha} n_j^{*\alpha}} \times \\ &\times \iint_{V^\beta} \left(n_{ij}^{*\beta}(\boldsymbol{r}^{(12)}) - n_i^{*\beta} n_j^{*\beta}\right) d\boldsymbol{r}^{(1)}\, d\boldsymbol{r}^{(2)}. \end{aligned}\right\} \tag{33.15}$$

Substitution of (33.13) to (33.15) into (33.6) leads immediately to

$$\left.\begin{aligned} A\Lambda^*(0,\ldots,1,\ldots,1,\ldots,0) &= \frac{1}{(1+\delta_{ij})\, \Gamma_i^* \Gamma_j^*} \iint_V \left(n_{ij}^*(\boldsymbol{r}^{(1)}, \boldsymbol{r}^{(2)}) - \hat{n}_{ij}^{*\alpha\beta}\right) d\boldsymbol{r}^{(1)}\, d\boldsymbol{r}^{(2)} - \\ &- \frac{1}{(1+\delta_{ij})\, \Gamma_i^* \Gamma_j^*} \iint_V \left(n_i^*(\boldsymbol{r}^{(1)})\, n_j^*(\boldsymbol{r}^{(2)}) - n_i^{*\alpha\beta} n_j^{*\alpha\beta}\right) d\boldsymbol{r}^{(1)}\, d\boldsymbol{r}^{(2)}, \end{aligned}\right\} \tag{33.16}$$

where $\hat{n}_{ij}^{*\alpha\beta}$ is defined as follows

$$\hat{n}_{ij}^{*\alpha\beta} = \left\{\begin{aligned} &n_{ij}^{*\alpha}(r^{(12)}) - n_i^{*\alpha} n_j^{*\alpha} + n_i^{*\alpha} n_j^{*\alpha\beta}(z^{(2)}) \qquad z^{(1)} \leq 0, \\ &n_{ij}^{*\beta}(r^{(12)}) - n_i^{*\beta} n_j^{*\beta} + n_i^{*\beta} n_j^{*\alpha\beta}(z^{(2)}) \qquad z^{(1)} > 0. \end{aligned}\right\} \tag{33.17}$$

It should be noted that $\hat{n}_{ij}^{\alpha\beta}$ is different from $n_{ij}^{\alpha\beta}$ defined by (26.20); otherwise the expression (33.16) is no longer valid.

On the other hand, according to (26.21), we have

$$\left.\begin{aligned} &\iint \left(n_i^*(\boldsymbol{r}^{(1)})\, n_j^*(\boldsymbol{r}^{(2)}) - n_i^{*\alpha\beta} n_j^{*\alpha\beta}\right) d\boldsymbol{r}^{(1)}\, d\boldsymbol{r}^{(2)} \\ &\quad = A^2 \Gamma_i^* \Gamma_j^* + \iint \left(n_i^*(\boldsymbol{r}^{(1)})\, n_j^{*\alpha\beta} + n_i^{*\alpha\beta} n_j^*(\boldsymbol{r}^{(1)}) - 2 n_i^{*\alpha\beta} n_j^{*\alpha\beta}\right) d\boldsymbol{r}^{(1)}\, d\boldsymbol{r}^{(2)}. \end{aligned}\right\} \tag{33.18}$$

Using the above relations we may rewrite (33.16) in the form[1]

$$\Lambda^*(0,\ldots,1,\ldots,1,\ldots,0) = \frac{1}{(1+\delta_{ij})} \iint \left(\frac{\Delta_{ij}^*}{\Gamma_i^* \Gamma_j^*} - 1\right) dx^{(12)}\, dy^{(12)}, \tag{33.19}$$

where

$$\left.\begin{aligned} \Delta_{ij}^* = \iint &\left(n_{ij}^*(\boldsymbol{r}^{(1)}, \boldsymbol{r}^{(2)}) - \hat{n}_{ij}^{*\alpha\beta} - n_i^*(\boldsymbol{r}^{(1)})\, n_j^{*\alpha\beta} - n_i^{*\alpha\beta} n_j^*(\boldsymbol{r}^{(2)}) + \right. \\ &\left. + 2 n_i^{*\alpha\beta} n_j^{*\alpha\beta}\right) dz^{(1)}\, dz^{(2)}. \end{aligned}\right\} \tag{33.20}$$

---

[1] It is easily seen that (33.19) corresponds to the expression for the cluster integral $b^{*\alpha}(0, \ldots, 1, \ldots, 1, \ldots, 0)$, for the homogeneous bulk phase α, given by $b^{*\alpha}(0, \ldots, 1, \ldots, 1, \ldots, 0)$ $= \dfrac{1}{(1+\delta_{ij})} \displaystyle\int \left(\frac{n_{ij}^*(r^{(12)})}{n_i^{*\alpha} n_j^{*\alpha}} - 1\right) d\boldsymbol{r}^{(12)}$.

Since the pair distribution function $n_{ij}^*(\mathbf{r}^{(1)}, \mathbf{r}^{(2)})$ reduces to the product of singlet distribution functions, $n_i^*(\mathbf{r}^{(1)})\, n_j^*(\mathbf{r}^{(2)})$, for values of $r^{(12)} = |\mathbf{r}^{(2)} - \mathbf{r}^{(1)}|$ sufficiently large compared with molecular dimensions, from (26.21), (33.16), (33.18) and (33.20) we obtain

$$\lim_{r^{(12)} \to \infty} \Delta_{ij}^* = \Gamma_i^* \Gamma_j^*. \tag{33.21}$$

Thus we may conclude that the superficial cluster integral (33.19) is independent of the area of the interface if the dimension of the container of the system can be taken sufficiently large compared with molecular dimensions. We can obtain similar equations for the more complicated superficial cluster integrals, expressed in terms of superficial multiplet densities and prove the results to be independent of the area A, but we shall not enter further into the details of those tedious calculations.

Differentiating (33.7) with respect to ζ_i, we obtain[1]

$$(\zeta_i - \zeta_i^*) \left(\frac{\partial \gamma}{\partial \zeta_i}\right)_{\zeta_j} = -kT \sum_{m_2=0}^{\infty} \cdots \sum_{m_\varkappa=0}^{\infty} m_i \Delta^*(m_2, \ldots, m_\varkappa) \prod_{l=2}^{\varkappa} \left((\zeta_l - \zeta_l^*) \frac{\Gamma_l^*}{\zeta_l^*}\right)^{m_l}. \tag{33.22}$$

In the limit that ζ_i^* tends to ζ_i, the above equation reduces, in accordance with (33.12), to [21]

$$\zeta_i \left(\frac{\partial \gamma}{\partial \zeta_i}\right)_{\zeta_j} = -kT\Gamma_i, \quad (i = 2, \ldots, \varkappa). \tag{33.23}$$

Making use of (31.2) we can rewrite (33.23) in the form

$$-\left(\frac{\partial \gamma}{\partial \mu_i}\right)_{\mu_j} = \Gamma_i, \quad (i = 2, \ldots, \varkappa). \tag{33.24}$$

Then we obtain from (33.24), for constant T and μ_1 [21],

$$-d\gamma = \sum_{i=2}^{\varkappa} \Gamma_i \, d\mu_i. \tag{33.25}$$

This is the Gibbs adsorption equation, (7.1), subject to the condition that μ is kept constant.

From (33.22) and (33.23) we obtain

$$\frac{\zeta_i - \zeta_i^*}{\zeta_i} \Gamma_i = \sum_{m_2=0}^{\infty} \cdots \sum_{m_\varkappa=0}^{\infty} m_i \Delta^*(m_2, \ldots, m_\varkappa) \prod_{j=2}^{\varkappa} \left((\zeta_j - \zeta_j^*) \frac{\Gamma_j^*}{\zeta_j^*}\right)^{m_j}. \tag{33.26}$$

If the reference state $\boldsymbol{\zeta}^*$ is the standard state $\boldsymbol{\zeta}^0$, on putting $\zeta_2^* = \zeta_3^* \cdots = \zeta_\varkappa^* = 0$, (33.26) reduces to

$$\Gamma_i = \sum_{m_2=0}^{\infty} \cdots \sum_{m_\varkappa=0}^{\infty} m_i \Delta^0(m_2, \ldots, m_\varkappa) \prod_{j=2}^{\varkappa} \left(\frac{\zeta_j}{f_j^0}\right)^{m_j}, \tag{33.27}$$

$$f_j^0 = \lim_{\zeta_2, \ldots, \zeta_\varkappa \to 0} \frac{\zeta_j}{\Gamma_j}, \tag{33.28}$$

[1] From the examination of the coefficients of the terms on the right-hand side of (33.7), we find those individual terms do depend on the choice of the location of the dividing surface, though the surface tension given as the sum of them is independent of this choice. Similarly the differential coefficient given by (33.23) depends on the choice of the dividing surface. But this dependence is only apparent, for the changes in ζ_i's or μ_i's are not independent of each other but subject to the condition (7.8), μ_1 being always kept constant in the osmotic solution.

where we may refer to f_j^0 as the superficial activity coefficient of solute species j in the infinitely dilute solution and Λ^0 corresponds to Λ^* in the infinitely dilute solution. When the surface tension of the pure solvent is denoted by γ^0, (33.7) is written

$$\gamma - \gamma^0 = -kT \sum_{m_2=0}^{\infty} \cdots \sum_{m_\varkappa=0}^{\infty} \Lambda^0(m_2, \ldots, m_\varkappa) \prod_{i=2}^{\varkappa} \left(\frac{\zeta_i}{f_i^0}\right)^{m_i}, \qquad (33.29)$$

which may be regarded as an integration of the Gibbs adsorption equation. Since the superficial cluster integral Λ^0 for a single molecule of any species is unity according to (33.12), we have, for sufficiently dilute solutions,

$$\gamma \approx \gamma^0 - kT\left[\frac{\zeta_2}{f_2^0} + \cdots + \frac{\zeta_\varkappa}{f_\varkappa^0}\right]. \qquad (33.30)$$

Since we have $-\Omega^\alpha - p^* V = (p - p^*) V$ for the homogeneous bulk phase α, from (5.15) we find

$$n_i^\alpha = \left(\frac{\partial(p-p^*)}{\partial \mu_i}\right)_{T, V^\alpha, \mu_j} \quad (i = 2, \ldots, \varkappa). \qquad (33.31)$$

Then by using (31.2), (33.1) and (33.4), (33.31) can be written as

$$\frac{\zeta_i - \zeta_i^*}{\zeta_i} n_i^\alpha = \sum_{m_2=0}^{\infty} \cdots \sum_{m_\varkappa=0}^{\infty} m_i b^{*\alpha}(m_2, \ldots, m_\varkappa) \prod_{j=2}^{\varkappa} \left\{(\zeta_j - \zeta_j^*) \frac{n_j^{*\alpha}}{\zeta_j^*}\right\}^{m_j}. \qquad (33.32)$$

If the reference state is the standard state ζ^0, (33.32) reduces to

$$n_i^\alpha = \sum_{m_2=0}^{\infty} \cdots \sum_{m_\varkappa=0}^{\infty} m_i b^{0\alpha}(m_2, \ldots, m_\varkappa) \prod_{j=2}^{\varkappa} \left(\frac{\zeta_j}{F_j^{0\alpha}}\right)^{m_j}, \qquad (33.33)$$

where $F_j^{0\alpha}$ is the ordinary activity coefficient of solute species i evaluated in the pure solvent:

$$F_j^{0\alpha} = \lim_{n_2^\alpha, \ldots, n_\varkappa^\alpha \to 0} \frac{\zeta_j}{n_j^\alpha}, \quad (j = 2, \ldots, \varkappa). \qquad (33.34)$$

We can obtain the adsorption isotherm, in principle, from (33.29) and (33.33) by eliminating $\zeta_2, \ldots, \zeta_\varkappa$ from these $\varkappa$ equations.

34. Gas adsorption at a solid surface. The treatments of the surface tension at a plane interface in the preceding sections can be applied to the problem of gas adsorption at a solid surface. Although the following results can be obtained as a special case of the corresponding results of the preceding sections by letting one of the phases be a solid adsorbent, we shall show an alternative and simpler method[1] [22], [37].

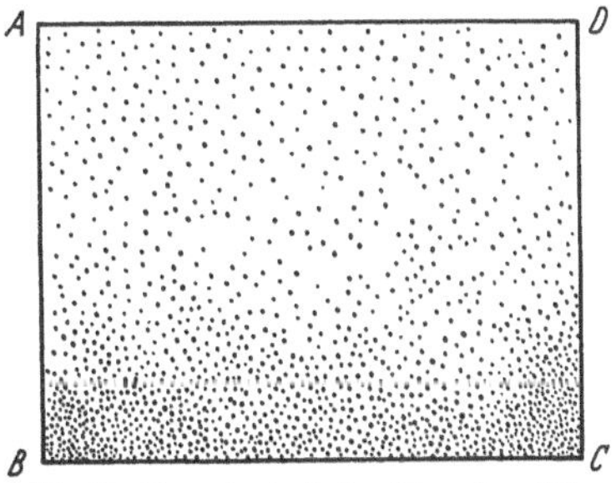

Fig. 12. Gas adsorbed at solid surface BC.

Let us consider a gas contained in a vessel as shown in Fig. 12. We assume that gas molecules interact with only one of the walls, say BC, which may be regarded as the solid adsorbent, and that they do not interact with the other walls[2]. For convenience let the volume of the vessel be V^α and α refer to the gas phase.

---

[1] See also T. L. HILL: Statistical Mechanics, pp. 424—425. New York: McGraw Hill 1956.

[2] In the present treatment, the solid surface need not be plane.

The grand partition function for this system is, according to (24.3) and (31.1),

$$\left.\begin{aligned}&\Xi(\zeta, T, V^\alpha)\\&\quad=\sum_{N_1=0}^{\infty}\cdots\sum_{N_\varkappa=0}^{\infty}\left(\prod_{i=1}^{\varkappa}\frac{\zeta_i^{N_i}}{N_i!}\right)\int\cdots\int\exp\left[-\frac{\Phi_N(\{\boldsymbol{r}_1\}_{N_1},\ldots,\{\boldsymbol{r}_\varkappa\}_{N_\varkappa})}{kT}\right]\prod_{i=1}^{\varkappa}d\{\boldsymbol{r}_i\}_{N_i},\end{aligned}\right\}\qquad(34.1)$$

$$\Phi_N=\Phi_N^\alpha+\sum_{i=1}^{\varkappa}\sum_{s=1}^{N_i}\psi_i(\boldsymbol{r}_i^{(s)}),\qquad(34.2)$$

$$\Phi_N^\alpha=\tfrac{1}{2}\sum_{i,j=1}^{\varkappa}\sum_{s=1}^{N_i}\sum_{t=1}^{N_j}\phi_{ij}(|\boldsymbol{r}_i^{(s)}-\boldsymbol{r}_j^{(t)}|),\qquad(34.3)$$

where Φ_N^α is the total intermolecular potential energy given by (25.3) and $\psi_i(\boldsymbol{r}_i^{(s)})$ is the potential energy of molecule i at $\boldsymbol{r}_i^{(s)}$ due to the interaction with the adsorbent.

Let us consider the same gas contained in the same vessel but under the condition that none of the walls interacts with the gas. We write the grand partition function for such a system as

$$\left.\begin{aligned}&\Xi^\alpha(\zeta, T, V^\alpha)\\&\quad=\sum_{N_1=0}^{\infty}\cdots\sum_{N_\varkappa=0}^{\infty}\left(\prod_{i=1}^{\varkappa}\frac{\zeta_i^{N_i}}{N_i!}\right)\int\cdots\int\exp\left[-\frac{\Phi_N^\alpha(\{\boldsymbol{r}_1\}_{N_1},\ldots,\{\boldsymbol{r}_\varkappa\}_{N_\varkappa})}{kT}\right]\prod_{i=1}^{\varkappa}d\{\boldsymbol{r}_i\}_{N_i}.\end{aligned}\right\}\qquad(34.4)$$

In the same way as in the case of (33.3), we expand $\log\Xi$ in powers of $\zeta_1^{m_1}\ldots\zeta_\varkappa^{m_\varkappa}$:

$$\log\Xi=\sum_{m_1=0}^{\infty}\cdots\sum_{m_\varkappa=0}^{\infty}V^\alpha b^0(m_1,\ldots,m_\varkappa)\,\zeta_1^{m_1}\ldots\zeta_\varkappa^{m_\varkappa},\qquad(34.5)$$

and similarly

$$\log\Xi^\alpha=\sum_{m_1=0}^{\infty}\cdots\sum_{m_\varkappa=0}^{\infty}V^\alpha b^{0\alpha}(m_1,\ldots,m_\varkappa)\,\zeta_1^{m_1}\ldots\zeta_\varkappa^{m_\varkappa},\qquad(34.6)$$

where $b^0(m_1,\ldots,m_\varkappa)$ and $b^{0\alpha}(m_1,\ldots,m_\varkappa)$ are the cluster integrals with and without the adsorbent, respectively[1].

Since $n_i^{*\alpha}/\zeta_i^*$ tends to unity at the infinite dilution, it is easily seen from (32.1), (33.1) and (33.3) that $b^0(m_1,\ldots,m_\varkappa)$ corresponds to $b^*(m_1,\ldots,m_\varkappa)$ at infinite dilution and that $b^{0\alpha}(m_1,\ldots,m_\varkappa)$ is identical with the cluster integral of MAYER[2] for a gaseous mixture.

According to (31.2) and (29.11), the average number of molecules of species i in the two cases are, respectively,

$$\overline{N_i}=\zeta_i\frac{\partial\log\Xi}{\partial\zeta_i}=\sum_{m_1=0}^{\infty}\cdots\sum_{m_\varkappa=0}^{\infty}V^\alpha m_i\,b^0(m_1,\ldots,m_\varkappa)\,\zeta_1^{m_1}\ldots\zeta_\varkappa^{m_\varkappa},\qquad(34.7)$$

$$\overline{N_i^\alpha}=\zeta_i\frac{\partial\log\Xi^\alpha}{\partial\zeta_i}=\sum_{m_1=0}^{\infty}\cdots\sum_{m_\varkappa=0}^{\infty}V^\alpha m_i\,b^{0\alpha}(m_1,\ldots,m_\varkappa)\,\zeta_1^{m_1}\ldots\zeta_\varkappa^{m_\varkappa}.\qquad(34.8)$$

Since the number of adsorbed molecules is defined as the average excess of molecules in V^α due to the presence of the adsorbent, the amount of the adsorbed gas for species i can be expressed with use of (34.7) and (34.8) as a function of activities of component gases as follows

$$\overline{N_i}-\overline{N_i^\alpha}=\sum_{m_1=0}^{\infty}\cdots\sum_{m_\varkappa=0}^{\infty}V^\alpha m_i\{b^0(m_1,\ldots,m_\varkappa)-b^{0\alpha}(m_1,\ldots,m_\varkappa)\}\,\zeta_1^{m_1}\ldots\zeta_\varkappa^{m_\varkappa}.\qquad(34.9)$$

---

[1] $b^0(m_1,\ldots,m_\varkappa)$ appearing in this section, is a special case of the cluster integrals discussed in the previous section.

[2] J.E. MAYER: J. Chem. Phys. **5**, 67 (1937). — J. Phys. Chem. **43**, 71 (1939). — See also in Vol. XII of this Encyclopedia.

As in the case of the coefficient $A^*(m_1, \ldots, m_\varkappa)$ of (33.7), we can show that $V^\alpha\{b^0(m_1, \ldots, m_\varkappa) - b^{0\alpha}(m_1, \ldots, m_\varkappa)\}$ is proportional to the area of the adsorbent.

We shall, however, restrict our detailed treatment to the case of one-component gas. Then (34.1) and (34.5) reduce, respectively, to

$$\left.\begin{aligned} &\Xi(\zeta, T, V^\alpha) \\ &= \sum_{N=0}^{\infty} \frac{\zeta^N}{N!} \int \cdots \int \exp\left[-\frac{1}{kT}\left\{\frac{1}{2}\sum_{s,t=1}^{N} \phi(r^{(st)}) + \sum_{s=1}^{N} \psi(\boldsymbol{r}^{(s)})\right\}\right] d\boldsymbol{r}^{(1)} \ldots d\boldsymbol{r}^{(N)}, \end{aligned}\right\} \tag{34.10}$$

$$\log \Xi = \sum_{m=0}^{\infty} V^\alpha b_m \zeta^m, \tag{34.11}$$

in which the subscript to indicate molecular species is omitted and the cluster integral is expressed by b_m. If we denote by b_m^α the cluster integral for the gas without the adsorbent, (34.9) can be rewritten as

$$\Gamma = \frac{V^\alpha}{A} \sum_{m=1}^{\infty} m(b_m - b_m^\alpha)\zeta^m, \tag{34.12}$$

Γ being the number of molecules adsorbed per unit area of the adsorbent.

Comparing the corresponding coefficients of (34.10) and (34.11) we obtain the following expressions for cluster integrals:

$$b_1 = 1 + \frac{1}{V^\alpha} \int (e^{-\psi(\boldsymbol{r}^{(1)})/kT} - 1)\, d\boldsymbol{r}^{(1)}, \tag{34.13}$$

$$\left.\begin{aligned} b_2 = \frac{1}{2V^\alpha} \iint &[(e^{-\phi(r^{(12)})/kT} - 1) + (e^{-\phi(r^{(12)})/kT} - 1)(e^{-\psi(\boldsymbol{r}^{(1)})/kT} - 1) + \\ &+ (e^{-\phi(r^{(12)})/kT} - 1)(e^{-\psi(\boldsymbol{r}^{(2)})/kT} - 1) + (e^{-\phi(r^{(12)})/kT} - 1) \times \\ &\times (e^{-\psi(\boldsymbol{r}^{(1)})/kT} - 1)(e^{-\psi(\boldsymbol{r}^{(2)})/kT} - 1)]\, d\boldsymbol{r}^{(1)}\, d\boldsymbol{r}^{(2)}. \end{aligned}\right\} \tag{34.14}$$

Likewise for the gas without adsorbent we have

$$b_1^\alpha = 1, \tag{34.15}$$

$$b_2^\alpha = \frac{1}{2V^\alpha} \iint (e^{-\phi(r^{(12)})/kT} - 1)\, d\boldsymbol{r}^{(1)}\, d\boldsymbol{r}^{(2)}. \tag{34.16}$$

In the case of rarefied gas, with use of (34.13) to (34.16), the leading terms of (34.12) are written as

$$\left.\begin{aligned} \Gamma \approx \frac{\zeta}{A} \int (e^{-\psi(\boldsymbol{r}^{(1)})/kT} - 1)\, d\boldsymbol{r}^{(1)} + \frac{\zeta^2}{A} \iint &[(e^{-\phi(r^{(12)})/kT} - 1) \times \\ \times \{(e^{-\psi(\boldsymbol{r}^{(1)})/kT} - 1) + (e^{-\psi(\boldsymbol{r}^{(2)})/kT} - 1) &+ (e^{-\psi(\boldsymbol{r}^{(1)})/kT} - 1) \times \\ \times (e^{-\psi(\boldsymbol{r}^{(2)})/kT} - 1)\}]\, d\boldsymbol{r}^{(1)}\, d\boldsymbol{r}^{(2)}. \end{aligned}\right\} \tag{34.17}$$

Since $e^{-\psi(\boldsymbol{r})/kT} - 1$ is zero except when $\boldsymbol{r}$ is within the molecular dimension from the surface of the adsorbent and $e^{-\phi(r)/kT} - 1$ is zero except when r is of the order of magnitude of molecular diameter, Γ given by (34.17) is asymptotically independent of A.

If we desire to express Γ as a function of the pressure, we have only to use the relation

$$p = kT\left[\zeta + \frac{\zeta^2}{2V^\alpha} \iint (e^{-\phi(r^{(12)})/kT} - 1)\, d\boldsymbol{r}^{(1)}\, d\boldsymbol{r}^{(2)} + \cdots\right], \tag{34.18}$$

which is easily obtained from $pV = kT \log \Xi^{\alpha}$ for the gas without adsorbent by using (34.15) and (34.16). Combining (34.17) and (34.18) we obtain

$$\left.\begin{aligned}\Gamma = \frac{1}{A}\Bigg[\frac{p}{kT}\int\limits_{V^{\alpha}}\left(e^{-\varphi(\boldsymbol{r}^{(1)})/kT}-1\right)d\boldsymbol{r}^{(1)} + \left(\frac{p}{kT}\right)^2\iint\limits_{V^{\alpha}}\left(e^{-\phi(r^{(12)})/kT}-1\right)\times\\ \times\Big\{\left(e^{-\varphi(\boldsymbol{r}^{(1)})/kT}-1\right)\left(e^{-\varphi(\boldsymbol{r}^{(2)})/kT}-1\right) + \tfrac{3}{4}\left(e^{-\varphi(\boldsymbol{r}^{(1)})/kT}-1\right)+\\ + \tfrac{3}{4}\left(e^{-\varphi(\boldsymbol{r}^{(2)})/kT}-1\right)\Big\}\,d\boldsymbol{r}^{(1)}\,d\boldsymbol{r}^{(2)} + \cdots\Bigg].\end{aligned}\right\} \quad (34.19)$$

Finally it will be seen that (34.17) is a special case of (33.27) and that the superficial activity coefficient f_0 defined by (33.28) reduces, according to (34.17), to

$$f^0 = \frac{A}{\int\limits_{V^{\alpha}}\left(e^{-\varphi(\boldsymbol{r}^{(1)})/kT}-1\right)d\boldsymbol{r}^{(1)}}. \quad (34.20)$$

II. Mechanical definition of surface tension.

35. Statistical expression for pressure tensor in fluids. The probability density function $P^{(N)}(\{\boldsymbol{r}_1\}, \ldots, \{\boldsymbol{r}_\varkappa\}; \{\boldsymbol{p}_1\}, \ldots, \{\boldsymbol{p}_\varkappa\})$ introduced in Sect. 24 can be defined as well in non-equilibrium systems and in such cases it changes, in general, with time according to the Liouville equation. The definition of the singlet distribution function $n_i(\boldsymbol{r}^{(1)})$, pair distribution function $n_{ij}(\boldsymbol{r}^{(1)}, \boldsymbol{r}^{(2)})$ and other distribution functions of higher order for a non-equilibrium system is just the same as for an equilibrium system.

It is the purpose of the present section to secure the expression for the pressure tensor at every point of a fluid in terms of distribution functions and the intermolecular potentials. We shall closely follow the method of IRVING and KIRKWOOD[1].

Let us take $d\boldsymbol{S}$ as a surface element located at point $\boldsymbol{r}$ in a fluid and divide the fluid into two parts by the plane tangent to $d\boldsymbol{S}$. Let us call the portion into which the vector $d\boldsymbol{S}$ points the outer fluid and another portion the inner fluid. We shall restrict ourselves to the case where the total potential energy of the fluid is the sum of pair potentials each of which is a function of the intermolecular distance alone as assumed in (25.3). Let us say that the force between a pair of molecules acts across $d\boldsymbol{S}$ only when the straight line joining the centers of the two molecules passes through $d\boldsymbol{S}$. For an infinitesimal $d\boldsymbol{S}$, this can happen only when two molecules are on opposite sides of $d\boldsymbol{S}$ (see also Fig. 13a on p. 212).

Let us next consider the force exerted by the molecules in the outer fluid on those in the inner fluid. A molecule of species i at $\boldsymbol{r}'$ in the inner fluid experiences a force $\frac{\boldsymbol{R}}{R}\phi'_{ij}(R)$ from a molecule of species j at $\boldsymbol{r}'+\boldsymbol{R}$ in the outer fluid. By convention the above intermolecular force is acting across $d\boldsymbol{S}$ only if the vector $\boldsymbol{r}'+\eta\boldsymbol{R}$ terminates on $d\boldsymbol{S}$ for a value of η between zero and one. If $\boldsymbol{R}$ is fixed, the volume element, over which the vector $\boldsymbol{r}'+\eta\boldsymbol{R}$ terminates on $d\boldsymbol{S}$ for η between η and $\eta+d\eta$, is $d\boldsymbol{S}\cdot\boldsymbol{R}\,d\eta$. The average number of molecular pairs, one member of which is a molecule of species i situated in the above volume element and the other species j in another volume element $d\boldsymbol{R}$ at $\boldsymbol{r}'+\boldsymbol{R}$, is given by

$$n_{ij}(\boldsymbol{r}', \boldsymbol{r}'+\boldsymbol{R})\,(d\boldsymbol{S}\cdot\boldsymbol{R}\,d\eta)\,d\boldsymbol{R} = n_{ij}(\boldsymbol{r}-\eta\boldsymbol{R}, \boldsymbol{r}-\eta\boldsymbol{R}+\boldsymbol{R})\,(d\boldsymbol{S}\cdot\boldsymbol{R}\,d\eta)\,d\boldsymbol{R}. \quad (35.1)$$

[1] J.H. IRVING and J.G. KIRKWOOD: J. Chem. Phys. 18, 817 (1950).

In the above equation $\boldsymbol{r}'+\eta\boldsymbol{R}$ is an arbitrary point on $d\boldsymbol{S}$ but the difference between $\boldsymbol{r}$ and $\boldsymbol{r}'+\eta\boldsymbol{R}$ is an infinitesimal of higher order.

We are now ready to derive an expression for the total force exerted by the outer molecules on the inner across $d\boldsymbol{S}$, which we shall denote by $\boldsymbol{F}^{(\Phi)}\,dS$. Multiplying (35.1) by $(\boldsymbol{R}/R)\,\phi'_{ij}(R)$, integrating the result over all $\boldsymbol{R}$ to make $\boldsymbol{R}\cdot d\boldsymbol{S}$ positive, i.e. over all the outer fluid, and further summing the result over molecular species i and j, we obtain

$$\boldsymbol{F}^{(\Phi)}dS = d\boldsymbol{S}\cdot\sum_{i=1}^{\varkappa}\sum_{j=1}^{\varkappa}\int\limits_{\boldsymbol{R}\cdot d\boldsymbol{S}>0} d\boldsymbol{R}\left\{\int_0^1 \frac{\boldsymbol{R}\boldsymbol{R}}{R}\,\phi'_{ij}(R)\,n_{ij}(\boldsymbol{r}-\eta\boldsymbol{R},\,\boldsymbol{r}-\eta\boldsymbol{R}+\boldsymbol{R})\,d\eta\right\}. \tag{35.2}$$

Putting $\xi = 1-\eta$, we find

$$\left.\begin{aligned}&\int_0^1 \frac{\boldsymbol{R}\boldsymbol{R}}{R}\,\phi'_{ij}(R)\,n_{ij}(\boldsymbol{r}-\eta\boldsymbol{R},\;\boldsymbol{r}-\eta\boldsymbol{R}+\boldsymbol{R})\,d\eta\\ &\qquad=\int_0^1 \frac{\boldsymbol{R}\boldsymbol{R}}{R}\,\phi'_{ij}(R)\,n_{ij}(\boldsymbol{r}+\xi\boldsymbol{R}-\boldsymbol{R},\;\boldsymbol{r}+\xi\boldsymbol{R})\,d\xi.\end{aligned}\right\} \tag{35.3}$$

Since $n_{ij}(\boldsymbol{r}^{(1)},\boldsymbol{r}^{(2)})$ must be equal to $n_{ji}(\boldsymbol{r}^{(2)},\boldsymbol{r}^{(1)})$ and $\phi'_{ij}(R)$ to $\phi'_{ji}(R)$ by definition, (35.3) can be rewritten in the form:

$$\left.\begin{aligned}&\int_0^1 \frac{\boldsymbol{R}\boldsymbol{R}}{R}\,\phi'_{ij}(R)\,n_{ij}(\boldsymbol{r}-\eta\boldsymbol{R},\boldsymbol{r}-\eta\boldsymbol{R}+\boldsymbol{R})\,d\eta\\ &\qquad=\int_0^1 \frac{\boldsymbol{R}\boldsymbol{R}}{R}\,\phi'_{ji}(R)\,n_{ji}(\boldsymbol{r}+\eta\boldsymbol{R},\boldsymbol{r}+\eta\boldsymbol{R}-\boldsymbol{R})\,d\eta.\end{aligned}\right\} \tag{35.4}$$

Consequently (35.2) can be written:

$$\boldsymbol{F}^{(\Phi)}\,dS = \frac{1}{2}\,d\boldsymbol{S}\cdot\sum_{i,j=1}^{\varkappa}\int d\boldsymbol{R}\left\{\int_0^1 \frac{\boldsymbol{R}\boldsymbol{R}}{R}\,\phi'_{ij}(R)\,n_{ij}(\boldsymbol{r}-\eta\boldsymbol{R},\boldsymbol{r}-\eta\boldsymbol{R}+\boldsymbol{R})\,d\eta\right\}, \tag{35.5}$$

where the integration over $\boldsymbol{R}$ extends all space.

On the other hand, from the macroscopic standpoint the pressure tensor $\mathbf{p}(\boldsymbol{r})$ in the fluid is given as a function of the spatial coordinates; the force acting across the surface element $d\boldsymbol{S}$ is $d\boldsymbol{S}\cdot\mathbf{p}(\boldsymbol{r})$. Henceforward, we restrict ourselves, for simplicity, to the case of a system at rest under no external field. Then the pressure tensor $\mathbf{p}(\boldsymbol{r})$ consists of the kinetic contribution and the contribution of intermolecular forces, $\mathbf{p}^{(k)}(\boldsymbol{r})$ and $\mathbf{p}^{(\Phi)}(\boldsymbol{r})$. When the fluid is at rest, $d\boldsymbol{S}\cdot\mathbf{p}^{(k)}(\boldsymbol{r})$ is equal to the momentum transferred per unit time across the area $d\boldsymbol{S}$ due to the thermal motion of molecules and $\mathbf{p}^{(k)}(\boldsymbol{r})$ is given by the familiar expression $kT\sum_{i=1}^{\varkappa} n_i(\boldsymbol{r})\,\mathbf{1}$, as in the case of dilute gases, for the off-diagonal parts vanish in fluids at rest. The contribution of intermolecular forces $-d\boldsymbol{S}\cdot\mathbf{p}^{(\Phi)}(\boldsymbol{r})$ is the force acting across $d\boldsymbol{S}$ due to the interaction of molecules on opposite sides of $d\boldsymbol{S}$, and is given by (35.5). Then we obtain the following statistical mechanical expression for the pressure tensor:

$$\mathbf{p} = \mathbf{p}^{(k)} + \mathbf{p}^{(\Phi)}, \tag{35.6}$$

$$\mathbf{p}^{(k)}(\boldsymbol{r}) = kT\sum_{i=1}^{\varkappa} n_i(\boldsymbol{r})\,\mathbf{1}, \tag{35.7}$$

$$\mathbf{p}^{(\Phi)}(\boldsymbol{r}) = -\frac{1}{2}\sum_{i,j=1}^{\varkappa}\int d\boldsymbol{R}\left\{\int_0^1 \frac{\boldsymbol{R}\boldsymbol{R}}{R}\,\phi'_{ij}(R)\,n_{ij}(\boldsymbol{r}-\eta\boldsymbol{R},\boldsymbol{r}-\eta\boldsymbol{R}+\boldsymbol{R})\,d\eta\right\}. \tag{35.8}$$

The pressure tensor defined above is suject to the condition for hydrostatic equilibrium

$$\nabla \cdot \mathbf{p} = 0. \tag{35.9}$$

In the case of thermodynamic equilibrium, the singlet and pair distribution functions should satisfy the generalized Born-Green-Yvon equations (25.6). Thus the pressure tensor (35.6) for a system in thermodynamic equilibrium, automatically satisfies the hydrostatic equilibrium condition (35.9) [26].

It should be noted that the force acting across $d\boldsymbol{S}$ at $\boldsymbol{r}$ and hence the pressure tensor as a function of the spatial coordinates $\boldsymbol{r}$, cannot be defined without arbitrariness and that definitions other than (35.8) are possible; but from the physical point of view all definitions should give identical results when integrated over a certain domain greater than the range of intermolecular forces. The last requirement cannot be fulfilled in the case of a droplet or bubble so small that the radius of the interface is not very large compared with the radius of the molecules.

36. Plane interface. We are now ready to translate what we have obtained in Sect. 14 from a purely mechanical point of view, into terms of intermolecular forces and the molecular distribution functions. As in Sect. 14 we define a rectangular coordinate system (x, y, z) with the dividing surface as the (x, y) plane and z axis normal to the dividing surface directed from phase α to phase β. Since in the case of a plane interface any physical property within the interface depends only on z, the singlet distribution function of species i can be expressed by $n_i(z^{(1)})$ and the pair distribution function by $n_{ij}(z^{(1)}; \boldsymbol{r}^{(12)})$ as in Sect. 26.

The pressure tensor $\mathbf{p}$ has been given by (14.2):

$$\mathbf{p} = p_T(z)\,(\boldsymbol{e}_x\boldsymbol{e}_x + \boldsymbol{e}_y\boldsymbol{e}_y) + p_N(z)\,\boldsymbol{e}_z\boldsymbol{e}_z, \tag{36.1}$$

where $p_T(z)$ and $p_N(z)$ are, respectively, the normal and tangential components of the pressure tensor. If the right-hand side of (36.1) is equated to the statistical expression (35.6), we can at once obtain

$$\left.\begin{aligned} p_T(z^{(1)}) &= kT\sum_{i=1}^{\varkappa} n_i(z^{(1)}) - \\ &\quad - \frac{1}{2}\sum_{i,j=1}^{\varkappa}\int \frac{(x^{(12)})^2}{r^{(12)}}\,\Phi'_{ij}(r^{(12)})\int_0^1 n_{ij}(z^{(1)} - \eta z^{(12)}; \boldsymbol{r}^{(12)})\,d\eta\,d\boldsymbol{r}^{(12)}, \end{aligned}\right\} \tag{36.2}$$

$$\left.\begin{aligned} p_N(z^{(1)}) &= kT\sum_{i=1}^{\varkappa} n_i(z^{(1)}) - \\ &\quad - \frac{1}{2}\sum_{i,j=1}^{\varkappa}\int \frac{(z^{(12)})^2}{r^{(12)}}\,\Phi'_{ij}(r^{(12)})\int_0^1 n_{ij}(z^{(1)} - \eta z^{(12)}; \boldsymbol{r}^{(12)})\,d\eta\,d\boldsymbol{r}^{(12)}, \end{aligned}\right\} \tag{36.3}$$

and an alternative expression for $p_T(z^{(1)})$ found by replacing $x^{(12)}$ in (36.2) by $y^{(12)}$. The normal component p_N must be equal to the uniform hydrostatic pressure p in the bulk phases α and β as required from the condition of hydrostatic equilibrium.

Substitution of (36.2) and (36.3) into (14.8) immediately leads to

$$\gamma = \frac{1}{2}\sum_{i,j=1}^{\varkappa}\int_{-\infty}^{\infty} dz^{(1)}\int d\boldsymbol{r}^{(12)}\int_0^1 \frac{(x^{(12)})^2 - (z^{(12)})^2}{r^{(12)}}\,\Phi'_{ij}(r^{(12)})\,n_{ij}(z^{(1)} - \eta z^{(12)}; \boldsymbol{r}^{(12)})\,d\eta. \tag{36.4}$$

We can rewrite (36.4) in the simpler form (see [26]).

$$\gamma = \frac{1}{2} \sum_{i,j=1}^{\varkappa} \int_{-\infty}^{\infty} d z^{(1)} \int n_{ij}(z^{(1)}; \boldsymbol{r}^{(12)}) \, \phi'_{ij}(r^{(12)}) \, \frac{(x^{(12)})^2 - (z^{(12)})^2}{r^{(12)}} \, d\boldsymbol{r}^{(12)}, \qquad (36.5)$$

making use of the relation

$$\int_{-\infty}^{\infty} d z^{(1)} \int_{0}^{1} n_{ij}(z^{(1)} - \eta z^{(12)}; \boldsymbol{r}^{(12)}) \, d\eta = \int_{-\infty}^{\infty} n_{ij}(z^{(1)}; \boldsymbol{r}^{(12)}) \, d z^{(1)}, \qquad (36.6)$$

which is obtained by interchanging the order of the integrations with respect to η and $z^{(1)}$. The expression (36.5) is identical with the thermodynamic expression (26.13).

For the pressures of the bulk phases α and β, we can write, from (36.2), the compact form

$$p = kT \sum_{i=1}^{\varkappa} n_i^{\alpha\beta} - \frac{1}{2} \sum_{i,j=1}^{\varkappa} \int \frac{(x^{(12)})^2}{r^{(12)}} \, \phi'_{ij}(r^{(12)}) \, n_{ij}^{\alpha\beta} \, d\boldsymbol{r}^{(12)}, \qquad (36.7)$$

where $n_i^{\alpha\beta}$ and $n_{ij}^{\alpha\beta}$ are given by (26.19) and (26.20), respectively. Substitution of (36.2) and (36.7) into (14.5), which expresses the surface tension as the excess of tangential stress, leads to [24],

$$\gamma = kT \sum_{i=1}^{\varkappa} \int_{-\infty}^{\infty} (n_i^{\alpha\beta} - n_i) \, d z^{(1)} - \frac{1}{2} \sum_{i,j=1}^{\varkappa} \int_{-\infty}^{\infty} d z^{(1)} \int (n_{ij}^{\alpha\beta} - n_{ij}) \, \phi'_{ij}(r^{(12)}) \, \frac{(x^{(12)})^2}{r^{(12)}} \, d\boldsymbol{r}^{(12)}. \qquad (36.8)$$

This is identical with (26.18).

As mentioned in the previous section, the statistical definition of the pressure tensor is not unique. It is, however, easily seen from (36.5) or by direct calculation from (36.2) and (36.3) that if the changes in n_i and n_{ij} with $z^{(1)}$ are negligible in the range of intermolecular forces, we may take, instead of (36.2) and (36.3), the alternative expressions

$$p_T(z^{(1)}) = kT \sum_{i=1}^{\varkappa} n_i(z^{(1)}) - \frac{1}{2} \sum_{i,j=1}^{\varkappa} \int \frac{(x^{(12)})^2}{r^{(12)}} \, \phi'_{ij}(r^{(12)}) \, n_{ij}(z^{(1)}; \boldsymbol{r}^{(12)}) \, d\boldsymbol{r}^{(12)}, \qquad (36.9)$$

$$p_N(z^{(1)}) = kT \sum_{i=1}^{\varkappa} n_i(z^{(1)}) - \frac{1}{2} \sum_{i,j=1}^{\varkappa} \int \frac{(z^{(12)})^2}{r^{(12)}} \, \phi'_{ij}(r^{(12)}) \, n_{ij}(z^{(1)}; \boldsymbol{r}^{(12)}) \, d\boldsymbol{r}^{(12)}, \qquad (36.10)$$

or in the tensorial form [25],

$$\mathbf{p}(z^{(1)}) = kT \sum_{i=1}^{\varkappa} n_i(z^{(1)}) \mathbf{1} - \frac{1}{2} \sum_{i,j=1}^{\varkappa} \int \frac{\boldsymbol{r}^{(12)} \boldsymbol{r}^{(12)}}{r^{(12)}} \, \phi'_{ij}(r^{(12)}) \, n_{ij}(z^{(1)}; \boldsymbol{r}^{(12)}) \, d\boldsymbol{r}^{(12)}, \qquad (36.11)$$

which is usually applied to the statistical mechanics of transport phenomena and is a very accurate approximation for a bulk fluid phase[1] [25].

Let us consider the physical implication of the expression for the tangential component (36.9) in more detail [24]. We consider an infinitesimal strip normal to the x axis, with unit width in the y direction and bounded between $z^{(1)}$ and $z^{(1)} + dz^{(1)}$ in the z direction. Let $F_x^{(\phi)} dz^{(1)}$ be the x component of the force exerted by all the molecules in the semi-infinite space with positive x on those in the semi-infinite column of fluid, with the cross section of the above strip, which is perpendicular to the (y, z) plane and extends from $x = 0$ to $-\infty$ (see Fig. 13b).

[1] J. H. Irving and J. G. Kirkwood: J. Chem. Phys. 18, 817 (1950).

The x component of the average intermolecular force exerted by molecules in a volume element $d\boldsymbol{r}^{(12)}$ at $\boldsymbol{r}^{(1)}+\boldsymbol{r}^{(12)}$ on those in a volume element $d\boldsymbol{r}^{(1)}$ at $\boldsymbol{r}^{(1)}$ is

$$-\sum_{i,j=1}^{\varkappa} \frac{\partial \phi_{ij}(r^{(12)})}{\partial x^{(1)}} n_{ij}(z^{(1)};\boldsymbol{r}^{(12)})\, d\boldsymbol{r}^{(1)}\, d\boldsymbol{r}^{(12)}. \tag{36.12}$$

Integrating over all configurations $\boldsymbol{r}^{(1)}$ within the column and over all configurations $\boldsymbol{r}^{(12)}$ such that $x^{(2)}>0$, we obtain the result

$$F_x^{(\Phi)}(z^{(1)})\, dz^{(1)} = \sum_{i,j=1}^{\varkappa} dz^{(1)} \int_{-\frac{1}{2}}^{\frac{1}{2}} dy^{(1)} \int_{-\infty}^{0} dx^{(1)} \int_{-\infty}^{\infty} dz^{(12)} \int_{-\infty}^{\infty} dy^{(12)} \int_{-x^{(1)}}^{\infty} dx^{(12)}\, n_{ij} \phi'_{ij} \frac{x^{(12)}}{r^{(12)}}. \tag{36.13}$$

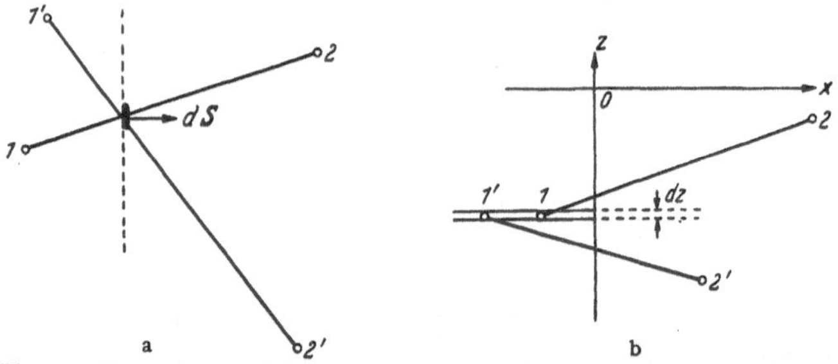

Fig. 13a and b. Diagrams illustrating the difference between two alternative definitions of tangential pressure p_T. (a) Diagram for (36.2). (b) Diagram for (36.9).

The interchange of the order of the integrations with respect to $x^{(12)}$ and $x^{(1)}$ leads, after some rearrangement taking into account the symmetry requirement, to [24],

$$-F_x^{(\Phi)}(z^{(1)}) = -\frac{1}{2}\sum_{i,j=1}^{\varkappa} \int_{-\infty}^{\infty} \frac{(x^{(12)})^2}{r^{(12)}} \phi'_{ij}(r^{(12)})\, n_{ij}(z^{(1)};\boldsymbol{r}^{(12)})\, d\boldsymbol{r}^{(12)}. \tag{36.14}$$

This can be regarded as the contribution of intermolecular forces to the pressure tensor component $p_{xx}(z^{(1)})$, i.e., the tangential component $p_T(z^{(1)})$. Hence adding the kinetic contribution $kT\sum_{i=1}^{\varkappa} n_i(z^{(1)})$ to $-F_x^{(\Phi)}$, we obtain the expression given by (36.9). Thus we find that p_T given by (36.9) is valid even in the plane interface region.

Since the definition of the tangential component of the pressure tensor given by (36.9) is in general different from the earlier one given by (36.2) (see Figs. 13a and b), it is evident that the statistical-mechanical definition of the pressure tensor is not unique. Furthermore, it will easily be seen that as regards $p_N(z^{(1)})$, a definition similar to the above with use of a strip perpendicular to the z axis, does not yield (36.10) but (36.3) based on the earlier definition. In fact the expression for p_N given by (36.10) is not valid in the interface region though both (36.3) and (36.10) lead to the same result when integrated over some domain sufficiently large compared with molecular dimensions, provided that edge effects may be neglected.

The location of the surface of tension z_s is given by (14.6). Substitution of (36.7) and (36.9) into (14.6) leads to

$$\left.\begin{aligned} z_s = \frac{1}{\gamma}\Bigg[&- kT\sum_{i=1}^{\varkappa} \int_{-\infty}^{\infty} (n_i - n_i^{\alpha\beta})\, z^{(1)}\, dz^{(1)} + \\ &+ \frac{1}{2}\sum_{i,j=1}^{\varkappa} \int d\boldsymbol{r}^{(12)} \frac{(x^{(12)})^2}{r^{(12)}} \phi'_{ij}(r^{(12)}) \int_{-\infty}^{\infty} (n_{ij} - n_{ij}^{\alpha\beta})\, z^{(1)}\, dz^{(1)}\Bigg]. \end{aligned}\right\} \tag{36.15}$$

In the case of a one component system, with use of (27.6), (27.7) and (27.11), (36.15) can be written as [24],

$$z_s = \frac{1}{2\gamma}\left[\int \frac{1}{r^{(12)}}\,\phi'(r^{(12)})\,\{(x^{(12)})^2\,[\Gamma_2]_1 + z^{(12)}\,[\Gamma_2]_2\}\,d\boldsymbol{r}^{(12)}\right]. \tag{36.16}$$

This equation is useful for the calculation of the distance between the surface of tension and the equimolecular dividing surface, δ, which plays an important role in the curvature dependence of surface tension as discussed in Sect. 12.

37. Spherical interface. In the case of a spherical interface the pressure tensor is given by (15.1). It is required from the symmetry of the system that $n_i(\boldsymbol{r}^{(1)})$ and $n_{ij}(\boldsymbol{r}^{(1)};\boldsymbol{r}^{(12)})$ can be expressed as $n_i(r^{(1)})$ and $n_{ij}(r^{(1)};\boldsymbol{r}^{(12)})$, respectively.

In a similar manner to the planar case, comparison of (35.6) and (15.1) immediately leads to

$$p_T(r^{(1)}) = kT\sum_{i=1}^{\varkappa} n_i(r^{(1)}) - \frac{1}{2}\sum_{i,j=1}^{\varkappa}\int \frac{(\boldsymbol{r}^{(12)}\cdot \boldsymbol{e}_\varphi^{(1)})^2}{r^{(12)}}\,\phi'_{ij}(r^{(12)})\,\tilde{n}_{ij}(r^{(1)};\boldsymbol{r}^{(12)})\,d\boldsymbol{r}^{(12)}, \tag{37.1}$$

$$\tilde{n}_{ij}(r^{(1)};\boldsymbol{r}^{(12)}) = \int_0^1 n_{ij}(|\boldsymbol{r}^{(1)} - \eta\,\boldsymbol{r}^{(12)}|;\boldsymbol{r}^{(12)})\,d\eta, \tag{37.2}$$

where the present notation $n_{ij}(\boldsymbol{r}^{(1)}-\eta\boldsymbol{r}^{(12)};\boldsymbol{r}^{(12)})$ corresponds to the earlier $n_{ij}(\boldsymbol{r}^{(1)}-\eta\boldsymbol{r}^{(12)},\ \boldsymbol{r}^{(1)}-\eta\boldsymbol{r}^{(12)}+\boldsymbol{r}^{(12)})$ used in (35.8).

For two homogeneous bulk phases α and β separated by a spherical dividing surface $r=a$, (37.1) reduces to

$$p^{\alpha\beta} = kT\sum_{i=1}^{\varkappa} n_i^{\alpha\beta} - \frac{1}{2}\sum_{i,j=1}^{\varkappa}\int \frac{(\boldsymbol{r}^{(12)}\cdot \boldsymbol{e}_\varphi^{(1)})^2}{r^{(12)}}\,\phi'_{ij}\,\tilde{n}_{ij}^{\alpha\beta}\,d\boldsymbol{r}^{(12)}, \tag{37.3}$$

where

$$\tilde{n}_{ij}^{\alpha\beta} = \tilde{n}_{ij}^{\alpha}(r^{(1)};\boldsymbol{r}^{(12)})\,[1 - A(r^{(1)}-a)] + \tilde{n}_{ij}^{\beta}(r^{(1)};\boldsymbol{r}^{(12)})\,A(r^{(1)}-a), \tag{37.4}$$

and $n_i^{\alpha\beta}$ is defined in a way similar to (28.20).

Substitution of (37.1) and (37.3) into (16.5) leads to an expression for the surface tension referred to a dividing surface with radius a (c.f. [26]),

$$\gamma = -\sum_{i=1}^{\varkappa}\Gamma_i kT + \frac{1}{2}\sum_{i,j=1}^{\varkappa}\int \tilde{\Gamma}_{ij}\,\phi'_{ij}(r^{(12)})\,\frac{(\boldsymbol{r}^{(12)}\cdot \boldsymbol{e}_\varphi^{(1)})^2}{r^{(12)}}\,d\boldsymbol{r}^{(12)}, \tag{37.5}$$

where, with $\xi = r - a$,

$$\Gamma_i = \int_{R^\alpha}^{R^\beta}(n_i - n_i^{\alpha\beta})\,\frac{(r^{(1)})^2}{a^2}\,dr^{(1)} = \int_{-\infty}^{\infty}(n_i - n_i^{\alpha\beta})\left(1+\frac{\xi}{a}\right)^2 d\xi, \tag{37.6}$$

$$\tilde{\Gamma}_{ij} = \int_{R^\alpha}^{R^\beta}(\tilde{n}_{ij} - \tilde{n}_{ij}^{\alpha\beta})\,\frac{(r^{(1)})^2}{a^2}\,dr^{(1)}. \tag{37.7}$$

We shall now show that in the case $\varkappa=1$, (37.5) is equivalent to (28.23), which is based on the thermodynamic definition. From (35.4), (37.2), (37.4) and (37.7), we have

$$\left.\begin{aligned}\sum_{i,j=1}^{\varkappa}\tilde{\Gamma}_{ij} = \sum_{i,j=0}^{\varkappa}\frac{1}{2}\int_0^1\int_{R^\alpha}^{R^\beta}[n_{ij}(|\boldsymbol{r}^{(1)}-\eta\boldsymbol{r}^{(12)}|;\boldsymbol{r}^{(12)}) - n_{ij}^{\alpha\beta}(|\boldsymbol{r}^{(1)}-\eta\boldsymbol{r}^{(12)}|;\boldsymbol{r}^{(12)}) +\\ + n_{ij}(|\boldsymbol{r}^{(1)}+\eta\boldsymbol{r}^{(12)}|;-\boldsymbol{r}^{(12)}) - n_{ij}^{\alpha\beta}(|\boldsymbol{r}^{(1)}+\eta\boldsymbol{r}^{(12)}|;-\boldsymbol{r}^{(12)})]\left(\frac{r^{(1)}}{a}\right)^2 dr^{(1)}\,d\eta.\end{aligned}\right\} \tag{37.8}$$

If the radius a of the dividing surface, which is now assumed to be located within the spherical interface, is sufficiently large compared with the range of intermolecular force, τ, so that $(\tau/a)^2$ can be neglected as compared with unity, we have, for such $r^{(12)} \sim \tau$

$$\left.\begin{aligned} &\int_{R^\alpha}^{R^\beta} n_{ij}(|\boldsymbol{r}^{(1)} \pm \eta \boldsymbol{r}^{(12)}|; \mp \boldsymbol{r}^{(12)})\,(r^{(1)})^2\, d r^{(1)} \\ &\quad\approx \int_{R^\alpha \pm \eta r^{(12)}\cos\psi^{(1)}}^{R^\beta \pm \eta r^{(12)}\cos\psi^{(1)}} n_{ij}(r^{(1)}; \mp \boldsymbol{r}^{(12)})\,(r^{(1)} \mp \eta r^{(12)}\cos\psi^{(1)})^2\, d r^{(1)}, \end{aligned}\right\} \tag{37.9}$$

$$\cos\psi^{(1)} = \frac{\boldsymbol{r}^{(1)}\cdot\boldsymbol{r}^{(12)}}{r^{(1)}\, r^{(12)}}. \tag{37.10}$$

Since the integrand of (37.8) vanishes outside the interface zone, with use of (37.9) and (37.8), we can rewrite the second term on the righthand side of (37.5) in the form

$$\left.\begin{aligned} &\frac{1}{2}\sum_{i,j=1}^{\varkappa}\int d\boldsymbol{r}^{(12)}\,\Phi'_{ij}(r^{(12)})\,\frac{(\boldsymbol{r}^{(12)}\cdot\boldsymbol{e}_\varphi^{(1)})^2}{r^{(12)}}\times \\ &\quad\times\int_{R^\alpha}^{R^\beta}\Big[\{n_{ij}(r^{(1)};\boldsymbol{r}^{(12)}) - n_{ij}^{\alpha\beta}(r^{(1)};\boldsymbol{r}^{(12)})\}\,\frac{(r^{(1)})^2 + \frac{1}{3}(r^{(12)}\cos\psi^{(1)})^2}{a^2} + \\ &\quad + \frac{1}{2}\{n_{ij}(r^{(1)};\boldsymbol{r}^{(12)}) - n_{ij}(r^{(1)}; -\boldsymbol{r}^{(12)})\}\,\frac{r^{(1)}\, r^{(12)}\cos\psi^{(1)}}{a^2}\Big]\, d r^{(1)}. \end{aligned}\right\} \tag{37.11}$$

By definition (see also the footnote 1 on p. 179) we have

$$n_{ij}(\boldsymbol{r}^{(1)}, \boldsymbol{r}^{(1)} - \boldsymbol{r}^{(12)}) = n_{ji}(\boldsymbol{r}^{(1)} - \boldsymbol{r}^{(12)}, \boldsymbol{r}^{(1)})$$

and hence, as in the case of (37.9),

$$\begin{aligned} &\int n_{ij}(r^{(1)}; -\boldsymbol{r}^{(12)})\, r^{(1)}\, r^{(12)}\cos\psi^{(1)}\, d r^{(1)} \\ &\quad\approx \int n_{ji}(r^{(1)}; \boldsymbol{r}^{(12)})\,(r^{(1)} + r^{(12)}\cos\psi^{(1)})\, r^{(12)}\cos\psi^{(1)}\, d r^{(1)}. \end{aligned}$$

Thus, neglecting terms of the order of magnitude of $(\tau/a)^2$, we can reduce (37.5) to

$$\gamma = -\sum_{i=1}^{\varkappa}\Gamma_i kT + \frac{1}{2}\sum_{i,j=1}^{\varkappa}\int \Gamma_{ij}\,\Phi'_{ij}(r^{12})\,\frac{(\boldsymbol{r}^{(12)}\cdot\boldsymbol{e}_\varphi^{(1)})^2}{r^{(12)}}\, d\boldsymbol{r}^{(12)}, \tag{37.12}$$

$$\Gamma_{ij} = \int_{-\infty}^{\infty}(n_{ij} - n_{ij}^{\alpha\beta})\left(1 + \frac{\xi^{(1)}}{a}\right)^2 d\xi^{(1)}. \tag{37.13}$$

Eq. (37.12) is valid down to the order of (τ/a) and, for a one-component case, reduces to (28.23).

Thus, we may conclude that the mechanical and thermodynamic expressions for the surface tension of a spherical interface are equivalent insofar as quantities of the order of magnitude of $(\tau/a)^2$ can be neglected.

In the case of a plane interface we have seen in Sect. 36 that two different expressions (36.2) and (36.9) are equally reasonable as the tangential component of the pressure tensor. By analogy with (36.9), let us assume, as an alternative expression for p_T given by (37.1), the following equation:

$$p_T(r^{(1)}) = kT\sum_{i=1}^{\varkappa} n_i(r^{(1)}) - \frac{1}{2}\sum_{i,j=1}^{\varkappa}\int\frac{(\boldsymbol{r}^{(12)}\cdot\boldsymbol{e}_\varphi^{(1)})^2}{r^{(12)}}\,\Phi'_{ij}(r^{(12)})\, n_{ij}(r^{(1)};\boldsymbol{r}^{(12)})\, d\boldsymbol{r}^{(12)}. \tag{37.14}$$

Then in conjunction with (37.14), we must use, instead of (37.3), the following:

$$p^{\alpha\beta} = kT \sum_{i=1}^{\varkappa} n_i^{\alpha\beta} - \frac{1}{2} \sum_{i,j=1}^{\varkappa} \int \frac{(\boldsymbol{r}^{(12)} \cdot \boldsymbol{e}_\varphi^{(1)})^2}{r^{(12)}} \phi'_{ij}(r^{(12)})\, n_{ij}^{\alpha\beta}\, d\boldsymbol{r}^{(12)}, \tag{37.15}$$

where

$$n_{ij}^{\alpha\beta} = n_{ij}^{\alpha}(r^{(12)})\,[1 - A(r^{(1)} - a)] + n_{ij}^{\beta}(r^{(12)})\, A(r^{(1)} - a). \tag{37.16}$$

Substitution of (37.14) and (37.15) into (16.5) immediately leads to (37.12). Although the physical implication of the expression (37.14) has not been visualized yet for the spherical case, in contrast to the planar case, the resultant effects integrated over domains sufficiently large compared with molecular dimensions, would be expected to be equivalent both for (37.14) and (37.1).

Thus, we may conclude that the statistical-mechanical expression for the tangential component of the pressure tensor at a spherical interface can be taken to be identical in form with that of a plane interface apart from the errors of the order of $(\tau/R)^2$ and that effectively (insofar as we are concerned with surface tension) the singlet and the pair distribution functions are solely responsible for the radius dependence of p_T if the *latter* exists in the range of R in which $(\tau/R)^2$ is negligible compared with unity.

From the above we may conclude that the statistical-thermodynamic expression for the surface tension, (28.23), agrees with the statistical-hydrostatic expression, (37.5), for the one-component case apart from the error of the order of $(\tau/R)^2$. This is also the case for multicomponent systems. Then the curvature dependence of the surface tension can be divided into two parts. One is due to the geometry of the spherical interface discussed in Sect. 16, i.e., the factor $(1 + a^{-1}\xi)^2$ appearing in (16.7). The other is that arising from the change in p_T itself, which results only from the modification of the singlet and pair distribution functions, and its evaluation will be impossible until we obtain some approximate solution of the Born-Green-Yvon equation in the surface layer to a sufficient accuracy.

If the radius of a droplet or bubble is comparable with the range of intermolecular potentials, a discrepancy between the two definitions of the surface tension would appear. It is, however, to be noted that the surface tension would lose its proper meaning for such small systems, because the interior phase would not have bulk properties even at its center. At the same time, the technique used in deriving the thermodynamic expression for the surface tension in Sects. 8 and 16 ceases to be valid because of edge effects. One of the most important features which prevent us from defining the surface tension as a thermodynamic quantity for such small drops or bubbles, is that the area of the interface cannot be defined as an external parameter. It would be impossible to control the shape of such a drop or bubble against the thermal agitation with use of an auxiliary field such as a weak central field considered in Sect. 28. This implies that the area of the interface never can be regarded as a controllable external parameter in a thermodynamic sense.

It should now be noted and emphasized that the equations for surface tension given in Sects. 36 and 37 are valid for non-equilibrium systems as well if the systems are at rest under no external field and their total potential energies are expressible as (25.3). Therefore, we may conclude that surface tension is expressed by a statistical-mechanical expression which is identical in form both for the equilibrium and non-equilibrium cases and that singlet and pair distribution functions only are responsible for the difference between surface tension in equilibrium and that out of equilibrium.

III. Numerical calculations.

a) Pure liquid.

38. Surface of discontinuity [24]. We have found in the foregoing that the calculation of surface tension needs a precise knowledge of the singlet and pair distribution functions $n_i(\mathbf{r}^{(1)})$ and $n_{ij}(\mathbf{r}^{(1)}, \mathbf{r}^{(2)})$ in the interface layer. For the case of a one-component homogeneous fluid in thermodynamic equilibrium, rather precise calculations of the pair distribution function have been carried out with the Born-Green-Yvon equation (25.10) based on the superposition approximation as mentioned in Sect. 25. But a similar calculation has not been carried out yet for the distribution functions in the interface layer because of computational difficulties. Thus we shall use here more drastic approximations, restricting our treatment to the case of a one-component system with a plane interface.

We shall assume that the system is composed of a homogeneous liquid phase α with a constant bulk density up to the dividing surface in contact with a vapor phase β so rarefied that it may be considered a vacuum. Then we have

$$\left.\begin{aligned} n(z^{(1)}) &= n^{\alpha}, \quad z^{(1)} \leqq 0, \\ &= 0, \quad\ \ z^{(1)} > 0, \end{aligned}\right\} \tag{38.1}$$

$$n_2(z^{(1)}; \mathbf{r}^{(12)}) = n(z^{(1)})\, n(z^{(1)} + z^{(12)})\, g_2^{\alpha}(r^{(12)}), \tag{38.2}$$

where $g_2^{\alpha}(r^{(12)})$ is the correlation function in the homogeneous liquid phase defined by (25.11). Then the dividing surface corresponds to the equimolecular dividing surface, i.e., $\Gamma = 0$. Substitution of (38.2) into (27.7) leads to

$$[\Gamma_2]_\nu = \left\{\begin{matrix} (-1)^{\nu+1} \dfrac{(z^{(12)})^{\nu+1}}{\nu+1} (n^{\alpha})^2 g_2^{\alpha}(r^{(12)}); & z^{(12)} \geqq 0, \\ 0; & z^{(12)} < 0. \end{matrix}\right\} \tag{38.3}$$

Thus, with use of (27.12), we obtain the desired approximation to the surface tension as follows:

$$\gamma = \frac{\pi (n^{\alpha})^2}{8} \int\limits_0^{\infty} \phi'(r)\, g_2^{\alpha}(r)\, r^4\, dr. \tag{38.4}$$

This expression was first derived by Fowler [27], based upon the same approximation to the pair density n_2 but on a method quite different from that described above.

Substitution of (38.3) into (27.16) yields the expression for the surface energy referred to the equimolecular dividing surface, to the above approximation, as follows:

$$v = -\frac{\pi (n^{\alpha})^2}{2} \int\limits_0^{\infty} \phi(r)\, g_2^{\alpha}(r)\, r^3\, dr. \tag{38.5}$$

From (36.16), (38.3) and (38.4), we obtain for the location of the surface of tension

$$\delta_\infty = -z_s = \frac{4}{15} \frac{\int\limits_0^{\infty} g_2^{\alpha}(r)\, \phi'(r)\, r^5\, dr}{\int\limits_0^{\infty} g_2^{\alpha}(r)\, \phi'(r)\, r^4\, dr}, \tag{38.6}$$

where z_s is the distance of the surface of tension measured from the equimolecular dividing surface toward the vapor phase β. The symbol δ_∞ denotes the value of δ, which is given by (12.8), for the planar case.

The hydrostatic pressure and the molar energy of vaporization, are respectively given by the familiar expression:

$$p = n^\alpha kT - \frac{2\pi}{3}(n^\alpha)^2 \int_0^\infty \phi'(r)\, g_2^\alpha(r)\, r^3\, dr, \tag{38.7}$$

$$\Delta U_v = -2\pi N_0 n^\alpha \int_0^\infty \phi(r)\, g_2^\alpha(r)\, r^2\, dr, \tag{38.8}$$

where N_0 is the Avogadro number. For the derivation of (38.7) we have used (26.24); and (38.8) is a direct consequence of (25.4) if the contribution of intermolecular forces is ignored for the vapor phase.

Table 5. *The surface tension, surface energy and the distance between the surface of tension and equimolecular dividing surface of argon at* 90° K.

	(dyne/cm) γ	(dyne/cm) v	(Å) δ_∞
KIRKWOOD and BUFF [24]	14.9	27.2	3.63
HILL [31b]	6.0	19.0	2.81
Experiment[1]	11.9	35.0	

We can now easily calculate the thermodynamic quantities γ, v and δ_∞ from (38.4), (38.5) and (38.6), respectively, if only we know the intermolecular potential $\phi(r)$ and the correlation function $g_2^\alpha(r)$ of the liquid. Such calculations for liquid argon at 90° K were carried out by KIRKWOOD and BUFF [24]. They employed the Lennard-Jonnes potential modified by RUSHBROOKE[2]

$$\phi(r) = \frac{8.62\times 10^{-8}}{r^{11.4}} - \frac{1.11\times 10^{-10}}{r^6}, \quad \begin{pmatrix} \phi(r) & \text{in erg} \\ r & \text{in Å} \end{pmatrix}. \tag{38.9}$$

and the analytic approximation

$$\left.\begin{aligned} g_2^\alpha(r) &= \left(\frac{a_1}{r}\right)^s \exp\left\{\left(\frac{a_m}{a_1}\right)^t - \left(\frac{a_m}{r}\right)^t\right\} \quad & a_1 > r \geqq 0, \\ &= 1 & a_1 < r, \end{aligned}\right\} \tag{38.10}$$

to the first peak of the experimental curve determined from the X-ray diffraction work of EISENSTEIN and GINGRICH[3]. Since only the first peak plays a significant role in such calculations, it is adequate to approximate $g_2^\alpha(r)$ by unity for distances greater than a_1. The parameters a_1 and t are determined from the experimental data. The remaining s and a_m are adjusted to satisfy (38.7) and (38.8) with the experimental values of p and ΔU_v. By this procedure they found for liquid argon at 90° K in equilibrium with the saturated vapor [24],

$$\left.\begin{aligned} a_1 &= 4.50\,\text{Å}; \quad & t &= 14, \\ a_m &= 3.55\,\text{Å}; & s &= 7.00. \end{aligned}\right\} \tag{38.11}$$

The calculated values[4] of γ, v and δ_∞ are presented in Table 5, together with the results of HILL's calculation to be mentioned in the next section and the experimental values[1].

The agreement between the theoretical and experimental values is good in view of the drastic approximation employed for the distribution functions.

---

[1] BALY and DONNAN: J. Chem. Soc. **81**, 907 (1902). — G. RUDORF: Ann. d. Phys. **29**, 751 (1909).

[2] G. S. RUSHBROOKE: Proc. Roy. Soc. Edinburgh **60**, 182 (1940).

[3] A. EISENSTEIN and N. S. GINGRICH: Phys. Rev. **62**, 261 (1942).

[4] According to a private communication from Dr. BUFF, the calculation was based on a set of more complete values $a_1 = 4.500$ Å; $t = 14.000$; $a_m = 3.554$ Å; $s = 7.007$. The figures in (38.11) are rounded off.

The positive value of δ_∞ implies that the surface of tension lies at a distance 3.63 Å from the equimolecular dividing surface on the liquid side and consequently, due to (12.10), the surface tension must decrease for a drop or increase for a bubble with decreasing radius of the interface[1]. The value of δ_∞ is of the same sign and magnitude as those estimated by TOLMAN [30] for other liquids on the basis of quasithermodynamics.

Let us further calculate the values of $p_T(z)$ and $p_N(z)$ separately by inserting (38.1) and (38.2) in (36.9) and (36.3) for a one-component system. Then we obtain the following expressions:

$$\left.\begin{aligned} p_T(z) &= p^\alpha + \frac{\pi}{3}(n^\alpha)^2 \int_{-z}^{\infty} \Phi'(r)\, g_2^\alpha(r)\, r^3\, dr - \\ &\quad - \frac{\pi}{6}(n^\alpha)^2 z^3 \int_{-z}^{\infty} \Phi'(r)\, g_2^\alpha(r)\, dr + \frac{\pi}{2}(n^\alpha)^2 z \int_{-z}^{\infty} \Phi'(r)\, g_2^\alpha(r)\, r^2\, dr, \end{aligned}\right\} \tag{38.12}$$

$$\left.\begin{aligned} p_N(z) &= p^\alpha + \frac{2\pi}{3}(n^\alpha)^2 \int_{-z}^{\infty} \Phi'(r)\, g_2^\alpha(r)\, r^3\, dr - \\ &\quad - \frac{\pi}{3}(n^\alpha)^2 z^3 \int_{-z}^{\infty} \Phi'(r)\, g_2^\alpha(r)\, dr + \pi(n^\alpha)^2 z \int_{-z}^{\infty} \Phi'(r)\, g_2^\alpha(r)\, r^2\, dr, \end{aligned}\right\} \tag{38.13}$$

where p^α is the hydrostatic pressure in the interior of the liquid phase α.

With use of (38.12) and (38.13) the numerical values of p_T and p_N as functions of z have been calculated by HARASIMA [29a], [29b] based on the same approximations as (38.9) to (38.11). His calculated results are illustrated in Fig. 14.

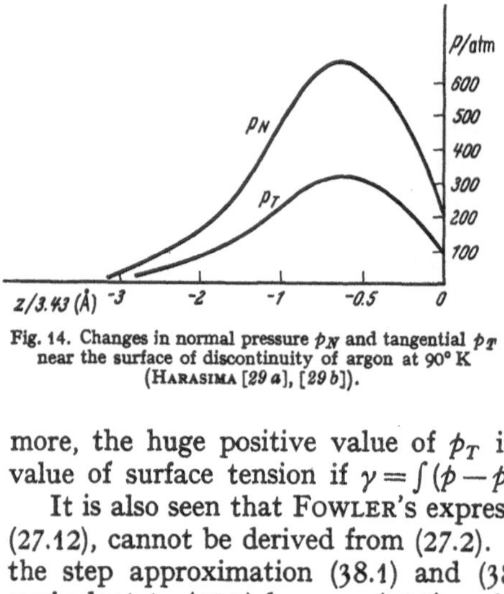

Fig. 14. Changes in normal pressure p_N and tangential p_T near the surface of discontinuity of argon at 90° K (HARASIMA [29a], [29b]).

It is observed from the figure that the normal pressure p_N varies greatly in the interface region taking the maximum value 660 atm at $z = -2.4$ Å. This contradicts the condition for hydrostatic equilibrium which requires a uniform value of p_N throughout the system and hence implies that an interface with such a discontinuous change in density as given by (38.1) and (38.2) is extraordinarily far from the state in mechanical equilibrium. Furthermore, the huge positive value of p_T in the interface cannot give a positive value of surface tension if $\gamma = \int (p - p_T)\, dz$ is used.

It is also seen that FOWLER's expression (38.4), which we have obtained from (27.12), cannot be derived from (27.2). This contradiction arises from the use of the step approximation (38.1) and (38.2) for which (27.12) does not become equivalent to (27.2) because (27.5) and hence (27.11) do not hold for (38.1) and (38.2).

Nevertheless, we can obtain a rather good value for the surface tension as shown in Table 5 when we use the relation $\gamma = \int (p_N - p_T)\, dz$. It seems that in this calculation the large errors in p_N and p_T are almost cancelled out, for the

[1] HILL [28] applied the step approximation (38.1) and (38.2) to spherical surfaces. But his values for the radius dependence of surface tension appear to be too small.

errors introduced by the step approximation in $\mathbf{p}$ seem to be almost independent of the direction of $\mathbf{p} \cdot \boldsymbol{e}$ in the interface layer (where $\boldsymbol{e}$ is a unit vector in an arbitrary direction), although this remains to be justified in the future.

39. Quasithermodynamic theory [*31a*], [*31b*]. The inconsistency encountered in the previous section seems to arise from the mathematical discontinuity of the density at the interface layer. Therefore we shall here place emphasis on the continuity of the change in the fluid density within the interface layer so that it may approximately satisfy the equilibrium condition.

The surface tension can be calculated directly from (14.5) or (20.5) if we know p_T as a function of z. For the purpose of obtaining p_T as a function of z, we shall start with some quasithermodynamic considerations of the same system as dealt with in Sects. 18 to 21.

From the quasithermodynamic Gibbs-Duhem relation (20.3) we have

$$\left(\frac{\partial \mu}{\partial p_T}\right)_T = \frac{1}{n}. \tag{39.1}$$

As discussed in Sect. 18, in quasithermodynamics we assume that p_T is uniquely determined as a function of T and n at each point. If we know the functional form $p_T(n, T)$, from (39.1) we can express the chemical potential μ in terms of n at every point and in turn μ is required to be uniform throughout the system at equilibrium.

Since it is difficult to know the exact form of p_T as a function of $n(z)$ and T, even if p_T is assumed to be so expressible, we shall calculate p_T by assuming that it obeys an equation of state usually employed to calculate the pressure p in bulk phases, e.g., the van der Waals equation of state for one mole [*30*], [*31a*],

$$\left(p + \frac{N_0^2 a}{V^2}\right)(V - N_0 b) = N_0 kT, \tag{39.2}$$

where N_0 is the Avogadro number, V the molar volume and a and b VAN DER WAALS' constants per molecule. This simple semi-empirical equation provides a rather good approximation to the observed p-V relationship even down to the volume of the liquid phase. Solving (39.2) for p we obtain

$$\frac{pV}{N_0 kT} = 1 + \left(b - \frac{a}{kT}\right)\left(\frac{N_0}{V}\right) + \cdots, \tag{39.3}$$

where $\left(b - \frac{a}{kT}\right)$ is the second virial coefficient.

According to the statistical-mechanical treatment[1], one of the simplest forms of intermolecular potential, which lead to (39.3) in the low density and high temperature limit, is given by

$$\left.\begin{aligned} \phi(r) &= -\varepsilon\left(\frac{r_0}{r}\right)^6 & r &> r_0, \\ &= +\infty & r &\leqq r_0. \end{aligned}\right\} \tag{39.4}$$

The molecules obeying the potential law (39.4) behave like hard spheres of radius $\frac{1}{2} r_0$. Then it is shown that the constants a and b are given by, respectively,

$$a = \varepsilon b, \tag{39.5}$$

$$b = 2\pi \frac{r_0^3}{3}. \tag{39.6}$$

---

[1] For example, J.E. MAYER and M.G. MAYER: Statistical Mechanics, pp. 266—269. New York: John Wiley & Sons, Inc. 1940.

For a system composed of molecules obeying the potential law (39.4), we use an approximate correlation function

$$g_2(r) = \begin{cases} 1 & r > r_0, \\ 0 & r \leq r_0, \end{cases} \tag{39.7}$$

which is valid in the high temperature limit.

We shall further assume that the tangential pressure p_T is given, even for the interface region, by the van der Waals equation:

$$(p_T + n^2 a)\left(\frac{1}{n} - b\right) = kT, \tag{39.8}$$

where the molar volume V is replaced by N_0/n and the tangential pressure p_T may assume negative values.

We can rewrite (39.8) in the convenient form

$$\frac{p_T b}{kT} = \frac{\vartheta}{1-\vartheta} - \frac{\alpha}{2}\vartheta^2, \tag{39.9}$$

$$\vartheta = n b, \tag{39.10}$$

$$\alpha = \frac{2a}{b\,kT} = \frac{2\varepsilon}{kT}, \tag{39.11}$$

where the last expression of (39.11) results from (39.5).

We integrate (39.1) with respect to n after substitution of (39.9), obtaining

$$\mu = \mu_0(T) + kT \log\left(\frac{kT}{b}\right) + kT\,\nu, \tag{39.12}$$

$$\nu = f(\vartheta) - \alpha\vartheta, \tag{39.13}$$

$$f(\vartheta) = \log\frac{\vartheta}{1-\vartheta} + \frac{\vartheta}{1-\vartheta}, \tag{39.14}$$

where the integration constant $\mu_0(T)$ is so determined that the chemical potential μ approaches (7.3) in the limit of infinite dilution and is a function of T only.

As seen from (25.4) the average intermolecular potential energy per molecule, $u^{(\phi)}$, in a homogeneous bulk fluid can be expressed in the form

$$u^{(\phi)} = 2\pi n \int_0^\infty g_2(r)\,\phi(r)\,r^2\,dr. \tag{39.15}$$

Substitution of (39.4) and (39.7) into (39.15) leads to

$$u^{(\phi)} = -2\pi\frac{r_0^3 n\varepsilon}{3}. \tag{39.16}$$

Multiplying (39.10) by (39.11) times $-\frac{1}{2}kT$ and using (39.6), we obtain the same expression as the right-hand side of (39.16). Thus we have, comparing with (39.15),

$$\alpha\vartheta kT = -4\pi n \int_0^\infty g_2(r)\,\phi(r)\,r^2\,dr. \tag{39.17}$$

This implies that $-\alpha\vartheta\,kT$ in (39.12) is the contribution of the potential energy per molecule to the chemical potential.

Thus we can use the van der Waals equation as a relation connecting the tangential pressure $p_T(z)$ at z in the interface layer to $n(z)$, the number density at the same point. On the other hand, the statistical mechanical expression (36.9) shows that in the interface layer, $p_T(z)$ is determined not only by $n(z)$

at z but also by effects arising from the changes in the number density in the neighborhood of this point. Hence $p_T(z)$ cannot be identical with the pressure of a bulk fluid with the same density $n(z)$.

To take account of the contribution of the spatial change in the number density of molecules to the internal energy in a rather exact manner, HILL [*31a*], [*31b*] has replaced the above crude expression $-\alpha\vartheta\, kT$ for the intermolecular potential by

$$kT\,\Psi(z^{(1)}) = \int n(z^{(1)} + z^{(12)})\, g_2(r^{(12)})\, \phi(r^{(12)})\, d\boldsymbol{r}^{(12)}, \tag{39.18}$$

which is more plausible than (39.17) from the statistical-mechanical standpoint.

For a system composed of molecules obeying the potential law (39.4) and having $g_2(r)$ given by (39.7), the above equation is written as

$$\Psi(s) = -\frac{3\alpha}{8}\left[\int_{-\infty}^{-1} \vartheta(s+t)\, t^{-4}\, dt + \int_{-1}^{1} \vartheta(s+t)\, dt + \int_{1}^{\infty} \vartheta(s+t)\, t^{-4}\, dt\right], \tag{39.19}$$

$$s = \frac{z^{(1)}}{r_0}; \qquad t = \frac{z^{(12)}}{r_0}; \qquad \vartheta(s) = b\, n(z^{(1)}). \tag{39.20}$$

Then in the expression for μ, (39.12), ν is given by

$$\nu = f(\vartheta(s)) + \Psi(s), \tag{39.21}$$

instead of (39.13).

According to the quasithermodynamic condition for equilibrium, (19.4), this chemical potential must be constant over the whole range of s. Thus (39.12) with (39.21) is a non-linear integral equation to determine the function $\vartheta(s)$. Solving the integral equation and using (39.20), we may obtain $n(z)$ as a function of z.

A further improvement in the above method was made by HILL [*31b*] in utilizing a more accurate equation of state than VAN DER WAALS'. He employed TONKS' equation[1]

$$\frac{p\,b'}{kT} = \frac{\vartheta(1 + 2.9619\,\vartheta + 5.483\,\vartheta^2)}{1 - 0.8517\,\vartheta^3 - 0.1483\,\vartheta^4}; \qquad \vartheta = n\,b'; \qquad b' = \frac{r_0^3}{\sqrt{2}}, \tag{39.22}$$

which describes the p-V-T behavior of the fluid of hard sphere molecules [i.e., $\varepsilon = 0$ in (39.4)] with accuracy up to the third virial coefficient.

For molecules obeying the potential law (39.4) with (39.7), TONK's equation may be modified as follows:

$$\frac{p_T\, b'}{kT} = \frac{\vartheta(1 + 2.9619\,\vartheta + 5.483\,\vartheta^2)}{1 - 0.8517\,\vartheta^3 - 0.1483\,\vartheta^4} - \frac{\alpha'}{2}\vartheta^2, \tag{39.23}$$

$$\alpha' = 2.9619\,\alpha, \tag{39.24}$$

where $\alpha = 2\varepsilon/kT$ is given by (39.11). The second term on the right-hand side of (39.23), which corresponds to $-\frac{\alpha}{2}\vartheta^2$ in (39.9), expresses the contribution to the internal energy due to the attractive forces. The modified TONK's equation (39.23) does not give the exact third virial coefficient for a system of molecules with the attractive potential (39.4), but still may be regarded as more accurate than VAN DER WAALS' since at the critical temperature (39.23) gives a value $p_c V_c / N_0 k T_c = 0.342$, which is in better agreement with the average of the experimental values for monatomic gases, 0.293, than the value 0.375 provided by VAN DER WAALS' equation.

[1] L. TONKS: Phys. Rev. **50**, 955 (1936).

After substituting (39.23) into (39.1), we integrate with respect to n and replace $-\alpha'\vartheta$ by the statistical-mechanical expression $\tilde{\Psi}(s)$ which is obtained from (39.19) by replacing α by α' given by (39.24).

Finally we obtain the following equations:

$$\mu = \mu_0(T) + kT \log \frac{kT}{b'} + kT\nu', \tag{39.25}$$

$$\nu' = f(\vartheta(s)) + \tilde{\Psi}(s), \quad \vartheta = n b', \tag{39.26}$$

$$\tilde{\Psi}(s) = -\frac{3\alpha'}{8}\left[\int_{-\infty}^{-1} \vartheta(s+t)\, t^{-4}\, dt + \int_{-1}^{1} \vartheta(s+t)\, dt + \int_{1}^{\infty} \vartheta(s+t)\, t^{-4}\, dt\right], \tag{39.27}$$

$$\left.\begin{aligned} f(\vartheta) = \log \frac{\vartheta}{(1-\vartheta)^3} - 0.99195 \log(1 + 0.173069\,\vartheta) + 1.49597 \log(1 + \\ + 0.826931\,\vartheta + 0.856884\,\vartheta^2) - 1.33238 \tan^{-1}(1.03462\,\vartheta + \\ + 0.49923) + \frac{1 + 2.9619\,\vartheta + 5.483\,\vartheta^2}{1 - 0.8517\,\vartheta^3 - 0.1483\,\vartheta^4} - 0.38307. \end{aligned}\right\} \tag{39.28}$$

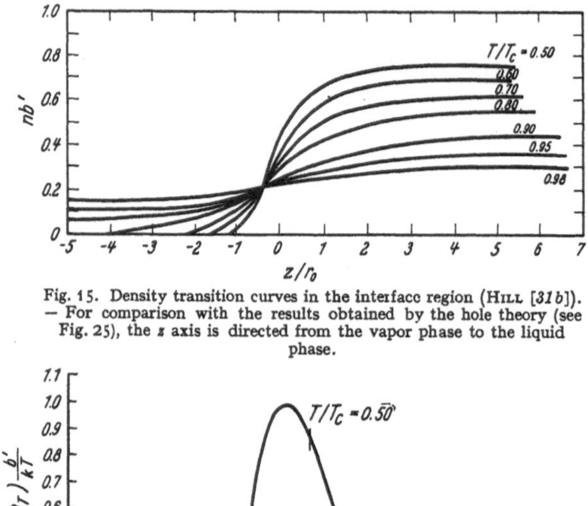

Fig. 15. Density transition curves in the interface region (Hill [31b]). — For comparison with the results obtained by the hole theory (see Fig. 25), the z axis is directed from the vapor phase to the liquid phase.

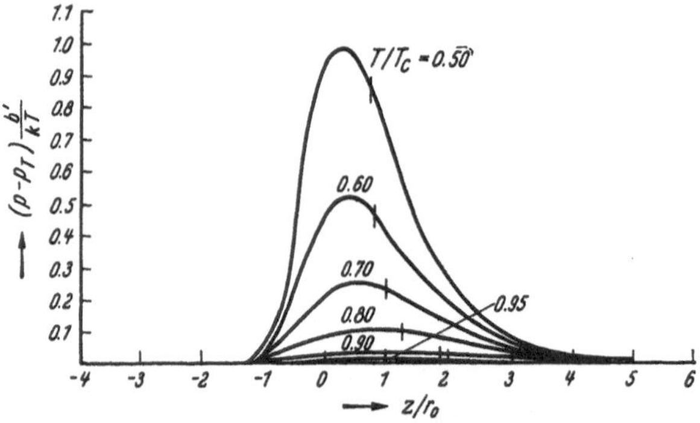

Fig. 16. Transition curves of tangential pressure p_T in the interface region and location of the surface of tension (Hill [31b]). — For comparison with the results obtained by the hole theory (see Fig. 27), the z axis is directed from the vapor phase to the liquid phase.

The non-linear integral equation (39.25) was numerically solved by Hill [31b] with the help of an I. B. M. punched card machine and the calculated values are shown in Table 5, Fig. 15 and Fig. 16 where T_c is the critical temperature given by $\varepsilon/kT_c = 2.44$.

The value of γ obtained from this method is not in better agreement with the experimental value than that calculated by Kirkwood and Buff [24] with use of the radial distribution function. This may be due to the crudeness of the quasithermodynamic treatment employed in Hill's method. It is, however, significant that this method, though it includes some inconsistency between assumptions concerning the contributions of energy and entropy to the chemical potential, provides a relatively simple means of calculating the density change in the interface layer and that p_T obtained from this method assumes negative values, in contrast with Kirkwood-Buff's approximation discussed in the previous section, and as is actually the case in the interface layer as seen [10], [9] from (14.5) or (20.5) (see also Fig. 27).

b) Electrolyte solution.

40. Surface tension of dilute solutions of strong electrolytes. The potential of average force in a very dilute electrolyte solution with a plane interface can be obtained by means of a modification of DEBYE and HÜCKEL's theory[1]. In the present section, from this potential of average force, an approximate expression for the surface tension of an electrolyte solution will be derived on the basis of the general theory developed in Sect. 32.

α) Homogeneous electrolyte solution. Before dealing with the modification of the average force between ions in solution due to the presence of the surface, we consider a homogeneous solution composed of $\varkappa - 1$ ionic species dissolved in a solvent regarded as a structureless medium with uniform dielectric constant D. Let R denote a distance from an ion of species i, which will be treated as a point charge with charge ε_i, $\varrho_i(R)$ the average electrostatic charge density at a distance R from a central ion of species i and $\psi_i(R)$ the average electrostatic potential at R. Then $\varrho_i(R)$ and $\psi_i(R)$ are related by the Poisson equation

$$\nabla^2 \psi_i(R) = -\frac{4\pi}{D} \varrho_i(R). \tag{40.1}$$

On the other hand, the average number of j ions per unit volume at distance R from the central ion i is $n_{ij}(R)/n_i$, where the pair distribution function, $n_{ij}(R)$, is a function of R alone due to the spherical symmetry of an ion and the isotropy of the homogeneous solution. If the potential of average force defined by (31.15) is used, we have

$$\frac{n_{ij}(R)}{n_i} = n_j \, \mathrm{e}^{-w_{ij}(R)/kT}, \tag{40.2}$$

where $w_{ij}(R)$ is the component potential of average force defined by (31.18) and w_i and w_j may be taken as zero in a homogeneous bulk solution.

If the component potential between solute ions, $w_{ij}(R)$, is used, the charge density $\varrho_i(R)$ is, with use of (40.2),

$$\varrho_i(R) = \sum_{j=2}^{\varkappa} n_j \, \varepsilon_j \, \mathrm{e}^{-w_{ij}(R)/kT}. \tag{40.3}$$

On the other hand, we shall assume that the solution is so dilute that the contributions of the short range repulsive forces and VAN DER WAALS' forces between ions to the w_{ij} would be very small compared with the electrostatic contribution. Then we may use the approximation

$$w_{ij}(R) = \varepsilon_j \psi_i(R). \tag{40.4}$$

Inserting (40.3) in (40.1) and using (40.4), we obtain the Poisson-Boltzmann equation:

$$\nabla^2 \psi_i(R) = -\frac{4\pi}{D} \sum_{j=2}^{\varkappa} n_j \, \varepsilon_j \, \mathrm{e}^{-\varepsilon_j \psi_i(R)/kT}. \tag{40.5}$$

The further approximation made by DEBYE and HÜCKEL is to assume that $\sum_{j=2}^{\varkappa} \varepsilon_j \psi_j(R)/kT$ is small compared with unity for all important values of R. Then

---

[1] P. DEBYE and E. HÜCKEL: Phys. Z. **24**, 185 (1923). — See also E. DARMOIS in Vol. XX of this Encyclopedia.

we can expand the exponential in (40.5) as a power series and note that the first term vanishes owing to the electroneutrality condition:

$$\sum_{j=2}^{\varkappa} n_j \varepsilon_j = 0. \tag{40.6}$$

Neglecting all terms higher than the second, we obtain the approximation first obtained by DEBYE and HÜCKEL[1]

$$\nabla^2 \psi_i(R) = K^2 \psi_i(R), \tag{40.7}$$

where K is the shielding constant given by

$$K^2 = 4\pi \sum_{j=2}^{\varkappa} n_j \frac{\varepsilon_j^2}{D\,kT}. \tag{40.8}$$

The solution of (40.7), which remains finite at a great distance from the central ion, is given by[2]

$$\psi_i(R) = \frac{\varepsilon_i}{D} \frac{e^{-KR}}{R}, \tag{40.9}$$

With use of the Poisson equation, the charge density $\varrho_i(R)$ is

$$\varrho_i(R) = -\frac{K^2 \varepsilon_i}{4\pi} \frac{e^{-KR}}{R}, \tag{40.10}$$

where the screening constant $1/K$ is often called the mean thickness of the ionic atmosphere.

Since the electrostatic potential due to the central ion is given by ε_i/DR, the average potential due to the remaining ions is

$$\tilde{\psi}_i(R) = \frac{\varepsilon_i}{DR} (e^{-KR} - 1). \tag{40.11}$$

The electrostatic potential energy of the central ion itself is expressed as the product of its charge and the potential (40.11) at $R=0$, $\tilde{\psi}_i(0)$:

$$\varepsilon_i \tilde{\psi}_i(0) = -\frac{\varepsilon_i^2 K}{D}. \tag{40.12}$$

However, if we apply this argument to every ion, we should be counting each ion twice: once as the central ion, and again as part of the surrounding ions. Thus the contribution to the energy of an ion of species i due to the ionic interaction is

$$\mu_i^{\mathrm{el}} = -\frac{\varepsilon_i^2 K}{2D}. \tag{40.13}$$

The chemical potential due to the pure mixing, i.e., the chemical potential of a perfect solution is given in the form, $\mu_i^0 + kT \log n_i$, and hence we have[3]

$$\mu_i = \mu_i^0 + kT \log n_i - \frac{\varepsilon_i^2 K}{2D}, \tag{40.14}$$

where μ_i^0 is independent of concentration.

[1] P. DEBYE and E. HÜCKEL: Phys. Z. **24**, 185 (1923). — See also R.A. ROBINSON and R.H. STOKES: Electrolyte Solutions, pp. 72—78. London: Butterworth's Scientific Publications 1955.

[2] Usually the mean distance of approach, a, is employed in the Debye-Hückel theory. This distance a represents the limit within which no other ion can approach the central ion. Then the solution, (40.9), is modified in the form $\psi_i(R) = \frac{\varepsilon_i}{D} \frac{e^{Ka}}{1+Ka} \frac{e^{-KR}}{R}$, the integration constant being determined from the electrical neutrality. The above equation (40.9) is obtained in the limit of vanishing a (see also [*35*]).

[3] E. SCHMUTZER [*35*] used, in his theory of surface tension, an improved equation applicable to some non-perfect solutions.

In the special case of a solution of a univalent ion pair, (40.8) and (40.14) reduce, respectively, to

$$K^2 = \frac{8\pi n \varepsilon^2}{D\,kT}, \tag{40.15}$$

and

$$\left.\begin{aligned} \mu_2 &= \mu_2^0 + kT\log n - \frac{\varepsilon^2 K}{2D}, \\ \mu_3 &= \mu_3^0 + kT\log n - \frac{\varepsilon^2 K}{2D}, \end{aligned}\right\} \tag{40.16}$$

where n is the number density of the positive or negative ions in the solution.

β) Solution with plane interface. Let us now consider a solution with a plane interface in contact with air. As in Sect. 26 we use the (x, y, z) coordinate system and take the dividing surface as the (x, y) plane. Though the solvent is assumed to be a structureless medium with uniform dielectric constant D filling the space from below up to the dividing surface, the concentrations of ions may vary with z due to the surface potential. For such a case, the average electrostatic potential at point $\boldsymbol{r}$ when an ion of species i is fixed at $\boldsymbol{r}^{(1)}$, depends not only on $\boldsymbol{r}-\boldsymbol{r}^{(1)}$, but also on $\boldsymbol{r}^{(1)}$ itself and hence may be denoted by $\psi_i(\boldsymbol{r}^{(1)}, \boldsymbol{r})$. Then the modification of (40.7) for this case may be expressed in the form [32], [33], [34]

$$\nabla^2 \psi(\boldsymbol{r}^{(1)}, \boldsymbol{r}) = K^2(z)\,\psi_i(\boldsymbol{r}^{(1)}, \boldsymbol{r}), \tag{40.17}$$

$$K^2(z) = \frac{4\pi \sum\limits_{j=2}^{\varkappa} n_j(z)\,\varepsilon_j^2}{D\,kT}; \quad z \leqq 0. \tag{40.18}$$

On the other hand, the vapor pressures of ions are assumed to be completely negligible and hence the electrostatic potential $\psi_i^0(\boldsymbol{r}^{(1)}, \boldsymbol{r})$ outside the ionic solution satisfies the Laplace equation:

$$\nabla^2 \psi_i^0(\boldsymbol{r}^{(1)}, \boldsymbol{r}) = 0; \quad z > 0. \tag{40.19}$$

According to the theory of electrostatics, the solutions of these equations should be subject to

$$\left.\begin{aligned} (\psi_i)_{z=0} &= (\psi_i^0)_{z=0}; & \left(\frac{\partial \psi_i}{\partial x}\right)_{z=0} &= \left(\frac{\partial \psi_i^0}{\partial x}\right)_{z=0}, \\ \left(\frac{\partial \psi_i}{\partial y}\right)_{z=0} &= \left(\frac{\partial \psi_i^0}{\partial y}\right)_{z=0}; & D\left(\frac{\partial \psi_i}{\partial z}\right)_{z=0} &= D^0\left(\frac{\partial \psi_i^0}{\partial z}\right)_{z=0}, \end{aligned}\right\} \tag{40.20}$$

where D^0 is the dielectric constant of the air.

Although the exact solution of these equations has been obtained by SCHMUTZER [35] with use of the modified Hankel functions, we make the following two approximations.

First, we replace the dielectric constant of the air by

$$D^0 = 0, \tag{40.21}$$

which was first employed by WAGNER [32]. It has also been shown by ONSAGER and SAMARAS [34] that the relative error in the potential of average force due to this approximation is less than $2/(D+D^0)$.

Secondly, we replace the shielding constant given by (40.18), by its value in the interior of the homogeneous ionic solution:

$$K(z) = K(-\infty) = K, \tag{40.22}$$

which was first adopted by OKA [33].

With the help of the approximations (40.21) and (40.22), we can solve the above differential equations (40.17) and (40.19) subject to the condition (40.20), obtaining the simple form [34],

$$\psi_i(\boldsymbol{r}^{(1)}, \boldsymbol{r}) = \varepsilon_i \frac{e^{-K|\boldsymbol{r}-\boldsymbol{r}^{(1)}|}}{D|\boldsymbol{r}-\boldsymbol{r}^{(1)}|} + \varepsilon_i \frac{e^{-K|\boldsymbol{r}-\boldsymbol{r}^{(1)}+2z^{(1)}\boldsymbol{e}_z|}}{D|\boldsymbol{r}-\boldsymbol{r}^{(1)}+2z^{(1)}\boldsymbol{e}_z|}, \tag{40.23}$$

where the second term arises from the surface charge. Since $\psi_i(\boldsymbol{r}^{(1)}, \boldsymbol{r})$ is the average electrostatic potential of $\boldsymbol{r}$ when an ion of species i is fixed at $\boldsymbol{r}^{(1)}$, the potential of average force between the ion of species i at $\boldsymbol{r}^{(1)}$ and another ion of species j at $\boldsymbol{r}^{(2)}$, $w_{ij}(\boldsymbol{r}^{(1)}; \boldsymbol{r}^{(2)}-\boldsymbol{r}^{(1)})$, is given by $\varepsilon_j \psi_i(\boldsymbol{r}^{(1)}, \boldsymbol{r}^{(2)})$ as in the case of (40.4), and hence we obtain, from (40.23), the expression [34],

$$w_{ij}(\boldsymbol{r}^{(1)}; \boldsymbol{r}^{(12)}) = \varepsilon_i \varepsilon_j \left\{ \frac{e^{-K r^{(12)}}}{D r^{(12)}} + \frac{e^{-K r_{im}^{(12)}}}{D r_{im}^{(12)}} \right\}, \tag{40.24}$$

where $r^{(12)}$ is the distance between the two ions and $r_{im}^{(12)} = |\boldsymbol{r}^{(2)} - \boldsymbol{r}^{(1)} + 2z^{(1)}\boldsymbol{e}_z|$, which has been given by (32.6), is equal to the distance between one of the ions and the image of the other as illustrated in Fig. 17.

Let us denote the average electrostatic potential at the center of an ion of species i at $\boldsymbol{r}^{(1)}$ due to the remaining ions and the surface charge by $\psi_i(\boldsymbol{r}^{(1)})$. Substracting $\frac{\varepsilon_i}{D|\boldsymbol{r}-\boldsymbol{r}^{(1)}|}$, the potential due to the ion i itself, from (40.23) and putting $\boldsymbol{r}$ equal to $\boldsymbol{r}^{(1)}$, we have

$$\psi_i(\boldsymbol{r}^{(1)}) = \varepsilon_i \frac{e^{2Kz^{(1)}}}{-2Dz^{(1)}} - \frac{\varepsilon_i K}{D}, \tag{40.25}$$

of which the first term is the electrostatic potential of the screened image force. The average electrostatic potential due to the induced surface charge depends on the distance between the ion i and its image point, $r_{im}^{(11)}$. Hence, the electrostatic force acting on this ion is given by [34],

$$\varepsilon_i \left.\frac{\partial \psi_i(\boldsymbol{r}^{(1)})}{\partial r_{im}^{(11)}}\right|_{r_{im}^{(11)}=-2z^{(1)}} = -\frac{1}{2}\varepsilon_i \frac{\partial \psi_i(\boldsymbol{r}^{(1)})}{\partial z^{(1)}}. \tag{40.26}$$

Then from (40.25) and (40.26) the potential of the average force exerted on the ion of the species i itself due to the remaining ions is [34], [35],

$$w_i(\boldsymbol{r}^{(1)}) = \frac{\varepsilon_i^2 e^{2Kz^{(1)}}}{-4Dz^{(1)}}, \tag{40.27}$$

where the integration constant is so chosen that it vanishes in the interior of the solution. This is called the adsorbing potential [32], [34].

With use of (31.20), the number density of molecules of species i at $\boldsymbol{r}$ is given by

$$n_i(\boldsymbol{r}) = n_i^\alpha \exp\left\{\frac{\varepsilon_i^2 e^{2Kz}}{4DzkT}\right\}; \quad z \leq 0. \tag{40.28}$$

This function has been calculated by Oka[1]. The curve for a solution of a univalent ion-pair at temperature 20° C and consentration 0.05 mole/liter, is illustrated in Fig. 18.

[1] S. Oka [33] used, however, as the adsorbing potential twice our potential (40.27) owing to the omisson of the factor $\frac{1}{2}$ on the right-hand side of (40.26).

Since the number densities of solutes in the vapor phase are assumed to be zero, substitution of (40.28) into (26.21) leads to the expression for the superficial number density

$$\Gamma_i = n_i^\alpha \int_{-\infty}^{0} \left[\exp\left(\frac{\varepsilon_i^2 e^{2Kz}}{4Dz\,kT}\right) - 1\right] dz. \tag{40.29}$$

We are now ready to calculate the surface tension difference between a very dilute electrolyte solution and the pure solvent by using (32.17) or (32.18).

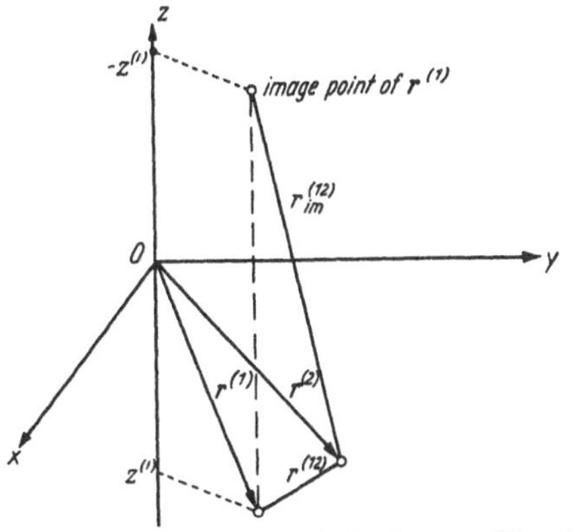

Fig. 17. Diagram showing the relation between $r^{(12)}$ and $r_{im}^{(12)}$ with regard to a point charge at $r^{(1)}$ and its image.

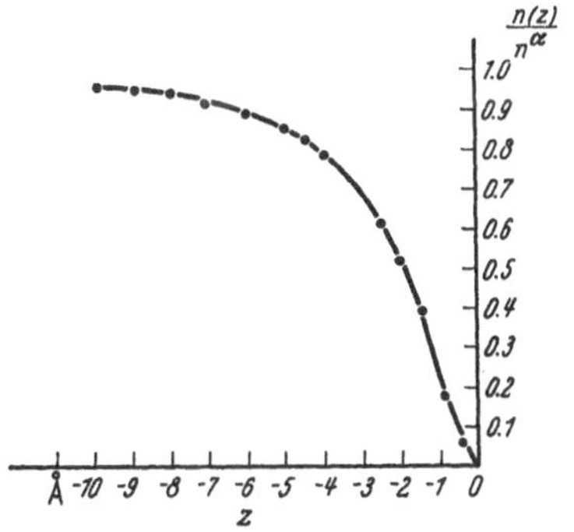

Fig. 18. Relative density distribution in the surface layer for uni-univalent electrolytes.

Since there is no screening effect in the pure solvent, we have $K=0$ in the limit of zero concentration. Then, (40.24) and (40.27) reduce, respectively, to

$$w_{ij}^0(\boldsymbol{r}^{(1)}, \boldsymbol{r}^{(2)}) = \varepsilon_i \varepsilon_j \left(\frac{1}{D r^{(12)}} + \frac{1}{D r_{im}^{(12)}}\right), \tag{40.30}$$

$$w_1^0(\boldsymbol{r}^{(1)}) = \frac{\varepsilon_i^2}{-4Dz^{(1)}}. \tag{40.31}$$

These potentials are due to the pure Coulomb force and the image force.

From (31.15), (31.18) and (31.20), we obtain

$$\left.\begin{aligned} n_{ij}(\boldsymbol{r}^{(1)}, \boldsymbol{r}^{(2)}) &= n_i^\alpha n_j^\alpha \exp\left\{-\frac{1}{kT}\left(w_i(\boldsymbol{r}^{(1)}) + w_j(\boldsymbol{r}^{(2)}) + w_{ij}(\boldsymbol{r}^{(1)}; \boldsymbol{r}^{(2)})\right)\right\} \\ &= n_i(\boldsymbol{r}^{(1)})\, n_j(\boldsymbol{r}^{(1)} + \boldsymbol{r}^{(12)})\, e^{-w_{ij}(\boldsymbol{r}^{(1)}; \boldsymbol{r}^{(12)})/kT}. \end{aligned}\right\} \tag{40.32}$$

Since all terms of the second order in w_{ij}/kT have been neglected in the present treatment, we may use, in place of (40.32), the linearized relation

$$n_{ij}(\boldsymbol{r}^{(1)}; \boldsymbol{r}^{(12)}) = n_i(\boldsymbol{r}^{(1)})\, n_j(\boldsymbol{r}^{(1)} + \boldsymbol{r}^{(12)}) \left(1 - \frac{w_{ij}(\boldsymbol{r}^{(1)}; \boldsymbol{r}^{(12)})}{kT}\right). \tag{40.33}$$

Substituting (40.30) and (40.33) into (32.17), we obtain, with use of the electroneutrality condition (40.6),

$$\gamma - \gamma^0 = \int_{-\infty}^{0} \left(\pi - \tilde{\pi}_T(z^{(1)})\right) dz^{(1)}, \tag{40.34}$$

where

$$\left.\begin{aligned} \tilde{\pi}_T(z^{(1)}) &= kT \sum_{i=2}^{\varkappa} n_i(z^{(1)}) - \frac{1}{2} \sum_{i,j=2}^{\varkappa} \int_{-\infty}^{-z^{(1)}} dz^{(12)} \int_{-\infty}^{\infty} dy^{(12)} \int_{-\infty}^{\infty} dx^{(12)} \times \\ &\times \frac{n_i n_j \varepsilon_i \varepsilon_j}{D\,kT} \left\{\frac{(x^{(12)})^2}{(r^{(12)})^3} + \frac{(x^{(12)})^2}{(r_{im}^{(12)})^3}\right\} w_{ij}(z^{(1)}; \boldsymbol{r}^{(12)}). \end{aligned}\right\} \tag{40.35}$$

The integration with respect to $z^{(12)}$ is, in the above equation, restricted in such a manner that $\boldsymbol{r}^{(2)}$ falls inside the solution. Since $r_{im}^{(12)}$ is the distance between $\boldsymbol{r}^{(2)}$ and the image point of $\boldsymbol{r}^{(1)}$, which always lies outside the solution as seen from Fig. 17, we may rewrite (40.35), with use of (40.24), in the form

$$\left.\begin{aligned}\tilde{\pi}_T(z^{(1)}) &= kT\sum_{i=2}^{\varkappa} n_i(z^{(1)}) - \\ &\quad -\frac{1}{2}\sum_{i,j=2}^{\varkappa}\int d\boldsymbol{r}^{(12)}\left[\left\{\frac{(x^{(12)})^2}{(r^{(12)})^3}+\frac{(x^{(12)})^2}{(r_{im}^{(12)})^3}\right\}\frac{e^{-Kr^{(12)}}}{r^{(12)}}\,\frac{n_i(\boldsymbol{r}^{(1)})\,n_j(\boldsymbol{r}^{(2)})\,\varepsilon_i^2\,\varepsilon_j^2}{D^2\,kT}\right],\end{aligned}\right\} \quad (40.36)$$

where the integration with respect to $\boldsymbol{r}^{(12)}$ extends over the whole space.

At a great distance from the interface, $\tilde{\pi}_T$ reduces to the osmotic pressure

$$\pi = kT\sum_{i=2}^{\varkappa} n_i^\alpha - \frac{1}{2}\sum_{i,j=2}^{\varkappa}\int d\boldsymbol{r}^{(12)}\frac{(x^{(12)})^2}{(r^{(12)})^3}\,\frac{e^{-Kr^{(12)}}}{r^{(12)}}\,\frac{n_i^\alpha\,n_j^\alpha\,\varepsilon_i^2\,\varepsilon_j^2}{D^2\,kT}, \quad (40.37)$$

as seen from (32.14).

In exactly the same way we obtain from (32.18)

$$\gamma-\gamma^0 = -\sum_{i=2}^{\varkappa}\int_{-\infty}^{\infty} dz^{(1)}\frac{dw_i^0(z^{(1)})}{dz^{(1)}}\,n_i(z^{(1)})\,z^{(1)} + \int_{-\infty}^{\infty} dz^{(1)}\left[\tilde{\pi}_N(z^{(1)})-\tilde{\pi}_T(z^{(1)})\right], \quad (40.38)$$

where

$$\left.\begin{aligned}\tilde{\pi}_N(z^{(1)}) &= kT\sum_{i=2}^{\varkappa} n_i(z^{(1)}) \\ &\quad -\frac{1}{2}\sum_{i,j=2}^{\varkappa}\int d\boldsymbol{r}^{(12)}\left[\left\{\frac{(z^{(12)})^2}{(r^{(12)})^3}+\frac{(z^{(1)}+z^{(2)})^2}{(r_{im}^{(12)})^3}\right\}\frac{e^{-Kr^{(12)}}}{r^{(12)}}\,\frac{n_i(\boldsymbol{r}^{(1)})\,n_j(\boldsymbol{r}^{(2)})\,\varepsilon_i^2\,\varepsilon_j^2}{D^2\,kT}\right].\end{aligned}\right\} \quad (40.39)$$

The notation $\tilde{\pi}_T$ and $\tilde{\pi}_N$ has been used because $\tilde{\pi}_T$ and $\tilde{\pi}_N$ may be considered in a sense as measures of the tangential osmotic pressure and of the normal osmotic pressure as will be surmised from the arguments concerning p_T and p_N discussed in Sect. 36 [*23*]. However, the exact mechanical expressions for the tangential osmotic pressure and the normal osmotic pressure, π_T and π_N, have not been obtained yet.

41. Numerical calculation for solution of monovalent ion-pair [*23*], [*34*]. We shall carry out the numerical calculation of the surface tension difference between a solution of a monovelent ion-pair and the pure solvent on the basis of the above mentioned theory of electrolyte solutions. Let $n(\boldsymbol{r}^{(1)})$ be the number density of ions of either species 2 or 3 and ε be the absolute value of the electronic charge. Then, (40.36) and (40.39) reduce, respectively, to the following expressions [*23*]:

$$\tilde{\pi}_T(z^{(1)}) = 2n(z^{(1)})\,kT - \frac{n(z^{(1)})\,\varepsilon^2 K^2}{4\pi D}\int\left\{\frac{(x^{(12)})^2}{(r^{(12)})^3}+\frac{(x^{(12)})^2}{(r_{im}^{(12)})^3}\right\}\frac{e^{-Kr^{(12)}}}{r^{(12)}}\,d\boldsymbol{r}^{(12)}, \quad (41.1)$$

$$\tilde{\pi}_N(z^{(1)}) = 2n(z^{(1)})\,kT - \frac{n(z^{(1)})\,\varepsilon^2 K^2}{4\pi D}\int\left\{\frac{(z^{(12)})^2}{(r^{(12)})^3}+\frac{(z^{(1)}+z^{(2)})^2}{(r_{im}^{(12)})^3}\right\}\frac{e^{-Kr^{(12)}}}{r^{(12)}}\,d\boldsymbol{r}^{(12)}, \quad (41.2)$$

where K, the screening constant defined by (40.18), assumes the form

$$K^2 = \frac{8\pi n(z^{(1)})\,\varepsilon^2}{D\,kT}. \quad (41.3)$$

According to the approximation (40.22), this is regarded as being independent of z and hence equal to (40.15).

The integrals in the above equations can be evaluated with use of dipolar coordinates with the result:

$$\tilde{\pi}_T = 2n\,(z^{(1)}) \left[kT - \frac{\varepsilon^2 K}{D}\left\{\frac{1}{6} + \frac{1}{u^3} - \frac{e^{-u}}{2u} - \frac{e^{-u}}{u^2} - \frac{e^{-u}}{u^3}\right\}\right], \tag{41.4}$$

$$\tilde{\pi}_N = 2n\,(z^{(1)}) \left[kT - \frac{\varepsilon^2 K}{D}\left\{\frac{1}{6} + \frac{1}{2u} - \frac{2}{u^3} + \frac{e^{-u}}{2u} + \frac{2e^{-u}}{u^2} + \frac{2e^{-u}}{u^3}\right\}\right], \tag{41.5}$$

$$u = -2K z^{(1)}. \tag{41.6}$$

On denoting the surface tension increase derived from (40.34) by $\Delta\gamma^{\mathrm{I}}$ and that according to (40.38) by $\Delta\gamma^{\mathrm{II}}$, these quantities can be expressed as follows [23]:

$$\left.\begin{aligned} \Delta\gamma^{\mathrm{I}} = \frac{n^\alpha\varepsilon^2}{D}\Big\{\Big(\frac{1}{6} - \frac{1}{2y}\Big)\int\limits_0^\infty\Big[\exp\Big(-\frac{y\,e^{-u}}{u}\Big) - 1\Big]du + \int\limits_0^\infty \exp\Big(-\frac{y\,e^{-u}}{u}\Big) \times \\ \times \Big[\frac{1}{u^3} - \frac{1}{u^2 y} - \frac{e^{-u}}{2u}\Big]du\Big\}, \end{aligned}\right\} \tag{41.7}$$

$$\Delta\gamma^{\mathrm{II}} = \frac{n^\alpha\varepsilon^2}{D}\int\limits_0^\infty \exp\Big(-\frac{y\,e^{-u}}{u}\Big)\Big[\frac{3}{u^3} - \frac{3}{u^2 y} - \frac{e^{-u}}{u}\Big]du, \tag{41.8}$$

$$y = \frac{\varepsilon^2 K}{2DkT}, \tag{41.9}$$

where $w^0(\boldsymbol{r}^{(1)})$ given by (40.31) has been used as the adsorbing potential in (40.38) and n^α is the number density in the bulk solution. After analytical evaluation of prototype integrals, BUFF and STILLINGER [23] derived the following expressions

$$\Delta\gamma^{\mathrm{I}} = \frac{n^\alpha\varepsilon^2}{D}F^{\mathrm{I}}(y), \tag{41.10}$$

$$F^{\mathrm{I}}(y) = L(y) + \sum_{n=1}^{\infty}\frac{y^n}{n!\,(n+2)!}\,(A_n^{\mathrm{I}}\log y + B_n^{\mathrm{I}}), \tag{41.11}$$

and

$$\Delta\gamma^{\mathrm{II}} = \frac{n^\alpha\varepsilon^2}{D}F^{\mathrm{II}}(y), \tag{41.12}$$

$$F^{\mathrm{II}}(y) = L(y) + \sum_{n=1}^{\infty}\frac{y^n}{n!\,(n+2)!}\,(A_n^{\mathrm{II}}\log y + B_n^{\mathrm{II}}), \tag{41.13}$$

where

$$L(y) = -\tfrac{1}{2}\log y - \gamma + \tfrac{3}{4}, \tag{41.14}$$

$$A_n^{\mathrm{I}} = (n+1)^n\,\frac{(n^2-2)}{2} - \frac{n^n}{6}\,(5n^2 - 3n - 2), \tag{41.15}$$

$$\left.\begin{aligned} B_n^{\mathrm{I}} = \frac{(n+1)^n}{2}\{(n^2-2)\,(\log(n+1)+1) - n(n+1)\,f(n+1) - (n-1)\times \\ \times (n+2)\,f(n+2) + 2(n+1)\,f(n+3)\} - \frac{n^n}{6}\{(5n^2-3n-2)\log n + \\ + (5n^2+3n-2) + 2(n+1)(n+2)\,f(n+1) - 6n^2(f(n) + f(n+3))\}, \end{aligned}\right\} \tag{41.16}$$

$$A_n^{\mathrm{II}} = (n+1)^{n+1}(n-1) - 3n^{n+2}, \tag{41.17}$$

$$\left.\begin{aligned} B_n^{\mathrm{II}} = (n+1)^{n+1}\{(n-1)\,(\log(n+1) + 1 - f(n+1)) - (n+2)\,f(n+2) + \\ + 3f(n+3)\} - 3n^{n+2}\Big(\log n + \frac{n+1}{n} - f(n) - f(n+3)\Big), \end{aligned}\right\} \tag{41.18}$$

$$f(n) = -\gamma + \sum_{m=1}^{n-1}\frac{1}{m}, \tag{41.19}$$

$$\gamma = \text{EULER's constant} = 0.5772157\ldots. \tag{41.20}$$

The values of $F^{\mathrm{I}}(y)$ and $F^{\mathrm{II}}(y)$ are given in Table 6.

Another method of obtaining the surface tension difference of a solution is to integrate the Gibbs adsorption formula (6.8). OKA [33] used an equation equivalent to (6.18), and ONSAGER and SAMARAS [34] and SCHMUTZER [35] (33.25). In the case of a solution of a univalent ion pair, (33.25) is written as

$$-d\gamma = \Gamma_2 d\mu_2 + \Gamma_3 d\mu_3. \tag{41.21}$$

From (40.29) we have

$$\Gamma = \Gamma_2 = \Gamma_3 = \frac{n^\alpha}{2K}\int\limits_0^\infty \left[\exp\left(-\frac{y\,e^{-u}}{u}\right) - 1\right] du. \tag{41.22}$$

Using the integral in the complex domain and LAGRANGE's theorem[1], ONSAGER and SAMARAS [34] obtained an expression for Γ given by (41.22) as follows[2]:

$$\Gamma = \frac{n^\alpha}{2K}\sum_{s=1}^{\infty}\frac{s^s y^s}{s!\,s!}(\log y + g_s), \tag{41.23}$$

$$g_s = \log s + 1 + 2\gamma - 2\sum_{n=1}^{s}\frac{1}{n}, \tag{41.24}$$

where y is given by (41.9) and γ EULER's constant given by (41.20).

From (40.15) and (40.16) we have for constant temperature

$$d\mu_2 = d\mu_3 = 2kT\frac{dK}{K} - \frac{\varepsilon^2}{2D}dK. \tag{41.25}$$

Substitution of (41.23) and (41.25) into (41.21) leads to

$$d\gamma = -\frac{K\,kT}{16\pi y}\sum_{s=1}^{\infty}\frac{s^s}{s!\,s!}(2-y)\,y^s(\log y + g_s)\,dK. \tag{41.26}$$

Integrating with respect to K we obtain

$$\Delta\gamma^{\mathrm{III}} = \frac{n^\alpha\varepsilon^2}{D}F^{\mathrm{III}}(y), \tag{41.27}$$

$$\left.\begin{aligned} F^{\mathrm{III}}(y) &= -\frac{1}{2y^2}\sum_{s=1}^{\infty}\frac{y^s}{s!\,s!}\times \\ &\times\left\{\frac{2y^{s+1}}{s+1}\left(\log y - \frac{1}{s+1} + g_s\right) - \frac{y^{s+2}}{s+2}\left(\log y - \frac{1}{s+2} + g_s\right)\right\}. \end{aligned}\right\} \tag{41.28}$$

The values of $F^{\mathrm{III}}(y)$ are also given in Table 6. The experimental values are determined by BUFF and STILLINGER so that the curve passes through SCHWENKER's value[3].

The same limiting law is obtained for cases I, II and III, which for a univalent electrolyte pair takes the identical form [34]

$$\Delta\gamma = -\frac{n^\alpha\varepsilon^2}{2D}\left(\log y - \frac{1}{2} + g_1\right) = -\frac{n^\alpha\varepsilon^2}{2D}\left(\log\frac{\varepsilon^2 K}{2DkT} - 0.34557\right), \tag{41.29}$$

[1] E. T. WHITTAKER and G. N. WATSON: A Course of Modern Analysis, 4th ed. § 7.32. London: Cambridge University Press 1935.

[2] ONSAGER and SAMARAS [34] have carried out the integration by expanding the exponential function in a power series. SCHMUTZER [35] evaluated the integral by taking account of the distance of closest approach of two ions, a, to avoid the expansion of the exponential function for an interval that makes the expansion meaningless.

[3] G. SCHWENKER: Ann. d. Phys. 11, 525 (1931).

Table 6. *Comparison of theoretical values of $F^{\mathrm{I}}(y)$, $F^{\mathrm{II}}(y)$, $F^{\mathrm{III}}(y)$ with experimental values for uni-univalent electrolytes in water.* (BUFF and STILLINGER [23].)

y^2	$F^{\mathrm{III}}(y)$	$F^{\mathrm{I}}(y)$	$F^{\mathrm{II}}(y)$	$F_{\mathrm{expt}}(y)$	c (mole/liter) at 18° C [1]
0.0063	1.486	1.492	1.549		0.0046
0.0126	1.325	1.333	1.405		0.0091
0.0189	1.233	1.242	1.323		0.0137
0.0315	1.117	1.130	1.223	1.25	0.0228
0.0378	1.077	1.090	1.189	1.22	0.0273
0.0630	0.966	0.981	1.095	1.18	0.0456
0.0944	0.880	0.899	1.023	1.14	0.0683
0.1259	0.819	0.839	0.974	1.12	0.0911
0.1574	0.772	0.794	0.937	1.06	0.1138
0.1889	0.735	0.758	0.908	1.03	0.1366
0.2014	0.722	0.745	0.897	1.02	0.1457
0.2266	0.697	0.722	0.879	1.00	0.1639

or [34], [23],

$$\Delta\gamma = \frac{80.00}{D}\, c \,\mathrm{Log}\, \frac{1.130\times 10^{-13}(D\,T)^3}{c}\,; \qquad \begin{pmatrix} c\text{: mole/liter} \\ \gamma\text{: dyne/cm} \end{pmatrix} \tag{41.30}$$

where c is the concentration of the solute in moles per liter[2].

In more concentrated solutions, the basic model begins to break down. Furthermore, the introduction of approximate distribution functions in rigorous formulae for the thermodynamic variables leads to an internal contradiction. The comparison of these values provides a criterion for the validity of the approximations. For example, the difference between $\Delta\gamma^{\mathrm{I}}$ and $\Delta\gamma^{\mathrm{II}}$ arises from the failure of the hydrostatic equilibrium condition. As shown by Table 6, $\Delta\gamma^{\mathrm{II}}$ is in better agreement with the experimental values. A similar self-contradiction has been pointed out in the case of the step approximation used in calculating the surface tension of liquid argon. In the present notation, γ^{II} corresponds to $\int (p_N - p_T)\,dz$ and γ^{I} to $\int (p - p_T)\,dz$. Then, as discussed in Sect. 38, γ^{II} is in a fairly good agreement with the experimental value but γ^{I} even takes a negative sign.

Although the Gibbs adsorption formula itself is valid, ONSAGER-SAMARAS' theory [34] based on it, also leads to some discrepancies as shown in Table 6. Some of these are attributed to the neglect of higher terms in expanding the exponential functions and of the distance of closest approach of two ions. Taking account of the distance of the closest approach, SCHMUTZER [35] calculated the surface tension difference. His results for solutions of sodium chloride are in good agreement with experiment up to relatively high concentration.

c) Gas adsorption on a solid surface.

42. Adsorption of monatomic gas molecules on solid surface. *a) General theory.* The number of molecules of species i in a system with an adsorbent is given by (34.7). In the case of a one-component system this equation reduces to

$$N = \zeta \frac{\partial \log \Xi}{\partial \zeta} = \sum_m V^\alpha m\, b_m \zeta^m, \tag{42.1}$$

[1] The dielectric constant of water has been calculated from Akerlof-Oshry equation [J. Amer. Chem. Soc. **72**, 2844 (1950)]: $D = 5321\,\frac{1}{T} + 233.76 - 0.9297\,T + 0.001417\,T^2 - 0.068292\,T^3$; $T = 273.1 + t$.

[2] The recent values of physical constants are taken from J. W. M. DUMOND and E. R. COHEN: Rev. Mod. Phys. **25**, 691 (1953).

where $\log \Xi$ is given by (34.11). If we retain terms only up to the second order in N/V^α, we obtain from (42.1)

$$\zeta = \frac{N}{V^\alpha b_1} - 2\frac{V^\alpha b_2 N^2}{(V^\alpha b_1)^3}. \tag{42.2}$$

Substituting (42.2) into (34.18) we have

$$p = \frac{N kT}{V^\alpha b_1}\left\{1 - \frac{N c_2}{(V^\alpha b_1)^2}\right\}, \tag{42.3}$$

$$c_2 = V^\alpha(2b_2 - b_1 b_2^\alpha), \tag{42.4}$$

where b_2^α is given by (34.16).

The dividing surface is defined as a mathematical plane passing through the centers of the molecules that make up the solid surface. Let the area of the surface be A and the distance of closest approach of the gas molecules to the solid molecules be D. For the present treatment we neglect the periodic structure of the crystal. Then a portion of V^α equal to AD is unavailable to the center of gas molecules.

It is convenient to express experimental results in terms of the apparent volume V^* of the system as defined by the perfect gas law, rather than the number of molecules N:

$$V^* = N\frac{kT}{p}. \tag{42.5}$$

Furthermore, we define the excess volume

$$V_{\text{ex}} = V^* - V^\alpha + AD. \tag{42.6}$$

Since $V^\alpha - AD$ is the volume available to the center of gas molecules, V_{ex} is the excess apparent volume due to the presence of the solid adsorbent and therefore expresses the quantity of adsorbed gas.

Substitution of N from (42.5) into (42.3) leads to

$$p = \frac{kT(V^\alpha b_1)^2}{V^* c_2} - \frac{kT(V^\alpha b_1)^3}{V^{*2} c_2}. \tag{42.7}$$

Differentiation of (42.7) gives the expression for the slope as follows:

$$\frac{dV^*}{dp} = \frac{V^{*3} c_2}{2kT(V^\alpha b_1)^3 - kT\,V^*(V^\alpha b_1)^2}. \tag{42.8}$$

Near the zero pressure limit, it is seen from (34.13) that

$$\lim_{p\to 0} b_1 = \lim_{V^\alpha\to\infty} b_1 = 1. \tag{42.9}$$

From (42.3) and (42.5) we have

$$\lim_{V^\alpha\to\infty} \frac{V^*}{V^\alpha b_1} = 1. \tag{42.10}$$

Thus near the zero pressure limit (42.8) reduces to

$$\left.\frac{dV^*}{dp}\right|_{p=0} = \frac{c_2}{kT}. \tag{42.11}$$

Using (34.13) and (34.16), we can write (34.14) in the form

$$\left.\begin{aligned} b_2 = b_2^\alpha + 2(b_1 - 1)\,b_2^\alpha + \frac{1}{2V^\alpha}\iint (e^{-\phi(r^{(12)})/kT} - 1) \times \\ \times (e^{-\varphi(\boldsymbol{r}^{(1)})/kT} - 1)(e^{-\varphi(\boldsymbol{r}^{(2)})/kT} - 1)\,d\boldsymbol{r}^{(1)}\,d\boldsymbol{r}^{(2)}. \end{aligned}\right\} \tag{42.12}$$

Then c_2 given by (42.4) can be expressed as

$$\left.\begin{aligned} c_2 = V^\alpha(3\, b_1 b_2^\alpha - 2 b_2^\alpha) + \iint (\mathrm{e}^{-\phi(r^{(12)})/kT} - 1)\, (\mathrm{e}^{-\psi(\boldsymbol{r}^{(1)})/kT} - 1) \times \\ \times (\mathrm{e}^{-\psi(\boldsymbol{r}^{(2)})/kT} - 1)\, d\boldsymbol{r}^{(1)}\, d\boldsymbol{r}^{(2)}. \end{aligned}\right\} \tag{42.13}$$

If the temperature is sufficiently high as compared with the critical temperature of the adsorbate and the gas pressure is low, the contribution of the interaction potential $\phi(r)$ between the gas molecules can be ignored. Then we obtain from (34.13) and (34.14)

$$b_1 = 1 + \frac{1}{V^\alpha} \int\limits_{V^\alpha} (\mathrm{e}^{-\psi(\boldsymbol{r}^{(1)})/kT} - 1)\, d\boldsymbol{r}^{(1)}, \tag{42.14}$$

$$b_2 = b_3 = \cdots = 0. \tag{42.15}$$

Thus, with use of (34.7) the number of molecules N is expressed as

$$N = \zeta \left\{ V^\alpha + \int\limits_{V^\alpha} (\mathrm{e}^{-\psi(\boldsymbol{r}^{(1)})/kT} - 1)\, d\boldsymbol{r}^{(1)} \right\}. \tag{42.16}$$

Since ζ is equal to p/kT for a perfect gas as seen from (34.18), we can rewrite (42.16) in terms of the apparent volume V^* defined by (42.5) as follows:

$$V^* = V^\alpha + \int\limits_{V^\alpha} (\mathrm{e}^{-\psi(\boldsymbol{r}^{(1)})/kT} - 1)\, d\boldsymbol{r}^{(1)}. \tag{42.17}$$

Then the excess volume (42.6) is given by

$$V_{\mathrm{ex}} - AD = \int\limits_{V^\alpha} (\mathrm{e}^{-\psi(\boldsymbol{r}^{(1)})/kT} - 1)\, d\boldsymbol{r}^{(1)}. \tag{42.18}$$

β) Crude model [*36*], [*37*]. At first we ignore the interaction between gas molecules [*36*]. If the attractive potential $\psi(r)$ between a gas molecule and the solid molecules is due to London forces, we have

$$\psi(r) = \left\{ \begin{matrix} \infty & r < D, \\ -\sum\limits_s \dfrac{C}{R_s^6} & r \geqq D, \end{matrix} \right\} \tag{42.19}$$

where R_s is the distance between the gas molecule and the s-th molecule of the crystalline solid adsorbent, and the summation is taken over all the solid molecules. If the crystal may be approximately replaced by a semi-infinite structureless solid with a plane surface[1], the summation can be replaced by integration and we have[2]

$$\psi(r) = \left\{ \begin{matrix} \infty & r < D, \\ -\psi_0 \dfrac{D^3}{r^3} & r \geqq D, \end{matrix} \right\} \tag{42.20}$$

$$\psi_0 = \frac{\pi C n^S}{6 D^3}, \tag{42.21}$$

where n^S is the number of molecules per unit volume of the solid.

If we insert (42.20) into (42.18), we obtain

$$V_{\mathrm{ex}} = A \int\limits_D^\infty \left\{ \exp\left(\frac{\psi_0 D^3}{kT r^3}\right) - 1 \right\} dr. \tag{42.22}$$

[1] Capillary and crystal lattice surfaces have been treated by W. A. Steele and G. D. Halsey: J. Phys. Chem. **59**, 57 (1955).

[2] T. L. Hill: J. Chem. Phys. **16**, 181 (1948).

After expanding the exponential in the power series, we carry out the integration, obtaining [36]

$$V_{ex} = AD \sum_{s=1}^{\infty} \frac{1}{s!(3s-1)} \left(\frac{\psi_0}{kT}\right)^s. \tag{42.23}$$

The values of $\log(V_{ex}/AD)$ calculated from (42.23) as a function of ψ_0/kT are plotted in Fig. 19 shown as a continuous curve.

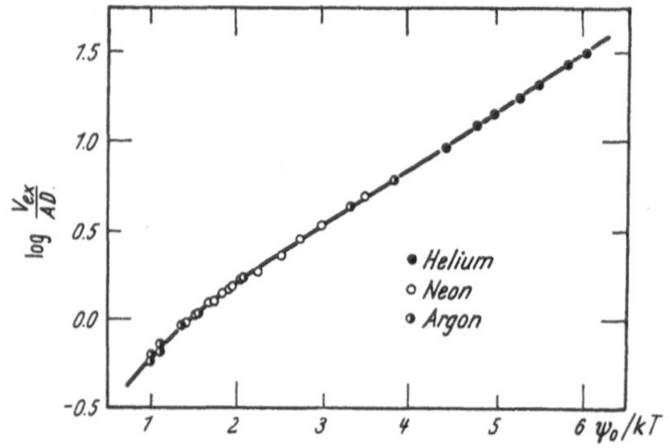

Fig. 19. Fitted values of excess volume for carbon black plotted against reciprocal temperature (in reduced units). Curve is calculated from (42.23). (Steele and Halsey [36].)

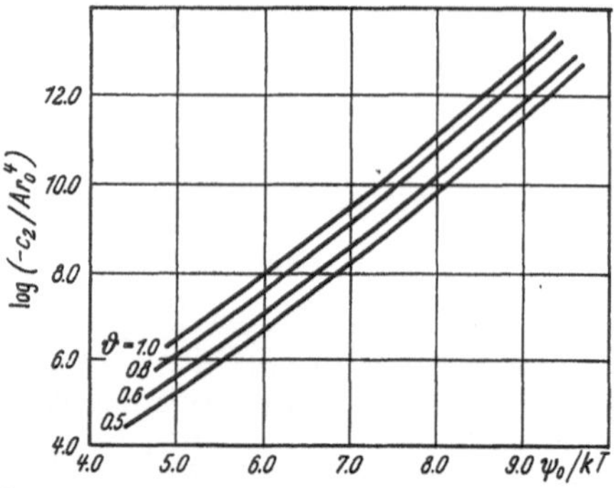

Fig. 20. Calculated values of $\log(-c_2/A r_0^4)$ for various values of $\vartheta = D/r_0$. (Freeman and Halsey [37].)

According to the Kirkwood-Müller equation[1], the London constant C in (42.21) is given by

$$C = 6mc^2 \alpha_G \alpha_S \left(\frac{\alpha_G}{\chi_G} + \frac{\alpha_S}{\chi_S}\right)^{-1}, \tag{42.24}$$

where mc^2 is the mass of electron times the square of the velocity of light, α and χ are polarizabilities and diamagnetic susceptibilities and the subscripts G and S refer to the gaseous and solid states, respectively.

Substitution of (42.24) into (42.21) leads to

$$\psi_0 = \frac{n^S mc^2 \pi \alpha_G \alpha_s}{D^3} \left(\frac{\alpha_G}{\chi_G} + \frac{\alpha_S}{\chi_S}\right)^{-1}. \tag{42.25}$$

Since the number density of molecules near the adsorbent surface may not be so low even when the gas itself in the absence of such a surface remains effectively ideal, we take the next higher order interaction into consideration [37].

If the bulk gas itself is perfect, b_2^α is negligible, and consequently we may omit the first term of (42.13):

$$c_2 = \iint \left(e^{-\phi(r^{(12)})/kT} - 1\right)\left(e^{-\psi(\boldsymbol{r}^{(1)})/kT} - 1\right)\left(e^{-\psi(\boldsymbol{r}^{(2)})/kT} - 1\right) d\boldsymbol{r}^{(1)} d\boldsymbol{r}^{(2)}. \tag{42.26}$$

The gas molecules are assumed to behave like rigid spheres of diameter r_0 with no attraction:

$$e^{-\phi(r)/kT} - 1 = \begin{Bmatrix} -1 & r < r_0 \\ 0 & r \geqq r_0 \end{Bmatrix}. \tag{42.27}$$

[1] A. Müller: Proc. Roy. Soc. Lond., Ser. A **154**, 624 (1936).

Inserting (42.20) and (42.27) into (42.26), and carrying out the analytical integrations, we obtain

$$\left.\begin{aligned}\frac{c_2}{A\,r_0^4} &= -\pi\int d\sigma^{(1)}\exp\left\{\frac{\psi_0}{kT}\left(\frac{\vartheta}{\sigma^{(1)}}\right)^3\right\}\times\\ &\times\int\exp\left\{\frac{\psi_0}{kT}\left(\frac{\vartheta}{\sigma^{(2)}}\right)^3\right\}\{1-(\sigma^{(1)}-\sigma^{(2)})^2\}\,d\sigma^{(2)},\end{aligned}\right\} \tag{42.28}$$

$$\vartheta=\frac{D}{r_0},\qquad \sigma^{(1)}=\frac{r^{(1)}}{r_0},\qquad \sigma^{(2)}=\frac{r^{(2)}}{r_0}, \tag{42.29}$$

where $\sigma^{(1)}$ goes from ϑ to $V^{\alpha}/A\,r_0$ and $\sigma^{(2)}$ goes from ϑ when ϑ is greater than $\sigma^{(1)}-1$ and otherwise from $\sigma^{(1)}-1$ to $\sigma^{(1)}+1$.

The integral (42.28) was evaluated graphically by FREEMAN and HALSEY [*37*]. The calculated values of $\log(-c_2/A\,r_0^4)$ are plotted as a function of ψ_0/kT for various values of the parameter ϑ in Fig. 20.

γ) Comparison with experimental results. Adsorption of rare gases on solids was carried out experimentally by STEELE and HALSEY [*36*]. They measured the apparent volumes of vessels containing high surface powders with rare gases at various high temperatures[1] and fitted the data to (42.23) by trial and error.

For these cases, V^{α} is the total volume of the vessel minus those of the enclosed powders, while the true dead space volume is $V^{\alpha}-AD$.

Table 7. *Experimental values of parameter ψ_0 determined from* (42.23); *distances of closest approaches D calculated from* (42.25) *and surface areas A per gram calculated from D and experimental values of AD.* (STEELE and HALSEY [*36*].)

Adsorbent	Gas	$N_0\,\psi_0$ kcal	D (Å)	A (m²/g)
Carbon black	Helium	0.60	2.47	462
	Neon	1.36	2.70	367
	Argon	4.34	2.75	262
Saran charcoal	Neon	1.33	2.72	774
	Argon	4.14	2.79	786
Porous glass	Neon	1.56	1.93	101
	Argon	3.79	2.21	71
Alumina	Argon	2.80	1.87	131
	Krypton	3.46	1.99	141

N_0 is the Avogadro number.

In the high temperature limit, V_{ex} vanishes according to (42.22) and then (42.6) reduces to

$$V^* = V^{\alpha} - AD, \tag{42.30}$$

which is the measured apparent volume. The value of the helium dead space at a sufficiently high temperature was used as the first trial value of $V^{\alpha}-AD$. The theoretical value of $\log(V_{\text{ex}}/AD)$ was obtained from (42.23) as a function of ψ_0/kT and is shown as a continuous curve in Fig. 19. The experimental data were then made to fit the theoretical curve by adjusting the scale. The experimental values of AD and ψ_0 were determined from the values of V_{ex} when $V_{\text{ex}}/AD=1$ and from those of $1/T$ when $\psi_0/kT=1$, respectively, as shown in Fig. 19. The values of D were calculated from ψ_0 and the observed values of polarizabilities and susceptibilities with use of (42.25). The values of A were calculated from these values of D and the above experimental values of AD. The experimental values of ψ_0 together with the calculated values of D and A are shown in Table 7.

[1] The temperature range is 78 to 303° K for helium, 153 to 492° K for neon, 302 to 492° K for argon and 332 to 491° K for krypton, which are all sufficiently high compared with their respective critical temperatures 5.3, 44, 151 and 230° K.

The large value of ψ_0 indicates that the rare gas atoms interact strongly with solid surfaces. The areas A for the various gases are rather close especially in the case of carbon black. The values of the area also agree with the B.E.T. areas[1]. Thus we can estimate the area of the adsorbent in the form of powders by this method as well.

The comparison of the slope dV^*/dp calculated from (42.11) with experiment was performed by FREEMAN and HALSEY [37] for rare gases on high surface area powders.

The values of $\log(-c_2/A r_0^4)$ calculated for various values of $\vartheta = D/r_0$ by means of (42.28) are plotted in Fig. 20. In Fig. 21, the observed values of argon on carbon black are compared with the calculated line by using the values given in Table 8, of which the values of r_0 are estimated from viscosity data and those of A and D are determined from (42.23) as shown in Table 7. The numerical comparison of the least-square slopes dV^*/dp for the experimental values and the values calculated from (42.11) and calculated values of c_2 are shown in Table 9. If we consider the crudeness of the model and the absence of any adjustable parameter, the close agreement between theory and experiment found here is satisfactory.

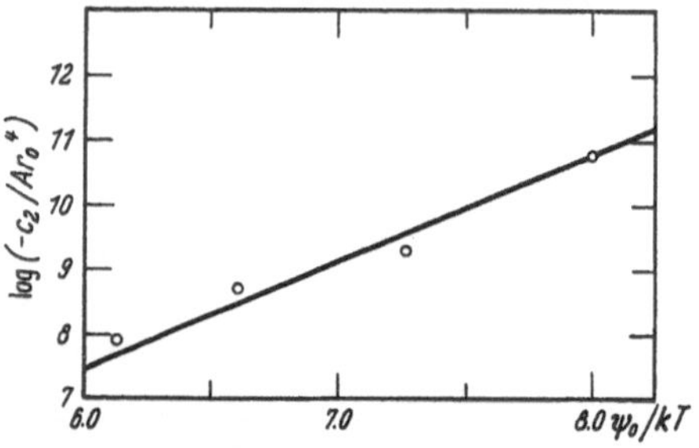

Fig. 21. Experimental data for argon on carbon black compared with theoretical line. (FREEMAN and HALSEY [37].)

Table 8. *Values of parameters used for calculating c_2 given by* (42.28). (FREEMAN and HALSEY [37].)

Adsorbent	Gas	$N_0\psi_0$ (kcal)	A (m^2/g)	r_0 (Å)	D (Å)
Carbon black	Argon	4.34	262	3.42	2.75
Alumina	Krypton	3.46	141	3.61	1.99
Saran S-85	Argon	3.66	1030	3.41	2.90
	Neon	1.28	1135	2.80	2.55

Table 9. *A comparison of theoretical and experimental slopes dV^*/dp.* (FREEMAN and HALSEY [37].)

Adsorbent	Gas	t (°C)	Slope (cm^3/g) mm Hg × 10^4 Calc.	Obs.
Carbon black	Argon	0.00	62.1	60.2
		27.71	13.1	16.4
		57.85	5.29	6.82
		84.3	2.25	2.70
Alumina	Krypton	0.00	1.51	2.17
		30.2	0.528	0.528
Saran S-85	Argon	0.00	34.0	32.1
		5.09	26.4	24.2
		31.0	10.1	13.6
	Neon	−195.7	806.0	538.0
		−194.7	679.0	645.0
		−182.1	87.3	83.4

[1] The area determined with use of the B.E.T. adsorption isotherm [S. BRUNAUER, P.H. EMMETT and E. TELLER: J. Amer. Chem. Soc. **60**, 309 (1938)]. Cf. S. BRUNAUER: Adsorption of Gases and Vapors. Princeton: Princeton University Press 1943. This method is most commonly used in determining the area of a powder. However, we shall restrict ourselves to an application of the general theory based upon statistical mechanics and shall not go into details of the theory of physical adsorption.

IV. Quantum statistical mechanics.

43. Quantum effects in surface tension [*42*]. Up to now we have dealt only with systems obeying classical mechanics. This would suffice for the theoretical treatment of most liquids. It is, however, found that quantum effects are quite prominent in the surface tension of lighter substances such as hydrogen and helium[1].

We shall now give an exact but rather formal quantum treatment of surface tension. Let us consider a system of N identical molecules confined to a cubic vessel with the edge length l (see Fig. 22). We assume that there exist two phases α and β meeting at a plane interface under a weak gravitational field which enables us to treat the area of the interface as an external parameter. We shall denote the energy eigenvalues of the system as a whole by $E_1, E_2, \ldots, E_\varrho, \ldots$. Then the quantum-mechanical partition function is given by

$$Z(T, V, A) = \sum_{\varrho} e^{-E_\varrho/kT}, \qquad (43.1)$$

where the sum extends over all energy states. If we take the summation over the energy levels, instead of the energy states, each term should be multiplied by the degeneracy of each level.

Since E_ϱ depends on V and A, we find from (43.1) that the partition function is a function of T, V and A.

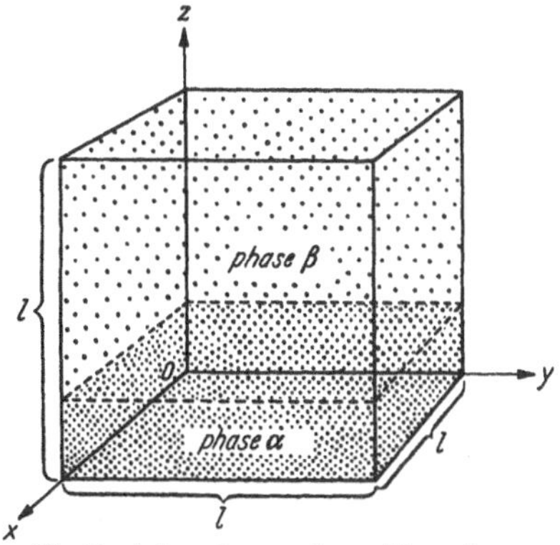

Fig. 22. A two-phase system with a plane interface in a cubic vessel.

As in the case of classical statistical mechanics, the probability of finding a system of the canonical ensemble in the energy state E_ϱ is $e^{-E_\varrho/kT}/\sum_{\varrho} e^{-E_\varrho/kT}$. Then the probability density in configuration space, $P(\boldsymbol{r}^{(1)}, \ldots, \boldsymbol{r}^{(N)})$, is given by

$$P(\boldsymbol{r}^{(1)}, \ldots, \boldsymbol{r}^{(N)}) = Z^{-1} \sum_{\varrho} e^{-E_\varrho/kT}\, \Psi_\varrho^*(\boldsymbol{r}^{(1)}, \ldots, \boldsymbol{r}^{(N)})\, \Psi_\varrho(\boldsymbol{r}^{(1)}, \ldots, \boldsymbol{r}^{(N)}). \qquad (43.2)$$

Here Ψ_ϱ's are a complete orthonormal set of symmetrized or antisymmetrized eigenfunctions.

The probability density corresponds to the diagonal part, with respect to coordinates, of the density matrix defined by[2]

$$\left.\begin{aligned} &\varrho(\boldsymbol{r}^{(1)}, \ldots, \boldsymbol{r}^{(N)}; \boldsymbol{r}'^{(1)}, \ldots, \boldsymbol{r}'^{(N)}) \\ &\quad = \sum_{\varrho} e^{-E_\varrho/kT}\, \Psi_\varrho^*(\boldsymbol{r}'^{(1)}, \ldots, \boldsymbol{r}'^{(N)})\, \Psi_\varrho(\boldsymbol{r}^{(1)}, \ldots, \boldsymbol{r}^{(N)}). \end{aligned}\right\} \qquad (43.3)$$

[1] The observed values of the surface tension of these substances deviate from the classical law of corresponding states as shown by DE BOER and BIRD [*16*]. Especially in helium below around the λ-point as shown in Fig. 6 the situation is much different from that predicated by the law of corresponding states.

To account for this peculiar temperature dependence of surface tension of helium II, ATKINS [*40*] assumed, following FRENKEL'S [*49*] classical treatment of ordinary surface tension, that this effect arises from the thermal energy associated with vibrational modes of the surface and applied to them a quantum mechanical procedure similar to that adopted in the Debye theory of solids. Before ATKIN'S theory, a pioneering work based on DEBYE'S theory was done by BORN and COURANT [*38*] to obtain a quantum mechanical expression for temperature dependence of surface tension. LOVEJOY [*43*] applied ATKIN'S theory [*40*] to the surface tension of He^3. Tentative theories have been proposed also by KOTHARI [*39*], TRIKHA *et al.* [*45*]. Quantum effects in surface energy have been treated by HARASIMA and SHIMURA [*29b*], [*44*] using PRIGOGINE and SARAGA'S cell method [*54*]. Their results give rather good agreement with experimental data but we shall leave to the references the further details of these crude models.

[2] For example, see D. TER HAAR: Elements of Statistical Mechanics, pp. 147—155. New York: Rinehart 1954.

The singlet and pair distribution functions are given respectively by

$$n(\boldsymbol{r}) = N \int\cdots\int \delta(\boldsymbol{r}-\boldsymbol{r}^{(1)})\, P(\boldsymbol{r}^{(1)},\ldots,\boldsymbol{r}^{(N)})\, d\boldsymbol{r}^{(1)}\ldots d\boldsymbol{r}^{(N)}, \tag{43.4}$$

$$n_2(\boldsymbol{r},\boldsymbol{r}') = N(N-1) \int\cdots\int \delta(\boldsymbol{r}-\boldsymbol{r}^{(1)})\,\delta(\boldsymbol{r}'-\boldsymbol{r}^{(2)})\, P(\boldsymbol{r}^{(1)},\ldots,\boldsymbol{r}^{(N)})\, d\boldsymbol{r}^{(1)}\ldots d\boldsymbol{r}^{(N)}, \tag{43.5}$$

corresponding to (25.1) and (25.2).

As in a classical system, the Helmholtz free energy is expressed by (24.1) in terms of the partition function Z and therefore the surface tension is given by (24.10). Thus inserting (43.1) in (24.10), we have

$$\gamma = Z^{-1} \sum_{\varrho} \left(\frac{\partial E_\varrho}{\partial A}\right)_V e^{-E_\varrho/kT}. \tag{43.6}$$

We calculate $(\partial E_\varrho(V,A)/\partial A)_V$ in the following manner. In general the Hamiltonian operator of the system, $\mathscr{H}$, is given by

$$\mathscr{H} = -\frac{\hbar^2}{2m}\sum_{s=1}^{N} \nabla^{(s)2} + \sum_{s<t}^{N} \phi(r^{(st)}), \tag{43.7}$$

where $\phi(r^{(st)})$ is the intermolecular potential.

The time-independent Schrödinger equation is

$$\mathscr{H}\Psi(\boldsymbol{r}^{(1)},\ldots,\boldsymbol{r}^{(N)}) = E\,\Psi(\boldsymbol{r}^{(1)},\ldots,\boldsymbol{r}^{(N)}), \tag{43.8}$$

where E is an energy eigenvalue of the system. The wave function Ψ must fall to zero at the walls of vessel, $x=0$, $y=0$, $z=0$, $x=l$, $y=l$ and $z=l$.

Let us shift outwards one of the vertical walls, perpendicular to the x axis, by a length εl and at the same time push inwards one of the horizontal walls, perpendicular to z axis, by the length εl. Then E and Ψ make some changes depending on ε, and therefore we may express them as $E(\varepsilon)$ and $\Psi(\boldsymbol{r}^{(1)},\ldots,\boldsymbol{r}^{(N)};\varepsilon)$. The boundary condition $\Psi(\varepsilon)=0$ is now imposed when $x^{(s)}=0$, $y^{(s)}=0$, $z^{(s)}=0$, $x^{(s)}=l\,(1+\varepsilon)$, $y^{(s)}=l$ and $z^{(s)}=l(1-\varepsilon)$ $(s=1,2,\ldots,N)$.

Since we have the area $A(\varepsilon)=l^2(1+\varepsilon)$ and the volume $V(\varepsilon)=l^3(1-\varepsilon^2)=V(0)\,(1-\varepsilon^2)$, the differentiation of $E(\varepsilon)$ with respect to ε gives

$$\frac{1}{l^2}\,\frac{\partial E(0)}{\partial \varepsilon} = \left(\frac{\partial E}{\partial A}\right)_V, \tag{43.9}$$

where $(\partial E(\varepsilon)/\partial\varepsilon)_{\varepsilon=0}$ is conventionally expressed by $\partial E(0)/\partial\varepsilon$.

By changing variables, $\bar{x}^{(s)}=x^{(s)}/(1+\varepsilon)$, $\bar{y}^{(s)}=y^{(s)}$, $\bar{z}^{(s)}=z^{(s)}/(1-\varepsilon)$, we may write the Hamiltonian in the form

$$\bar{\mathscr{H}}(\varepsilon) = -\frac{\hbar^2}{2m}\sum_{s=1}^{N}\left(\frac{\partial^2}{(1+\varepsilon)^2\,\partial\bar{x}^{(s)2}} + \frac{\partial^2}{\partial\bar{y}^{(s)2}} + \frac{\partial^2}{(1-\varepsilon)^2\,\partial\bar{z}^{(s)2}}\right) + \sum_{s<t}^{N}\phi(\bar{r}^{(st)}(\varepsilon)), \tag{43.10}$$

$$\bar{r}^{(st)}(\varepsilon) = \left[(1+\varepsilon)^2(\bar{x}^{(t)}-\bar{x}^{(s)})^2 + (\bar{y}^{(t)}-\bar{y}^{(s)})^2 + (1-\varepsilon)^2(\bar{z}^{(t)}-\bar{z}^{(s)})^2\right]^{\frac{1}{2}}, \tag{43.11}$$

and the wave function in the form

$$\bar{\Psi}(\varepsilon) = \Psi\big((1+\varepsilon)\,\bar{x}^{(1)}, \bar{y}^{(1)}, (1-\varepsilon)\,\bar{z}^{(1)}, \ldots, (1+\varepsilon)\,\bar{x}^{(N)}, \bar{y}^{(N)}, (1-\varepsilon)\,\bar{z}^{(N)}\big). \tag{43.12}$$

Then (43.8) can be rewritten in the form

$$\bar{\mathscr{H}}(\varepsilon)\,\bar{\Psi}(\varepsilon) = E(\varepsilon)\,\bar{\Psi}(\varepsilon), \tag{43.13}$$

where the boundary condition $\overline{\Psi}(\varepsilon)=0$ is imposed at $\bar{x}^{(s)}=0$, $\bar{y}^{(s)}=0$, $\bar{z}^{(s)}=0$, $\bar{x}^{(s)}=l$, $\bar{y}^{(s)}=l$, and $\bar{z}^{(s)}=l\ (s=1,2,\ldots,N)$, and hence $\overline{\Psi}(\varepsilon)$ has the same boundary conditions irrespective of the value of ε.

Then differentiating (43.13) with respect to ε and putting $\varepsilon=0$ in the result, we obtain

$$\overline{\mathscr{H}}(0)\,\frac{\partial\overline{\Psi}(0)}{\partial\varepsilon}+\frac{\partial\overline{\mathscr{H}}(0)}{\partial\varepsilon}\,\overline{\Psi}(0)=E(0)\,\frac{\partial\overline{\Psi}(0)}{\partial\varepsilon}+\frac{\partial E(0)}{\partial\varepsilon}\,\overline{\Psi}(0), \tag{43.14}$$

where $\partial\overline{\Psi}(0)/\partial\varepsilon$ has a meaning similar to $\partial E(0)/\partial\varepsilon$ in (43.9). Then multiplying (43.14) by $\overline{\Psi}^*(0)$ and integrating the result over the whole configuration space inside the vessel, we obtain

$$\left.\begin{aligned}\int\overline{\Psi}^*(0)\overline{\mathscr{H}}(0)\,\frac{\partial\overline{\Psi}(0)}{\partial\varepsilon}\,d\boldsymbol{r}^{(1)}\ldots d\boldsymbol{r}^{(N)}+\int\overline{\Psi^*(0)}\,\frac{\partial\overline{\mathscr{H}}(0)}{\partial\varepsilon}\,\overline{\Psi}(0)\,d\boldsymbol{r}^{(1)}\ldots d\boldsymbol{r}^{(N)}&\\=E(0)\int\overline{\Psi}^*(0)\,\frac{\partial\overline{\Psi}(0)}{\partial\varepsilon}\,d\boldsymbol{r}^{(1)}\ldots d\boldsymbol{r}^{(N)}+\frac{\partial E(0)}{\partial\varepsilon}.&\end{aligned}\right\} \tag{43.15}$$

From the Hermitian character of the operator $\mathscr{H}(0)$, we obtain

$$\left.\begin{aligned}\int\overline{\Psi}^*(0)\overline{\mathscr{H}}(0)\,\frac{\partial\overline{\Psi}(0)}{\partial\varepsilon}\,d\boldsymbol{r}^{(1)}\ldots d\boldsymbol{r}^{(N)}=\int\left(\overline{\mathscr{H}}(0)\,\overline{\Psi}(0)\right)^*\frac{\partial\overline{\Psi}(0)}{\partial\varepsilon}\,d\boldsymbol{r}^{(1)}\ldots d\boldsymbol{r}^{(N)}&\\=E(0)\int\overline{\Psi}^*(0)\,\frac{\partial\overline{\Psi}(0)}{\partial\varepsilon}\,d\boldsymbol{r}^{(1)}\ldots d\boldsymbol{r}^{(N)}.&\end{aligned}\right\} \tag{43.16}$$

Hence (43.15) can be expressed as

$$\frac{\partial E(0)}{\partial\varepsilon}=\int\overline{\Psi}^*(0)\,\frac{\partial\overline{\mathscr{H}}(0)}{\partial\varepsilon}\,\overline{\Psi}(0)\,d\boldsymbol{r}^{(1)}\ldots d\boldsymbol{r}^{(N)}. \tag{43.17}$$

Using (43.10), we can rewrite (43.17) in the form

$$\left.\begin{aligned}\frac{\partial E(0)}{\partial\varepsilon}=\int\Psi^*\left[-\frac{\hbar^2}{m}\sum_{s=1}^{N}\left(\frac{\partial^2}{\partial z^{(s)2}}-\frac{\partial^2}{\partial x^{(s)2}}\right)+\sum_{s<t}^{N}\frac{(x^{(st)})^2-(z^{(st)})^2}{r^{(st)}}\,\phi'(r^{(st)})\right]\times&\\\times\,\Psi\,d\boldsymbol{r}^{(1)}\ldots d\boldsymbol{r}^{(N)},&\end{aligned}\right\} \tag{43.18}$$

$\overline{\Psi}(0)$ being nothing but the original wave function Ψ.

From (43.6), (43.9) and (43.18), we obtain the following expression for the surface tension:

$$\left.\begin{aligned}\gamma=Z^{-1}l^{-2}\sum_{\varrho}e^{-E_\varrho/kT}\int\Psi_\varrho^*\left[-\frac{\hbar^2}{m}\sum_{s=1}^{N}\left(\frac{\partial^2}{\partial z^{(s)2}}-\frac{\partial^2}{\partial x^{(s)2}}\right)+\right.&\\\left.+\sum_{s<t}^{N}\frac{(x^{(st)})^2-(z^{(st)})^2}{r^{(st)}}\,\phi'(r^{(st)})\right]\Psi_\varrho\,d\boldsymbol{r}^{(1)}\ldots d\boldsymbol{r}^{(N)},&\end{aligned}\right\} \tag{43.19}$$

where Ψ_ϱ is the eigenfunction corresponding to the state with energy E_ϱ.

The quantum-statistical average of the momentum operator $\boldsymbol{p}^{(s)}$ of molecule s is given by

$$\overline{\boldsymbol{p}^{(s)}}=Z^{-1}\sum_{\varrho}e^{-E_\varrho/kT}\int\Psi_\varrho^*\left(\frac{\hbar}{i}\,\nabla^{(s)}\right)\Psi_\varrho\,d\boldsymbol{r}^{(1)}\ldots d\boldsymbol{r}^{(N)}. \tag{43.20}$$

Then using (43.5) and (43.20), we can rewrite (43.19) in the form:

$$\gamma=\frac{N}{Am}\left(\overline{p_z^2}-\overline{p_x^2}\right)+\frac{1}{2A}\iint\frac{(x^{(12)})^2-(z^{(12)})^2}{r^{(12)}}\,\phi'(r^{(12)})\,n_2(\boldsymbol{r}^{(1)},\boldsymbol{r}^{(2)})\,d\boldsymbol{r}^{(1)}d\boldsymbol{r}^{(2)}. \tag{43.21}$$

The second term may be written in a form identical with (27.1) in the classical case[1].

The first term is characteristic of quantum statistics and does not necessarily cancel out even in the case of equilibrium [*41*], [*42*], while it evidently vanishes in the classical case due to the isotropy of the Maxwell distribution in momentum space.

In addition, since the pair distribution function in quantum statistics differs in general from that in classical statistics, quantum effects may arise also from the second term although the quantum mechanical modification of the pair distribution function has not been calculated yet.

With use of the density matrix (43.3), we can express (43.19) in the form [*42*]

$$\left.\begin{aligned}\gamma = \frac{1}{l^2}\Bigg[2\,\mathrm{trace}\left\{-\frac{\hbar^2}{2m}\sum_{s=1}^{N}\left(\frac{\partial^2}{\partial z^{(s)2}}-\frac{\partial^2}{\partial x^{(s)2}}\right)\varrho_N\left(\boldsymbol{r}^{(1)},\ldots,\boldsymbol{r}^{(N)};\ \boldsymbol{r}'^{(1)},\ldots,\boldsymbol{r}'^{(N)}\right)\right\}\times\\ \times\,(\mathrm{trace}\,\varrho_N)^{-1}+\frac{1}{2}\iint\frac{(x^{(12)})^2-(z^{(12)})^2}{r^{(12)}}\,\phi'\left(r^{(12)}\right)n_2\left(\boldsymbol{r}^{(1)},\boldsymbol{r}^{(2)}\right)d\boldsymbol{r}^{(1)}\,d\boldsymbol{r}^{(2)}\Bigg].\end{aligned}\right\}\quad (43.2$$

C. Lattice theory approaches.

As was discussed in Chap. B, if we know the intermolecular forces, all the thermodynamic properties of a fluid can, in principle, be calculated according to statistical thermodynamics, with the help of the partition function or the molecular distribution functions. But an exact calculation is difficult except for perfect gases or such rarefied gases that the virial expansion is valid. To carry out the numerical calculation of the thermodynamic quantities of a bulk fluid, the Born-Green-Yvon equation can be used successfully on the basis of the superposition approximation, as mentioned briefly in Sect. 25. It seems, however, extremely difficult to solve such an integro-differential equation in an interface region even with the help of a high speed electronic computing machine. The distribution function methods for the numerical calculation of surface tension discussed in the foregoing, are based on rather crude approximations such as the step approximation and the quasithermodynamic theory.

Some models have proved useful for obtaining approximate partition functions for liquids and, they are, of course, applicable to the calculation of surface tension as well. We shall here discuss the application to a plane interface of lattice theories which have been used successfully to a certain extent to calculate thermodynamic properties of pure liquids and solutions of non-electrolytes. Lattice models seem, however, to be inconvenient to treat a curved interface.

I. Pure liquids.

a) Free volume theory.

44. Cell method for calculating partition function[2] [*52*]. We shall consider a classical one-component system made up of N particles which belongs to a canonical ensemble. Then the thermodynamic properties of the system are calculated

[1] Recently a numerical calculation for the surface tensions of He^4 and He^3 has been carried out with use of (43.21) by R. BROUT and M. NAUENBERG [Phys. Rev. **112**, 1452 (1958)], based on similar assumptions to those of KIRKWOOD and BUFF for liquid argon [see (38.1) and (38.2)]. The first term of (43.21) has been neglected. They have employed the Slater-Kirkwood potential between two helium atoms and an approximate radial distribution function determined from experimental data in a way similar to the Kirkwood-Buff method in Sect. 38. The calculated values show rather good agreement with experiment.

[2] J. G. KIRKWOOD: J. Chem. Phys. **18**, 380 (1950). — See also J. O. HIRSCHFELDER, C. F. CURTIES and R. B. BIRD: [*51*], pp. 271—310.

from the partition function given by (24.2). Let us now divide the coordinate space into L cells; $\Delta_1, \ldots, \Delta_L$. These cells are assumed to have the same shape and the same volume Δ. The configuration integral (24.3) can then be written as a sum of terms in which the coordinates of each molecule are confined to a partic ular cell:

$$Q_N = (N!)^{-1} \sum_{l_1=1}^{L} \cdots \sum_{l_N=1}^{L} \int_{\Delta_{l_1}} \cdots \int_{\Delta_{l_N}} \exp\{-\Phi(\boldsymbol{r}^{(1)}, \ldots, \boldsymbol{r}^{(N)})/kT\} \prod_{s=1}^{N} d\boldsymbol{r}^{(s)}. \quad (44.1)$$

Since the molecules are indistinguishable, the value of each term remains unaltered by a permutation of the molecules. Let us take $Q_N(m_1, m_2, \ldots, m_L)$ as various terms in (44.1) which correspond to m_1 molecules in cell 1, m_2 molecules in cell 2 ..., m_L molecules in cell L. Owing to the permutation of N molecules, for fixed $\{m\} = m_1, m_2, \ldots, m_L$, there are $\left(N! \big/ \prod_{s=1}^{L} m_s!\right)$ ways of placing N molecules in L cells. Consequently the configuration integral is given in the form

$$Q_N = \sum_{\{m\}} Q_N(m_1, \ldots, m_L) \Big/ \prod_{s=1}^{L} m_s!, \quad (44.2)$$

where the summation is extended over all sets of $\{m\}$ subject to the restriction $\sum_{s=1}^{L} m_s = N$.

In general L is not necessarily equal to N, but for the present we restrict ourselves to the case $L = N$. Then (44.2) becomes

$$Q_N = \sum_{\{m\}} Q_N(m_1, \ldots, m_N) \Big/ \prod_{s=1}^{N} m_s!, \quad (44.3)$$

and the volume per cell is equal to the volume per molecule:

$$v = V/N. \quad (44.4)$$

At sufficiently high densities, all the $Q_N(m_1, \ldots, m_N)$'s but $Q_N(1, \ldots, 1)$ vanish, since the strong repulsive forces between molecules prevent multiple occupancy. In such a case (44.3) reduces to

$$Q_N = Q_N(1, 1, \ldots, 1) \equiv Q_N^{(1)}. \quad (44.5)$$

For low densities, we have to take into account the terms corresponding to double and multiple occupancy. Thus in general we may write the configuration integral as

$$Q_N = \eta^N Q_N^{(1)}. \quad (44.6)$$

Comparing (44.3) with (44.6) we obtain an expression for η:

$$\eta^N = \sum_{\{m\}} \left(\prod_{s=1}^{N} m_s!\right)^{-1} Q_N(m_1, \ldots, m_N)/Q_N^{(1)}. \quad (44.7)$$

At sufficiently high densities all $Q_N(m_1, \ldots, m_N)$'s vanish except $Q_N^{(1)}$, so that

$$\eta = 1 \quad \text{(high density limit)}. \quad (44.8)$$

For an extremely dilute gas there is effectively no contribution due to interaction between the molecules so that the terms on the right hand side of (44.1) are all equal to each other, while the number of such terms is equal to N^N

corresponding to N^N ways of placing N molecules in N cells. Hence from (44.1) and (44.6) combined with STIRLING's formula for $N!$ we find

$$\eta = \mathrm{e} \quad \text{(low density limit)}. \tag{44.9}$$

45. Free volume and communal entropy. We shall evaluate the configuration integral $Q_N^{(1)}$ corresponding to the configuration in which each cell is occupied by one and only one molecule. Let the potential energy of the system, $\Phi(\boldsymbol{r}^{(1)}, \ldots, \boldsymbol{r}^{(N)})$, be expressed by a sum of $\frac{1}{2}N(N-1)$ intermolecular potential terms:

$$\Phi(\boldsymbol{r}^{(1)}, \ldots, \boldsymbol{r}^{(N)}) = \tfrac{1}{2} \sum_{s,r} \phi(r^{(sr)}), \tag{45.1}$$

which corresponds to (25.3). When each of N molecules is at the center of its cell, (45.1) reduces to

$$\Phi_0 = \tfrac{1}{2} \sum_{s,r} \phi(|\boldsymbol{a}^{(r)} - \boldsymbol{a}^{(s)}|), \tag{45.2}$$

$\boldsymbol{a}^{(s)}$ being the position vector of the center of cell s. The contribution to the total potential energy due to the interaction of molecule s at the location $\boldsymbol{R}^{(s)}$ with respect to the origin of the s-th cell with the remaining $N-1$ molecules located at the origins of their respective cells is

$$\Psi(\boldsymbol{R}^{(s)}) = \sum_{r \neq s}^{N} \{\phi(|\boldsymbol{a}^{(r)} - \boldsymbol{a}^{(s)} - \boldsymbol{R}^{(s)}|) - \phi(|\boldsymbol{a}^{(r)} - \boldsymbol{a}^{(s)}|)\}. \tag{45.3}$$

As a first approximation to the total potential energy given by (45.1) we assume

$$\Phi(\boldsymbol{r}^{(1)}, \ldots, \boldsymbol{r}^{(N)}) = \Phi_0 + \sum_{s=1}^{N} \Psi(\boldsymbol{R}^{(s)}), \tag{45.4}$$

where Φ_0 is the zeroth approximation given by (45.2).

If $\boldsymbol{a}^{(1)}, \ldots, \boldsymbol{a}^{(N)}$ constitute a spatial lattice, Φ_0 may be regarded as the lattice energy of a crystal lattice, and $\Psi(\boldsymbol{R}^{(s)})$ as the potential energy increase of the s-th molecule due to the deviation $\boldsymbol{R}^{(s)}$ from its equilibrium point. This implies that the liquid may be regarded as a distorted crystal with one molecule at or near each lattice point. By such an intuitive consideration LENNARD-JONES and DEVONSHIRE first developed a cell theory and hence this model is usually called the Lennard-Jones and Devonshire model[1].

If the approximate potential given by (45.4) is used, we can readily obtain from (44.1)

$$Q_{(N)}^{(1)} = \mathrm{e}^{-\Phi_0/kT} v_f^N, \tag{45.5}$$

in which v_f is defined by

$$v_f = \int_{\Delta} \mathrm{e}^{-\Psi(\boldsymbol{R})/kT} d\boldsymbol{R}. \tag{45.6}$$

The partition function (45.5) is equivalent to the crystalline partition function for the Einstein model in which each molecule moves freely in the volume v_f under the field of a constant potential Φ_0. Hence v_f is called the free volume.

In this article we use the term *free volume theory* for a lattice theory of liquids in which the number of cells is chosen equal to the number of molecules. This is to be distinguished from the *hole theory* to be described in Sect. 49.

Using (44.6) and (45.5), we obtain an approximate expression for the Helmholtz free energy given by (24.1):

$$F = -N\,kT \log(\lambda^{-3} j(T)) - N\,kT \log(\eta\, v_f) + \Phi_0. \tag{45.7}$$

[1] J.E. LENNARD-JONES and A.F. DEVONSHIRE: Proc. Roy. Soc. Lond., Ser. A **163**, 53 (1937); Ser. A **165**, 1 (1938).

At sufficiently low density, η is, according to (44.9), taken as e, Φ_0 as zero and v_f as the cell volume v. Hence in the low density limit (45.7) reduces to

$$F = -N\,kT\log\left(\lambda^{-3}j(T)\right) - N\,kT\log\frac{V}{N} - N\,kT, \tag{45.8}$$

which is identical with the expression for the Helmholtz free energy of a perfect gas.

In this case $-N\,kT\log\eta$ furnishes a term $-N\,kT$ to the Helmholtz free energy and hence Nk to the entropy. This additional entropy is attributed to the free migration of molecules over the entire volume and hence called the communal entropy. On the other hand, $-N\,kT\log\eta$ is zero in the high density limit. Since the Helmholtz free energy is a continuous function of density, it is generally reasonable to assume that η is also a continuous function of the density. But the exact evaluation of the density dependence of η seems to be extremely difficult. As an approximation, it is often assumed that the communal entropy appears on melting and that the liquid state has the complete communal entropy[1]. According to this view, the entropy of fusion is expected to be equal to 1.98 cal/deg.mole. Although the experimental values for most monatomic substances satisfy this condition, the problem of communal entropy seems to be still open to question, as emphasized by RICE[2].

46. Smeared free volume. For evaluating the Helmholtz free energy provided by (45.7) we have to estimate the value of the free volume v_f. But the calculation of the exact free volume from (45.6) is very difficult because of the geometrical complexity of the cell and the dependence of the potential $\Psi(\boldsymbol{R})$ on the direction of the vector $\boldsymbol{R}$ as well as on R.

It is customary in the free volume calculation to employ a smeared or sphericalized free volume. This is achieved by replacing the angle-dependent potential $\Psi(\boldsymbol{R})$ in (45.6) by an angle average $\overline{\Psi}(R)$ given by

$$\overline{\Psi}(R) = \sum_{r\neq s}^{N}\frac{1}{4\pi}\iint\left\{\phi\left(|\boldsymbol{a}^{(r)}-\boldsymbol{a}^{(s)}-\boldsymbol{R}|\right)-\phi\left(|\boldsymbol{a}^{(r)}-\boldsymbol{a}^{(s)}|\right)\right\}\sin\vartheta\,d\vartheta\,d\varphi, \tag{46.1}$$

where ϑ and φ together with R form a spherical polar coordinate system. If all the lattice sites are equivalent, $\Psi(R)$ does not depend on s.

Let there be c nearest-neighbors, located at a distance a from an arbitrary site and c_2 second neighbors, at distance a_2. If we neglect effects arising from third and more distant neighbors, an expression for the sphericalized potential (46.1) is

$$\left.\begin{aligned}\overline{\Psi}(R) = c\left[\frac{1}{4\pi}\iint\phi\left(|\boldsymbol{a}-\boldsymbol{R}|\right)\sin\vartheta\,d\vartheta\,d\varphi-\phi(a)\right]+\\ +c_2\left[\frac{1}{4\pi}\iint\phi\left(|\boldsymbol{a}_2-\boldsymbol{R}|\right)\sin\vartheta\,d\vartheta\,d\varphi-\phi(a_2)\right].\end{aligned}\right\} \tag{46.2}$$

Substituting (46.2) into (45.6), we obtain an expression for the sphericalized or smeared free volume $\bar{v}_f$:

$$\bar{v}_f = 4\pi\int_0^{\bar{a}/2} e^{-\overline{\Psi}(R)/kT}R^2\,dR, \tag{46.3}$$

where $\bar{a}/2$ is the radius of a sphericalized cell. However, the choice of the upper limit for the integral appears rather arbitrary.

---

[1] J.O. HIRSCHFELDER, D.P. STEVENSON and H. EYRING: J. Chem. Phys. 5, 896 (1937).
[2] O.K. RICE: J. Chem. Phys. 6, 476 (1938).

The smeared free volume for the Lennard-Jones potential (23.4) with the consideration of nearest neighbors only is given by

$$\bar{v}_f = 2\pi a^3 g, \tag{46.4}$$

$$g(v^*, T^*) = \int_0^{y_0} \sqrt{y} \exp\left\{-\frac{c}{T^*}\left(\frac{l(y)}{v^{*4}} - \frac{2m(y)}{v^{*2}}\right)\right\} dy, \tag{46.5}$$

$$l(y) = (1 + 12y + 25.2y^2 + 12y^3 + y^4)(1-y)^{-10} - 1, \tag{46.6}$$

$$m(y) = (1+y)(1-y)^{-4} - 1, \tag{46.7}$$

$$y = (R/a)^2, \tag{46.8}$$

where v^* and T^* are, respectively, the reduced cell size and the reduced temperature given by (27.18).

If the lattice points form a face-centered cubic lattice of the closest packing structure, we have $y_0 = (3/4\pi\sqrt{2})^{\frac{2}{3}}$. The calculation can be improved by taking account of second and more distant neighbors.

The quantity Φ_0 given by (45.2) can be calculated from (23.4). The result for the face-centered cubic lattice is[1]

$$\Phi_0 = \frac{6NkT}{T^*}\left(\frac{1.0110}{v^{*4}} - \frac{2.4090}{v^{*2}}\right), \tag{46.9}$$

where the fractions 0.0110 and 0.4090 appearing on the right-hand side are due to the contributions of more distant neighbors than the nearest. Substitution of (46.4), (46.9) and $\eta = e$ into (45.7) yields the desired explicit expression for the Helmholtz free energy. Then using the relation $p = -(\partial F/\partial V)_T$, we obtain the equation of state. The equation obtained exhibits a loop of the van der Waals type below a reduced temperature given by[2]

$$T_c^* = 1.30.$$

The temperature at which the loop disappears, is usually identified with the critical temperature and denoted by T_c.

47. Surface energy. The free-volume theory is applied here to the numerical computation of the surface energy of liquids. We assume now that the vapor density is so low that it may be regarded as a vacuum for calculating the surface energy, as in Sect. 38, and that the liquid and its surface forms a column of lattice like structure which exposes one of its crystal planes. The lattice points of the surface plane exposed to the vapor are assumed to be shifted slightly towards the main body of the liquid, while the lattice points of the other planes are unaffected by the presence of the surface. We choose the location of the dividing surface HK just above and parallel to the surface layer. Then the plane HK is an equimolecular dividing surface, for the vapor phase has been assumed to be a vacuum. Let there be c_1 nearest neighbors at distance a_1, c_2 second neighbors at a_2, c_3 third neighbors at a_3 and so on.

Neglecting surface effects, we assume at first that all the N molecules below the plane HK can be considered as being in the interior of the liquid phase α.

---

[1] R. H. WENTORF, R. J. BUEHLER, J. O. HIRSCHFELDER and C. F. CURTISS: J. Chem. Phys. **18**, 1484 (1950).

[2] In the original treatment, LENNARD-JONES and DEVONSHIRE neglected the contribution of distant neighbors to the free volume and obtained $T_c^* = 1.33$. — See R. H. FOWLER and E. A. GUGGENHEIM: Statistical Thermodynamics, 2nd Impression, p. 345. London: Cambridge University Press 1949.

According to (45.2), the expression for the potential energy of a homogeneous bulk liquid of N molecules is given by

$$\Phi_0^\alpha = \sum_{s<r}^{N} \phi(|\boldsymbol{a}^{(r)} - \boldsymbol{a}^{(s)}|) = \tfrac{1}{2} N \sum_{m=1}^{\infty} c_m \phi(a_m), \tag{47.1}$$

where the replacement of the finite sum over all points inside the liquid by an infinite series, is allowed because of the rapid convergence of $\sum_m c_m \phi(a_m)$.

On the other hand, in the actual liquid bounded by a plane surface, any molecule in or near the surface layer has fewer neighboring molecules. We shall refer to the surface layer composed of the lattice points on the lattice plane exposed to the vapor phase as the first layer, and to the successive layers of lattice points parallel to the first as the second, ..., t-th, ..., w-th layers. Each layer is assumed to have the same area A, the same thickness d and the same number of lattice points per unit area, ν.

Let us consider a molecule situated at a lattice point in the t-th layer and denote by $c_m^{(t)}$ the number of its neighbors at distance a_m. Then $c_m^{(t)}$ differs, in general, from c_m due to the presence of the surface. Thus in analogy to (47.1) we have the expression for the potential energy of a lattice liquid with a plane surface exposed to a vacuum

$$\Phi_0 = \tfrac{1}{2} A \nu \sum_{t=1}^{w} \sum_{m=1}^{\infty} c_m^{(t)} \phi(a_m), \tag{47.2}$$

where w is the total number of lattice layers parallel to the surface and the shift of the first layer is neglected for the time being. We may replace N by $wA\nu$ in (47.1), obtaining

$$\Phi_0^\alpha = \tfrac{1}{2} w A \nu \sum_{m=1}^{\infty} c_m \phi(a_m). \tag{47.3}$$

From (47.2) and (47.3) we obtain the superficial density of potential energy

$$\upsilon^{(\phi)} = \frac{\Phi_0 - \Phi_0^\alpha}{A} = \frac{1}{2} \nu \sum_{t=1}^{\infty} \sum_{m=1}^{\infty} (c_m^{(t)} - c_m) \phi(a_m), \tag{47.4}$$

where the upper limit of the summation with respect to t may be taken as ∞, because $c_m^{(t)} - c_m$ rapidly approaches zero as t increases.

In a face-centered cubic lattice, there are three principal planes. The area per molecule is $2^{-1} 3^{\frac{1}{2}} a^2$ for a (111)-plane and a^2 for a (100)-plane, while a is equal to $2^{\frac{1}{6}} v^{\frac{1}{3}}$. Then we have

$$\nu = b v^{-\frac{2}{3}}, \tag{47.5}$$

$$b = \begin{cases} 3^{-\frac{1}{2}} 2^{\frac{2}{3}} & \text{for the (111)-plane,} \\ 2^{-\frac{1}{3}} & \text{for the (100)-plane.} \end{cases} \tag{47.6}$$

The thickness per lattice layer, d, is

$$d = \nu v = b v^{\frac{1}{3}}. \tag{47.7}$$

If the contributions of the deviation of molecules from their own lattice points and the shift of the first layer are neglected, the superficial density of internal energy υ defined by (3.2) and (6.7), becomes equal to (47.4), for the kinetic energy does not contribute to the surface energy insofar as the classical law of equipartition is valid. From this stand-point, HARASIMA [29b], [47] derived the following expression for the superficial energy density of pure liquids composed of molecules interacting according to the Lennard-Jones potential (23.4) and being arranged in a face-centered cubic lattice:

$$v = \frac{1.43\,\varepsilon}{v^{*\frac{2}{3}} D^2} \left(\frac{2.70}{v^{*2}} - \frac{1}{v^{*4}} \right), \tag{47.8}$$

where v^* is the reduced cell volume $v^* = v/D^3$. HARASIMA's calculation was carried out for the (111)-plane as the exposed face and coefficients 2.70 and 1 in (47.8) are obtained from a rather rough estimation. The calculated values of the surface energy are shown in Table 10 together with experimental values determined from the observed surface tension with use of (6.18). In his calculation the values of v^* are determined from the experimental liquid densities at melting point.

Table 10. *Theoretical and experimental surface energy* (HARASIMA [*29b*], [*47*])[1]. Calculated values of v_{calc} are based on (47.8).

Substance	$1/v^*$	v_{calc}(erg/cm²)	v_{exper}(erg/cm²)
He	0.39	0.6	0.6
H_2	0.56	4.4	5.4
D_2	0.66	6.4	8.2
Ne	0.75	9.1	15.0
N_2	0.96	22.4	26
A	0.86	26.4	35

HARASIMA considered that the choice of which face as exposed is immaterial due to the roughness of the experimental values. The careful calculation by CORNER with the help of (47.4), however, shows that the (111)-plane corresponds to the lowest potential energy. CORNER [*50*] calculated the numerical values of $v^{(\Phi)}$ for the (111)-face with use of the Lennard-Jones potential (23.4) by summing the contributions of 428 lattice points which are not more than $5a$ from a lattice point in the surface layer and approximating the contributions of all those molecules at distances greater than $5a$.

Table 11. *Contribution of intermolecular potential energy to surface energy* (CORNER [*50*]).

$\frac{1}{v^{*2}}$	$\frac{\Delta z_0}{(a\sqrt{3}/2)}$ †	$-\frac{\chi(\Delta z_0)}{\varepsilon}$	$-\frac{\chi(0)}{\varepsilon}$	$\frac{(\Phi_0(\Delta z_0)-\Phi_0^{\alpha})}{A}\frac{D^2}{\varepsilon}$	$v\frac{D^2}{\varepsilon} = \frac{1.43}{v^{*\frac{2}{3}}} \times \left(\frac{2.70}{v^{*2}} - \frac{1}{v^{*4}}\right)$ ††
0.7	0.075	10.6524	10.1361	1.550	1.778
0.8	0.054	11.1648	10.8585	1.990	2.018
0.9	0.036	11.5496	11.3994	2.377	2.237
1.0	0.020	11.8077	11.7589	2.703	2.431

† $\frac{a\sqrt{3}}{2}$: distance between (111)-planes.
†† From (47.8).

CORNER [*50*] further took into account the displacement, Δz, of the first layer toward the interior of the liquid and calculated the potential energy of a surface molecules, $\chi(\Delta z)$, as a function of Δz, assuming the molecules in the other lattice layers are held fixed. The equilibrium value of the displacement of the first layer, Δz_0, was determined so as to minimize $\chi(\Delta z)$. Some results of these calculations are given in Table 11. First the values of Δz_0 are shown. The potential energy of a surface molecule, $\chi(\Delta z_0)$, when the surface layer is displaced by Δz_0 and that in the absence of such a displacement are given together. The

[1] For the Lennard-Jones constants of He and H_2 the quantum mechanical ones (Qu) in Table 2 were used.

superficial density of potentail energy, $v^{(\Phi)} = (\Phi_0(\Delta z_0) - \Phi_0^\alpha)/A$, corresponds to (47.8) which was considered as being approximately equal to the surface energy by HARASIMA [29b]. In the last column, the values of v calculated from (47.8) are added for comparison.

48. Surface tension. If we use $wA\nu$ to express the total number of the molecules, instead of N, the expression for the Helmholtz free energy of the homogeneous liquid phase α, (45.7), is written as

$$F^\alpha = wA\nu kT\log(\lambda^3 j^{-1}) - wA\nu kT\log(\eta v_f) + \Phi_0^\alpha, \tag{48.1}$$

where v_f is the free volume defined by (45.6) and w is the number of lattice layers.

We shall next calculate the Helmholtz free energy F of a liquid, bounded by a plane surface, based on the free-volume theory [46]. The excess potential energy $\Psi^\dagger(\boldsymbol{R})$ of a surface molecule due to the deviation $\boldsymbol{R}$ from its equilibrium point, must differ from $\Psi(\boldsymbol{R})$ in (45.6) for an interior molecule of the liquid, because of the difference in the number of the neighbors. Thus, assuming the difference in composition from that of the bulk is confined to the first layer, the total potential energy provided by (45.4) takes the form

$$\Phi(\boldsymbol{r}^{(1)}, \ldots, \boldsymbol{r}^{(N)}) = \Phi_0(\Delta z_0) + \sum_{i=1}^{A\nu} \Psi^\dagger(\boldsymbol{R}^{(i)}) + \sum_{j=A\nu+1}^{wA\nu} \Psi(\boldsymbol{R}^{(j)}), \tag{48.2}$$

where $\Phi_0(\Delta z_0)$ is the potential energy when each molecule is located at its lattice point and the surface layer is displaced by Δz_0 toward the main body of the liquid as discussed in the previous section, the numbers 1 to $A\nu$ being assigned to the surface molecules. The modification of the excess potential energy of molecules in the second and distant layers is ignored.

If we use the approximate potential (48.2) instead of (45.4), we obtain, in analogy with (45.5), the following expression for $Q_N^{(1)}$

$$Q_N^{(1)} = e^{-\Phi_0(\Delta z_0)/kT} v_f'^{A\nu} v_f^{(w-1)A\nu}, \tag{48.3}$$

where v_f' is the free volume of a surface molecule given by

$$v_f' = \int_{\Delta'} e^{-\Psi^\dagger(\boldsymbol{R})/kT} d\boldsymbol{R}, \tag{48.4}$$

Δ' being any cell in the surface layer.

Since the surface molecules can migrate more easily than interior ones, (44.6) may be modified for the present case as follows:

$$Q_N = \eta'^{A\nu} \eta^{(w-1)A\nu} Q_N^{(1)}, \tag{48.5}$$

the factor η'/η corresponding to the contribution per surface molecule due to the probable increase in the probability of multiple occupancy. From (24.1), (24.2), (48.3) and (48.5), we obtain, in analogy with (48.1),

$$F = wA\nu kT\log(\lambda^3 j^{-1}) - (w-1)A\nu kT\log(\eta v_f) - A\nu\log(\eta' v_f') + \Phi_0(\Delta z_0). \tag{48.6}$$

Substitution of (48.1) and (48.6) into (5.9) leads to the following expression for the surface tension:

$$\gamma = \frac{F - F^\alpha}{A} = v^{(\Phi)} - \nu kT\log\frac{v_f'}{v_f} - \nu kT\log\frac{\eta'}{\eta}. \tag{48.7}$$

The free volume v_f may be replaced by the smeared free volume $\bar{v}_f$ given by (46.3). For surface molecules we may replace $\Psi^\dagger(\boldsymbol{R})$ to a good approximation by a cylindricalized one, instead of a sphericallized one, in the case of hexagonal packing. Then we may replace v_f' by $\bar{v}_f'$, a smeared free volume, calculated from (46.3) by using the cylindricalized potential $\bar{\Psi}^\dagger(\sqrt{R^2-z^2})$, in place of $\bar{\Psi}^\dagger(\boldsymbol{R})$, i.e. an average over the azimuth. If we neglect the difference between η and η', (48.7) reduces to

$$\gamma = v^{(\Phi)} - \nu\, kT \log \frac{\bar{v}_f'}{\bar{v}_f}, \tag{48.8}$$

where the contribution of molecular rotation would be more important than that due to the difference in η in the cases of di- and polyatomic molecules.

CORNER used (48.8) to calculate the surface tension as a function of T. His results for $\gamma^* = \gamma D^2/\varepsilon$ and $\log \bar{v}_f'/\bar{v}_f$ are given as a function of $v^* = v/D^3$ in Table 12. To obtain numerical values of the surface tension we need further a relation between the density and temperature. CORNER [*50*] adopted

$$\frac{1}{v^*} = \frac{1}{3.05}\left[1 + \frac{3}{4}\left(1 - \frac{T^*}{1.29}\right) + \frac{7}{4}\left(1 - \frac{T^*}{1.29}\right)^{\frac{1}{3}}\right], \tag{48.9}$$

where v^* and T^* are the reduced volume and the reduced temperature defined by (27.18). This relation was obtained by substituting the empirial values $kT_c/\varepsilon = 1.29$ and $v_c/D^3 = 3.05$ into GUGGENHEIM's empirical equation for simple liquids [*15*],

$$\frac{v_c}{v} = 1 + \frac{3}{4}\left(1 - \frac{T}{T_c}\right) + \frac{7}{4}\left(1 - \frac{T}{T_c}\right)^{\frac{1}{3}}, \tag{48.10}$$

where v is the volume per molecule in liquids and the subscript c corresponds to the critical point.

For liquid argon at 90° K, where T/T_c is 0.6, from Table 12 and the values of ε and D for argon in Table 2, the surface tension becomes 23.1 dyne/cm which is considerably greater than the experimental value 11.9 dyne/cm [*51*].

Table 12. *Surface tension based on free-volume theory* (CORNER [*50*]).

$\frac{1}{v^{*2}}$	$\frac{T}{T_c}$	$\frac{1}{T^*}$	$\log\frac{\bar{v}_f'}{\bar{v}_f}$	$\gamma^* = \gamma D^2/\varepsilon$
0.7	0.620	1.2495	0.009	1.545
0.8	0.521	1.4871	0.153	1.902
0.9	0.421	1.8432	0.295	2.236
1.0	0.319	2.4280	0.427	2.542

Since the dividing surface is equimolecular in the present case, (6.9) assumes the form

$$\gamma = v - T\sigma. \tag{48.11}$$

If we accept to identify $v^{(\Phi)}$ on the right-hand side of (48.8) with the surface energy as was done by HARASIMA [*47*], [*29b*] (see also Sect. 47), the second term $\nu kT \log(\bar{v}_f'/\bar{v}_f)$ will be regarded as the surface entropy multiplied by T according to (48.11) and the value of σ will be calculated from the values of $\log(\bar{v}_f'/\bar{v}_f)$ given in Table 12. Thus, for the surface entropy of liquid argon at 90° K, the calculated value of σ is 0.012 erg/cm² deg, which is too small compared with the experimental value $\sigma = 0.25$ erg/cm² deg, obtained from the observed value of γ and its temperature dependence.

On the other hand, with use of (47.5), we can rewrite (48.7) in the form

$$\gamma = v^{(\Phi)} - \frac{b\,kT}{v^{\frac{2}{3}}}\left(\log\frac{\bar{v}_f'}{\bar{v}_f} + \log\frac{\eta'}{\eta}\right). \tag{48.12}$$

It seems to be unreasonable to put $\eta = \eta'$ as in (48.8), because contribution of the communal entropy of surface molecules may be larger than that of internal molecules due to the disordered arrangement of molecules in the interface layer [*48*], which will be treated from the viewpoint of the hole theory of liquids in the subsequent sections. If we would take account of the above difference, the calculated value of surface tension would be improved, and, at the same time, a more reasonable value would be expected for the surface entropy.

b) Hole theory.

49. General survey of hole theory of liquids[1] [*52*]. We have hitherto considered the number of cells L as equal to the number of molecules, N. We may choose, however, L somewhat greater than N to allow the presence of vacant cells. We assume that each cell is sufficiently small that the probability of multiple occupancy is negligible and at the same time sufficiently large that the interactions of molecules beyond immediately adjacent cells are regarded as only a small correction. Both of these conditions can be satisfied only in the case of sufficiently short-ranged intermolecular forces such as of non-polar neutral molecules. Since L is larger than N, there exist vacant cells which we call *holes*. Then a liquid or a gas can be regarded as if it were composed of molecules and holes, and consequently a lattice theory in which the number of cells is chosen larger than the number of molecules, is often called a *hole theory*, and is to be distinguished from the *free volume theory* discussed in the preceding sections.

Let us begin with the configuration integral expressed in the form (44.1). Though the sum contains L^N terms corresponding to the number of possible arrangements of N molecules in L cells, any term corresponding to multiple occupancy of a cell is neglected according to the above assumption. Then (44.1) reduces to

$$Q_N = \sum \int_{\Delta_{l_1}} \cdots \int_{\Delta_{l_N}} \exp[-\Phi(\boldsymbol{r}^{(1)}, \ldots, \boldsymbol{r}^{(N)})\, kT] \prod_{s=1}^{N} d\boldsymbol{r}^{(s)}, \tag{49.1}$$

where the summation is over all the $L!/N!(L-N)!$ arrangements which differ by more than a permutation.

The configuration integral expressed in the from of (49.1) is valid for gases as well as for liquids. In the present case, the volume of a cell, q, is different from $v = V/N$:

$$q = V/L. \tag{49.2}$$

As in the free volume theory, the evaluation of the integral (49.1) depends on the choice of the lattice structure and the shape and orientation of the cells. We shall consider a space-lattice such that for each lattice site there are c nearest neighbors at a distance a. Let us denote by $\omega^{(s)}$ the fraction of the nearest neighbor sites of the s-th molecule which are vacant, for a given configuration. Then, if the total number of pairs of nearest-neighbor cells, one member of which is occupied and the other vacant, is denoted by cX, we obtain

$$X = \sum_{s=1}^{N} \omega^{(s)}. \tag{49.3}$$

When all of the molecules are at the origins of their own cells, potential energy of the s-th molecule is $c(1-\omega^{(s)})\,\phi(a)$, and the total potential energy of the system is given by

$$\Phi_0 = \frac{c}{2}(N - X)\,\phi(a), \tag{49.4}$$

[1] F. Cernuschi and H. Eyring: J. Chem. Phys. **7**, 547 (1939).

where $\phi(r)$ is the intermolecular potential, the contributions from second and more distant neighbors being neglected. When the molecules deviate from the origins of their own cells, we may approximate the total potential energy by

$$\Phi(\boldsymbol{r}^{(1)}, \ldots, \boldsymbol{r}^{(N)}) = \Phi_0 + \sum_{s=1}^{N} \Psi_s(\boldsymbol{R}^{(s)}), \tag{49.5}$$

$$\Psi_s(\boldsymbol{R}^{(s)}) = \sum_{r \neq s}^{N} \{\phi(|\boldsymbol{a}^{(r)} - \boldsymbol{a}^{(s)} - \boldsymbol{R}^{(s)}|) - \phi(|\boldsymbol{a}^{(r)} - \boldsymbol{a}^{(s)}|)\}. \tag{49.6}$$

These two equations correspond to (45.4) and (45.3) in the free-volume theory, respectively, but $\Psi_s(\boldsymbol{R}^{(s)})$ depends, not only on $\boldsymbol{R}^{(s)}$, but on the number and arrangement of $c(1-\omega^{(s)})$ neighbor molecules around the s-th molecule.

In analogy with (46.2) we replace the angle-dependent $\Psi_s(\boldsymbol{R}^{(s)})$ by the sphericalized potential $\overline{\Psi}(R, \omega^{(s)})$, expressed in the form

$$\overline{\Psi}(R, \omega) = c(1-\omega)\left[\frac{1}{4\pi}\iint \phi(|\boldsymbol{a} - \boldsymbol{R}|)\sin\vartheta\, d\vartheta\, d\varphi - \phi(a)\right], \tag{49.7}$$

which depends, not only on the location of the molecule, $\boldsymbol{R}$, but on the number of the nearest-neighbor molecules. For intermediate values of ω between 0 and 1, $\Psi_s(\boldsymbol{R}, \omega^{(s)})$ may be less spherical.

If the sphericalized potential (49.7) is utilized, substitution of (49.5) into (49.1) leads to

$$Q_N = e^{-Nc\phi(a)/2kT} \sum \{v_f(\omega^{(1)})\, v_f(\omega^{(2)}) \ldots v_f(\omega^{(N)})\}\, e^{Xc\phi(a)/2kT}, \tag{49.8}$$

where

$$v_f(\omega) = \int_{\Delta} e^{-\overline{\Psi}(R,\omega)/kT}\, 4\pi R^2\, dR. \tag{49.9}$$

The function $v_f(\omega)$ is a generalized free volume. When all the adjacent cells are occupied, ω is equal to zero and $v_f(0)$ becomes exactly the same as the sphericalized free volume defined by (46.3). When all the adjacent cells are vacant, i.e. ω equal to unity, $v_f(1)$ is just the cell volume q. Thus we find

$$v_f(0) = \bar{v}_f, \quad v_f(1) = q. \tag{49.10}$$

From (49.8) and (49.10), the limiting form of the configuration integral at very high and low densities are given by, respectively,

$$Q_N = \bar{v}_f^N\, e^{-Nc\phi(a)/2kT}, \qquad \text{(high density limit)}, \tag{49.11}$$

$$Q_N = q^N \frac{L!}{N!\,(L-N)!} \approx e^N \left(\frac{V}{N}\right)^N, \qquad \text{(low density limit)}. \tag{49.12}$$

The equation for the high density limit is identical with the Einstein approximation for the configuration integral of a crystal. The equation for the low density limit is exactly equal to the configuration integral of a perfect gas. Since the factor e^N appears in (49.12) as a natural consequence of the theory, the hole theory is more satisfactory than the free volume theory in that, without use of the questionable concept of communal entropy, it allows a continuous transition from liquids to gases.

Since $v_f(\omega)$ depends on ω in a complicated manner, we need to assume some simplified functional dependence for $v_f(\omega)$ to carry out the summation in (49.8). The only calculations which have been carried out, are limited to the linear approximation. Of the four linear approximations[1-3] [52] to $\log v_f(\omega)$, we shall use only the following [52]:

$$\log v_f(\omega) = \log v_f(0) + \omega \log \frac{v_f(1)}{v_f(0)}. \tag{49.13}$$

Substitution of the linear approximation (49.13) into (49.8) leads to

$$Q_N = \mathrm{e}^{-Nc\phi(a)/2kT} \left(v_f(0)\right)^N \sum_X G(N, L, X)\, \mathrm{e}^{-X\zeta/kT}, \tag{49.14}$$

$$\zeta = -\frac{c}{2}\phi(a) - kT \log \frac{v_f(1)}{v_f(0)}, \tag{49.15}$$

where $G(N, L, X)$ is the number of the distinguishable arrangements of N molecules in L cells for a specified value X.

This number $G(N, L, X)$ has not been evaluated exactly, but its approximate value has been obtained by means of the quasi-chemical approximation[4] in the theory of binary alloys. Applications of the quasichemical method to homogeneous fluids are rather successful, but its application to an interface is rather cumbersome[5]. Consequently we employ a more crude approximation equivalent to BRAGG-WILLIAMS', for treating an interface. The probability that a cell of the homogeneous fluid is occupied by a molecule is given by

$$x = \frac{N}{L} = \frac{q}{v}. \tag{49.16}$$

Let us consider the probability that out of two cells adjacent to each other, one is occupied and the other unoccupied. If the mixing of molecules and holes is completely random, this probability is given by $x(1-x)$. We shall use the approximation of random mixing, which is equivalent to BRAGG-WILLIAMS'. To this approximation, the average total number, $c\overline{X}$, of nearest-neighbor pairs, one member being an occupied cell and the other a hole, is given by

$$c\overline{X} = cL\,x(1-x). \tag{49.17}$$

Replacing X in (49.14) by $\overline{X}$ given by (49.17), we obtain the approximate configuration integral expressed in the form

$$Q_N = \mathrm{e}^{-Nc\phi(a)/2kT} \left(v_f(0)\right)^N \frac{L!}{N!(L-N)!} \exp\left\{-\frac{Lx(1-x)\zeta}{kT}\right\}, \tag{49.18}$$

where we have used the relation

$$\sum_X G(N, L, X) = \frac{L!}{N!(L-N)!}, \tag{49.19}$$

corresponding to the ways of arranging N molecules among the L cells.

We immediately obtain, from (49.18), with use of (24.1) and (24.2), the Helmholtz free energy:

$$\left.\begin{aligned} F = \frac{Nc\phi(a)}{2} - NkT\log\left(\lambda^{-3} j(T)\, v_f(0)\right) + \\ + LkT\{x\log x + (1-x)\log(1-x)\} + Lx(1-x)\zeta. \end{aligned}\right\} \tag{49.20}$$

---

[1] F. CERNUSCHI and H. EYRING: J. Chem. Phys. 7, 547 (1939).

[2] H.M. PEEK and T.L. HILL: J. Chem. Phys. 18, 1252 (1950).

[3] J.S. ROWLINSON and C.F. CURTISS: J. Chem. Phys. 19, 1519 (1951).

[4] For example, R.H. FOWLER and E.A. GUGGENHEIM: Statistical Thermodynamics, 2nd Impression, pp. 246–253, 361–366. London: Cambridge University Press 1949.

[5] Actually the hole theory presented in this article was generalized by KURATA [53] to the quasi-chemical approximation. The difference between the results based on the two approximations is only a few percent.

Then the familiar equations for the pressure p and the chemical potential μ are, respectively, expressed in the forms:

$$p = -\left(\frac{\partial F}{\partial V}\right)_{T,N} = -\frac{1}{q}\left(\frac{\partial F}{\partial L}\right)_{T,N}, \tag{49.21}$$

$$\mu = \left(\frac{\partial F}{\partial N}\right)_{T,V} = \frac{1}{L}\left(\frac{\partial F}{\partial x}\right)_{L,T}. \tag{49.22}$$

Substituting (49.20) into (49.21), we obtain

$$p = -\frac{L\,kT}{V}\left\{\log(1-x) + x^2\frac{\zeta}{kT}\right\} = -\frac{kT}{q}\left\{\log\left(1-\frac{q}{v}\right) + \left(\frac{q}{v}\right)^2\frac{\zeta}{kT}\right\}, \tag{49.23}$$

and similarly

$$\mu = \frac{c}{2}\phi(a) - kT\log\left(\lambda^{-3} j\, v_f(0)\right) + kT\log\frac{x}{1-x} + (1-2x)\zeta. \tag{49.24}$$

Since the cell volume q is considered as being constant, (49.23) is the equation of state. It can be shown that this equation of state exhibits the isothermals with loops of the van der Waals type for temperatures below the critical temperature T_c given by

$$kT_c = \frac{-\frac{c}{2}\phi(a)}{2 + \log(v_f(1)/v_f(0))}. \tag{49.25}$$

The above-mentioned property of the isothermals may be regarded to indicate the coexistence of the two phases, α and β, at equilibrium, below T_c. Let us denote the values of x for the phases α and β, respectively, by x^α and x^β, of which the larger one corresponds to a liquid and the smaller to a vapor. Since the pressure and the chemical potential of the phase α must be the same as those of the phase β, if the two phases coexist at equilibrium, we obtain from (49.23) and (49.24) the condition for equilibrium in the following forms:

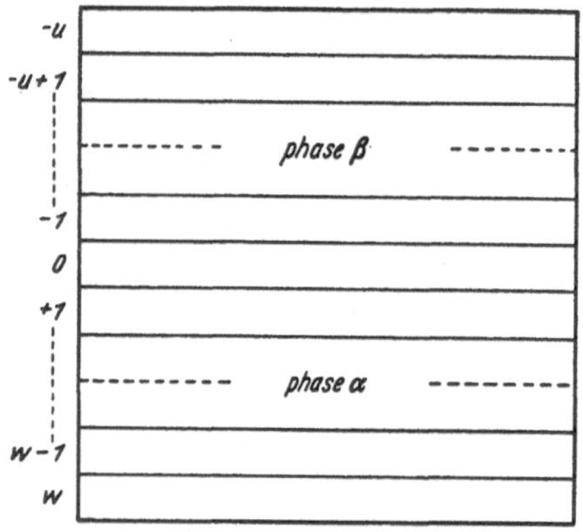

Fig. 23. A two-phase system with a plane interface. The system is assumed to consist of lattice layers parallel to the interface.

$$\frac{\zeta}{kT}(2x^\alpha - 1) = \log\frac{x^\alpha}{1-x^\alpha}, \tag{49.26}$$

$$x^\alpha + x^\beta = 1. \tag{49.27}$$

50. Free energy of two-phase system [52]. The hole theory developed in the previous section is extended here to an inhomogeneous system consisting of two phases separated by a plane interface. We shall consider a system confined to a rectangular parallelepiped vessel placed so that one of its edges points in the direction of the gravitational force. We assume that one of the principal planes in the lattice which composes the system, is perpendicular to the gravitational force. We may refer to these lattice planes as equipotential lattice planes. For abbreviation we refer to one of the equipotential lattice planes as the zeroth layer and give successively numbers $-1, -2, \ldots, -u$ upwards to the layers above the zeroth and $1, 2, \ldots, w$, downwards to those below the zeroth (Fig. 23). The location of the zeroth layer will be chosen conveniently as described in Sect. 52.

We shall indicate the properties referred to the t-th layer by superscript t, e.g. the number of molecules in the t-th layer is indicated by $N^{(t)}$. These layers are assumed to have the same area. As in the free volume theory, d and A denote the thickness and area of each layer, respectively.

The expression for the configuration integral (49.8) is valid whether the system is homogeneous or not, but not convenient for the present case of an inhomogeneous system, because the number density of molecules whithin the interface region may change appreciably along the vertical direction. We shall express the configuration integral Q_N as the sum of terms which will be denoted by $Q_N(N^{(-u)}, \ldots, N^{(-1)}, N^{(0)}, \ldots, N^{(w)})$ corresponding to the given set of numbers $N^{(-u)}, \ldots, N^{(0)}, \ldots, N^{(w)}$:

$$Q_N = \sum_{N^{(-u)}} \cdots \sum_{N^{(w)}} Q_N(N^{(-u)}, \ldots, N^{(0)}, \ldots, N^{(w)}), \tag{50.1}$$

$$\sum_{t=-u}^{w} N^{(t)} = N, \tag{50.2}$$

where

$$Q_N(N^{(-u)}, \ldots, N^{(0)}, \ldots, N^{(w)}) = \mathrm{e}^{-Nc\phi(a)/2kT} \sum \{v_f(\omega^{(1)}) \ldots v_f(\omega^{(N)})\}\, \mathrm{e}^{Xc\phi(a)/2kT}, \tag{50.3}$$

summation extending over all configurations consistent with the given set of $N^{(-u)}, \ldots, N^{(0)}, \ldots, N^{(w)}$.

If we insert the linear approximation (49.13) in (50.3), we have

$$Q_N(N^{(-u)}, \ldots, N^{(w)}) = \mathrm{e}^{-Nc\phi(a)/2kT} (v_f(0))^N \sum_X g(N^{(-u)}, \ldots, N^{(w)}; X)\, \mathrm{e}^{-X\zeta/kT}, \tag{50.4}$$

where $g(N^{(-u)}, \ldots, N^{(0)}, \ldots, N^{(w)}; X)$ is the number of the ways of arranging $N^{(-u)}, \ldots, N^{(0)}, \ldots, N^{(w)}$ molecules respectively in $A\nu$ cells of each of the corresponding layers with the given value of X. If the functions $g(N^{(-u)}, \ldots, N^{(0)}, \ldots, N^{(w)}; X)$ are summed with respect to X, we obtain the number of the above arrangements without restriction:

$$\sum_X g(N^{(-u)}, \ldots, N^{(0)}, \ldots, N^{(w)}; X) = \prod_{t=-u}^{w} \left\{ \frac{(A\nu)!}{N^{(t)}!\,(A\nu - N^{(t)})!} \right\}, \tag{50.5}$$

which corresponds to (49.19).

Let us consider an arbitrary cell in the t-th layer. Let the number of its adjacent cells within this layer be lc and the number of its adjacent cells in one of the next layers be mc. It is evident that the total number of cells adjacent to a central cell is $(l+2m)c$ and we have by definition

$$l + 2m = 1. \tag{50.6}$$

On the other hand, the probability that an arbitrary cell in the t-th layer is occupied by a molecule, is given by

$$x^{(t)} = \frac{N^{(t)}}{A\nu}. \tag{50.7}$$

Let us consider the total number of nearest-neighbor pairs consisting of a hole and an occupied cell, to which we shall hereafter refer, for brevity, as the hole-molecule contacts. If we assume a completely random configuration of molecules in each layer, the average number of hole-molecule contacts within the t-th layer is, by analogy with (49.17),

$$c\,\overline{X}^{(t,t)} = c\,A\,\nu\,l\,x^{(t)}(1 - x^{(t)}). \tag{50.8}$$

In the same way, the average number $c\overline{X}^{(t,t+1)}$ of hole-molecule contacts between the $(t+1)$-th and t-th layers is

$$c\overline{X}^{(t,t+1)} = \frac{cA\nu}{2} m\left[x^{(t)}(1-x^{(t+1)}) + x^{(t+1)}(1-x^{(t)})\right]. \tag{50.9}$$

Then for any given set $x^{(-u)}, \ldots, x^{(0)}, \ldots, x^{(w)}$, the average of the total number of hole-molecule contacts in this system, $c\overline{X}$, is

$$\left.\begin{aligned} c\overline{X} = \frac{cA\nu}{2}\sum_{t=-u}^{w}\{&2l\,x^{(t)}(1-x^{(t)}) + m\,x^{(t)}(1-x^{(t+1)}) + \\ &+ m\,x^{(t+1)}(1-x^{(t)}) + m\,x^{(t)}(1-x^{(t-1)}) + m\,x^{(t-1)}(1-x^{(t)})\},\end{aligned}\right\} \tag{50.10}$$

where edge effects are ignored.

As in the previous section, replacing X in (50.4) by $\overline{X}$ given by (50.10) and making use of (50.5), we obtain the following approximation to the configuration integral given by (50.3),

$$\left.\begin{aligned} Q_N(x^{(-u)}, \ldots, x^{(0)}, \ldots, x^{(w)}) &= (v_f(0))^N \prod_{t=-u}^{w}\left\{\frac{(A\nu)!}{(A\nu x^{(t)})!\,(A\nu - A\nu x^{(t)})!}\right\}\times \\ \times\exp\Big[-\frac{Nc\phi(a)}{2kT} &- \frac{A\nu\zeta}{2kT}\sum_{t=-u}^{w}\{2l\,x^{(t)}(1-x^{(t)}) + m\,x^{(t)}(1-x^{(t+1)}) + \\ + m\,x^{(t+1)}(1-x^{(t)}) &+ m\,x^{(t)}(1-x^{(t-1)}) + m\,x^{(t-1)}(1-x^{(t)})\}\Big].\end{aligned}\right\} \tag{50.11}$$

The number of terms contained in Q_N provided by (50.1) is less than $(A\nu)^{u+w+1}$, the number for the case where the total number of molecules is not restricted. Thus, for the maximum term Q_N^* of (50.1) we obtain the inequality

$$0 < \log Q_N - \log Q_N^* < (u+w+1)\log(A\nu). \tag{50.12}$$

Under ordinary conditions, $(u+w+1)$ is of the order of $N^{\frac{1}{3}}$ and hence the contribution of the difference between $\log Q_N$ and $\log Q_N^*$ to the Helmholtz free energy of the system, is of the order of magnitude of $N^{\frac{1}{3}}kT$, which may be neglected even as compared with the surface contribution. Thus we may replace $\log Q$ by $\log Q^*$ to a sufficiently good approximation. Hence by using (24.1) and (24.2) we have

$$F = -NkT\log(\lambda^{-3}j(T)) - kT\log Q_N^*. \tag{50.13}$$

Inserting (50.11) in (50.13) and using the Stirling formula we have

$$\left.\begin{aligned} F = \frac{cN}{2}\phi(a) - NkT\log(\lambda^{-3}j(T)\,v_f(0)) &+ A\nu kT\sum_{t=-u}^{w}\{x^{(t)}\log x^{(t)} + \\ +(1-x^{(t)})\log(1-x^{(t)})\} + \frac{A\nu\zeta}{2}&\sum_{t=-u}^{w}\{2l\,x^{(t)}(1-x^{(t)}) + m\,x^{(t)}(1-x^{(t+1)}) + \\ + m\,x^{(t+1)}(1-x^{(t)}) + m\,x^{(t)}&(1-x^{(t-1)}) + m\,x^{(t-1)}(1-x^{(t)})\},\end{aligned}\right\} \tag{50.14}$$

where the set of the fractions $(x^{(-u)}, \ldots, x^{(0)}, \ldots, x^{(w)})$ is determined so as to maximize $Q_N(x^{(-u)}, \ldots, x^{(0)}, \ldots, x^{(w)})$ given by (50.11) under the restriction

$$A\nu\sum_{t=-u}^{w}x^{(t)} = N. \tag{50.15}$$

Using the Stirling formula for factorials, we obtain, from (50.11) and (50.15), the following condition for the maximization:

$$\left.\begin{array}{c}\zeta\{m(1-2x^{(t-1)})+l(1-2x^{(t)})+m(1-2x^{(t+1)})\}+kT\log\frac{x^{(t)}}{1-x^{(t)}}=\Lambda,\\ (t=-u,\ldots,0,\ldots,w),\end{array}\right\}\tag{50.16}$$

where Λ is LAGRANGE'S undetermined multiplier.

It is seen from (50.13) that the set of fractions, $x^{(-u)}, \ldots, x^{(0)}, \ldots, x^{(w)}$ determined from (50.16) and (50.15) minimizes the Helmholtz free energy given by (50.14) as a function of $x^{(-u)}, \ldots, x^{(0)}, \ldots, x^{(w)}$. Since the number density of molecules in the t-th layer is $x^{(t)}/q$, the above implies that the equilibrium form of the local density is determined by the condition that F is minimized under the restraint (50.15), just as for the equilibrium condition in quasithermodynamics, as discussed in Sect. 19.

51. Calculation of equilibrium form of local density [*52*]. The equilibrium values of the density $\frac{1}{q}x^{(-u)}, \ldots, \frac{1}{q}x^{(0)}, \ldots, \frac{1}{q}x^{(w)}$, and the undetermined multiplier Λ can be, in principle, determined by the set of $u+w+2$ equations, (50.16) and (50.15). It is, however, not necessary to carry out the actual calculation of the density distribution for all layers, since we may anticipate that all x's are equal to x^{α} or x^{β} except for those within or near the interface layer. For a system composed of liquid and vapor phases, from the macroscopic standpoint, layers of sufficiently large positive t lie in the interior of the liquid phase α and those of sufficiently large negative t in the interior of the vapor phase β, i.e.,

$$\lim_{t\to\infty} x^{(t)} = x^{\alpha}, \quad \lim_{t\to-\infty} x^{(t)} = x^{\beta}. \tag{51.1}$$

Consequently for large positive values of t, (50.16) reduces to [see (50.6)]

$$\zeta(1-2x^{\alpha}) + kT\log\frac{x^{\alpha}}{1-x^{\alpha}} = \Lambda. \tag{51.2}$$

By comparison with (49.26) we immediately find

$$\Lambda = 0. \tag{51.3}$$

Let us introduce the following notation for the differences:

$$\Delta x^{(t)} = x^{(t+1)} - x^{(t)}; \qquad \Delta^2 x^{(t)} = \Delta x^{(t+1)} - \Delta x^{(t)}. \tag{51.4}$$

Then, using (50.6), (51.3) and (51.4) we can rewrite (50.16) in a more compact form

$$\Delta^2 x^{(t-1)} = \frac{1}{m}\left(\frac{kT}{2\zeta}\log\frac{x^{(t)}}{1-x^{(t)}} + \frac{1}{2} - x^{(t)}\right), \tag{51.5}$$

which is a non-linear difference equation. We can solve it numerically by exactly the same method as for numerical integration of an ordinary differential equation. That is, we can obtain a solution of this difference equation if the values of $x^{(t)}$ and $\Delta x^{(t)}$ for a certain t, say $x^{(1)}$ and $\Delta x^{(1)}$ are given. The boundary conditions for our problem are given by (51.1).

As with the case of boundary value problems for ordinary differential equations, the boundary condition can be satisfied only for certain definite values of $x^{(1)}$ and $\Delta x^{(1)}$. However, it is not so easy here to find the solution with proper behavior by adjusting two parameters $x^{(1)}$ and $\Delta x^{(1)}$. But, in our case, we can eliminate either of these parameters with the help of the symmetry property that (51.5) is invariant under the transformation of t to $1-t$ and $x^{(t)}$ to $1-x^{(t)}$.

If $x^{(t)}$ satisfies the difference equation (51.5) subject to (51.1), then so does $\bar{x}^{(t)} = 1 - x^{(1-t)}$ also, as (51.5) is invariant under the above transformation, and at the same time, the boundary condition (51.1) is satisfied due to (49.27). If there is only one solution with the proper boundary values which changes the sign of $x^{(t)} - \frac{1}{2}$ between the zeroth and first layers, we have

$$x^{(1-t)} + x^{(t)} = 1. \tag{51.6}$$

For $t = 1$ this reduces to

$$x^{(0)} + x^{(1)} = 1, \tag{51.7}$$

from which we immediately obtain $\Delta x^{(1)} = x^{(1)} - x^{(0)} = 2x^{(1)} - 1$. Consequently we have only to adjust the value of a single parameter $x^{(1)}$ to obtain the solution

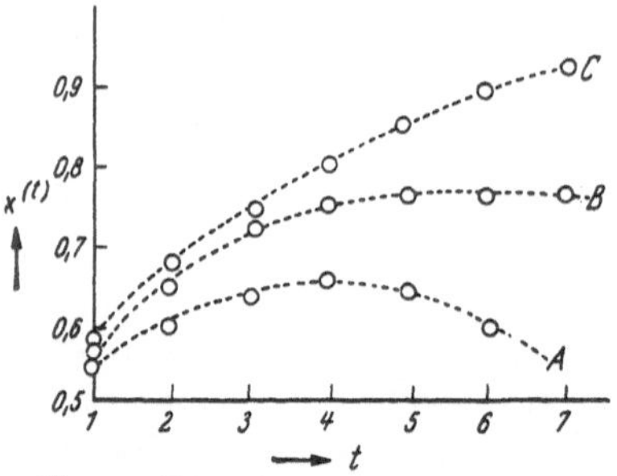

Fig. 24. Trial solutions of difference equation (51.5) (ONO [52]). Out of three curves, B is the desired solution, for its value approaches a constant x^{α} as t increases.

Table 13. *Density distribution in the surface layer for $\zeta/kT = 5.0$.*

t	(100)	(111)	(110)
1	0.836	0.903	0.760
2	0.988	0.991	0.970
3	0.993	0.993	0.992
4	0.993	0.993	0.993
...	...	...	...
∞	0.993	0.993	0.993

The values for the (100)-plane were calculated by ONO [52], and those for (111) and (110) by KURATA [53], a modified equation having been used in the case of the (110)-plane.

which satisfies the boundary condition. It is easily proved from the symmetric property of (51.5) that the solution which satisfies (51.7) always satisfies (51.6).

We shall assume for simplicity that the lattice is a face-centered cubic array. The behaviors of some solutions of (51.5) with $2kT/\zeta = 1.05$ for the case of the

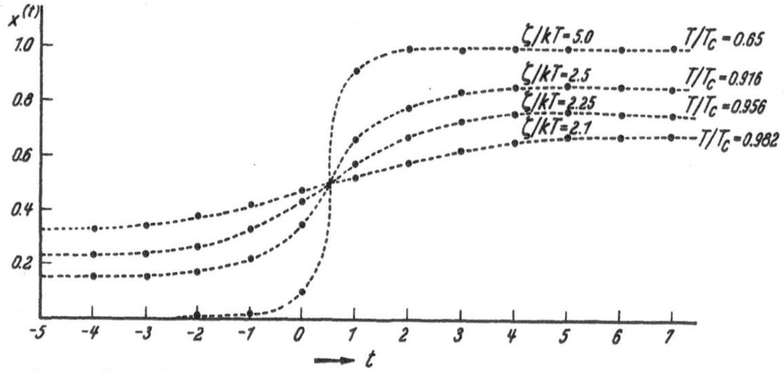

Fig. 25. Change in density around the interface calculated from the hole theory; (111) face.

(100)-plane, for which $m = \frac{1}{3}$, calculated for varying values of $x^{(1)}$, are shown in Fig. 24. Out of these curves, only Curve B with $x^{(1)} = 0.525$ satisfies the boundary condition (51.1). The numerical calculations of the solutions of (51.5) were carried out by ONO [52] and KURATA [53][1]. Some of the results are shown in Table 13 and Fig. 25.

[1] In the case of the (100)-plane modification of (51.5) is necessary, since some of the adjacent cells are found in the second neighbor layer, whereas we have merely to put $m = \frac{1}{4}$ for the (111)-plane.

By using (49.25) we can eliminate $\phi(a)$ from (49.15) to obtain

$$\frac{\zeta}{2kT} = \frac{T_c}{T}\left\{1+\frac{1}{2}\left(1-\frac{T}{T_c}\right)\log\frac{v_f(1)}{v_f(0)}\right\}. \tag{51.8}$$

Elimination of ζ from (51.8) with use of (49.26) and (49.27) yields

$$\frac{T}{T_c} = \frac{\{2+\log(v_f(1)/v_f(0))\}(1-2x^\beta)}{\log\frac{1-x^\beta}{x^\beta}+(1-2x^\beta)\log\frac{v_f(1)}{v_f(0)}}. \tag{51.9}$$

The values of ζ/kT given in Fig. 25 are calculated from (51.8) for the given value $\log v_f(1)/v_f(0)=3.55$ which is determined so as to fit (51.9) to the observed change in orthobaric densities x^β of nearly perfect liquids with varying temperature [*52*]. We shall see that the density transition in the interface zone at $T/T_c=0.65$ is essentially completed in two molecular layers and that the thickness of the transition zone increases as T approaches the critical temperature.

Furthermore, we find that the density versus depth curves shown in Fig. 25 strongly resemble those in Fig. 15 calculated by HILL based upon the quasi-thermodynamic treatment[1]. This point will be discussed again in Sect. 53.

52. Surface tension and surface energy. Let us place the dividing surface between the 0th and 1st layers. Then from (51.6) and (50.2) we have

$$N = N^\alpha + N^\beta, \tag{52.1}$$

i.e., this dividing surface is the equimolecular dividing surface, $\Gamma=0$.

Then, according to (3.1) and (5.8), the surface tension γ is given by

$$\gamma = \frac{F-F^\alpha-F^\beta}{A}, \tag{52.2}$$

where F^α and F^β are, respectively, the Helmholtz free energies of the bulk liquid and vapor which are strictly homogeneous right up to the dividing surface. According to (49.20) the free energies F^α and F^β are, respectively, given by

$$\left.\begin{aligned} F^\alpha &= \frac{N^\alpha c}{2}\phi(a) - N^\alpha kT\log\left(\lambda^{-3}j(T)\,v_f(0)\right) + \\ &\quad + L^\alpha kT\{x^\alpha\log x^\alpha+(1-x^\alpha)\log(1-x^\alpha)\} + L^\alpha x^\alpha(1-x^\alpha)\zeta, \end{aligned}\right\} \tag{52.3}$$

$$\left.\begin{aligned} F^\beta &= \frac{N^\beta c}{2}\phi(a) - N^\beta kT\log\left(\lambda^{-3}j(T)\,v_f(0)\right) + \\ &\quad + L^\beta kT\{x^\beta\log x^\beta+(1-x^\beta)\log(1-x^\beta)\} + L^\beta x^\beta(1-x^\beta)\zeta, \end{aligned}\right\} \tag{52.4}$$

where

$$L^\alpha = w\,A\,v; \quad L^\beta = (u+1)\,A\,v, \tag{52.5}$$

$$L = L^\alpha + L^\beta. \tag{52.6}$$

---

[1] An attempt was made by J.W. CAHN and J.E. HILLIARD [J. Chem. Phys. **28**, 258 (1958)] to express the local free energy at an interface as the sum of two contributions which are functions of the local composition and the composition derivatives, respectively. The calculation of composition versus depth curve was carried out for the interface between two coexisting binary regular solutions. Their curve resembles those of Fig. 25. The resemblance will be accounted for by the fact that the hole theory is a theory of the regular solution of holes and molecules. Since their theory is simpler, though less rigorous, than ONO's and HILL's, they could show that the thickness of the interface increases with temperature and becomes infinite at the critical temperature T_c, and that at a temperature T just below T_c the surface tension is proportional to $(T-T_c)^{\frac{3}{2}}$. The agreement of the theory with experimental data is very good.

Substitution of (50.14), (52.3) and (52.4) into (52.2) yields the expression for the surface tension:

$$\left.\begin{aligned}\gamma = \frac{\nu\zeta}{2}\sum_{t=-u}^{w}\{&2l\,x^{(t)}(1-x^{(t)}) + m\,x^{(t)}(1-x^{(t+1)}) + m\,x^{(t+1)}(1-x^{(t)}) + \\ &+ m\,x^{(t)}(1-x^{(t-1)}) + m\,x^{(t-1)}(1-x^{(t)}) - 2x^{\alpha}(1-x^{\alpha})\} + \\ &+ \nu kT\sum_{t=-u}^{w}\{x^{(t)}\log x^{(t)} + (1-x^{(t)})\log(1-x^{(t)}) - x^{\alpha}\log x^{\alpha} - \\ &- (1-x^{\alpha})\log(1-x^{\alpha})\} = \nu\sum_{t=-u}^{w}\{\zeta\,x^{(t)}(m\,x^{(t-1)} + l\,x^{(t)} + m\,x^{(t+1)}) + \\ &+ kT\log(1-x^{(t)}) - \zeta(x^{\alpha})^2 - kT\log(1-x^{\alpha})\}\end{aligned}\right\} \tag{52.7}$$

where we have used the relations (49.27), (50.16) with (51.3) and (49.26).

Next, replacing x^{1-t} by $1-x^{t}$ according to (51.6) we can rewrite (52.7) in the following form

$$\left.\begin{aligned}\gamma = \nu\zeta\sum_{t=1}^{\infty}[&\{2l\,x^{(t)}(1-x^{(t)}) + m\,x^{(t)}(1-x^{(t+1)}) + m\,x^{(t+1)}(1-x^{(t)}) + \\ &+ m\,x^{(t)}(1-x^{(t-1)}) + m\,x^{(t-1)}(1-x^{(t)})\} - 2x^{\alpha}(1-x^{\alpha})] + \\ &+ 2\nu kT\sum_{t=1}^{\infty}[x^{(t)}\log x^{(t)} + (1-x^{(t)})\log(1-x^{(t)}) - x^{\alpha}\log x^{\alpha} - \\ &- (1-x^{\alpha})\log(1-x^{\alpha})],\end{aligned}\right\} \tag{52.8}$$

where the upper limit of the summations has been replaced by infinity because the terms in both sums rapidly approach zero for sufficiently large values of t.

The number of cells ν per unit surface area of each layer in the free volume theory is given by (47.5) but we must replace v by the cell volume q in the hole theory:

$$\nu = b\,q^{-\frac{2}{3}}, \tag{52.9}$$

where b is given by (47.6). Then, we may rewrite (52.8) in the following form more convenient for practical calculation:

$$\gamma = 2^{-\frac{1}{3}}q^{-\frac{2}{3}}\frac{\zeta}{3}P\left(\frac{\zeta}{kT}\right) - 2^{\frac{2}{3}}q^{-\frac{2}{3}}kT\,D\left(\frac{\zeta}{kT}\right), \tag{52.10}$$

where

$$\left.\begin{aligned}P\left(\frac{\zeta}{kT}\right) = 3\cdot 2^{\frac{1}{3}}b\sum_{t=1}^{\infty}\{&x^{(t)}(1-m\,x^{(t-1)} - l\,x^{(t)} - m\,x^{(t+1)}) + \\ &+ (1-x^{(t)})(m\,x^{(t-1)} + l\,x^{(t)} + m\,x^{(t+1)}) - 2x^{\alpha}(1-x^{\alpha})\},\end{aligned}\right\} \tag{52.11}$$

$$\left.\begin{aligned}D\left(\frac{\zeta}{kT}\right) = 2^{\frac{1}{3}}b\sum_{t=1}^{\infty}\{&x^{(t)}\log x^{(t)} + (1-x^{(t)})\log(1-x^{(t)}) - \\ &- x^{\alpha}\log x^{\alpha} - (1-x^{\alpha})\log(1-x^{\alpha})\}.\end{aligned}\right\} \tag{52.12}$$

The calculated values of P and D as functions of ζ/kT are shown in Table 14. From (52.7), (50.16) and (51.3), we can easily prove

$$\left(\frac{\partial\gamma}{\partial x^{(t)}}\right)_{T,\,x^{(s\neq t)},\,x^{\alpha}} = 0. \tag{52.13}$$

In the same way, using (49.26), we have

$$\left(\frac{\partial \gamma}{\partial x^{\alpha}}\right)_{T,\, x_{-u},\, \ldots,\, x_w} = 0. \tag{52.14}$$

Since the dividing surface is equimolecular in the present case, we can calculate the surface energy from (52.10) and (6.18). Thus, using (52.13) and (52.14), we can readily obtain the expression for the surface energy

$$v = 2^{-\frac{1}{3}} q^{-\frac{2}{3}} \left(\zeta - T\frac{d\zeta}{dT}\right) \frac{P\left(\frac{\zeta}{kT}\right)}{3}. \tag{52.15}$$

Similarly from (6.17) and (52.10) we obtain

$$\sigma = -2^{-\frac{1}{3}} q^{-\frac{2}{3}} \left(\frac{d\zeta}{dT}\,\frac{P(\zeta/kT)}{3} - 2kT\,D\left(\frac{\zeta}{kT}\right)\right). \tag{52.16}$$

Also, from (49.15) and (49.10) we have

$$\frac{d\zeta}{dT} = -k\log\frac{q}{\bar{v}_f} - kT\,\frac{d\log(q/\bar{v}_f)}{dT}, \tag{52.17}$$

$$\zeta - T\frac{d\zeta}{dT} = -\frac{1}{2}c\,\phi(a) + kT^2\,\frac{d\log(q/\bar{v}_f)}{dT}. \tag{52.18}$$

Table 14. *Function $P(\zeta/kT)$ and $D(\zeta/kT)$ for the (111) face.*

ζ/kT	P	D
6.0	0.990	0.226
5.5	1.090	0.281
5.0	1.148	0.331
4.5	1.217	0.382
4.0	1.298	0.441
3.5	1.360	0.480
3.0	1.394	0.499
2.5	1.272	0.447
2.25	1.100	0.380
2.1	0.753	0.255

In general, if we know $q/\bar{v}_f$ as a function of temperature, we can calculate the value of the surface tension as a function of T from (52.10), (51.8) and the values of P and D tabulated in Table 14. In a similar way we can calculate the values of the surface energy and surface entropy.

If we use the empirical value $\log(q/\bar{v}_f) = 3.55$ mentioned in Sect. 51 and neglect the temperature dependence of $\log(q/\bar{v}_f)$, we obtain from (52.10) the calculated value $\gamma = 11.2$ dyne/cm for liquid argon at 90° K with molar volume 28.1 cm³/mol. This value is in very good agreement with the experimental one $\gamma = 11.9$ dyne/cm. However, the value of the surface energy calculated from (52.15) is 25 erg/cm² which is not in good agreement with the experimental value, 35 erg/cm². This discrepancy seems to arise from the neglect of thermal expansion of the cell volume.

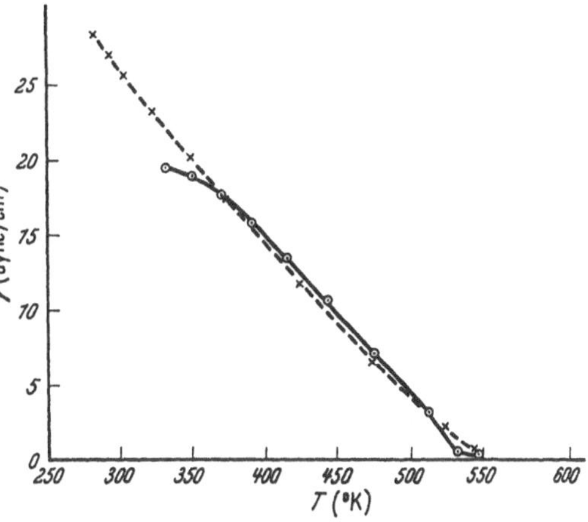

Fig. 26. A comparison of theoretical (for the (111)-face) (⊙) and experimental (×) surface tension for carbon tetrachloride. (The experimental data are from the International Critical Tables.)

Neglecting $d\log(q/v_f)/dT$ and eliminating ζ from (52.10) with use of (51.8), we obtain

$$\gamma = 2^{-\frac{1}{3}} q^{-\frac{2}{3}} k T_c \left[\left(2 + \log\frac{q}{\bar{v}_f}\right)\frac{P\left(\frac{\zeta}{kT}\right)}{3} - \frac{T}{T_c}\left\{\frac{P\left(\frac{\zeta}{kT}\right)}{3}\log\frac{q}{\bar{v}_f} + 2D\left(\frac{\zeta}{kT}\right)\right\}\right]. \tag{52.19}$$

The temperature dependence of the surface tension of carbon tetrachloride calculated from (52.19), is shown together with the experimental values in Fig. 26. In this figure we have adopted the empirical value $\log(q/\bar{v}_f) = 4.0$ determined from the temperature dependence of the orthobaric density of carbon tetrachloride and calculated q from $T_c = 556.3°$ K and the molar volume 94.18 cm³/mole at the melting point. If we consider that there are no adjustable parameters, the agreement between the theoretical and the experimental values is surprisingly good.

The values of γ calculated for the (100)- and (110)-faces are higher than that for the (111)-face. Therefore the calculation of the surface tension in the hole theory should be made for the (111)-face.

Prigogine and Saraga [54] developed an alternative theory of surface tension. Their theory is based on a crude model of a square-well intermolecular potential, instead of the Lennard-Jones potential (23.4), and on a monomolecular layer model with two parameters: β (to adjust the excess free volume of a surface molecule over the free volume of an interior molecule) and the above $x^{(1)}$ (i.e. the fraction of occupied cells in the surface layer). Their results for $\beta = 0.25$ and $x^{(1)} = 0.7$, are in very good agreement with observed values especially for surface entropy; e.g. $\gamma_{\text{calc}} = 12.5$ dyne/cm ($\gamma_{\text{obs}} = 13.2$ dyne/cm) and $\sigma_{\text{calc}} = 0.27$ erg/deg. cm² ($\sigma_{\text{obs}} = 0.26$ erg/deg. cm²) for liquid argon at 85° K.

53. Relation to quasithermodynamic theory[1]. The expression for the Helmholtz free energy given by (50.14) may be rewritten after some rearrangement as follows:

$$F = A\nu \sum_{t=-u}^{w} x^{(t)} f^{(t)}, \tag{53.1}$$

$$\left.\begin{aligned} f^{(t)} &= \frac{c}{2}\phi(a) - kT\log\left(\lambda^{-3} j(T)\, v_f(0)\right) + kT\left\{\log\frac{x^{(t)}}{1-x^{(t)}} + \frac{1}{x^{(t)}}\log\left(1-x^{(t)}\right)\right\} + \\ &\quad + \zeta\left\{m\left(1-x^{(t-1)}\right) + l\left(1-x^{(t)}\right) + m\left(1-x^{(t+1)}\right)\right\}. \end{aligned}\right\} \tag{53.2}$$

Since $A\nu x^{(t)}$ is the number of molecules in the t-th layer, $f^{(t)}$ may be regarded as the Helmholtz free energy per molecule in the t-th layer, and corresponds to $f(z)$ in the quasithermodynamic theory discussed in Sect. 18. Nevertheless, it must be noted that a unique definition of $f^{(t)}$ cannot be given. On the other hand, the chemical potential of a molecule in the t-th layer may be given as, according to (50.14),

$$\left.\begin{aligned} \mu^{(t)} &= \frac{1}{A\nu}\frac{\partial F}{\partial x^{(t)}} = \frac{c}{2}\phi(a) - kT\log\left(\lambda^{-3} j(T)\, v_f(0)\right) + kT\log\frac{x^{(t)}}{1-x^{(t)}} + \\ &\quad + \zeta\left\{m\left(1-2x^{(t-1)}\right) + l\left(1-2x^{(t)}\right) + m\left(1-2x^{(t+1)}\right)\right\}. \end{aligned}\right\} \tag{53.3}$$

The condition for equilibrium, (50.16), is expressed in terms of $\mu^{(t)}$ in the following form:

$$\mu^{(t)} = \mu, \tag{53.4}$$

where μ is the chemical potential of the bulk phase. This corresponds to the quasithermodynamic condition for equilibrium, (19.4). Furthermore, this condition is derived by minimization of the Helmholtz free energy as a function of the local densities just as the fundamental postulate of quasithermodynamics discussed in Sect. 18.

To obtain an expression for the tangential pressure we shall change the area of the plane interface. Then the change in the Helmholtz free energy given by (50.14) is

$$\left.\begin{aligned} \left(\frac{\partial F}{\partial A}\right)_N &= \nu kT\sum_{t=-u}^{w}\left\{x^{(t)}\log x^{(t)} + \left(1-x^{(t)}\right)\log\left(1-x^{(t)}\right)\right\} + \frac{\nu\zeta}{2}\sum_{t=-u}^{w}\left\{2l\,x^{(t)}\left(1-x^{(t)}\right) + \right. \\ &\quad \left. + m\,x^{(t)}\left(1-x^{(t+1)}\right) + m\,x^{(t+1)}\left(1-x^{(t)}\right) + m\,x^{(t)}\left(1-x^{(t-1)}\right) + m\,x^{(t-1)}\left(1-x^{(t)}\right)\right\} \\ &= \nu kT\sum_{t=-u}^{w}\left\{x^{(t)}\log x^{(t)} + \left(1-x^{(t)}\right)\log\left(1-x^{(t)}\right)\right\} + \\ &\quad + \nu\zeta\sum_{t=-u}^{w}\left\{m\,x^{(t)}\left(1-x^{(t-1)}\right) + l\,x^{(t)}\left(1-x^{(t)}\right) + m\,x^{(t)}\left(1-x^{(t+1)}\right)\right\}, \end{aligned}\right\} \tag{53.5}$$

[1] S. Ono: Suppl. Nuovo Cim. 9, 166 (1958).

where the second expression can be obtained by using the fact that the changes in the $x^{(t)}$'s due to the change in area do not contribute according to (50.16) with (51.3) as in the case of (52.13).

Meanwhile, from (50.16) and (51.3) we obtain

$$kT\log x^{(t)} = kT\log(1-x^{(t)}) - \zeta\{m(1-2x^{(t-1)}) + l(1-2x^{(t)}) + m(1-2x^{(t+1)})\}. \quad (53.6)$$

Substituting (53.6) into the second expression of (53.5), we have

$$\left(\frac{\partial F}{\partial A}\right)_N = \nu kT \sum_{t=-u}^{w} \log(1-x^{(t)}) + \nu\zeta \sum_{t=-u}^{w} x^{(t)}(m x^{(t-1)} + l x^{(t)} + m x^{(t+1)}). \quad (53.7)$$

On the other hand, the elementary work done, due to the increase δA in area, at constant T, by the portion of the system located between z and $z+dz$, is given by (18.4). Then, in the case of the hole theory, if the tangential pressure for the t-th layer is denoted by $p_T^{(t)}$, the work done by the t-th layer of the system is

$$\delta w^{(t)} = p_T^{(t)}(\delta A)\, d. \quad (53.8)$$

Thus, the change in the Helmholtz free energy due to the change in the area, δA, may be expressed as

$$\left.\begin{aligned} \delta F &= -\sum_t \delta w^{(t)} \\ &= -(\delta A)\sum_t p_T^{(t)} d. \end{aligned}\right\} \quad (53.9)$$

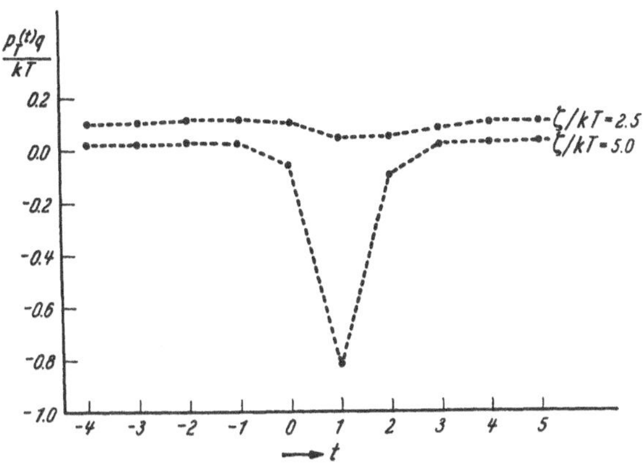

Fig. 27. Spatial variation of tangential pressure calculated from the hole theory. ($\zeta/kT = 2.0$ corresponds to the critical temperature.)

Comparing (53.9) with (53.7), we obtain an expression for the tangential pressure[1]

$$p_T^{(t)} = -\frac{1}{q}\{kT\log(1-x^{(t)}) + \zeta x^{(t)}(m x^{(t-1)} + l x^{(t)} + m x^{(t+1)})\}. \quad (53.10)$$

In the interior of the bulk phases (53.10) reduces to the ordinary hydrostatic pressure given by (49.23).

From (53.1) to (53.3) and (53.10) we readily obtain the relation

$$\mu = \mu^{(t)} = f^{(t)} + p_T^{(t)}\left(\frac{q}{x^{(t)}}\right). \quad (53.11)$$

Since $q^{-1}x^{(t)}$ is the number density of molecules in the t-th layer, (53.11) is identical in form with the quasithermodynamic condition for thermodynamic equilibrium given by (19.4).

Furthermore, we obtain from (49.23), (49.16), (52.7), and (53.10)

$$\gamma = \sum_{t=-u}^{w}(p - p_T^{(t)})\, d, \quad (53.12)$$

which corresponds to the mechanical definition of surface tension (14.5) or (20.5).

Thus, it is seen that in the hole theory of surface tension we have relations identical in form with those of quasithermodynamics. The chemical potential $\mu^{(t)}$ given by (53.3) depends, however, not only on $x^{(t)}$ but on $x^{(t+1)}$ and $x^{(t-1)}$.

[1] Definition of the tangential pressure is not unique. — See also Sect. 36.

In our quasithermodynamic theory (see Sects. 17 to 20), the chemical potential $\mu^{(t)}$ is a function of $x^{(t)}$ but depends neither on $x^{(t+1)}$ nor $x^{(t-1)}$. According to this hypothesis $\mu^{(t)}$ may be obtained from (49.24) when x is replaced by $x^{(t)}$. Consequently we can see that the hole theory does not correspond to the crude form of our quasithermodynamics but rather to Hill's modification [31a], [31b] discussed in Sect. 39 (see also Tolman's treatment [11]). The resemblance between the density versus depth curves given in Fig. 15 and Fig. 25 seems to be due to this fact.

The variation of the tangential pressure (53.10) within the interface, calculated for the (111)-plane, is shown in Fig. 27. In the case of liquid argon, the minimum value of the tangential pressure is -280 atm for $\zeta/kT = 5.0$, which corresponds to $T = 104°$ K.

II. Solutions of non-electrolytes.

a) Regular solutions.

54. Application of cell method to solutions. Let us now consider a mixture of molecules of two species A and B, and evaluate the partition function by dividing the space into cells in a manner similar to the case of pure liquids discussed in the preceding sections. Let us assume that the size of a molecule of component species A is nearly equal to that of B, since otherwise some modifications must be introduced. The cell size is chosen so that multiple occupancy is excluded and at the same time the interaction of molecules beyond immediately adjacent cells may be regarded as only a very small correction. The molecules are assumed to be spherical or nearly spherical.

The same cell methods discussed in the foregoing are applicable to fluid mixtures as well, but we shall restrict ourselves to a more simplified treatment of binary solutions. Let the number of cells be equal to the total number of molecules. First we assume that there are no vacant cells and that molecules of both species pack so that their centers form a face-centered cubic lattice. Secondly, we assume that the free volumes v_{fA} and v_{fB} of species A and B do not depend on the arrangement of the neighboring molecules and that they are equal to those in their respective pure liquids. Although this restriction appears to be too severe, we can take account of the dependence of free volumes on the ratio of the number of molecules of species A to that of species B in nearest-neighbor cells, without any other essential alteration, by employing the linear approximation[1] analogous with (49.13) in the hole theory of pure liquids. We shall refer to solutions which satisfy the above assumptions together with the assumption concerning the cell size, as the regular solutions[2].

Let χ_{AA}, χ_{AB}, and χ_{BB} denote the interaction potential energies of pairs of molecules A and A, A and B, and B and B at the centers of adjacent cells, to which we shall refer hereafter as AA, AB and BB contacts, respectively. Symbols N_A and N_B denote the total numbers of molecules A and B in the system, respectively. Then the partition function of the solution is, according to (24.2),

$$Z_{N_A N_B} = (\lambda_A^{-3} j_A)^{N_A} (\lambda_B^{-3} j_B)^{N_B} \eta_A^{N_A} \eta_A^{N_B} Q_{N_A N_B}^{(1)}, \tag{54.1}$$

$$\lambda_A^{-1} = \frac{\sqrt{2\pi m_A kT}}{h}; \qquad \lambda_B^{-1} = \frac{\sqrt{2\pi m_B kT}}{h}, \tag{54.2}$$

[1] S. Ono: Mem. Fac. Engng., Kyushu Univ. **12**, 201 (1950).

[2] This definition of a regular solution is somewhat different from that of J.H. Hildebrand [J. Amer. Chem. Soc. **51**, 66 (1929)], but essentially identical with the strictly regular solution of Fowler and Guggenheim (cf. R.H. Fowler and E.A. Guggenheim: Statistical Thermodynamics, 2nd impression, § 814. London: Cambridge University Press 1949).

where $\eta_A^{N_A}\eta_B^{N_B}Q_{N_A N_B}^{(1)}$ is the configuration integral and η_A and η_B are the factors corresponding to the communal entropy as defined by (44.7). The configuration integral $Q_{N_A N_B}^{(1)}$ corresponds to the state in which each cell has one and only one molecule of either species [see (48.3) and (48.14)]:

$$Q_{N_A N_B}^{(1)} = v_{fA}^{N_A} v_{fB}^{N_B} \exp\{-(N_A c\chi_{AA} + N_B c\chi_{BB})/2kT\} \sum_X G(N_A, N_B, X)\, \mathrm{e}^{-cX\chi/kT}, \quad (54.3)$$

$$\chi = \chi_{AB} - \frac{\chi_{AA}+\chi_{BB}}{2}, \quad (54.4)$$

where cX is the total number of AB contacts and $G(N_A, N_B, X)$ the number of the distinguishable arrangements for a specified value of X.

As in the case of pure liquids in Sect. 44, we assume the low density limit (44.9) is also applicable to η_A and η_B:

$$\eta_A = \eta_B = \mathrm{e}. \quad (54.5)$$

We shall evaluate the configuration integral based on the approximation of random mixing[1] as in Sect. 49 for pure liquids. Let x be the mole fraction of component A and N be the total number of molecules:

$$x = \frac{N_A}{N}, \qquad 1 - x = \frac{N_B}{N}; \quad (54.6)$$

$$N = N_A + N_B. \quad (54.7)$$

Using the assumption of random mixing, the average value of cX is, by analogy with (49.17), expressed as

$$c\overline{X} = cNx(1-x). \quad (54.8)$$

Then, replacing X in (54.3) by $\overline{X}$ and using the relation $\sum_X G(N_A, N_B, X) = N!/N_A!\,N_B!$, we obtain

$$Q_{N_A N_B}^{(1)} = \mathrm{e}^{-(N_A c\chi_{AA}+N_B c\chi_{BB})/2kT}\, v_{fA}^{N_A} v_{fB}^{N_B} \frac{N!}{N_A!\,N_B!} \exp\{-Nc\chi(1-x)\,x/kT\}. \quad (54.9)$$

Recalling (24.1) and using the Stirling formula we obtain from (54.1), (54.5) and (54.9) the following expression for the Helmholtz free energy of the solution:

$$\left.\begin{aligned} F = \frac{c}{2}\{N_A\chi_{AA} + N_B\chi_{BB}\} - N_A kT\log(\lambda_A^{-3} j_A e v_{fA}) - N_B kT\log(\lambda_B^{-3} j_B e v_{fB}) + \\ + NkT\{x\log x + (1-x)\log(1-x)\} + Ncx(1-x)\chi. \end{aligned}\right\} \quad (54.10)$$

Then the difference between the chemical potentials μ_A and μ_B of molecules of species A and B is given by

$$\left.\begin{aligned} \mu_A - \mu_B = \frac{1}{N}\left(\frac{\partial F}{\partial x}\right)_{V,T,N} = \frac{c}{2}(\chi_{AA} - \chi_{BB}) + kT\log\left(\frac{\lambda_B^{-3} j_B v_{fB}}{\lambda_A^{-3} j_A v_{fA}}\right) + \\ + kT\log\frac{x}{1-x} + c(1-2x)\chi. \end{aligned}\right\} \quad (54.11)$$

To obtain the chemical potential μ_A and μ_B separately, we neglect the difference between the Helmholtz free energy and Gibbs free energy. This is usually legitimate under a pressure such as the orthobaric pressure. Then we have

$$\mu_A = \left(\frac{\partial G}{\partial N_A}\right)_{p,T,N_B} \approx \left(\frac{\partial F}{\partial N_A}\right)_{p,T,N_B}. \quad (54.12)$$

---

[1] See also R. H. FOWLER and E. A. GUGGENHEIM: Statistical Thermodynamics, 2nd impression, §§ 814–818. London: Cambridge University Press 1949.

Thus, using (54.10) we immediately obtain

$$\mu_A = \frac{c}{2}\chi_{AA} - kT\log(\lambda_A^{-3} j_A\, e\, v_{fA}) + kT\log x + c(1-x)^2\chi, \tag{54.13}$$

$$\mu_B = \frac{c}{2}\chi_{BB} - kT\log(\lambda_B^{-3} j_B\, e\, v_{fB}) + kT\log(1-x) + c\,x^2\chi. \tag{54.14}$$

When the system is composed of two phases, say α and β, each of the chemical potentials μ_A and μ_B must have the same value in both phases. From this we obtain the following condition for equilibrium:

$$\frac{c\chi}{kT}(2x^\alpha - 1) = \log\frac{x^\alpha}{1-x^\alpha}, \tag{54.15}$$

$$x^\alpha + x^\beta = 1, \tag{54.16}$$

which correspond to (49.26) and (49.27), respectively. It is seen from the above that two phases can coexist in equilibrium only when χ is positive.

55. Regular solution with plane interface. Let us consider a regular solution separated by a plane interface of area A from the vapor phase. For convenience we consider the system confined to a parallelepiped vessel as shown in Fig. 28. We assume that the temperature is so low that we may regard the vapor phase as a vacuum. In the case when we must take account of the vapor density, we also cannot neglect the presence of holes in the solution[1]. Choosing the dividing surface HK just above and parallel to the exposed lattice plane, we refer to the successive layers of lattice cells parallel to the exposed lattice plane, as the 1st, 2nd, ..., w-th layers (see Fig. 28).

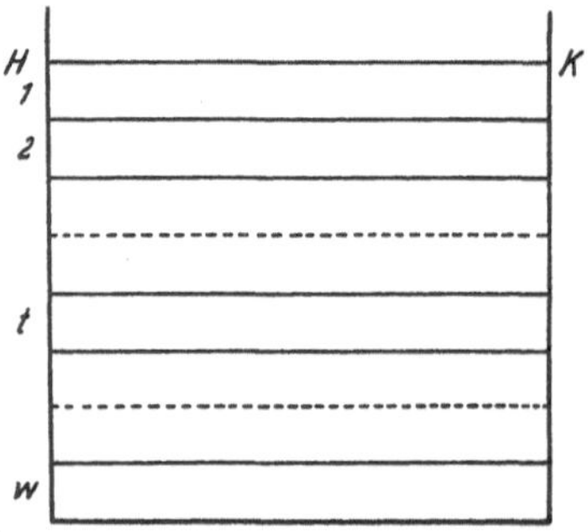

Fig. 28. A solution with plane interface HK exposed to its vapor phase of near vacuum.

Let $N_A^{(t)}$ and $N_B^{(t)}$ denote the numbers of molecules of species A and B in the t-th layer, respectively. Then, in analogy with (50.1), we can write the configuration integral in the form:

$$Q_{N_A N_B}^{(1)} = \sum_{N_A^{(1)}} \cdots \sum_{N_A^{(w)}} Q(N_A^{(1)}, N_A^{(2)}, \ldots, N_A^{(w)}), \tag{55.1}$$

where $Q(N_A^{(1)}, N_A^{(2)}, \ldots, N_A^{(w)})$ is the term corresponding to the set of numbers $N_A^{(1)}, N_A^{(2)}, \ldots, N_A^{(w)}$. These terms are expressed in a form analogous with (50.4), but we must take into consideration the following modifications.

First we shall consider the change in the total number of AA contacts due to the presence of the surface. As in the hole theory, we use the notation l and m defined in Sect. 50. Among the cN_A nearest neighbors of N_A molecules of species A, cX neighbors are molecules of species B and $mcN_A^{(1)}$ neighbors of the $N_A^{(1)}$ surface molecules of species A are vacant cells in the vapor phase. Since the number of the remaining pairs, $cN_A - cX - mcN_A^{(1)}$, is twice the number of AA contacts and likewise $(cN_B - cX - mcN_B^{(1)})$ is twice that of BB contacts:

$$\left.\begin{aligned} &\text{the number of } AA \text{ contacts: } \frac{cN_A - cX - mcN_A^{(1)}}{2},\\ &\text{the number of } BB \text{ contacts: } \frac{cN_B - cX - mcN_B^{(1)}}{2}. \end{aligned}\right\} \tag{55.2}$$

[1] Actually, Kurata [60] applied the hole theory given in Sects. 49 to 51, to a binary regular solution by treating it as a triple-component system of molecules of species A and B and holes.

Secondly, we shall take account of modification of the free volumes due to the surface. Let v'_{fA} and v'_{fB} denote the free volumes of surface molecules of species A and B, respectively. Then we have, instead of (50.4),

$$\left.\begin{aligned} Q(N_A^{(1)},\ldots,N_A^{(w)}) &= v_{fA}^{N_A}\, v_{fB}^{N_B}\left(\frac{v'_{fA}}{v_{fA}}\right)^{N_A^{(1)}}\left(\frac{v'_{fB}}{v_{fB}}\right)^{N_B^{(1)}}\times \\ &\times \exp\left[-\frac{1}{2kT}\{(N_A c\,\chi_{AA}+N_B c\,\chi_{BB})-(N_A^{(1)} mc\,\chi_{AA}+N_B^{(1)} mc\,\chi_{BB})\}\right]\times \\ &\times \sum_X g(N_A^{(1)},\ldots,N_B^{(w)};X)\,\mathrm{e}^{-Xc\chi/kT}, \end{aligned}\right\} \quad (55.3)$$

where $g(N_A^{(1)}, N_A^{(2)}, \ldots, N_A^{(w)}; X)$ is the number of the configurations for a given set of numbers $N_A^{(1)}, N_A^{(2)}, \ldots, N_A^{(w)}$, and X.

Let $x^{(t)}$ be the mole fraction of species A in the t-th layer. Then we have

$$x^{(t)} = \frac{N_A^{(t)}}{A\nu}; \qquad 1 - x^{(t)} = \frac{N_B^{(t)}}{A\nu}, \qquad (55.4)$$

where as in (52.9) ν denotes the number of cells per unit area of each layer. Then the probability of finding a molecule of species A in a cell in the t-th layer is equal to $x^{(t)}$, and therefore, using the assumption of random mixing, we obtain, exactly in analogy with (50.10), the average number of AB contacts as follows:

$$\left.\begin{aligned} c\overline{X} = \frac{A\nu c}{2}\Big[\sum_{t=2}^{w}\{&2l\,x^{(t)}(1-x^{(t)}) + m\,x^{(t)}(1-x^{(t+1)}) + \\ &+ m\,x^{(t+1)}(1-x^{(t)}) + m\,x^{(t)}(1-x^{(t-1)}) + m\,x^{(t-1)}(1-x^{(t)})\} + \\ &+ 2l\,x^{(t)}(1-x^{(1)}) + m\,x^{(1)}(1-x^{(2)}) + m\,x^{(2)}(1-x^{(1)})\Big]. \end{aligned}\right\} \quad (55.5)$$

Replacing X in (55.3) by the above $\overline{X}$, we obtain

$$\sum_X g(N_A^{(1)}, N_A^{(2)}, \ldots, N_A^{(w)}; X)\,\mathrm{e}^{-Xc\chi/kT} = \mathrm{e}^{-\overline{X}c\chi/kT}\prod_{t=1}^{w}\frac{(A\nu)!}{(A\nu x^{(t)})!\,(A\nu - A\nu x^{(t)})!}. \qquad (55.6)$$

Just as in Sect. 50, we may replace (55.1) by its maximum term with sufficient accuracy. Thus substituting (55.6) into (55.3) and recalling (24.1) and (54.1) together with (54.5), we obtain the Helmholtz free energy for a regular solution with a plane interface in the following form [*58*],

$$\left.\begin{aligned} F = \frac{c}{2}(N_A\chi_{AA}+N_B\chi_{BB}) &- \frac{A\nu mc}{2}\{x^{(1)}\chi_{AA} + (1-x^{(1)})\chi_{BB}\} - \\ &- N_A kT\log(\lambda_A^{-3} j_A\, e v_{fA}) - N_B kT\log(\lambda_B^{-3} j_B\, e v_{fB}) - \\ &- A\nu x^{(1)} kT\log\frac{v'_{fA}}{v_{fA}} - A\nu(1-x^{(1)})\,kT\log\frac{v'_{fB}}{v_{fB}} + \\ &+ A\nu kT\sum_{t=1}^{w}\{x^{(t)}\log x^{(t)} + (1-x^{(t)})\log(1-x^{(t)})\} + \\ &+ \frac{A\nu c\chi}{2}\Big[\sum_{t=2}^{w}\{2l\,x^{(t)}(1-x^{(t)}) + m\,x^{(t)}(1-x^{(t+1)}) + \\ &+ m\,x^{(t+1)}(1-x^{(t)}) + m\,x^{(t)}(1-x^{(t-1)}) + m\,x^{(t-1)}(1-x^{(t)})\} + \\ &+ 2l\,x^{(1)}(1-x^{(1)}) + m\,x^{(1)}(1-x^{(2)}) + m\,x^{(2)}(1-x^{(1)})\Big], \end{aligned}\right\} \quad (55.7)$$

where the set of $x^{(1)}, x^{(2)}, \ldots, x^{(w)}$ should be chosen so as to maximize $Q(N_A^{(1)}, \ldots, N_A^{(w)})$, i.e. minimize F at given values of N_A and N_B. This particular set of $x^{(1)}, \ldots, x^{(w)}$ is regarded as the set of equilibrium values of the local composition, and the condition to determine these values is given by

$$\left.\begin{aligned} &-kT\log\frac{v'_{fA}}{v_{fA}} + kT\log\frac{v'_{fB}}{v_{fB}} - \frac{mc}{2}(\chi_{AA}-\chi_{BB}) + \\ &\qquad + \{l(1-2x^{(1)}) + m(1-2x^{(2)})\}c\chi + kT\log\frac{x^{(1)}}{1-x^{(1)}} = \mu^*, \\ &\{m(1-2x^{(t-1)}) + l(1-2x^{(t)}) + m(1-2x^{(t+1)})\}c\chi + \\ &\qquad + kT\log\frac{x^{(t)}}{1-x^{(t)}} = \mu^*; \quad 2 \leq t \leq w, \end{aligned}\right\} \tag{55.8}$$

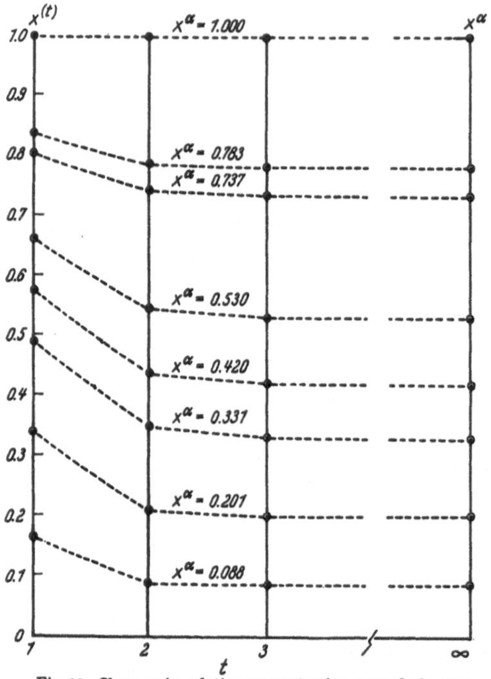

Fig. 29. Changes in relative concentration around the surface for a binary solution with $c\chi = 450$ cal/mole at 15° C.

Lagrange's multiplier μ^* being determined from the constraint

$$A\nu\sum_{t=1}^{w} x^{(t)} = N_A. \tag{55.9}$$

Since $x^{(t)}$ becomes identical with x^α in the interior of the solution, we have the condition

$$\lim_{t\to\infty} x^{(t)} = x^\alpha. \tag{55.10}$$

At the same time, with the help of (50.6) and the second relation of (55.8) we have

$$\left.\begin{aligned} &c\chi(1-2x^\alpha) + \\ &\qquad + kT\log\frac{x^\alpha}{1-x^\alpha} = \mu^*, \end{aligned}\right\} \tag{55.11}$$

from which we can conveniently determine μ^*.

Using (55.11), (50.6) and the notation of differences defined by (51.4), we can rewrite (55.8) in the following forms:

$$\left.\begin{aligned} \Delta x^{(1)} = \frac{1}{2\nu mc\chi}(\gamma_A-\gamma_B) + \frac{1}{2m}\{l(1-2x^{(1)}) + m(1-2x^{(1)}) - (1-2x^\alpha)\} + \\ + \frac{kT}{2mc\chi}\left(\log\frac{x^{(1)}}{1-x^{(1)}} - \log\frac{x^\alpha}{1-x^\alpha}\right), \end{aligned}\right\} \tag{55.12}$$

$$\Delta^2 x^{(t)} = \frac{1}{m}(x^\alpha - x^{(t)}) + \frac{kT}{2mc\chi}\left\{\log\frac{x^{(t)}}{1-x^{(t)}} - \log\frac{x^\alpha}{1-x^\alpha}\right\}; \quad 2 \leq t, \tag{55.13}$$

where γ_A and γ_B are the surface tensions of pure liquids A and B given respectively by

$$\gamma_A = -\frac{\nu mc}{2}\chi_{AA} - \nu kT\log\frac{v'_{fA}}{v_{fA}}, \tag{55.14}$$

$$\gamma_B = -\frac{\nu mc}{2}\chi_{BB} - \nu kT\log\frac{v'_{fB}}{v_{fB}}. \tag{55.15}$$

The above equations (55.14) and (55.15) correspond to the expression for the surface tension of pure liquids, (48.8), based on the free volume theory. Eqs. (55.12) and (55.13) were first obtained and solved numerically by ONO [58][1].

Some solutions of (55.12) and (55.13) for $c\chi = 450$ cal/mole (see Sect. 58) are shown in Fig. 29.

56. Surface tension and Gibbs adsorption formula [61]. If the exposed face of the lattice of a regular solution is chosen as the dividing surface and the vapor density is neglected, we obtain from (54.6) and (55.4) the superficial density of molecules of species A and B, according to the definition given by (2.1) and (6.5),

$$\Gamma_A = \nu \sum_{t=1}^{w} (x^{(t)} - x^{\alpha}), \quad \Gamma_B = \nu \sum_{t=1}^{w} \{(1 - x^{(t)}) - (1 - x^{\alpha})\}. \tag{56.1}$$

Then we can see immediately that

$$\Gamma_A + \Gamma_B = 0. \tag{56.2}$$

Multiplying the first equation of (56.1) by (54.11) in which x is put equal to x^{α}, and rearranging the result, we have the relation

$$\left.\begin{aligned}(\mu_A - \mu_B)\Gamma_A = \frac{\nu c}{2} \sum_{t=1}^{w} \big[&(x^{(t)} - x^{\alpha})\chi_{AA} + \{(1 - x^{(t)}) - (1 - x^{\alpha})\}\chi_{BB} - \\ &-(x^{(t)} - x^{\alpha})\, kT \log(\lambda_A^{-3} j_A\, e\, v_{fA}) - \{(1 - x^{(t)}) - (1 - x^{\alpha})\}\, kT \times \\ &\times \log(\lambda_B^{-3} j_B\, e\, v_{fB})\big] + \nu kT \sum_{t=1}^{w} \big[(x^{(t)} - x^{\alpha}) \log x^{\alpha} + \\ &+ \{(1 - x^{(t)}) - (1 - x^{\alpha})\} \log(1 - x^{\alpha})\big] + \nu c \chi \sum_{t=1}^{w} (x^{(t)} - x^{\alpha})(1 - 2x^{\alpha}).\end{aligned}\right\} \tag{56.3}$$

On the other hand, the Helmholtz free energy of the bulk phase which remains strictly homogeneous right up to the dividing surface, is, according to (54.10), expressed as

$$\left.\begin{aligned}F^{\alpha} = \frac{c}{2}(N_A^{\alpha}\chi_{AA} + N_B^{\alpha}\chi_{BB}) - N_A^{\alpha} kT \log(\lambda_A^{-3} j_A\, e\, v_{fA}) - N_B^{\alpha} kT \log(\lambda_B^{-3} j_B\, e\, v_{fB}) + \\ + N kT\{x^{\alpha} \log x^{\alpha} + (1 - x^{\alpha}) \log(1 - x^{\alpha})\} + N c\, x^{\alpha}(1 - x^{\alpha})\chi,\end{aligned}\right\} \tag{56.4}$$

where N_A^{α} and N_B^{α} are, respectively, the total number of molecules A and B, in the above hypothetical bulk phase α. Using (56.2) we can reduce the expression (5.8) to

$$\gamma = \frac{F - F^{\alpha}}{A} - (\mu_A - \mu_B)\Gamma_A. \tag{56.5}$$

Substitution of (55.7), (56.3) and (56.4) into (56.5) yields

$$\left.\begin{aligned}\gamma = x^{(1)}\gamma_A + (1 - x^{(1)})\gamma_B + \nu kT \sum_{t=1}^{w} \left\{x^{(t)} \log \frac{x^{(t)}}{x^{\alpha}} + (1 - x^{(t)}) \log \frac{1 - x^{(t)}}{1 - x^{\alpha}}\right\} - \\ - \frac{\nu c \chi}{2}\Big[2l(x^{(1)} - x^{\alpha})^2 + m\{4(x^{(1)} - x^{\alpha})^2 - (x^{(2)} - x^{(1)})(1 - 2x^{(1)}) + \\ + 2x^{(1)}(1 - x^{(1)})\} + \sum_{t=2}^{w} \{2l(x^{(t)} - x^{\alpha})^2 + \\ + 4m(x^{(t)} - x^{\alpha})^2 - m(x^{(t-1)} + x^{(t+1)} - 2x^{(t)})(1 - 2x^{(t)})\}\Big],\end{aligned}\right\} \tag{56.6}$$

[1] ONO obtained these equations for the case of a regular solution in contact with a solid wall regarded as an adsorbent. But there is no essential alteration in the above treatment except that the solid wall is replaced by the vapor phase which is practically regarded as a vacuum. He obtained the adsorption isotherm on the basis of the difference equation method [S. ONO: J. Phys. Soc. Japan 5, 232 (1950)].

where γ_A and γ_B are the surface tensions of pure liquids A and B, respectively, given by (55.14) and (55.15).

The local composition of molecules $x^{(1)}, x^{(2)}, \ldots, x^{(t)}, \ldots, x^{(w)}$ changes with the value of μ_A and μ_B. Let us change $x^{(1)}, x^{(2)}, \ldots, x^{(t)}, \ldots, x^{(w)}$ in such a way that they are subject to the condition for equilibrium given by (55.8), but not to (55.9), i.e., the number of molecules is not restricted. Then, from (55.7) and (56.4), we obtain, with use of (54.11) and (55.11),

$$\frac{1}{A} d(F - F^{\alpha}) = (\mu_A - \mu_B)\, d\Gamma_A. \tag{56.7}$$

From (56.5), (56.7) and (56.2) we find

$$-d\gamma = \Gamma_A\, d(\mu_A - \mu_B) = \Gamma_A\, d\mu_A + \Gamma_B\, d\mu_B. \tag{56.8}$$

This is the Gibbs adsorption equation given by (7.1). Thus, the present theory is self-consistent with respect to the thermodynamic relations of the surface.

Furthermore, we have from (54.11) the following relation

$$\frac{d(\mu_A - \mu_B)}{d x^{\alpha}} = -2c\chi + \frac{kT}{x^{\alpha}(1 - x^{\alpha})}. \tag{56.9}$$

Combining with (56.8) we obtain Hildebrand's equation[1], i.e. the exact Gibbs adsorption formula for a regular solution:

$$\frac{d\gamma}{d x^{\alpha}} = -\frac{\Gamma_A}{x^{\alpha}(1 - x^{\alpha})} \{kT - 2c\, x^{\alpha}(1 - x^{\alpha})\, \chi\}. \tag{56.10}$$

For a perfect solution, which is a regular solution with the propertiy that $\chi = 0$, the equilibrium condition (55.8) reduces simply, with use of (55.14) and (55.15), to

$$\gamma_A - \gamma_B + \nu\, kT \log \frac{x^{(1)}}{1 - x^{(1)}} = \nu\, kT \log \frac{x^{\alpha}}{1 - x^{\alpha}}, \tag{56.11}$$

$$x^{(2)} = x^{(3)} = \cdots = x^{\alpha}. \tag{56.12}$$

This implies that, in the case of a perfect solution, the concentration $x^{(t)}$ in the t-th layer is the same as that in the bulk, except for the first layer. Then the assumption that the difference in the composition of the surface from the bulk is confined to a unimolecular layer, as used in the early thories of the surface tension of solutions [55], [56], [57], is legitimate in the case of a perfect solution.

From (56.6) we can readily obtain the following expression for the surface tension of a perfect solution

$$\gamma = x^{(1)} \gamma_A + (1 - x^{(1)})\, \gamma_B + \nu\, kT \left\{ x^{(1)} \log \frac{x^{(1)}}{x^{\alpha}} + (1 - x^{(1)}) \log \frac{1 - x^{(1)}}{1 - x^{\alpha}} \right\}. \tag{56.13}$$

Using (56.11) we can rewrite (56.13) in the form

$$\gamma = \gamma_A + \nu\, kT \log \frac{x^{(1)}}{x^{\alpha}} = \gamma_B + \nu\, kT \log \frac{1 - x^{(1)}}{1 - x^{\alpha}}, \tag{56.14}$$

which is equivalent to Guggenheim's mixing law [57] for surface tension of perfect solutions

$$e^{-\gamma/\nu kT} = x^{\alpha} e^{-\gamma_A/\nu kT} + (1 - x^{\alpha})\, e^{-\gamma_B/\nu kT}. \tag{56.15}$$

[1] J.H. Hildebrand: Solubility, p. 190. New York: Reinhold 1939.

57. Monolayer model for interface. As seen from (56.11), the *monolayer model*, in which the difference in composition from that of the bulk phase is confined to a single layer of molecules at the interface, is accurate for a perfect solution. Furthermore, we observe from Fig. 29 that, even in the more general case of a regular solution, the difference in the composition is small except for the first layer. Thus the monolayer model seems to be approximately valid and to be applicable to a non-perfect regular solutions [*57*].

In the case of the monolayer model, in which all the $x^{(t)}$'s except for $t=1$ are equal to the concentration of the bulk phase, (56.6) reduces to

$$\left.\begin{aligned}\gamma = x^{(1)}\gamma_A + (1-x^{(1)})\,\gamma_B + \nu\, kT\left\{x^{(1)}\log\frac{x^{(1)}}{x^\alpha} + (1-x^{(1)})\log\frac{1-x^{(t)}}{1-x^\alpha}\right\} - \\ - \nu c\chi\{(l+m)\,(x^{(1)}-x^\alpha)^2 + m\,x^{(1)}(1-x^{(1)})\}.\end{aligned}\right\} \tag{57.1}$$

The condition for equilibrium (55.12) becomes

$$\left.\begin{aligned}\gamma_A - \gamma_B + \nu\{l(1-2x^{(1)}) + m(1-2x^\alpha)\}\,c\chi + \nu\, kT\log\frac{x^{(1)}}{1-x^{(1)}} \\ = \nu(1-2x^\alpha)\,c\chi + \nu\, kT\log\frac{x^\alpha}{1-x^\alpha},\end{aligned}\right\} \tag{57.2}$$

by means of which we can determine the equilibrium value of $x^{(1)}$ for the monolayer model. Using (57.2), we may rewrite (57.1) in GUGGENHEIM's form [*57*],

$$\left.\begin{aligned}\gamma &= \gamma_A + \nu\, kT\log\frac{x^{(1)}}{x^\alpha} + \nu l c\chi\{(1-x^{(1)})^2 - (1-x^\alpha)^2\} - \nu m c\chi(1-x^\alpha)^2 \\ &= \gamma_B + \nu\, kT\log\frac{1-x^{(1)}}{1-x^\alpha} + \nu l c\chi\{(x^{(1)})^2 - (x^\alpha)^2\} - \nu m c\chi(x^\alpha)^2.\end{aligned}\right\} \tag{57.3}$$

Nevertheless, the assumption of a unimolecular layer is not self-consistent and incompatible with the Gibbs adsorption formula as pointed out by DEFAY and PRIGOGINE [*59*]. According to the monolayer model, from (56.1) the superficial density of molecules of species A is given by

$$\Gamma_A = \nu(x^{(1)} - x^\alpha). \tag{57.4}$$

We obtain from (57.3) and (57.2)

$$\frac{d\gamma}{dx^\alpha} = \nu(x^{(1)} - x^\alpha)\left[-\frac{kT}{x^\alpha(1-x^\alpha)} + 2c(l+m)\chi\right]. \tag{57.5}$$

This can be rewritten, by means of (57.4), in the form

$$\frac{d\gamma}{dx^\alpha} = -\frac{\Gamma_A}{x^\alpha(1-x^\alpha)}\left[kT - 2c(l+m)\,x^\alpha(1-x^\alpha)\,\chi\right]. \tag{57.6}$$

This cannot be identical with HILDEBRAND's equation (56.10) except for the case of a perfect solution, for which $\chi=0$. In other words, the monolayer model is incompatible with the Gibbs adsorption formula for non-perfect regular solutions. This arises from the fact that the density distribution such that $x^{(2)} = x^{(3)} = \cdots = x^\alpha$ does not satisfy the second equation of the condition (55.8), and cosequently can never be in equilibrium.

Though it has been shown by DEFAY and PRIGOGINE [*59*] that the deviation from the Gibbs formula becomes practically inappreciable in the case of the two-layer model, any theory based upon a model of a finite number of layers, contradicts, in principle, the Gibbs formula [*61*].

58. Comparison with experiment. There are many kinds of solutions that may be regarded as approximately regular solutions. To compare calculated values based on cell theory with experimental data for approximately regular solutions we first need to estimate the mixing energy χ defined by (54.4).

Since the vapor in equilibrium with a regular solution is very dilute in the present treatment, it may be regarded with sufficient accuracy as a mixture of perfect gases. Then as given by (7.3) the chemical potentials become

$$\mu_A = \varphi_A + kT \log p_A; \qquad \mu_B = \varphi_B + kT \log p_B, \tag{58.1}$$

where p_A and p_B are the partial pressures of the components A and B, respectively. The chemical potentials, μ_A^0 and μ_B^0, of the pure liquids with vapor pressures p_A^0 and p_B^0 at the same temperature, are, respectively, given by

$$\mu_A^0 = \varphi_A + kT \log p_A^0; \quad \mu_B^0 = \varphi_B + kT \log p_B^0. \tag{58.2}$$

From (58.1) and (58.2) we obtain the following relation:

$$kT \log \left\{ \frac{p_A}{p_B} \frac{(1-x)\, p_B^0}{x p_A^0} \right\} = \{\mu_A - \mu_A^0 - kT \log x\} - \{\mu_B - \mu_B^0 - kT \log (1-x)\}. \tag{58.3}$$

Comparing with (54.11) we obtain

$$(1-2x)\, c\chi = kT \log \left(\frac{p_A}{p_B} \frac{(1-x)\, p_B^0}{x p_A^0} \right). \tag{58.4}$$

For the system of ether-acetone at 288° K, the value of $c\chi$ calculated from the vapor pressure, is $c\chi = 450$ cal/mole[1]. Then we obtain from (55.12) and (55.13) the curve for the concentration change of ether and acetone near the interface as shown in Fig. 29. The numerical calculation has been carried out for the (111)-plane as the exposed face, i.e. $l = \frac{1}{2}$, $m = \frac{1}{4}$, with use of the average ν given by

$$\frac{1}{\nu} = \frac{x}{\nu_A} + \frac{1-x}{\nu_B}, \tag{58.5}$$

where ν_A and ν_B are the values of ν for pure liquids A and B, respectively. The curve for the surface tension shown in Fig. 30 is calculated from (56.6) by using the above value of $c\chi$ and the experimental values of surface tension of pure ether and acetone. The values calculated by Prigogine and Narbond[2] with use of Guggenheim's formula (57.3) based on the monolayer model, are equal to those given in Fig. 30 with an accuracy of the order of one percent[3]. Thus we may conclude that the monolayer model gives a good concentration dependence of the surface tension of regular solutions, while the relative adsorption is significantly modified as seen from Fig. 29.

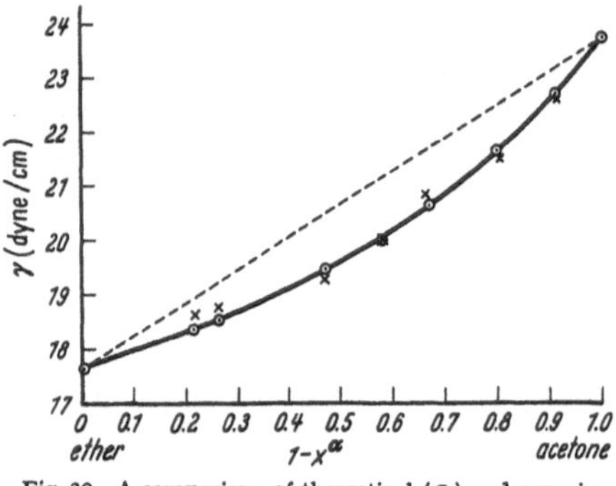

Fig. 30. A comparison of theoretical (⊙) and experimental[1] (×) surface tension for ether-acetone solutions at 15° C as a function of mole fraction of acetone, $1 - x^\alpha$.

[1] I. Prigogine and J. Narbond: Trans. Faraday Soc. **44**, 628 (1948).

[2] I. Prigogine: Trans. Faraday Soc. **44**, 626 (1948). — I. Prigogine and J. Narbond: Trans. Faraday Soc. **44**, 628 (1948).

[3] However, the best agreement was obtained in the case of simple cubic lattice, which seems hardly appropriate for regular solutions.

59. Dynamic surface tension. The concentration in the interface layer of a solution at thermodynamic equilibrium is in general different from that in the bulk phase as discussed in the foregoing. Immediately after a fresh interface has been formed, however, it may have almost the same concentration as that in the interior of the bulk phase. The surface tension of such an interface is called the *dynamic surface tension* [62]. The evaluation of such a dynamic surface tension of a regular solution is rather easy as will be shown below, although the evaluation of the surface tension of a system in an intermediate stage is difficult [63], [64]. The above interface is apparently out of thermodynamic equilibrium, so the dynamic surface tension should be calculated from equations based on thermodynamics of irreversible processes as given in Sect. 22.

Since the concentration in the fresh interface mentioned above is assumed to be exactly the same as that in the interior phase, the superficial number density is zero, i.e., $\Gamma_A = \Gamma_B = 0$ and hence (22.35) yields

$$\gamma_d = \frac{F - F^\alpha}{A}, \tag{59.1}$$

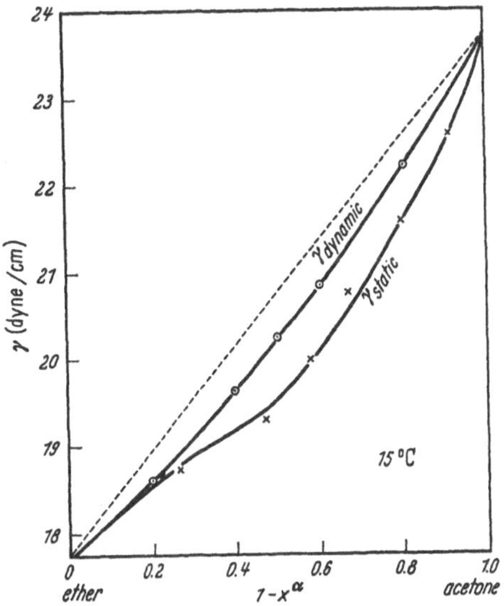

Fig. 31. A comparison of theoretical dynamic surface tension (⊙) and experimental static surface tension (×) for ether acetone solutions as a function of mole fraction of acetone (Prigogine and Defay [63]).

where the contribution of the vapor phase is neglected as before. Meanwhile, the Helmholtz free energy is given by

$$\left.\begin{aligned} F = {} & \frac{c}{2}\{N_A \chi_{AA} + N_B \chi_{BB}\} - \frac{A \nu m c}{2}\{x^\alpha \chi_{AA} + (1 - x^\alpha)\chi_{BB}\} - \\ & - N_A kT \log(\lambda_A^{-3} j_A e v_{fA}) - N_B kT \log(\lambda_B^{-3} j_B e v_{fB}) - A \nu x^\alpha kT \log \frac{v'_{fA}}{v_{fA}} - \\ & - A\nu(1 - x^\alpha) kT \log \frac{v'_{fB}}{v_{fB}} + N kT\{x^\alpha \log x^\alpha + (1 - x^\alpha)\log(1 - x^\alpha)\} + \\ & + N c x^\alpha (1 - x^\alpha)\chi - A \nu c m x^\alpha (1 - x^\alpha)\chi, \end{aligned}\right\} \tag{59.2}$$

which is obtained from (55.7) by putting all the $x^{(i)}$'s equal to x^α. Substitution of (59.2) and (56.4) into (59.1) yields

$$\left.\begin{aligned} \gamma_d = {} & -\frac{\nu c m}{2}\{x^\alpha \chi_{AA} + (1 - x^\alpha)\chi_{BB}\} - \nu x^\alpha kT \log \frac{v'_{fA}}{v_{fA}} - \\ & - \nu(1 - x^\alpha) kT \log \frac{v'_{fB}}{v_{fB}} - \nu c m x^\alpha (1 - x^\alpha)\chi. \end{aligned}\right\} \tag{59.3}$$

By means of the expressions for γ_A and γ_B given by (55.14) and (55.15), (59.3) can be rewritten in the form [63]

$$\gamma_d = x^\alpha \gamma_A + (1 - x^\alpha)\gamma_B - \nu c m x^\alpha (1 - x^\alpha)\chi. \tag{59.4}$$

The comparison of the theoretical dynamic surface tension based on (59.4) with the experimental static surface tension has been carried out by Prigogine and Defay [63] for the ether-acetone system mentioned in Sect. 58, as shown in Fig. 31, where m is chosen to be $\frac{1}{6}$, corresponding to a simple cubic lattice. In the calculation of these theoretical values they have utilized the static surface tensions γ_A and γ_B for pure liquids because the relaxation time for the establishment of mechanical equilibrium is relatively short compared with the time required for the interface of the solution to reach thermodynamic equilibrium through diffusion of substances from the interior of the solution to the surface.

Furthermore, in the case of a perfect solution (59.4) reduces to [63]

$$\gamma_d = x^\alpha \gamma_A + (1 - x^\alpha)\,\gamma_B \quad \text{(perfect solution)}, \tag{59.5}$$

which is evidently different from the expression for the static surface tension (56.13).

In general, the expression for surface tension has a form identical both for the equilibrium and non-equilibrium cases (see Sect. 37). Then, (59.4) is only a special case of (56.6) in which all the $x^{(t)}$'s are put equal to x^α.

PRIGOGINE and DEFAY [63] calculated also the lateral chemical potentials given by (22.19) and (22.20) from the values of the dynamic surface tension by means of the relation (22.31) which reduces, for the present case, to

$$x^\alpha \varepsilon_A^\alpha + (1 - x^\alpha)\,\varepsilon_B^\alpha = 0. \tag{59.6}$$

It has been shown by DEFAY [14] that this condition is actually the case for the dynamic surface tension, based upon the assumption that the change at the fresh surface is purely hydrodynamic with no irreversible production of entropy. However, this assumption does not seem so convincing to the authors.

b) Polymer solutions.

60. Lattice theory for athermal polymer solutions. Although the applications of the lattice theory to solutions have thus far been restricted to the case of solutions composed of molecules of nearly equal sizes, this method is also applicable to solutions composed of molecules of quite different sizes such as polymer solutions. For the present we shall restrict ourselves to a solution composed of molecules of solvent species A which occupy one cell per molecule and rigid-rod-like molecules[1] of solute species B which occupy r cells per molecule to deduce simple explicit formulae which are convenient for a qualitative discussion of surface tension of polymer solutions.

Among cr nearest-neighbor cells of the r cells occupied by a molecule B, $2(r-1)$ cells are occupied by the central molecule itself. Hence the number, qc, of the nearest-neighbor cells occupied by other molecules around a rigid-rod-like molecule of species B satisfies

$$c(r - q) = 2(r - 1). \tag{60.1}$$

Let us now regard any molecule of species B as being divided into r segments each occupying just one cell. Then the configuration integral for a homogeneous bulk solution composed of N_A molecules of species A and N_B molecules of species B is, in analogy with (54.3),

$$Q^{(1)}_{N_A N_B} = v_{fA}^{N_A} v_{fB}^{r N_B} \exp\left\{-\frac{1}{2kT}(N_A c \chi_{AA} + N_B c q \chi_{BB})\right\} \sum_X G(N_A, N_B, X)\,e^{-X c \chi/kT},$$

$$\chi = \chi_{AB} - \frac{\chi_{AA} + \chi_{BB}}{2}, \tag{60.2}$$

where cX and χ_{AB} denote the number of pairs of a molecule of species A and a segment of a molecule of species B and the interaction potential between them;

[1] For more general cases when polymer (trimer) molecules are flexible, see [66].

χ_{AA} and χ_{BB} are the intermolecular potential between two molecules of species A and the potential energy of interaction between two adjacent segments belonging to different molecules of species B, v_{fB} being the free volume per segment. The number of configurations $G(N_A, N_B, X)$ is different from that for regular solutions. When χ vanishes, the solution is called an athermal polymer solution, which corresponds to the perfect solution dealt with in the foregoing sections in connection with regular solutions.

The sum of $G(N_A, N_B, X)$ over X is the total number of ways of arranging N_A molecules of species A and N_B molecules of species B. According to FLORY and HUGGINS' method[1], we have

$$\sum_X G(N_A, N_B, X) = \frac{c^{N_B}}{(N_A + r N_B)^{N_B(r-1)}} \frac{1}{2^{N_B}} \frac{(N_A + r N_B)!}{N_A!\, N_B!}, \tag{60.3}$$

where $1/2^{N_B}$ results from the symmetry of polymer molecules. Thus (60.2) reduces for an athermal solution to

$$\left.\begin{aligned} Q^{(1)}_{N_A N_B} &= v_{fB}^{N_A} v_{fB}^{r N_B} \exp\left\{-\frac{1}{2kT}(N_A c \chi_{AA} + N_B c q \chi_{BB})\right\} \times \\ &\times \frac{c^{N_B}}{(N_A + r N_B)^{N_B(r-1)}} \frac{1}{2^{N_B}} \frac{(N_A + r N_B)!}{N_A!\, N_B!}. \end{aligned}\right\} \tag{60.4}$$

Then we can obtain in analogy with (54.10) the Helmholtz free energy F for the bulk athermal solution as follows:

$$\left.\begin{aligned} F &= \frac{c}{2}(N_A \chi_{AA} + N_B q \chi_{BB}) - N_A kT \log(\lambda_A^{-3} j_A \eta_A v_{fA}) - \\ &- N_B kT \log(\lambda_B^{-3} j_B \eta_B^r v_{fB}^r) - N_B kT \log\frac{c}{2} + \\ &+ (r-1) N_B kT - N_A kT \log\frac{N_A + r N_B}{N_A} - N_B kT \log\frac{N_A + r N_B}{N_B}, \end{aligned}\right\} \tag{60.5}$$

where η_B is the communal entropy per segment of a molecule B.

61. Surface tension of athermal polymer solutions [*68*]. We are now ready to consider an athermal polymer solution separated from its vapor by a plane interface. In general, rigid-rod molecules consisting of r segments each, may take various orientations with respect to the surface plane. It is extremely complicated to deal with such a system by taking account of the possible orientations of these molecules.

We shall, therefore, restrict ourselves to a simplified model in which only the configuration of rods parallel to the interface, is taken into account, as done by PRIGOGINE and MARÉCHAL [*68*]. The calculations for more general cases have been carried out by PRIGOGINE and his coworkers [*65*], [*66*], [*67*] and KURATA [*69*].

Let us consider an athermal polymer solution confined to a parallelepiped vessel and separated from its vapor phase (approximately a vacuum) by a dividing surface HK, which is located just above the exposed face of the lattice of the solution, as shown in Fig. 28 on p. 264. We give the numbers from 1 to w to the successive cell layers parallel to the surface as in Sect. 55 and use the same notation l and m as before.

[1] For example, see A. R. MILLER: The Theory of Solutions of High Polymers. London: Oxford University Press 1948.

If the orientations of rigid-rod molecules are confined only to the direction parallel to the surface, the number of configurations of molecules of species A and B, $\sum_X G(N_A, N_B, X)$, is diminished to

$$\sum_X G(N_A, N_B, X) = \frac{(l c)^{N_B}}{(N_A + r N_B)^{N_B(r-1)}} \frac{1}{2^{N_B}} \frac{(N_A + r N_B)!}{N_A! \, N_B!}. \tag{61.1}$$

Let us now consider such a hypothetical system that has the same bulk properties as in the interior of the above solution right up to the dividing surface. Then, with use of (61.1), we have, instead of (60.5),

$$\left.\begin{aligned} F^\alpha = {} & \frac{c}{2}(N_A^\alpha \chi_{AA} + N_B^\alpha q' \chi_{BB}) - N_A^\alpha kT \log(\lambda_A^{-3} j_A \eta_A v_{fA}) - \\ & - N_B^\alpha kT \log(\lambda_B^{-3} j_B \eta_B^r v_{fB}^r) - N_B^\alpha kT \log\left(\frac{l c}{2}\right) + \\ & + (r-1) N_B^\alpha kT + N_A^\alpha kT \log\frac{N_A^\alpha}{N_A^\alpha + r N_B^\alpha} + N_B^\alpha kT \log\frac{N_B^\alpha}{N_A^\alpha + r N_B^\alpha}; \end{aligned}\right\} \tag{61.2}$$

where superscript α denotes the bulk liquid phase α and $q' = l q + 2 m r$.

The chemical potentials μ_A and μ_B satisfy the following condition, in analogy with (54.11),

$$\left.\begin{aligned} \mu_A - \frac{1}{r}\mu_B = {} & \frac{kT}{r}\log\left(\frac{l c}{2}\right) - \frac{kT}{r}\log\frac{N_B^\alpha}{N_A^\alpha + r N_B^\alpha} + kT \log\frac{N_A^\alpha}{N_A^\alpha + r N_B^\alpha} - \\ & - kT \log(\lambda_A^{-3} j_A \eta_A v_{fA}) + \frac{kT}{r}\log(\lambda_B^{-3} j_B \eta_B^r v_{fB}^r) + \frac{c}{2}\left(\chi_{AA} - \frac{q'}{r}\chi_{BB}\right). \end{aligned}\right\} \tag{61.3}$$

Let the number of molecules of species A and B in the t-th layer of the actual system be $N_A^{(t)}$ and $N_B^{(t)}$, respectively. No molecules of species B extend over two or more layers according to the assumption. Then the configuration integral for the athermal polymer solution with a plane interface can be written, in analogy with (55.1) and (55.3),

$$Q_{N_A N_B}^{(1)} = \sum_{N_A^{(1)}} \cdots \sum_{N_A^{(w)}} Q(N_A^{(1)}, \ldots, N_A^{(w)}), \tag{61.4}$$

$$\left.\begin{aligned} Q(N_A^{(1)}, \ldots, N_A^{(w)}) = {} & v_{fA}^{N_A} v_{fB}^{r N_B} \exp\left\{-\frac{1}{2kT}(N_A c \chi_{AA} + N_B c q' \chi_{BB}) + \right. \\ & \left. + \frac{1}{2kT}(N_A^{(1)} m c \chi_{AA} + N_B^{(1)} m c r \chi_{BB})\right\} \left(\frac{v_{fA}'}{v_{fA}}\right)^{N_A^{(1)}} \left(\frac{v_{fB}'}{v_{fB}}\right)^{r N_B^{(1)}} \times \\ & \times \prod_{t=1}^{w} \left[\frac{(l c)^{N_B^{(t)}}}{(A \nu) N_B^{(t)} (r-1)} \frac{1}{2^{N_B^{(t)}}} \frac{(N_A^{(t)} + r N_B^{(t)})!}{N_A^{(t)}! \, N_B^{(t)}!}\right]. \end{aligned}\right\} \tag{61.5}$$

We shall use the abbreviations

$$\varphi_A^{(t)} = \frac{N_A^{(t)}}{N_A^{(t)} + r N_B^{(t)}}; \qquad \varphi_B^{(t)} = \frac{r N_B^{(t)}}{N_A^{(t)} + r N_B^{(t)}}, \tag{61.6}$$

which are the volume fractions of components A and B, respectively. We have

$$A \nu = N_A^{(t)} + r N_B^{(t)}, \tag{61.7}$$

where ν and A denote the number of cells per unit area and the area of each lattice layer, respectively. From (61.7) we have, instead of (56.2), for the athermal polymer (r-mer) solution

$$\Gamma_A + r \Gamma_B = 0. \tag{61.8}$$

Then, corresponding to (60.5) we obtain from (61.4) and (61.5), the expression for the Helmholtz free energy

$$\left.\begin{aligned}F=&\frac{c}{2}(N_A\chi_{AA}+N_B q'\chi_{BB})-\frac{A\nu m c}{2}(\varphi_A^{(1)}\chi_{AA}+\varphi_B^{(1)}\chi_{BB})-N_A kT\log(\lambda_A^{-3} j_A\eta_A v_{fA})-\\&-N_B kT\log(\lambda_B^{-3} j_B\eta_B' v_{fB}')-A\nu\varphi_A^{(1)} kT\log\left(\frac{v_{fA}'}{v_{fA}}\right)-A\nu\varphi_B^{(1)} kT\log\left(\frac{v_{fB}'}{v_{fB}}\right)-\\&-N_B kT\log\left(\frac{lc}{2}\right)+\sum_{t=1}^{w}A\nu\left[(r-1)\left(\frac{\varphi_B^{(t)}}{r}\right)kT+\varphi_A^{(t)} kT\log\varphi_A^{(t)}+\left(\frac{\varphi_B^{(t)}}{r}\right)kT\log\left(\frac{\varphi_B^{(t)}}{r}\right)\right].\end{aligned}\right\}\quad(61.9)$$

In the above equation $\varphi_A^{(t)}$ and $\varphi_B^{(t)}$ should be determined by the condition to minimize (61.9) at given values of N_A and N_B. That is,

$$\left.\begin{aligned}-\frac{mc}{2kT}(\chi_{AA}-\chi_{BB})-\log\left(\frac{v_{fA}'}{v_{fA}}\right)+\log\left(\frac{v_{fB}'}{v_{fB}}\right)+\left(\log\varphi_A^{(1)}-\frac{1}{r}\log\frac{\varphi_B^{(1)}}{r}\right)\\=\log\varphi_A^{(2)}-\frac{1}{r}\log\frac{\varphi_B^{(2)}}{r}=\cdots=\log\varphi_A^{\alpha}-\frac{1}{r}\log\frac{\varphi_B^{\alpha}}{r}.\end{aligned}\right\}\quad(61.10)$$

This set of equations shows that $\varphi_A^{(t)}$ and $\varphi_B^{(t)}$ except $\varphi_A^{(1)}$ and $\varphi_B^{(1)}$ are all equal, respectively, to

$$\varphi_A^{\alpha}=\frac{N_A^{\alpha}}{N_A^{\alpha}+rN_B^{\alpha}};\qquad\varphi_B^{\alpha}=\frac{rN_B^{\alpha}}{N_A^{\alpha}+rN_B^{\alpha}}.\qquad(61.11)$$

The above implies that the monolayer model is accurate for athermal polymer solutions insofar as the orientation of rigid-rod molecules is confined to the direction parallel to the surface[1].

If we use (61.8), (5.8) is written as

$$\gamma=\frac{F-F^{\alpha}}{A}-\left(\mu_A-\frac{1}{r}\mu_B\right)\Gamma_A.\qquad(61.12)$$

Since $\Gamma_A=\frac{1}{A}\left(\sum_{t=1}^{w}N_A^{(t)}-N_A^{\alpha}\right)$, we obtain, with use of (61.2), (61.3), (61.9) and (61.10), from (61.12)

$$\left.\begin{aligned}\gamma=\varphi_A^{(1)}\gamma_A+\varphi_B^{(1)}\gamma_B+\nu kT\left(\varphi_A^{(1)}\log\frac{\varphi_A^{(1)}}{\varphi_A^{\alpha}}+\frac{\varphi_B^{(1)}}{r}\log\frac{\varphi_B^{(1)}}{\varphi_B^{\alpha}}\right)+\\+\nu kT(r-1)\frac{\varphi_B^{(1)}-\varphi_B^{\alpha}}{r},\end{aligned}\right\}\quad(61.13)$$

$$\gamma_A=-\frac{\nu mc}{2}\chi_{AA}-\nu kT\log\frac{v_{fA}'}{v_{fA}};\qquad\gamma_B=-\frac{\nu mc}{2}\chi_{BB}-\nu kT\log\frac{v_{fB}'}{v_{fB}},\qquad(61.14)$$

where γ_A and γ_B are the surface tensions of pure liquids of A and B. With use of the equilibrium condition (61.10) combined with (61.14) we can rewrite (61.13) in a simpler form

$$\left.\begin{aligned}\gamma&=\gamma_A+\nu kT\log\frac{\varphi_A^{(1)}}{\varphi_A^{\alpha}}+\nu kT(r-1)\frac{\varphi_B^{(1)}-\varphi_B^{\alpha}}{r}\\&=\gamma_B+\frac{\nu kT}{r}\log\frac{\varphi_B^{(1)}}{\varphi_B^{\alpha}}+\nu kT(r-1)\frac{\varphi_B^{(1)}-\varphi_B^{\alpha}}{r}.\end{aligned}\right\}\quad(61.15)$$

This becomes for $r=1$ identical with the expression for the surface tension of perfect solutions given by (56.14).

[1] When we need to take into account the orientation of polymer molecules normal to the surface, the monolayer model breaks down even for athermal solutions of rigid-rod molecules [65], [66], [69].

The equilibrium condition (61.10) can be rewritten by (61.14) in an alternative form

$$e^{(\gamma_A-\gamma_B)/\nu kT}\,\frac{(\varphi_B^{\alpha})^{1/r}}{1-\varphi_B^{\alpha}}=\frac{(\varphi_B^{(1)})^{1/r}}{1-\varphi_B^{(1)}}. \tag{61.16}$$

Eqs. (61.15) and (61.16) were first obtained by PRIGOGINE and MARÉCHAL [68]. If r is sufficiently large and neither φ_A^{α} nor φ_B^{α} is zero, we have

$$(\varphi_B^{\alpha})^{1/r}\approx(\varphi_B^{(1)})^{1/r}\approx 1. \tag{61.17}$$

Then, from (61.16) combined with (61.17) we can reduce (61.15) to

$$\gamma-\gamma_A=-(\gamma_A-\gamma_B)+\gamma\,kT\,\frac{r-1}{r}\,(1-e^{-(\gamma_A-\gamma_B)/\nu kT})\,\varphi_A^{\alpha}. \tag{61.18}$$

Furthermore, if $|\gamma_A-\gamma_B|/(\nu kT)\ll 1$, then (61.18) becomes

$$\gamma=\gamma_A\,\varphi_A^{\alpha}+\gamma_B\,\varphi_B^{\alpha}. \tag{61.19}$$

The surface tension γ for dibenzyl-benzene calculated from (61.15) with $r=2$ is shown in Fig. 32 as a function of the mole fraction of benzene in the bulk phase

$$x_A=\frac{N_A^{\alpha}}{N_A^{\alpha}+N_B^{\alpha}}. \tag{61.20}$$

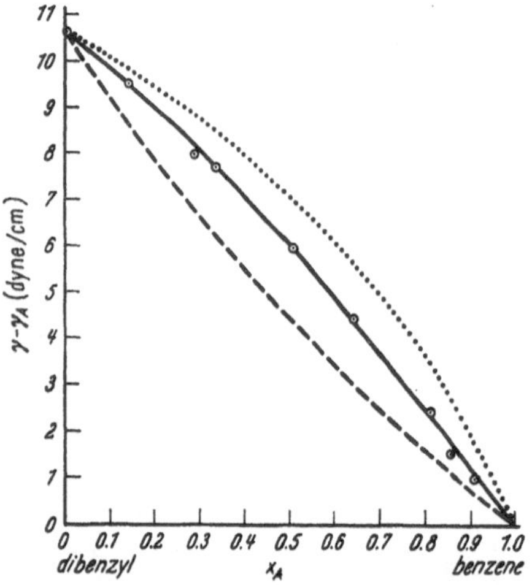

Fig. 32. A comparison of experimental surface tension of dibenzyl-benzene with values calculated from three different theoretical equations. 1. ⊙: Experimental values (MARÉCHAL [67], DEFAY and PRIGOGINE [13]). 2. Solid curve: from exact equation (61.15). 3. Dotted curve: from approximate equation (61.19). 4. Dashed curve: from equation for perfect solution (56.15). Calculation was made for $r=2$; $v^{\frac{2}{3}}=28.94$ Å^2 [67], [13] and for the (111)-plane, i.e. $\nu=2^{\frac{2}{3}}3^{-\frac{1}{2}}v^{-\frac{2}{3}}$ [see (47.6)].

The dotted curve represents the approximate equation (61.19) and the surface tension calculated from (56.15) for perfect solutions is shown by the dashed curve for comparison. The agreement between the theory and experimental data is satisfactory.

Traube's rule[1] [69]. Since it is difficult to solve (61.16) for $\varphi_B^{(1)}$ in general, we restrict ourselves to dilute solutions. Assuming that φ_B^{α} is small, we write (61.16) in the form

$$\left.\begin{aligned}\varphi_B^{(1)}=&(e^{(\gamma_A-\gamma_B)/(\nu kT)})^r\,\varphi_B^{\alpha}+\\&+(\text{higher terms in }\varphi_B^{\alpha}).\end{aligned}\right\} \tag{61.21}$$

On the other hand, from (61.6) and (61.11) we have $\varphi_A^{(1)}=1-\varphi_B^{(1)}$ and $\varphi_A^{\alpha}=1-\varphi_B^{\alpha}$. Hence, we can rewrite the first formula of (61.15) in the form

$$\gamma_A-\gamma=\nu kT\left[\frac{1}{r}(\varphi_B^{(1)}-\varphi_B^{\alpha})+\sum_{s=2}^{\infty}\frac{(\varphi_B^{(1)})^s-(\varphi_B^{\alpha})^s}{s}\right]. \tag{61.22}$$

Substituting (61.21) into (61.22) and neglecting higher terms in $\varphi_B^{(1)}$ and φ_B^{α}, we obtain

$$\gamma_A-\gamma=\nu kT\left[\left(e^{\frac{\gamma_A-\gamma_B}{\nu kT}}\right)^r-1\right]\left(\frac{\varphi_B^{\alpha}}{r}\right). \tag{61.23}$$

As seen from (61.21), if the adsorption of solute B is strong, we may assume $\left(e^{\frac{\gamma_A-\gamma_B}{\nu kT}}\right)^r\gg 1$. Then (61.23) reduces to

$$\gamma_A-\gamma=\nu kT\left(e^{\frac{\gamma_A-\gamma_B}{\nu kT}}\right)^r\left(\frac{\varphi_B^{\alpha}}{r}\right). \tag{61.24}$$

[1] Thanks are due to Professor KURATA, Kyoto University, for bringing this point to the notice of the authors. SZYSZKOWSKI's rule has also been derived by KURATA [69].

Since φ_B^α/r is practically equal to the mole fraction of the solute for small φ_B^α, the limiting expression (61.24) for surface tension is identical with TRAUBE's rule[1]. For very dilute solutions of a solute, the depression of surface tension caused by the solute is proportional to the concentration and the proportionality constant increases exponentially with increasing length of the solute polymer molecules[2].

Acknowledgements.

The authors are greatly indebted to Professor T. L. HILL of the University of Oregon, who has been kind enough to read the whole manuscript and make useful suggestions. They are very grateful to him for correcting many errors of expression and language. They want also to thank Professor M. KURATA at Kyoto University for reviewing the chapters on lattice theory approaches and for offering useful modifications; and Mr. N. MATSUDAIRA at Hosei University, Tokyo and Professor T. MURAKAMI at Kyushu University, Fukuoka for checking many chapters of this article.

Some of the figures or tables have by permission been copied from various journals and other books. The authors are grateful for such permission to the authors concerned and to the publishers of Journal of Chemical Physics, Transactions of Faraday Society, Journal of Physical Chemistry, Journal de Chimie Physique; and Wiley and Sons, Inc.

References.

A. Thermodynamics and quasithermodynamics.

I. Thermodynamics.

a) Thermodynamic quantities of interface layer.

[1] GIBBS, J.W.: Collected Works, Vol. 1. New Haven: Yale University Press 1948.

b) Plane interface.

[2] GUGGENHEIM, E.A., and N.K. ADAM: The Thermodynamics of Adsorption at the Surface of Solutions. Proc. Roy. Soc. Lond., Ser. A **139**, 218 (1933).

[3] GUGGENHEIM, E.A.: The Thermodynamics of Interfaces in Systems of Several Components. Trans. Faraday Soc. **37**, 397 (1940).

[4] GUGGENHEIM, E.A.: Thermodynamics. Amsterdam: North Holland Publishing Co. 1950.

c) Spherical interface.

[5] TOLMAN, R.C.: The Effect of Droplet Size on Surface Tension. J. Chem. Phys. **17**, 333 (1949).

[6] KOENIG, F.O.: On the Thermodynamic Relation between Surface Tension and Curvature. J. Chem. Phys. **18**, 449 (1950).

[7] BUFF, F.P.: The Spherical Interface. I. Thermodynamics. J. Chem. Phys. **19**, 1591 (1951).

[8] HILL, T.L.: Statistical Thermodynamics of the Transition Region between Two Phases. I. Thermodynamics and Quasithermodynamcis. J. Phys. Chem. **56**, 526 (1952).

[9] KONDO, S.: Thermodynamical Fundamental Equation for Spherical Interface. J. Chem. Phys. **25**, 662 (1956).

---

[1] I. TRAUBE: Liebigs Annalen **265**, 27 (1891): TRAUBE showed that, in homologous series of fatty acids and alcohols, the concentrations which gave equal lowering of surface tension, in dilute aqueous solutions, diminished three-fold for each additonal CH_2 group in the hydrocarbon chain (for example, see N.K. ADAM: The Physics and Chemistry of Surfaces, 3rd edn. London: Oxford University Press 1941, p. 121).

[2] Recently, R.H. ARANOW and L. WITTEN [J. Chem. Phys. **28**, 405 (1958)] have pointed out that the difference of the work done in bringing a molecule from the interior to the water surface for malonic acid and that for succinic acid is much smaller than the increment to this work for each additional CH_2 group of long chain hydro-carbons with one end group attracted by water. Since both of these acids have two attracting ends, their molecules are tightly held to water at both ends. The authors have concluded that the above discrepancy is attributable to upward orientation of the hydro-carbon molecules at the water surface, and that the conventional arguments supposing that chain molecules lie parallel to the surface for dilute solution are unconvincing. Thus they have proposed a crude theory to account for TRAUBE's rule based on the assumption of random hindered rotations of surface hydrocarbon molecules.

II. Hydrostatic approach.

[10] BAKKER, G.: Kapillarität und Oberflächenspannung. In: WIEN-HARMS' Handbuch der Experimentalphysik VI. Leipzig: Akademische Verlagsgesellschaft 1928.

III. Quasithermodynamics.

[11] TOLMAN, R.C.: Consideration of the Gibbs Theory of Surface Tension. J. Chem. Phys. **16**, 758 (1948).

IV. Thermodynamics of irreversible processes.

[12] DEFAY, R.: Étude Thermodynamique de la Tension Superficielle, Vol. 1. Paris: Gauthier-Villars 1934. — Groupement de Communication à l'Académie Royale de Belgique (Classe des Sciences). Années 1929 à 1934.

[13] DEFAY, R., et I. PRIGOGINE: Tension Superficielle et Adsorption. Liège: Editions Desoer 1951.

[14] DEFAY, R.: Du Choix Arbitraire de la Surface de Division dans les Couches Capillaires non en Equilibre. Bull. Acad. Roy. Belg., Cl. Sci. **41**, 138 (1955).

V. Empirical equations for temperature dependence of surface tension.

[15] GUGGENHEIM, E.A.: The Principle of Corresponding States. J. Chem. Phys. **13**, 253 (1945).

[16] DE BOER, J., and R.B. BIRD: Quantum Theory and the Equation of State. In: Molecular Theory of Gases and Liquids by J.O. HIRSCHFELDER, C.F. CURTISS and R.B. BIRD. New York: John Wiley & Sons, Inc. 1954.

B. Statistical mechanics.

I. Statistical thermodynamic method.

a) Canonical ensemble.

[17] BUFF, F.P.: Some Considerations of Surface Tension. Z. Elektrochem. **56**, 311 (1952).

[18] MCLELLAN, A.G.: A Statistical-Mechanical Theory of Surface Tension. Proc. Roy. Soc. Lond., Ser. A **213**, 274 (1952).

[19] HARASIMA, A.: Statistical Mechanics of Surface Tension. J. Phys. Soc. Japan **8**, 343 (1953).

[20] KONDO, S.: A Statistical-Mechanical Theory of Surface Tension of Curved Surface Layer I. J. Phys. Soc. Japan **10**, 381 (1955).

b) Grand canonical ensemble.

[21] SAITÔ, N.: Statistical Thermodynamics of Surface Layer. Bull. Kobayasi Inst. Phys. Res. Japan **1**, 157 (1951) [in Japanese].

[22] ONO, S.: Statistical Mechanics of Adsorption. Mem. Fac. Engng, Kyushu Univ. **12**, 9 (1950). — Application of URSELL and MAYER's Treatment for Imperfect Gases to Adsorption. J. Chem. Phys. **18**, 397 (1950). — Statistical Mechanics of Adsorption from Multicomponent Systems I. J. Phys. Soc. Japan **6**, 10 (1951).

[23] BUFF, F.P., and F.H. STILLINGER: Surface Tension of Ionic Solutions. J. Chem. Phys. **25**, 312 (1956).

II. Hydrostatic treatments of surface tension.

[24] KIRKWOOD, J.G., and F.P. BUFF: The Statistical Mechanical Theory of Surface Tension. J. Chem. Phys. **17**, 338 (1949).

[25] GREEN, H.S.: The Molecular Theory of Fluids. Amsterdam: North Holland Publishing Co. 1952.

[26] BUFF, F.P.: Spherical Interface. II. Molecular Theory. J. Chem. Phys. **23**, 419 (1955).

III. Numerical calculations.

a) Pure liquid: Surface of discontinuity.

[27] FOWLER, R.H.: A Tentative Statistical Theory of MACLEOD's Equation for Surface Tension, and the Parachor. Proc. Roy. Soc. Lond Ser. A **159**, 229 (1937).

[28] HILL, T.L.: Concerning the Dependence of the Surface Energy and Surface Tension of Spherical Drops and Bubbles on Radius. J. Amer. Chem. Soc. **72**, 3923 (1950).

[29a] HARASIMA, A.: Statistical Mechanics of Surface Tension. In: Proceedings of the International Conference of Theoretical Physics, Tokyo: Science Council of Japan 1954.

[29b] HARASIMA, A.: Molecular Theory of Surface Tension. In: Advances in Chemical Physics, Vol. 1 (ed. I. PRIGOGINE). New York: Interscience Publ. 1958.

Quasithermodynamic theory.

[30] TOLMAN, R.C.: The Superficial Density of Matter at a Liquid-Vapor Boundary. J. Chem. Phys. **17**, 118 (1949).

[31a] HILL, T.L.: Liquid-Vapor Transition Region and Physical Adsorption according to VAN DER WAALS' Equation. J. Chem. Phys. **19**, 261 (1951).

[31b] HILL, T.L.: Statistical Thermodynamics of the Transition Region Between Two Phases. II. One Component System with a Plane Interface. J. Chem. Phys. **20**, 141 (1952).

b) Electrolyte solution.

[32] WAGNER, C. Die Oberflächenspannung verdünnter Elektrolytlösungen. Phys. Z. **25**, 474 (1924).

[33] OKA, S.: Das quantitative Grenzgesetz der Oberflächenspannung starker binärer Elektrolyte. Proc. Phys.-Math. Soc. Japan **14**, 233 (1932). — Zur Theorie der Oberflächenspannung beliebiger verdünnter Elektrolyte mit einer Untersuchung über die Struktur der Kapillarschicht. Proc. Phys.-Math. Soc. Japan **14**, 649 (1932).

[34] ONSAGER, L., and N.N.T. SAMARAS: The Surface Tension of Debye-Hückel Elektrolytes. J. Chem. Phys. **2**, 528 (1934).

[35] SCHMUTZER, E.: Zur Theorie der Oberflächenspannung von Lösungen. Z. phys. Chem. **204**, 131 (1955).

c) Gas adsorption on a solid surface.

[36] STEELE, W.A., and G.D. HALSEY: The Interaction of Rare Gas Atoms with Surfaces. J. Chem. Phys. **22**, 979 (1954).

[37] FREEMAN, M.P., and G.D. HALSEY: The Interaction of Pairs of Gas Atoms with Surfaces. J. Phys. Chem. **59**, 181 (1955).

IV. Quantum statistical mechanics.

[38] BORN, M., u. R. COURANT: Zur Theorie des Eötvösschen Gesetzes. Phys. Z. **14**, 731 (1913).

[39] KOTHARI, D.S., and F.C. AULUCK: Surface Tension of Nuclear Matter and the Enumeration of Eigenfunctions of an Enclosed Particle. Nature, Lond. **159**, 204 (1947).

[40] ATKINS, K.R.: The Surface Tension of Liquid Helium. Canad. J. Phys. **31**, 1165 (1953).

[41] MCLELLAN, A.G.: The Stress Tensor, Surface Tension and Viscosity. Proc. Roy. Soc. Lond., Ser. A **217**, 92 (1953).

[42] TODA, M.: On the Theory of Quantum Liquids. I. Surface Tension and Stress. J. Phys. Soc. Japan **10**, 512 (1955).

[43] LOVEJOY, D.R.: Surface Tension of He^3. Canad. J. Phys. **33**, 49 (1955).

[44] HARASIMA, A., and Y. SHIMURA: Quantum Effects in the Theory of Surface Energy. J. Phys. Soc. Japan **11**, 14 (1956).

[45] TRIKHA, S.K., and O.P. RUSTGI: Surface Tension of Liquid He^4 and Liquid He^3. Progr. Theoret. Phys. **15**, 296 (1956). — The Surface Tension of Liquid He^3. Progr. Theoret. Phys. **17**, 303 (1957).

C. Lattice theory approaches.

I. Pure liquid.

a) Free volume theory.

[46] LENNARD-JONES, J.E., and J. CORNER: The Calculation of Surface Tension from Intermolecular Forces. Trans. Faraday Soc. **36**, 1156 (1940).

[47] HARASIMA, A.: Calculation of the Surface Energies of Several Liquids. Proc. Phys.-Math. Soc. Japan **22**, 825 (1940).

[48] HARASIMA, A.: Surface Tension of Liquids. Proc. Phys.-Math. Soc. Japan **23**, 983 (1941).

[49] FRENKEL, J.: Kinetic Theory of Liquids. Oxford: Clarendon Press 1946.

[50] CORNER, J.: The Calculation of Surface Tension from Intermolecular Forces. II. Numerical Results. Trans. Faraday Soc. **44**, 1036 (1948).

[51] HIRSCHFELDER, J.O., C.F. CURTISS and R.B. BIRD: Molecular Theory of Gases and Liquids. New York: John Wiley & Sons, Inc. 1954.

b) Hole theory.

[*52*] ONO, S.: Statistical Thermodynamics of Critical and Surface Phenomena. Mem. Fac. Engng, Kyushu Univ. **10**, 195 (1947).

[*53*] KURATA, M.: BETHE's Approximation for Surface Layers. Busseiron Kenkyu No. **39**, 77 (1951), [in Japanese].

[*54*] PRIGOGINE, I., et L. SARAGA: Sur la Tension Superficielle et le Modèle Cellulaire de l'Etat Liquide. J. Chim. phys. **49**, 399 (1952).

II. Solutions of non-electrolytes.

a) Regular solutions.

[*55*] BELTON, J.W., and M.G. EVANS: Studies of the Molecular Forces Involved in Surface Formation. I. The Surface Energies of Pure Liquids. Trans. Faraday Soc. **37**, 1 (1941). — Studies in the Molecular Forces Involved in Surface Formation. II. The Surface Free Energies of Simple Liquid Mixtures. Trans. Faraday Soc. **41**, 1 (1945).

[*56*] ZHUKHOVITSKII, A.A.: Surface Tension of Solutions. J. Phys. Chem. USSR. **17**, 313 (1943). — Acta physicochim. USSR. **19**, 176, 508 (1944).

[*57*] GUGGENHEIM, E.A.: Statistical Thermodynamics of the Surface of a Regular Solution. Trans. Faraday Soc. **41**, 150 (1945).

[*58*] ONO, S.: Statistical Thermodynamics of Critical and Surface Phenomena. II. Statistical Theory of Adsorption from Regular Solution. Mem. Fac. Engng, Kyushu Univ. **12**, 1 (1950).

[*59*] DEFAY, R., and I. PRIGOGINE: Surface Tension of Regular Solutions. Trans. Faraday Soc. **46**, 199 (1950).

[*60*] KURATA, M.: On the Gibbs Adsorption at the Surface of a Regular Solution. Busseiron Kenkyu No. **27**, 37 (1950) [in Japanese].

[*61*] MURAKAMI, T., S. ONO, M. TAMURA and M. KURATA: On the Theory of Surface Tension of Regular Solution. J. Phys. Soc. Japan **6**, 309 (1951).

b) Dynamic surface tension.

[*62*] RICE, O.K.: Dynamic Surface Tension and the Structure of Surfaces. J. Phys. Chem. **31**, 207 (1927).

[*63*] PRIGOGINE, I., et R. DEFAY: Tension Superficielle Dynamique des Solutions Régulières. J. Chim. phys. **46**, 367 (1949).

[*64*] DEFAY, R.: Calcul de la Tension Superficielle en dehors de l'Equilibre. J. Chim. phys. **51**, 299 (1954).

c) Polymer solutions.

[*65*] PRIGOGINE, I.: Sur la Tension Superficielle des Solutions de Molécules de Dimensions Différentes. J. Chim. phys. **47**, 33 (1950).

[*66*] PRIGOGINE, I., and L. SAROLEA: Sur la Tension Superficielle des Solutions de Molécules de Dimensions Différentes. Seconde Communication. J. Chim. phys. **47**, 807 (1950).

[*67*] MARÉCHAL, J.: The Influence of Molecular Size on the Surface Tension of Hydrocarbon Mixtures, Part 3. Trans. Faraday Soc. **48**, 601 (1952).

[*68*] PRIGOGINE, I., and J. MARÉCHAL: The Influence of Differences in Molecular Size on the Surface Tension of Solutions. IV. J. Coll. Sci. **7**, 122 (1952).

[*69*] KURATA, M.: On the Surface Tension of Solutions of Chain Molecules. I. Liquids and Solutions of Rod-like Molecules. Busseiron Kenkyu No. **56**, 60 (1952) [in Japanese].

The Theory of Capillarity.

By

FRANK P. BUFF.

With 2 Figures.

Introduction.

It is the purpose of this chapter to examine the foundations of the theory of capillarity and to present some of its applications. From an historical point of view, it is noted that the classical investigations of LAPLACE[1] had already led to the proper descriptive formalism. However, the underlying theories proposed by him, YOUNG[2], and GAUSS[3] were unsatisfactory due to the use of naive molecular models as well as to the lack of distinction between energy and free energy. The rigorous molecular formulation of planar and spherical interfaces has recently been carried out with help of the statistical mechanical theory of inhomogeneous fluids[4,5,6]. When this approach is supplemented by hydrostatic concepts [*1*] to [*3*], the thermodynamic description due to GIBBS [*4*] may be analyzed and extended.

Although the detailed examination of a system in terms of its molecular structure is frequently desirable, the phenomenological description of surface phenomena may be derived without recourse to the general apparatus of statistical mechanics. The hydrostatic treatment to be presented is intermediate between the thermodynamic and molecular approaches. It leads to the classical phenomenological equations and to their first order correction terms. Since these corrections enter into the asymptotic expansion of the free energy with respect to the geometrical parameters of external force, they yield criteria for the range of validity of the classical treatment. At the end of the analysis, this approach also provides the explicit formulas for the thermodynamic parameters, which it is the task of molecular theories to evaluate numerically.

In order to carry out this program, we shall first examine a detailed picture of inhomogeneous fluids. This model will be used to derive the equilibrium conditions and work elements of the theory of capillarity. These results are then transcribed into thermodynamic terms and are finally summarized in the form of a variational principle. Following this survey of the foundations, some applications are considered. They include discussions of the planar interface, the concept of spreading, simple solutions of the Laplace equation, spherical droplet formation, and the electrocapillary curve.

A. General theory.

1. Molecular model. We shall now consider the hydrostatic model [*1*] which yields the basic equations of capillarity and provides criteria for the range of applicability of the conventional formulas. An outline of the statistical mechanical

[1] P. S. LAPLACE: Mécanique Céleste, Suppl. to Vol. X, Paris, 1806.

[2] T. YOUNG: Phil. Trans. Roy. Soc. Lond. **65** (1805).

[3] C. F. GAUSS: Comment. Soc. Reg. Sci., Göttingen **7** (1830).

[4] J. G. KIRKWOOD and F. P. BUFF: J. Chem. Phys. **17**, 338 (1949), (Molecular Theory of Surface Stresses).

[5] F. P. BUFF: Z. Elektrochem. **56**, 311 (1952), (Canonical Ensemble Theory).

[6] F. P. BUFF: J. Chem. Phys. **23**, 419 (1955), (Grand Canonical Ensemble Theory; Virial Theorem).

theory that is encountered when one investigates the distribution of matter in an external field will provide a convenient starting point for this formulation. We consider a macroscopic multiphase system which is represented by a statistical grand ensemble. In general, one seeks to evaluate the number densities of sets of molecules the members of which are located in the various regions of the system under consideration. For the present problem the singlet density of component i, $\varrho_i^{(1)}(\boldsymbol{R})$, is of chief interest since $\varrho_i^{(1)}(\boldsymbol{R})\,dv$ specifies the number of molecules of component i in the volume element dv surrounding the point $\boldsymbol{R}$. Although the expressions relating the number densities to molecular variables may be exhibited readily[1], the various chains of integral equations implied by these expressions are not easily amenable to explicit evaluation. However, this detailed information is not of particular interest for a liquid within a capillary tube of arbitrary dimension. From a practical point of view it suffices to relate the macroscopic distribution of matter in a two-phase region, separated by an interfacial region of small curvatures, to the distribution of a similar system with a planar interface. In order to treat this more restricted problem, we shall only concern ourselves with the predominant average properties of regions that are small on a macroscopic scale but large compared with the range of intermolecular forces. This permits a minimal set of assumptions concerning the small region under consideration, and, in view of the fact that it is postulated to contain a large number of particles, thermodynamic arguments may be employed to interpret the results of the theory.

In accordance with the foregoing discussion, we divide the whole system into small cells and specify their detailed structure depending upon their location in a bulk phase, a two-phase transition zone or the confluence of three phases. We first note [*1*] that whenever the small region is well within a bulk phase, the singlet densities will be constant apart from negligible contribution arising from local mass fluctuations. When the subregion encloses the non-homogeneous film, the singlet densities will vary rapidly between the values attained in the respective bulk phases as the transition zone is traversed. Following Gibbs we shall span the transition zone by a family of parallel dividing surfaces, Gibbs dividing surfaces, located in such a manner as to be "similarly situated with respect to the condition of adjacent matter" [*4*]. Expressed more explicitly, it will be assumed that, although there will be an appreciable variation of the singlet densities in the direction normal to the parallel surfaces, within the small cell under consideration the singlet densities will be effectively constant on each surface. When generalized coordinates u, v are introduced, the current point $\boldsymbol{r}$ of a given surface s may be represented parametrically by $\boldsymbol{r}=\boldsymbol{r}(u,v)$, while the current point $\boldsymbol{R}$ of the parallel surface S, located at a constant distance λ along the normal from surface s, is related to the corresponding point $\boldsymbol{r}$ by

$$\boldsymbol{R}=\boldsymbol{r}(u,v)+\lambda\,\boldsymbol{N}(u,v)\,;\qquad \boldsymbol{N}=\frac{\boldsymbol{r}_u\times\boldsymbol{r}_v}{|\boldsymbol{r}_u\times\boldsymbol{r}_v|}\,, \tag{1.1}$$

where $\boldsymbol{N}$ is the common unit normal to the surfaces s and S. Consequently our assumption concerning the spatial dependence of the number densities, within a given small region, may be expressed in the form

$$\varrho_i^{(1)}=\varrho_i^{(1)}(\lambda)\,. \tag{1.2}$$

Finally we consider a fluid lens embedded in two fluid phases and label the phases in order of decreasing density by I, II and III. Although the dividing surface

[1] F. P. Buff: Proceedings on the Conference on Interfacial Phenomena. Geophys. Res. Papers, No. 37, **2**, 51 (1955). — F. P. Buff and F. H. Stillinger jr.: J. Chem. Phys. **25**, 312 (1956).

concept is physically applicable in interfacial regions far removed from the three-phase confluent zone, this concept breaks down as the region of intersection is approached. This difficulty may be resolved [2] by selecting appropriate dividing surfaces for the two faces of the lens and extrapolating these surfaces to their intersection along the space curve C^* which is located in the confluent zone. We similarly extrapolate the set of parallel I—III surfaces and for each small region we select the one that intersects the faces of the lens along C^*. It will be seen subsequently that this extrapolatory procedure is characteristic for surface problems and will give rise to excess thermodynamic functions.

The spatial variation of the density is ultimately determined by both external and intermolecular forces. Since the equations representing the local balance of intermolecular and external forces may be transformed into the equation of hydrostatics[1], the use of the latter is more convenient. It is thus of interest to remark that although the results of the theory may be expressed in terms of properties of the stress tensor σ, the latter may be established as a function of molecular variables at any stage of the analysis. Upon restriction to mass dependent external forces, the equation of hydrostatics is given by

$$\left.\begin{aligned} \nabla\cdot\sigma &= \varrho\,\nabla\psi;\\ \varrho &= \sum_{i=1}^{n} M_i\,\varrho_i^{(1)};\\ \psi = g\,z - \frac{\omega^2 r^2}{2}; &\qquad r = |\boldsymbol{R}\times\boldsymbol{k}| \end{aligned}\right\} \tag{1.3}$$

where ϱ is the mass density, M_i is the molecular weight of component i, ψ is the potential corresponding to both a uniform gravitational field and a centrifugal field characterized by angular velocity ω and $\boldsymbol{k}$ is the unit vector directed along the space-fixed z-axis.

In view of the fact that we shall only attempt to elucidate the main features of very small macroscopic regions, it is both necessary and convenient to carry out a spatial smoothing of the parameters appearing in Eq. (1.3). The results of this smoothing operation have already been postulated in the case of number densities. The corresponding specification of the stress tensor will again depend on the location of the region to be treated. To the orders of magnitude with which we shall be concerned, for regions located in the interior of bulk phase α it is assumed that the stress tensor is isotropic

$$\sigma_\alpha = -\,p_\alpha\,\mathbf{1}, \tag{1.4}$$

where $\mathbf{1}$ is the three-dimensional unit tensor and p_α is the pressure at the point considered. For regions enclosing the two-phase transition zone it is assumed that the stress tensor takes the form

$$\sigma = \sigma_{T_1}\,\boldsymbol{e}_1\,\boldsymbol{e}_1 + \sigma_{T_2}\,\boldsymbol{e}_2\,\boldsymbol{e}_2 + \sigma_N\,\boldsymbol{N}\,\boldsymbol{N} \tag{1.5}$$

where $\boldsymbol{N}$ is the unit normal and $\boldsymbol{e}_1$ and $\boldsymbol{e}_2$ are the unit vectors along the lines of curvature of the dividing surface under consideration. Finally within a three-phase confluent zone it is assumed that one of the principal stresses is directed along the space curve C^*. It should be emphasized that any errors inherent in this general representation will lead to deviations which are at most comparable with the resulting first-order corrections to the classical equations.

In preparation for later results, it will be convenient to designate the unit normal directed from the denser phase α to phase β by $\boldsymbol{N}_{\alpha\beta}$. Furthermore, it is

[1] See for example, J.M. Irving and J.G. Kirkwood: J. Chem. Phys. 18, 817 (1950).

noted that the invariant mean and Gaussian curvatures of the surface $\boldsymbol{r} = \boldsymbol{r}(u, v)$ may be expressed in terms of the coefficients of the first and second fundamental forms:

$$\frac{c_1 + c_2}{2} = \frac{2fF - Eg - eG}{2(EG - F^2)}, \tag{1.6}$$

$$c_1 c_2 = \frac{eg - f^2}{EG - F^2}, \tag{1.7}$$

where the following standard abbreviations have been used:

$$E = r_u^2, \qquad F = \boldsymbol{r}_u \cdot \boldsymbol{r}_v, \qquad G = r_v^2, \tag{1.8}$$

$$e = \boldsymbol{r}_{uu} \cdot \boldsymbol{N}, \quad f = \boldsymbol{r}_{uv} \cdot \boldsymbol{N}, \quad g = \boldsymbol{r}_{vv} \cdot \boldsymbol{N}. \tag{1.9}$$

When the unit normal appearing in Eq. (1.9) is directed from the concave to the convex side of elliptic surfaces, the sign convention of Eq. (1.6) leads to positive normal curvatures. Upon introduction of the two-parametric gradient

$$\nabla_2 = \frac{1}{|\boldsymbol{r}_u \times \boldsymbol{r}_v|} \left[\boldsymbol{r}_v \times \boldsymbol{N} \frac{\partial}{\partial u} + \boldsymbol{N} \times \boldsymbol{r}_u \frac{\partial}{\partial v}\right], \tag{1.10}$$

one obtains the useful relation

$$c_1 + c_2 = \nabla_2 \cdot \boldsymbol{N}, \tag{1.11}$$

and the following[1] two-dimensional analogue of the divergence theorem. Let C be any closed curve drawn on a surface. At any point of this curve the sense of its unit tangent $\boldsymbol{t}$ is determined by the requirement that $\boldsymbol{t} \times \boldsymbol{N}$ be drawn outward from the region enclosed by C. Then for any tensor $\boldsymbol{F}$

$$\iint \nabla_2 \cdot \boldsymbol{F}\, ds = \oint \boldsymbol{F} \cdot \boldsymbol{t} \times \boldsymbol{N}\, dl + \iint \boldsymbol{F} \cdot \boldsymbol{N}\, \nabla_2 \cdot \boldsymbol{N}\, ds \tag{1.12}$$

where ds is an element of area and dl is an element of length of the curve C whose positive sense is the sense of $\boldsymbol{t}$.

2. Phenomenological equations. The foregoing considerations suffice for the derivation of extended forms of the equations of capillarity. Although we must refer to references [*1*] and [*2*] for details, it will be of interest to indicate the important physical concepts that are encountered. It is first recognized that, while the pressure in the interior of a bulk phase varies slowly, the components of stress undergo drastic change as a two-phase transition zone is traversed. This complication is circumvented by the extrapolation of bulk properties into the transition zone, the theory then giving rise to convergent integrals whose integrands involve the difference between the true state of affairs and extrapolated properties. These integrals, whose numerical values depend on the particular parallel reference surface selected, will subsequently be shown to represent the thermodynamic functions required for the phenomenological description of surface phenomena. Thus the interfacial variation of the stress and density finally makes its appearance in the form of thermodynamic parameters. It should be recognized that, although the individual functions depend on the reference surface selected, those combinations which are operationally meaningful must be invariant to this arbitrary choice. For the description of two-phase properties

[1] C.E. WEATHERBURN: Differential Geometry, Vol. I, p. 238. Cambridge: Cambridge University Press 1930.

it is found that the following integrals appear:

$$\gamma_i = \int_{-\lambda_\alpha}^{\lambda_\beta} (\sigma_{T_i} - \sigma_{\alpha\beta}) \left[1 + (c_1 + c_2)\,\lambda\right] d\lambda; \qquad i = 1, 2; \tag{2.1}$$

$$\frac{C_i}{s} = \int_{-\lambda_\alpha}^{\lambda_\beta} (\sigma_{T_i} - \sigma_{\alpha\beta})\,\lambda\, d\lambda; \qquad i = 1, 2; \tag{2.2}$$

$$\Gamma = \int_{-\lambda_\alpha}^{\lambda_\beta} (\varrho - \varrho_{\alpha\beta})\, d\lambda \tag{2.3}$$

where

$$\sigma_{\alpha\beta} = \sigma_\alpha\left[1 - A(\lambda)\right] + \sigma_\beta A(\lambda);$$

$$\varrho_{\alpha\beta} = \varrho_\alpha\left[1 - A(\lambda)\right] + \varrho_\beta A(\lambda);$$

$$A(\lambda) = \begin{cases} 0, & \lambda < 0 \\ 1, & \lambda \geqq 0 \end{cases}$$

and the distances λ_α and λ_β are located in regions where bulk properties subsist. From Eq. (2.3) it is seen that Γ is the superficial excess of mass associated with the reference surface, while on the basis of their properties, it will be shown that γ is the generalized exess surface free energy and C/s is the intensive curvature term.

The subtractive procedure involving extrapolated quantities is illustrated even better in connection with calculations involving the three-phase confluent zone. Here the extrapolated bulk pressures are first subtracted from the relevant stress component, followed by a subtraction of extrapolated two-phase properties, to yield finally the thermodynamic parameters conjugate to the space curve C^*, the length parameter $\mathscr{L}^*$ and the linear mass excess Γ_l^*. The formulas for the correction terms $\mathscr{L}^*$ and Γ_l^* are exhibited in reference [2].

These considerations then lead to the following phenomenological relations for the spatial distribution of a multiphase fluid system at equilibrium in an external field. Within a homogeneous phase α the bulk pressure variation is determined by:

$$\nabla_3 p_\alpha = -\varrho_\alpha \nabla_3 \psi. \tag{2.4}$$

Interfacial regions, characterized by the arbitrary dividing surface s, are described by a two-dimensional analogue of the equation of hydrostatics:

$$\nabla_2 \cdot \boldsymbol{\sigma}_2 + (p_\alpha - p_\beta)\,\boldsymbol{N}_{\alpha\beta} - \Gamma \nabla_3 \psi = 0 \tag{2.5}$$

where the surface stress $\boldsymbol{\sigma}_2$ is given by

$$\boldsymbol{\sigma}_2 = \sum_{i=1}^{2} \left[\gamma_i \boldsymbol{e}_i \boldsymbol{e}_i - \frac{C_i}{s} \boldsymbol{e}_i \boldsymbol{e}_i \cdot \nabla_2 \boldsymbol{N}_{\alpha\beta}\right]. \tag{2.6}$$

p_α and p_β are the extrapolated bulk pressures evaluated at the surface s, and γ_i and C_i/s are given by Eqs. (2.1) and (2.2). Finally the three-phase confluent regions of fluid lenses, characterized by the space curve C^* with unit tangent $\boldsymbol{t}^*$, are described by a one-dimensional analogue of the equation of hydrostatics

$$\boldsymbol{\sigma}_2^{\mathrm{I\,II}} \cdot \boldsymbol{t}^* \times \boldsymbol{N}_{\mathrm{I\,II}} + \boldsymbol{\sigma}_2^{\mathrm{II\,III}} \cdot \boldsymbol{t}^* \times \boldsymbol{N}_{\mathrm{II\,III}} - \boldsymbol{\sigma}_2^{\mathrm{I\,III}} \cdot \boldsymbol{t}^* \times \boldsymbol{N}_{\mathrm{I\,III}} + \Gamma_l^* \nabla_3 \psi + \nabla_1 \cdot \boldsymbol{\sigma}_1^* = 0 \tag{2.7}$$

where

$$\nabla_1 = \boldsymbol{t}^* \frac{\partial}{\partial l^*}, \qquad \boldsymbol{\sigma}_1^* = \mathscr{L}^* \boldsymbol{t}^* \boldsymbol{t}^*. \tag{2.8}$$

The interpretation of Eq. (2.5) follows with use of the two-dimensional divergence theorem. Integration of this equation over a region bounded by the curve C leads to

$$\oint \sigma_2 \cdot \boldsymbol{t} \times \boldsymbol{N}_{\alpha\beta}\, dl + \iint [(p_\alpha - p_\beta)\, \boldsymbol{N}_{\alpha\beta} - \Gamma \nabla_3 \psi]\, ds = 0. \tag{2.9}$$

Consequently the dividing surface behaves as a membrane under the influence of the normal pressures, an external surface force and a peripheral tension. Here $\sigma_2 \cdot \boldsymbol{t} \times \boldsymbol{N}\, dl$ may be taken as the surface force acting across the element dl of C. In the same manner the meaning of Eq. (2.7) may be established. In this case the space curve behaves as a spring under the influence of the surface stresses, an external linear force and a linear tension. From a practical point of view, Eq. (2.5) in conjunction with Eq. (2.4) leads to the Laplace differential equation, the basic relation of capillarity, while Eq. (2.7) reduces to the Neumann triangle which provides the natural boundary condition for the Laplace equation at the confluence of three liquid phases.

In order to obtain the thermodynamic interpretation of these equilibrium conditions, the work elements must be specified. Here it is necessary to differentiate between two kinds of displacements which, to the orders of magnitude here considered, are not always identical in form. The first corresponds to the virtual increase in the system achieved by comparing neighboring regions. This variation, designated by $\bar{\delta}$, leads to the earlier equilibrium relations in the thermodynamic formulation. The second involves the actual physical increase in the boundaries of the system. This variation, designated by δ_b, leads to the desired free energy properties.

In the case of a homogeneous fluid region α, whose volume v is bounded by the surface S, the stress tensor (1.4) leads to the following expression for the actual work done by the system corresponding to the displacement $\delta_b \boldsymbol{x}$:

$$\delta_b W^{(3)} = \iint p_\alpha\, \delta_b \boldsymbol{x} \cdot d\boldsymbol{S} = \delta_b \iiint p_\alpha\, dv \tag{2.10}$$

where p_α is the pressure acting across the surface S. Since $\delta_b W^{(3)}$ corresponds to the creation of a new region, p_α represents the mechanical contribution to the free energy density of the homogeneous phase α.

For the description of the two-phase region, an arbitrary Gibbs dividing surface s bounded by the closed curve C is again located in the transition zone. The surface s precisely separates the total volume into the volume v_α of phase α and the volume v_β of phase β. The stress tensor (1.5) then implies the following expression for the work done by the system corresponding to an actual increase in extent of the region under consideration:

$$\delta_b W^{(2)} = \delta_b \iiint p_\alpha\, dv_\alpha + \delta_b \iiint p_\beta\, dv_\beta - \oint \delta_b \boldsymbol{x} \cdot \sum_{i=1}^{2} \gamma_i\, \boldsymbol{e}_i\, \boldsymbol{e}_i \cdot \boldsymbol{t} \times \boldsymbol{N}_{\alpha\beta}\, dl. \tag{2.11}$$

Here $\boldsymbol{t}$ is the unit tangent along C, $\boldsymbol{t} \times \boldsymbol{N}_{\alpha\beta}$ is drawn outward from the region and γ_i is given by Eq. (2.1). Upon noting that to within first order correction terms, $\gamma_1 = \gamma_2 = \gamma$, the circumstance that $\delta_b W^{(2)}$ involves the creation of a new interface demonstrates that γ is the mechanical contribution to the *excess* surface free energy density.

Finally the actual increase of a section of a fluid lens leads to the work element

$$\left.\begin{aligned} \delta_b W^{(1)} &= \delta_b \left[\sum_{\alpha=\mathrm{I}}^{\mathrm{III}} \iiint p_\alpha\, dv_\alpha - \sum_{\alpha<\beta=\mathrm{I}}^{\mathrm{III}} \iint \gamma_{\alpha\beta}\, ds_{\alpha\beta} \right] + \mathscr{L}^* \delta_b l^*; \\ \delta_b l^* &= \boldsymbol{t}_1^* \cdot \delta_b \boldsymbol{x} \end{aligned}\right\} \tag{2.12}$$

so that the small correction term $\mathscr{L}^*$ possesses free energy properties.

3. Laplace equation and the Neumann triangle. In order to derive these classical results of capillarity, we shall have to examine Eqs. (2.4) and (2.5) in greater detail. The former may be readily integrated to within first order correction terms. In the presence of a uniform gravitational field, the pressure $p(z)$ at z is related to the pressure $p(z_0)$ at z_0 by

$$p(z) = p(z_0) - \varrho(z_0)\, g(z - z_0)\left[1 - \frac{g(z-z_0)\,\varepsilon^0}{z} + \cdots\right] \tag{3.1}$$

where

$$\varepsilon^0 = \frac{\langle \Delta M^2\rangle^0}{RT\langle M\rangle^0}.$$

R is the gas constant, T is the absolute temperature and $\langle M^2\rangle$ is the local mass fluctuation. For the usual experimental conditions, the fluctuation correction is smaller than the dominant term by a factor of 10^7 and is thus completely negligible.

The integration of the tangential component of Eq. (2.5) leads to the curvature dependence of γ. To within first order correction terms, for all dividing surfaces, $\gamma_1 = \gamma_2 = \gamma$,

$$\gamma \sim \gamma_\infty + \left[\frac{C}{s} - \frac{\Gamma\gamma}{\varrho_\alpha - \varrho_\beta}\right]_\infty (c_1 + c_2) + \cdots. \tag{3.2}$$

Since the coefficients γ_∞ and $\left[\frac{C}{s} - \frac{\Gamma\gamma}{\varrho_\alpha - \varrho_\beta}\right]_\infty$ are only to be evaluated at a planar interface, they may be further simplified with use of the equation of hydrostatics. Apart from negligible contributions arising from the gravitational field, its application to the planar interface shows that here the normal component of the stress tensor is constant throughout. Thus with z-axis perpendicular to the interface, $\sigma_{\alpha\beta}(z) = \sigma_N(z)$. This relation permits the reduction of Eqs. (2.1) and (2.2) and yields the following final expressions for the basic thermodynamic parameters of the theory:

$$\gamma_\infty = \int_{-\infty}^{\infty} [\sigma_T(z) - \sigma_N(z)]\, dz; \tag{3.3}$$

$$\left(\frac{C}{s}\right)_\infty = \int_{-\infty}^{\infty} z\, [\sigma_T(z) - \sigma_N(z)]\, dz. \tag{3.4}$$

These relations demonstrate the invariance of γ and of $\left[\frac{C}{s} - \frac{\Gamma\gamma}{\varrho_\alpha - \varrho_\beta}\right]_\infty$ with respect to choice of dividing surface. They also emphasize the two most convenient choices of dividing surface that should be made. The first is the selection $\Gamma'' = 0$. This corresponds to physical intuition since the mass of the heterogeneous system may here be computed in terms of bulk densities only. The second selection, $(C/s)' = 0$, the surface of tension, leads to an *apparent* simplification of the formulas and was employed by GIBBS throughout his thermodynamic treatment of curved surfaces. It is convenient to exhibit the coefficient of the mean curvature correction term in the alternative forms

$$\left[\frac{C}{s} - \frac{\Gamma\gamma}{\varrho_\alpha - \varrho_\beta}\right]_\infty = \left(\frac{C}{s}\right)'' = -\,\delta_\infty \gamma_\infty \tag{3.5}$$

where δ_∞ is the distance from the surface of tension to the $\Gamma'' = 0$ surface. The form of Eq. (3.4), as well as approximate calculation[1] with use of molecular theories, shows that δ is roughly equal to the range of intermolecular forces. Thus the correction term in Eq. (3.2) is smaller than the dominant term by a

[1] J. G. KIRKWOOD and F. P. BUFF: J. Chem. Phys. 17, 338 (1949).

factor of 10^8 in the usual macroscopic applications of the theory. These considerations provide a quantitative basis for the usual neglect of these corrections on intuitive grounds. The further implication of these corrections must be deferred to the discussion of the total free energy of the system.

It is of interest to remark that Eq. (3.3) is the starting point for statistical mechanical theories of surface tension. When the stresses are expressed in terms of molecular variables, this formula may be employed to calculate the surface tension of representative monatomic liquids[1] and of solutions of electrolytes[2]. With use of statistical mechanical techniques which fall outside the scope of the present article, Eq. (3.3) may be used to verify the Gibbs adsorption equation. Furthermore, since this relation is also quantum mechanically exact, it may be employed in the calculation of high temperature quantum corrections.

After having effected these simple integrations we now turn to the derivation of the Laplace equation. One first recognizes that the normal component of Eq. (2.5) yields a generalized form of the Gibbs-Kelvin formula

$$p_\alpha - p_\beta = \sum_{i=1}^{2} \left(\gamma_i - c_i \frac{C_i}{s}\right) c_i + \Gamma g \, \boldsymbol{N}_{\alpha\beta} \cdot \boldsymbol{k}. \tag{3.6}$$

This expression relates the extrapolated bulk pressure difference across the dividing surface to the thermodynamic surface parameters and the principal curvatures of the surface. With use of Eqs. (3.1) and (3.2), the Gibbs-Kelvin formula reduces to the desired Laplace differential equation for the surface

$$p_\alpha(z_0) - p_\beta(z_0) - (\varrho_\alpha - \varrho_\beta)\, g\,(z - z_0) = \gamma_\infty (c_1 + c_2). \tag{3.7}$$

Here z is evaluated at any surface in the transition zone and it has been recognized that for macroscopic applications, the correction terms, exhibited in reference [*1*], are completely negligible. The integration of this nonlinear equation for a variety of geometrical shapes constitutes one of the largest chapters in the classical theory of capillarity. These solutions find their main application in the analysis of apparatus designed for surface tension determinations.

In order to obtain the boundary condition for the Laplace equation at the intersection of fluid lenses, we turn to Eq. (2.7). Closer examination of the thermodynamic length parameter $\mathscr{L}^*$ reveals that $\mathscr{L}^* = \lambda\gamma$, where λ is a distance comparable to the range of intermolecular forces. Again the first order correction terms, exhibited in reference [*2*], are entirely negligible in macroscopic applications. With their neglect one obtains the classical Neumann formula[3]

$$\gamma_\infty^{\mathrm{I\,II}}\, \boldsymbol{t}^* \times \boldsymbol{N}_{\mathrm{I\,II}} + \gamma_\infty^{\mathrm{II\,III}}\, \boldsymbol{t}^* \times \boldsymbol{N}_{\mathrm{II\,III}} - \gamma_\infty^{\mathrm{I\,III}}\, \boldsymbol{t}^* \times \boldsymbol{N}_{\mathrm{I\,III}} = 0. \tag{3.8}$$

This result shows that the possibility of construction of a triangle whose sides are numerically equal to the interfacial tensions of the coexisting phases is necessary for fluid lens formation. We observe that the Neumann triangle finds its most important application in the establishment of criteria for the spreading of liquids.

The functional form of Eq. (3.8) illustrates the dual nature of γ for fluid interfaces. On the one hand it possesses the excess surface free energy property and it also appears in mechanical tension relations. This feature accounts for its interchangeable designation as surface free energy or as surface tension.

---

[1] J. G. Kirkwood and F. P. Buff: J. Chem. Phys. **17**, 338 (1949).
[2] F. P. Buff and F. H. Stillinger jr.: J. Chem. Phys. **25**, 312 (1956).
[3] F. Neumann (first communicated in P. du Bois-Reymond's dissertation, Berlin, 1859).

The foregoing treatment has been entirely restricted to the description of fluid phases and interfaces. The complete discussion of the Laplace equation requires an additional boundary condition at the intersection of two fluid phases with a solid phase. It is provided by the classical Young equation. Since for solid-fluid interfaces the surface free energy-surface tension duality is no longer clear, the conventional discussion of the Young formula is postponed to section 5.

4. GIBBS' thermodynamic formulation. The preceding discussion will now be contrasted with the thermodynamic formulation which received its final form at the hands of GIBBS [4]. It will be recalled that the specification of a set of fundamental equations is the starting point of GIBBS' treatment. These equations summarize the first and second laws of thermodynamics for open systems. The contributions arising from the work done by the system will naturally differ when homogeneous or heterogeneous regions are described.

The treatment of small regions contained within homogeneous phases α is straightforward. Here the intrinsic energy E^α depends on the entropy S^α, composition $\{N_i^\alpha\}$, and volume v_α:

$$E^\alpha = E^\alpha(S^\alpha, \{N_i^\alpha\}, v_\alpha). \tag{4.1}$$

With these molar variables, reversible changes in state are then described by the fundamental equation

$$dE^\alpha = T\,dS^\alpha + \sum_{i=1}^{n} \mu_i^\alpha\, dN_i^\alpha - p_\alpha\, dv_\alpha \tag{4.2}$$

where T is the thermodynamic temperature and μ_i^α is the local chemical potential of component i. The other terms have their earlier interpretation. It is convenient to transform Eq. (4.2), which contains extensive variables, to an equivalent form in intensive variables only. Upon designating the local intrinsic energy density by E_v^α, the entropy density by S_v^α, and the bulk concentration of component i by c_i^α, Eq. (4.2) takes the form

$$dE_v^\alpha = T\,d\,S_v^\alpha + \sum_{i=1}^{n} \mu_i^\alpha\, d\,c_i^\alpha - \left[E_v^\alpha - T\,S_v^\alpha + p_\alpha - \sum_{i=1}^{n} \mu_i^\alpha c_i^\alpha\right] \frac{dv}{v}. \tag{4.3}$$

Since p_α has been shown to possess the free energy property, the terms in the bracket vanish. The desired relation finally takes the form

$$dE_v^\alpha = T\,dS_v^\alpha + \sum_{i=1}^{n} \mu_i^\alpha\, d\,c_i^\alpha, \tag{4.4}$$

while the free energy equation is given by

$$E_v^\alpha - T\,S_v^\alpha - \sum_{i=1}^{n} \mu_i^\alpha c_i^\alpha - p_\alpha. \tag{4.5}$$

The corresponding treatment of a very small macroscopic region which encloses a surface of discontinuity requires further geometrical specification. The region is considered to terminate at points within the phases α and β where bulk properties subsist. The transition zone is spanned by an initially arbitrary set of parallel dividing surfaces which separate the total volume into volumes v_α and v_β. For the Gibbs dividing surface of area s and principal curvatures c_1 and c_2, the intrinsic energy is then assumed to depend on the entropy, composition, volumes v_α and v_β, and area and principal curvatures of dividing surface:

$$E = E\left(S, \{N_i\}, v_\alpha, v_\beta, s, c_1, c_2\right). \tag{4.6}$$

With this set of molar variables, the following fundamental equation is postulated for reversible changes in state

$$dE = T\,dS + \sum_{i=1}^{n} \mu_i\,dN_i - p_\alpha\,dv_\alpha - p_\beta\,dv_\beta + \gamma\,ds + C_1\,dc_1 + C_2\,dc_2. \tag{4.7}$$

Here p_α and p_β are the bulk pressures for phases α and β, γ will be identified with the surface tension and C_1 and C_2 are designated as the thermodynamic principal curvature terms. GIBBS next transforms the last two terms as follows:

$$\left.\begin{aligned} C_1\,dc_1 + C_2\,dc_2 &= C\,d(c_1+c_2) + \left(\frac{C_1-C_2}{2}\right) d(c_1-c_2) \\ \text{with}\qquad C &= \frac{C_1+C_2}{2}. \end{aligned}\right\} \tag{4.8}$$

He observes that the coefficient of the term involving the differences in principal curvatures is of second order and it is thus deleted from further consideration. GIBBS finally selects that particular dividing surface, the surface of tension, which makes the surviving mean curvature term C vanish. His manipulation of the curvature terms has long been controversial, but the detailed analysis of Sect. 2 fully confirms the possibility of this approach. It will, however, be more useful to keep the surviving curvature term since more physically intuitive conclusions may be drawn by its retention. On the basis of these arguments, the fundamental equation, to within first order terms, takes the final form

$$dE = T\,dS + \sum_{i=1}^{n} \mu_i\,dN_i - p_\alpha\,dv_\alpha - p_\beta\,dv_\beta + \gamma\,ds + C\,d(c_1+c_2). \tag{4.9}$$

In order to extract from this equation the properties pertaining to the surface only, it is first necessary to subtract the bulk phase contributions. This is accomplished by extrapolating the bulk properties of the adjacent bulk phases up to the dividing surface, without regard to the existence of the surface of discontinuity. The fundamental equations applying to the matter contained within v_α and v_β respectively are given by

$$dE^j = T\,dS^j + \sum_{i=1}^{n} \mu_i^j\,dN_i^j - p_j\,dv_j; \quad j = \alpha, \beta. \tag{4.10}$$

When Eqs. (4.10) are subtracted from Eq. (4.9), the basic relation satisfied by the excess thermodynamic functions is obtained:

$$\left.\begin{aligned} dE^s &= T\,dS^s + \sum_{i=1}^{n} \mu_i^s\,dN_i^s + \gamma\,ds + C\,d(c_1+c_2); \\ E^s &= E - E^\alpha - E^\beta; \\ S^s &= S - S^\alpha - S^\beta; \\ N_i^s &= N_i - N_i^\alpha - N_i^\beta. \end{aligned}\right\} \tag{4.11}$$

For example, E^s is the difference between the actual energy of the region and that which would be computed on the basis of bulk properties only. Similarly N_i^s is the difference between the actual composition of component i and that found in volumes v_α and v_β with the help of the respective bulk densities. The manipulation of the superscript in the local chemical potential may be justified with use of the conditions of equilibrium.

It is again convenient to transform this equation for the excess thermodynamic functions into one containing intensive variables only. Upon designating the intrinsic excess surface energy density by E_s, the excess surface entropy density by S_s and the surface density of component i by Γ_i, Eq. (4.11) takes the form

$$\left.\begin{aligned} dE_s &= T\,dS_s + \sum_{i=1}^{n} \mu_i^s\,d\Gamma_i + \frac{C}{s}\,d(c_1+c_2) - \left[E_s - T\,S_s - \sum_{i=1}^{n}\mu_i^s\Gamma_i - \gamma\right]\frac{ds}{s}\,; \\ E_s &= \frac{E^s}{s}\,; \qquad S_s = \frac{S^s}{s}\,; \qquad \Gamma_i = \frac{N_i^s}{s}\,. \end{aligned}\right\} \tag{4.12}$$

Since γ has been previously shown to possess the free energy property, the terms in the bracket vanish. The desired relation finally takes the form

$$dE_s = T\,dS_s + \sum_{i=1}^{n}\mu_i^s\,d\Gamma_i + \frac{C}{s}\,d(c_1+c_2) \tag{4.13}$$

while the excess free energy equation is given by

$$E_s - T\,S_s = \sum_{i=1}^{n}\mu_i^s\,\Gamma_i + \gamma\,. \tag{4.14}$$

The thermodynamic description of regions corresponding to the intersection of three bulk phases along curve C^* is completely analogous. With the geometrical specification given in Sect. 2, bulk phase and surface properties are here extrapolated into the three-phase confluent region. The final equation for the local linear excess energy density E_l^*, etc., is given by

$$dE_l^* = T\,dS_l^* + \sum_{i=1}^{n}\mu_i^*\,d\Gamma_{il}^* \tag{4.15}$$

while the thermodynamic length parameter $\mathcal{L}^*$ enters into the relevant free-energy relation

$$E_l^* - T\,S_l^* = \sum_{i=1}^{n}\mu_i^*\,\Gamma_{il}^* - \mathcal{L}^*\,. \tag{4.16}$$

The equilibrium relations these fundamental equations imply follow from the Gibbs criterion of thermodynamic equilibrium

$$(\bar{\delta}E_{\text{tot}})_{S,\,V,\,\{N_i\}} = 0\,; \qquad E_{\text{tot}} = E_{\text{int}} + E_{\text{ext}}\,. \tag{4.17}$$

Here E_{tot} is the total energy of the three-phase fluid system under consideration. It consists of contributions from local intrinsic properties, E_{int}, and those arising from the external fields, E_{ext}. Their explicit representation is given by

$$\left.\begin{aligned} E_{\text{int}} &= \int E_v^\alpha\,dv_\alpha + \int E_v^\beta\,dv_\beta + E_s\,ds\,; \\ E_{\text{ext}} &= \int \varrho_\alpha\,\psi^\alpha\,dv_\alpha + \int \varrho_\beta\,\psi^\beta\,dv_\beta + \int \Gamma\,\psi^s(u,v)\,ds\,; \\ \Gamma &= \sum_{i=1}^{n} M_i\,\Gamma_i \end{aligned}\right\} \tag{4.18}$$

where the integrations are extended over the whole system. The corresponding expressions for total entropy S and total composition N_i are

$$\left.\begin{aligned} S &= \int S_v^\alpha\,dv_\alpha + \int S_v^\beta\,dv_\beta + \int S_s\,ds\,; \\ N_i &= \int c_i^\alpha\,dv_\alpha + \int c_i^\beta\,dv_\beta + \int \Gamma_i\,ds\,; \qquad i = 1, \ldots, n\,. \end{aligned}\right\} \tag{4.19}$$

The variational calculation dictated by the Gibbs criterion of equilibrium is conveniently carried out with use of Lagrangian multipliers. Thus

$$\bar{\delta}E_{\text{tot}} - T\,\bar{\delta}S - \sum_{=1}^{n}\bar{\mu}_i\,\bar{\delta}N_i = 0\,. \tag{4.20}$$

With use of Eqs. (4.2) and (4.13) it is immediately found that the entropy condition leads to constancy of temperature T, while the composition condition leads to the constancy of the partial potentials $\bar{\mu}_i$

$$\mu_i^\alpha + M_i \psi^\alpha = \mu_i^\beta + M_i \psi^\beta = \mu_i^s + M_i \psi^s = \bar{\mu}_i; \quad i = 1, \ldots, n. \tag{4.21}$$

The virtual displacements pertaining to the geometrical parameters may be calculated with either the theory of parallel surfaces or by means of the calculus of variations. Since the detailed analysis is reviewed in reference [*1*] and a related discussion will be presented in the next section, only the results need be summarized. To within first order correction terms, the only ones of significance in this approach, the previous extended forms of the Gibbs-Kelvin equation and Neumann triangle are obtained. Consequently, the detailed hydrostatic model fully confirms the validity of the present thermodynamic treatment. For example, the normal displacements to the dividing surface lead to the following relation

$$p_\alpha - p_\beta = \gamma (c_1 + c_2) - (c_1^2 + c_2^2) \frac{C}{s} + \Gamma \nabla_3 \psi \cdot \boldsymbol{N}. \tag{4.22}$$

In view of the earlier curvature dependence of surface tension, to be rederived by the thermodynamic route, this result is identical with Eq. (3.6).

Further important consequences of the theory follow from equations of the Gibbs-Duhem type. These are obtained from the fundamental equations by differentiating the free energy relations, Eqs. (4.5) and (4.14), and equating the respective results to Eqs. (4.4) and (4.13). In the case of bulk phases, the familiar Gibbs-Duhem equation is obtained

$$\sum_{i=1}^{n} c_i \, d\bar{\mu}_i + S_v \, dT - dp - \varrho \, d\psi = 0. \tag{4.23}$$

For the treatment of surface phenomena, this procedure leads to an extended form of the Gibbs adsorption equation

$$\sum_{i=1}^{n} \Gamma_i \, d\bar{\mu}_i + S_s \, dT + d\gamma - \Gamma \, d\psi - \frac{C}{s} \, d(c_1 + c_2) = 0. \tag{4.24}$$

Finally in the case of three-phase confluent regions, the following one-dimensional analogue of the preceding relations applies

$$\sum_{i=1}^{n} \Gamma_{il}^* \, d\bar{\mu}_i + S_l^* \, dT - d\mathcal{L}^* - \Gamma_l^* \, d\psi = 0. \tag{4.25}$$

The spatial variation of p, γ, and $\mathcal{L}^*$ may be obtained from these equations. Since at equilibrium, T and $\{\bar{\mu}_i\}$ are constant, they reduce to the following set of relations

$$dp = -\varrho \, d\psi, \tag{4.26}$$

$$d\gamma = \frac{C}{s} \, d(c_1 + c_2) + \Gamma \, d\psi, \tag{4.27}$$

$$d\mathcal{L}^* = -\Gamma_l^* \, d\psi. \tag{4.28}$$

Since Eq. (4.26) is equivalent to Eq. (2.4), its integration leads to the earlier result for the variation of pressure, Eq. (3.1). In the case of the spherical interface, GIBBS and TOLMAN[1] considered an equation similar to Eq. (4.27), and carried out its integration to molecular dimension. Their procedure is not justi-

[1] R. C. TOLMAN: J. Chem. Phys. **17**, 333 (1949).

fied since the thermodynamic approach can only yield valid first-order correction terms to the classical treatment. To within these first-order terms, the integration of Eq. (4.27) again yields the same curvature dependence of surface tension found in the hydrostatic approach, Eq. (3.2). The remaining relation for the spatial dependence of the length parameter $\mathcal{L}^*$ may be integrated in an identical manner.

We conclude this section by observing that the Gibbs criterion for equilibrium, Eq. (4.17) does not yield conditions concerning the stability of the system. For example, the equilibrium between spherical droplets of a one-component system in contact with supersaturated vapor is unstable. Although this case will be considered in more detail in connection with the subsequent discussion of nucleation, the interested reader is referred to GIBBS' monograph for the general treatment.

5. Variational formulation. The general thermodynamic theory of Sect. 4 may be summarized by means of a variational principle. It is obtained by applying consequences of the theory to the free energy relations. Since it has been shown that at equilibrium T and $\{\bar{\mu}_i\}$ are constant, it is convenient to carry out a Legendre transformation from the variables entropy and composition to the variables temperature and partial potentials. This transformation shows that for the equilibrium treatment of surface phenomena it suffices to consider the free energy Ω appropriate to the variables T and $\{\bar{\mu}_i\}$. The T, μ work function Ω is equal to the difference between the Helmholtz free energy $E_{\text{tot}} - TS$ and the Gibbs free energy $\sum_{i=1}^{n} N_i \bar{\mu}_i$:

$$E_{\text{tot}} - TS - \sum_{i=1}^{n} N_i \bar{\mu}_i = \Omega. \tag{5.1}$$

Consequently, it is of interest to note that Ω is connected with the grand partition function, i.e., G.P.F. $= \exp\left(-\frac{\Omega}{kT}\right)$; where k is the Boltzmann constant.

The explicit representation of Ω follows from Eqs. (4.5), (4.14), and (4.16)

$$\Omega \sim -\sum_{\alpha=\mathrm{I}}^{\mathrm{III}} \iiint p_\alpha \, dv_\alpha + \sum_{\alpha<\beta=\mathrm{I}}^{\mathrm{III}} \iint \gamma_{\alpha\beta} \, ds_{\alpha\beta} - \int \mathcal{L}^* \, dl^* + \cdots \tag{5.2}$$

where from the earlier results,

$$\left.\begin{aligned} p_\alpha(z) &= p_\alpha(z_0) - \varrho_\alpha(z_0)\, \Delta\psi \left[1 - \frac{\varepsilon_\alpha^0 \Delta\psi}{z} + \cdots\right], \\ \gamma &\sim \gamma_\infty + \left(\frac{C}{s}\right)''_\infty (c_1 + c_2), \\ \mathcal{L}^*(z) &\sim \mathcal{L}^*(z_0) - \Gamma_l^* \, \Delta\psi. \end{aligned}\right\} \tag{5.3}$$

For the interpretation of this result it is recognized that this representation of Ω provides the asymptotic expansion of the free energy with respect to the geometrical parameters of external force. Thus the detailed hydrodynamic model and its transcription into thermodynamic terms yield first-order asymptotic corrections to the classical representation. It is clear from the previous analysis, that these first-order terms lead to explicit ratios of microscopic distances to the dimensions of the system. This demonstrates that in the application of the theory to macroscopic systems (except possibly for coexisting phases of nearly identical density) it is entirely sufficient to employ only the leading terms of the theory. The correction terms thus provide concrete criteria for the range of validity of the conventional formalism. Their order of magnitude shows that the extrapolation of macroscopic thermodynamic concepts into the molecular domain, employed in current nucleation theories, is only approximate and that here more cumbersome techniques must be employed in free energy calculations.

We now turn to the general variation of Ω. Here it is again necessary to differentiate between the actual displacements of the system δ_b and the virtual displacements $\bar{\delta}$. The former provide the expressions for the work done by the system, $\delta_b W$, when it undergoes actual displacements under the condition of constant T and $\{\bar{\mu}_i\}$:

$$\delta_b W = -\,\delta_b \Omega. \tag{5.4}$$

The latter displacement corresponds to the shift from one set of dividing surfaces and space curves to their neighbors. Since Ω must be invariant to the choice of dividing surface, this procedure leads to the desired principle of virtual work

$$(\bar{\delta}\Omega)_{T,\{\bar{\mu}_i\},v} = 0. \tag{5.5}$$

The mathematical implications of these displacements may be calculated with use of the calculus of variations[1]. With its application, the following relations are obtained for the required general variations:

$$\left.\begin{aligned}
\delta \iiint \psi\, dv &= \iint \psi (\bar{\delta}\boldsymbol{x} + \delta_b \boldsymbol{x}) \cdot d\boldsymbol{s};\\
\delta \iint \psi\, ds &= \iint (\nabla_3 \psi \cdot \boldsymbol{N} + \psi \nabla_2 \cdot \boldsymbol{N})\, \bar{\delta}\boldsymbol{x} \cdot d\boldsymbol{s} + \oint \psi\, \boldsymbol{t} \times \boldsymbol{N} \cdot (\bar{\delta}\boldsymbol{x} + \delta_b \boldsymbol{x})\, dl;\\
\delta \iint (c_1 + c_2)\, dS &= z \iint c_1 c_2\, \bar{\delta}\boldsymbol{x} \cdot d\boldsymbol{s} + \\
&\quad + \oint [\bar{\delta}\boldsymbol{N} - \nabla_2 \boldsymbol{N} \cdot \bar{\delta}\boldsymbol{x} + \nabla_2 \cdot \boldsymbol{N} (\bar{\delta}\boldsymbol{x} + \delta_b \boldsymbol{x})] \cdot \boldsymbol{t} \times \boldsymbol{N}\, dl;\\
\delta \int \psi\, dl &= \Delta (\psi\, \boldsymbol{t} \cdot \delta \boldsymbol{x}) + \int \left[\nabla_3 \psi - \boldsymbol{t}\boldsymbol{t} \cdot \nabla_3 \psi - \psi \frac{d\boldsymbol{t}}{dl} \right] \cdot \bar{\delta}\boldsymbol{x}\, dl
\end{aligned}\right\} \tag{5.6}$$

where ψ is a known function of the coordinates $\boldsymbol{x}$.

The substitution of these results into Eq. (5.4) immediately leads to the earlier expressions for the work done by the system for actual displacements of its boundaries.

$$\delta_b W = \sum_{\alpha=\mathrm{I}}^{\mathrm{III}} \iint p_\alpha\, \delta_b \boldsymbol{x} \cdot d\boldsymbol{s}_\alpha - \sum_{\alpha<\beta=\mathrm{I}}^{\mathrm{III}} \int \gamma_{\alpha\beta}\, \boldsymbol{t}_{\alpha\beta} \times \boldsymbol{N}_{\alpha\beta} \cdot \delta_b \boldsymbol{x}\, dl_{\alpha\beta} + \Delta (\mathscr{L}^* \boldsymbol{t}^* \cdot \delta_b \boldsymbol{x}). \tag{5.7}$$

With future neglect of the correction terms, the application of the virtual displacements appearing in the variational principle leads to a relation of the type

$$\iint [-(p_\alpha - p_\beta) + \gamma_{\alpha\beta} \nabla_2 \cdot \boldsymbol{N}_{\alpha\beta}]\, \bar{\delta}\boldsymbol{x} \cdot d s_{\alpha\beta} = 0, \qquad \text{etc.} \tag{5.8}$$

so that the Kelvin (Laplace) equation and the Neumann triangle follow, i.e.

$$p_\alpha(z_0) - p_\beta(z_0) = \gamma_{\alpha\beta}(c_1 + c_2) + (\varrho_\alpha - \varrho_\beta)\, g\, (z - z_0); \tag{5.9}$$

and

$$\gamma_{\mathrm{I\,II}}\, \boldsymbol{t}^* \times \boldsymbol{N}_{\mathrm{I\,II}} + \gamma_{\mathrm{II\,III}}\, \boldsymbol{t}^* \times \boldsymbol{N}_{\mathrm{II\,III}} - \gamma_{\mathrm{I\,III}}\, \boldsymbol{t}^* \times \boldsymbol{N}_{\mathrm{I\,III}} = 0. \tag{5.10}$$

It will now be of interest to examine a further feature of the theory pertaining to Eqs. (5.9) and (5.10). Eq. (5.9) provides the partial differential equation, which, when solved subject to appropriate boundary conditions, determines the shapes of fluid interfaces. Eq. (5.10) supplies this boundary condition at the confluence of three fluid phases. The corresponding condition at the juncture

[1] R. COURANT and D. HILBERT: Methoden der mathematischen Physik, Vol. I, Chap. 4. Berlin: Springer 1931.

of two liquids and a solid is not as well definable because of familiar effects such as hysteresis, film formation, etc. The conventional phenomenological approach supplements Eq. (5.2) with terms of the form $\iint \gamma_{sf}\, ds_{sf}$ where γ_{sf} is a solid-fluid "free-energy". The variation analogous to Eq. (5.5) must now be subjected to the condition that $\bar{\delta}\boldsymbol{x}$ also satisfy the relation $\boldsymbol{N}_s \cdot \bar{\delta}\boldsymbol{x} = 0$, where $\boldsymbol{N}_s$ is the unit normal to the smooth surface. A sufficient condition for this restriction is provided by the relation

$$\boldsymbol{N}_s \times [\boldsymbol{t}^* \times \boldsymbol{N}_s \gamma_{s\alpha} + \boldsymbol{t}^* \times \boldsymbol{N}_{\alpha\beta} \gamma_{\alpha\beta} - \boldsymbol{t}^* \times \boldsymbol{N}_s \gamma_{s\beta}] = 0. \tag{5.11}$$

This relation is, of course, identical with the classical Young formula

$$\gamma_{s\beta} - \gamma_{s\alpha} = \boldsymbol{N}_s \cdot \boldsymbol{N}_{\alpha\beta} \gamma_{\alpha\beta}. \tag{5.12}$$

For the derivation of an equivalent relation involving alternative definitions of the solid free energies, the interested reader is referred to GIBBS' original treatment [*4*].

If one accepts the validity of Eq. (5.12), the Laplace equations may thus be solved to yield the shapes of fluid phases in a closed container of specified shape when the masses of the phases are known. This latter information is available under the conditions of DUHEM's theorem[1] which states that "whatever the number of phases, of components or of chemical reactions, the equilibrium state of a closed system, for which we know the initial masses, is completely determined by two independent variables". By the preceding theory, DUHEM's theorem may be extended by the further remark that specification of the shape of the container also provides the shapes of the fluid phases. Unfortunately, this mathematical result only has a most restricted experimental applicability due to the aforementioned difficulties of obtaining conditions under which the Young formula is applicable.

B. Applications.

6. Planar interface. The study of the thermodynamic properties of planar interfaces is the most important application of the theory of surface phenomena. With the help of the Gibbs adsorption equation, it relates surface tension changes to changes in the thermodynamic variables characterizing the bulk phases. As a preliminary to this study, it will be noted that in the planar case the Laplace equation reduces to

$$p_\alpha - p_\beta = g\,\Gamma_\infty \tag{6.1}$$

while the surface tension simplifies to

$$\gamma_\infty = \int_{-\infty}^{\infty} (\sigma_T - \sigma_N)\, dz - \int_{-\infty}^{\infty} (\varrho - \varrho_{\alpha\beta})\, g z\, dz. \tag{6.2}$$

Consequently, the gravitational field contributions are entirely negligible, and the bulk pressures are equal, $p_\alpha = p_\beta = p$. Similarly, the surface tension is now invariant to choice of dividing surface. These properties permit alternative treatments of the planar case [*5*] which avoid the use of Gibbs dividing surfaces altogether.

Since the gravitational field contributions are insignificant, the adsorption equation now takes the form

$$\sum_{i=1}^{n} \Gamma_i\, d\mu_i + S_s\, dT = -\, d\gamma. \tag{6.3}$$

---

[1] I. PRIGOGINE and R. DEFAY: Chemical Thermodynamics, p. 187. London: Longmans Green & Company 1954.

By comparison with the analogous Gibbs-Duhem equations for the adjacent bulk phases α and β

$$\sum_{i=1}^{n} c_i^j \, d\mu_i + S_v^j \, dT = dp_j; \quad j = \alpha, \beta \tag{6.4}$$

it is also immediately clear that Eq. (6.3) is invariant to shifts in dividing surfaces. This invariance also applies to the surface free energy relation

$$E_s - T S_s = \sum_{i=1}^{n} \Gamma_i \mu_i + \gamma, \tag{6.5}$$

since here the analogous bulk phase relations are given by

$$E_v^j - T S_v^j = \sum_{i=1}^{n} c_i^j \mu_i - p_j; \quad j = \alpha, \beta. \tag{6.6}$$

Any subsequent algebraic transformation on these basic surface relations will, of course, maintain this invariance property. The actual choice of dividing surface will in general be dictated by the application that is desired.

At this point it will be convenient to subject the adsorption equation to a change in intensive variables which will lead to the explicit disappearance of entropy and to the appearance of intrinsic energy. The resulting relation, completely equivalent to the original adsorption equation, is also more directly connected with the variables that appear in the statistical mechanical formulations of surface phenomena. Division of Eq. (6.3) by T and subsequent application of Eq. (6.5) leads directly to the desired relation

$$\sum_{i=1}^{n} \Gamma_i \, d\frac{\mu_i}{T} - E_s \, d\frac{1}{T} + d\frac{\gamma}{T} = 0. \tag{6.7}$$

The corresponding expression for the bulk phases is obtained in an identical manner and is given by

$$\sum_{i=1}^{n} c_i \, d\frac{\mu_i}{T} - E_v \, d\frac{1}{T} - d\frac{p}{T} = 0. \tag{6.8}$$

We shall now illustrate these relations by the consideration of one- and two-component systems. In the one-component system it is most convenient to select the dividing surface which makes the superficial density of matter vanish. This choice of $\Gamma'' = 0$ not only simplifies the formulas but also permits the calculation of the mass of the system in terms of bulk densities only. The free energy relation reduces to

$$E_s'' - T S_s'' = \gamma \tag{6.9}$$

and shows that γ is equal to the excess Helmholtz free energy computed with respect to the $\Gamma'' = 0$ surface. Similarly, the adsorption equation immediately leads to the surface analogue of the Gibbs-Helmholtz equation

$$\left(\frac{\partial \frac{\gamma}{T}}{\partial \frac{1}{T}} \right)_{\Delta\mu = 0} = \frac{d\frac{\gamma}{T}}{d\frac{1}{T}} = E_s''. \tag{6.10}$$

Eq. (6.3) directly leads to a formula for the calculation of the excess surface entropy:

$$S_s'' = -\frac{d\gamma}{dT}. \tag{6.11}$$

In connection with these formulas which relate temperature derivatives of γ to other thermodynamic functions, it is useful to observe that the reduced experimental temperature dependence of γ is well approximated by the following van der Waals-type expression:

$$\gamma \propto \left(1 - \frac{T}{T_c}\right)^{1+r} \tag{6.12}$$

where T_c is the critical temperature. GUGGENHEIM[1] has shown that the selection $r = \frac{2}{9}$ provides an excellent fit for simple liquids and has compared this result with the equally satisfactory Katayama formula

$$\gamma \propto \left(\frac{c_{\text{liq}} - c_{\text{vap}}}{c_{\text{crit}}}\right)^{\frac{2}{3}} \left(1 - \frac{T}{T_c}\right) \tag{6.13}$$

where c_{liq} and c_{vap} are the densities of coexisting liquid and vapor and c_{crit} is the critical density. In conclusion we observe that this type of empirical dependence on density for γ is employed in connection with the concept of the parachor [6].

The treatment of the two phase binary system is more complex since it is bivariant from the point of view of the phase rule. In order to transform the adsorption equation from the chemical potential variables to experimentally observed composition variables, it is necessary to employ the familiar thermodynamic relation

$$d\frac{\mu_i}{T} = \bar{v}_i\, d\frac{p}{T} + \bar{E}_i\, d\frac{1}{T} + \left(\frac{\partial \frac{\mu_i}{T}}{\partial x_2}\right)_{T,p} dx_2 \tag{6.14}$$

where x_i is the mol fraction of component i and $\bar{v}_i$ and $\bar{E}_i$ are its respective partial molar volume and energy. Since in the general case the derivatives describing the equilibrium phase lines are required, we shall derive these first. Recognition that $\mu_i^\alpha = \mu_i^\beta = \mu_i$, substitution of Eq. (6.14) into Eq. (6.4), and use of the Gibbs-Duhem equation for phase α

$$\sum_{i=1}^{2} x_i^\alpha \left(\frac{\partial \mu_i}{\partial x_2^\alpha}\right)_{T,p} = 0, \tag{6.15}$$

leads to the desired equation for the phase lines:

$$\sum_{i=1}^{2} x_i^\beta (\bar{E}_i^\beta - \bar{E}_i^\alpha)\, d\frac{1}{T} + \sum_{i=1}^{2} x_i^\beta (\bar{v}_i^\beta - \bar{v}_i^\alpha)\, d\frac{p}{T} + \frac{x_1^\alpha - x_1^\beta}{T x_1^\alpha} \left(\frac{\partial \mu_2^\alpha}{\partial x_2^\alpha}\right)_{T,p} dx_2^\alpha = 0. \tag{6.16}$$

The partial molar volumes and partial molar energies of phase β appear upon recognition that E and V are extensive.

The desired transformation of the adsorption equation again proceeds with use of Eqs. (6.14), (6.15) and (6.3). The final result is

$$\left.\begin{aligned} d\frac{\gamma}{T} = {} & \left[E_s - \sum \Gamma_i \bar{E}_i^\alpha - \left(\sum \Gamma_i \bar{v}_i^\alpha\right) \left(\frac{\partial \frac{p}{T}}{\partial \frac{1}{T}}\right)_{x_2^\alpha}\right] d\frac{1}{T} - \\ & - \frac{1}{T}\left[\left(\Gamma_2 - \frac{x_2^\alpha}{x_1^\alpha}\Gamma_1\right)\left(\frac{\partial \mu_2}{\partial x_2^\alpha}\right)_{T,p} + \left(\sum \Gamma_i \bar{v}_i^\alpha\right)\left(\frac{\partial p}{\partial x_2^\alpha}\right)_T\right] dx_2^\alpha \end{aligned}\right\} \tag{6.17}$$

where the pressure derivatives are obtained from Eq. (6.16). In this form the equation is still too complicated to be applied readily. It can be somewhat simplified by proper choice of dividing surfaces, and for this we refer to KOENIG's

[1] E. A. GUGGENHEIM: J. Chem. Phys. 13, 259 (1945).

exhaustive review [7]. For example, the selection $\sum \Gamma_i \bar{v}_i^\alpha = 0$ completely deletes the pressure derivatives from the general result. However, in applications to liquids in contact with dilute gases these pressure terms are negligible and the basic relation is here given by

$$d\frac{\gamma}{T} = \left(E_s - \sum \Gamma_i \bar{E}_i^\alpha\right) d\frac{1}{T} - \frac{1}{T}\left(\Gamma_2 - \frac{x_2^\alpha \Gamma_1}{x_1^\alpha}\right)\left(\frac{\partial \mu_2^\alpha}{\partial x_2^\alpha}\right)_{T,p} dx_2^\alpha, \tag{6.18}$$

where it is convenient to take α as the liquid phase. This formula shows that only the invariant combinations appearing on the right-hand side of Eq. (6.18) are amenable to experimental determination by thermodynamic means. This aspect has not only been emphasized by GUGGENHEIM, but he has also shown that non-thermodynamic selections of the dividing surface can lead to fruitful results [5].

For dilute solutions of solute 2, the Gibbs selection $\Gamma_1 = 0$ is convenient. Here the invariant composition combination reduces to $\Gamma_{2(1)}$, the superficial density of the solute computed with respect to the $\Gamma_1 = 0$ convention. For solutions which are sufficiently dilute so that they satisfy HENRY's law,

$$\left(\frac{\partial \mu_2^\alpha}{\partial x_2^\alpha}\right)_{T,p} = \frac{RT}{x_2^\alpha}, \tag{6.19}$$

the following relation is obtained:

$$\Gamma_{2(1)} \approx -\frac{x_2^\alpha}{RT}\frac{\partial \gamma}{\partial x_2^\alpha} \approx -\frac{c_2^\alpha}{RT}\frac{\partial \gamma}{\partial c_2^\alpha}. \tag{6.20}$$

It follows from this formula that for $\Gamma_{2(1)} > 0$, addition of solute leads to a decrease in γ (e.g. soap solutions) and that for $\Gamma_{2(1)} < 0$, addition of solute leads to an increase in γ (e.g. strong electrolyte solutions). For a more complete coverage of the Gibbs adsorption equation, the interested reader is referred to ADAMS' monograph [8].

7. The Neumann triangle and spreading. A more detailed examination of the Neumann triangle will illustrate the applications of this boundary condition of the Laplace equation at fluid boundaries. We again consider a fluid lens imbedded in two fluid phases and label the phases in order of decreasing density by I, II and III. For example, a completely liquid system can be obtained from water, aniline and hexane. Subsequent addition of detergent lowers the interfacial tensions sufficiently so that real solutions of the triangle relation are obtained and the middle aniline-rich phase collapses into a liquid lens. The two sides of the lens are respectively described analytically by the two-dimensional analogue of the equation of hydrostatics, Eq. (2.5).

$$\nabla_2 \cdot \sigma_2^{\alpha\beta} + (p_\alpha - p_\beta)\, \boldsymbol{N}_{\alpha\beta} = 0; \qquad \alpha < \beta. \tag{7.1}$$

The normal components of these two equations yield the respective Laplace equations which determine the shape of the lens. Even with cylindrically symmetrical lenses these equations can only be numerically integrated in the general case. However, the forces that act on the total lens can be easily determined since the two-dimensional divergence theorem, Eq. (1.12), leads to a simple first integral of the equations.

The desired modified form of ARCHIMEDES' principle is obtained by integration of Eq. (7.1) over both faces of the lens. The result is

$$\oint \gamma_{\alpha\beta}\, \boldsymbol{t}^* \times \boldsymbol{N}_{\alpha\beta}\, dl^* + \iint (p_\alpha - p_\beta)\, \boldsymbol{N}_{\alpha\beta}\, ds_{\alpha\beta} = 0; \qquad \alpha < \beta. \tag{7.2}$$

Upon addition, these relations may be exhibited in the form

$$\boldsymbol{F}_{\text{buoy}}+\boldsymbol{F}_{\text{ext}}+\boldsymbol{F}_{\text{per}}=0, \tag{7.3}$$

where $\boldsymbol{F}_{\text{buoy}}$ is the total normal force exerted by phases I and III on the lens

$$\boldsymbol{F}_{\text{buoy}}=\iint p_{\text{I}}\,\boldsymbol{N}_{\text{I II}}\,ds_{\text{I II}}+\iint p_{\text{III}}\,\boldsymbol{N}_{\text{III II}}\,ds_{\text{II III}}, \tag{7.4}$$

and $\boldsymbol{F}_{\text{ext}}$ is the total normal force exerted by the lens on phases I and III

$$\boldsymbol{F}_{\text{ext}}=\iint p_{\text{II}}\,\boldsymbol{N}_{\text{II I}}\,ds_{\text{I II}}+\iint p_{\text{II}}\,\boldsymbol{N}_{\text{II III}}\,ds_{\text{II III}}. \tag{7.5}$$

With use of GREEN's theorem and the equation of hydrostatics pertaining to phase II, it is seen to be the gravitational force acting on the lens:

$$\boldsymbol{F}_{\text{ext}}=\iiint \nabla p_{\text{II}}\,dv_{\text{II}}=-\boldsymbol{k}\,g\,\boldsymbol{M}_{\text{II}}. \tag{7.6}$$

The peripheral force may also be simplified when the Neumann equation, Eq. (3.8), is utilized so that the ARCHIMEDES' principle finally reduces to

$$\boldsymbol{F}_{\text{buoy}}-\boldsymbol{k}\,g\,\boldsymbol{M}_{\text{II}}+\oint \gamma^{\text{I III}}\,\boldsymbol{t}\times\boldsymbol{N}_{\text{I III}}\,dl=0 \tag{7.7}$$

where the last term is directed away from the lens. Entirely analogous considerations apply to similar force calculations for liquids contained within solid containers.

For most problems more naive considerations suffice. Restricting ourselves to macroscopic surfaces of revolution, the Neumann triangle implies that lens formation is only possible when the largest interfacial tension is less than the sum of the smaller ones. In this case, the interior angle of the lens ϑ_{int} is given by

$$\cos\vartheta_{\text{int}}=\frac{\gamma_{\text{I III}}^2-\gamma_{\text{I II}}^2-\gamma_{\text{II III}}^2}{2\gamma_{\text{I II}}\,\gamma_{\text{II III}}}, \tag{7.8}$$

a relationship which has been checked experimentally by MILLER[1]. In those cases where the above interfacial tension inequality is violated, so that lens formation is impossible, two limiting cases arise. When $\gamma_{\text{I II}}>\gamma_{\text{I II}}+\gamma_{\text{II III}}$, the middle phase spreads out as a film, while for $\gamma_{\text{II III}}>\gamma_{\text{I II}}+\gamma_{\text{I III}}$, the middle phase forms a globule which floats by buoyancy. Although these simple relations summarize the basic criteria, a large literature is available on more detailed aspects of spreading and film formation. It is reviewed in BURDON's monograph [*9*] and in HARKINS' review article [*10*].

8. Simple solutions of the Laplace equation. Since some of the leading mathematicians of the last century were concerned with finding solutions to the Laplace equation, an enormous literature is available on this topic. The most important results that have been obtained are exhaustively reviewed in BAKKER's monograph [*11*]. In addition to this reference, the interested reader is referred to the summaries of ADAMS [*8*] and HARKINS [*12*], which are primarily concerned with the mathematical analysis of equipment designed for the experimental determination of surface tension. Due to the availability of these reviews, we restrict ourselves to some simple illustrations.

When α designates the denser phase, it is convenient to exhibit the Laplace equation in the form:

$$\left.\begin{aligned} a^2\,\nabla_3\cdot\boldsymbol{N}_{\alpha\beta}+2(z-z_0)&=2\,\frac{[p_\alpha(z_0)-p_\beta(z_0)]}{(\varrho_\alpha-\varrho_\beta)\,g},\\ a^2&=\frac{2\gamma}{(\varrho_\alpha-\varrho_\beta)\,g},\end{aligned}\right\} \tag{8.1}$$

[1] N.F. MILLER: J. Phys. Chem. **45**, 1025 (1941).

where

$$\nabla_3 \cdot \boldsymbol{N}_{\alpha\beta} = c_1 + c_2. \tag{8.2}$$

The following features of these relations should be noted: (1), the "capillary constant" a^2 has dimensions distance squared; (2), the right-hand side of Eq. (8.1) vanishes when the origin of z is referred to a planar interface; and (3), Eq. (8.2) usually provides a simple way for the calculation of the mean curvature of the surface. In view of property (1), division of Eq. (8.1) by the parameter a leads to the dimensionless form of the Laplace equation, which is utilized in HARKINS' principle of similitude [*12*].

Two representative examples of cylindrical interfaces $z = z(x)$ will now be examined. In case (a), illustrated by Fig. 1, the liquid is infinitely extended to the left and meets a vertical flat plate at an angle of contact Θ. In case (b), illustrated by Fig. 2, we consider the liquid interface between two parallel flat plates which dip into the liquid. They are assumed identical and vertical to the horizontal. When ψ designates the angle which the tangent to the curve $z(x)$ makes with the positive x-axis, the Laplace equation takes the form

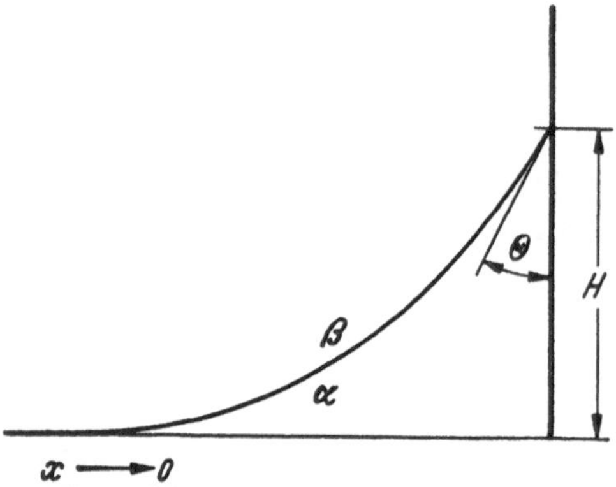

Fig. 1. Profile of liquid interface at flat vertical plate.

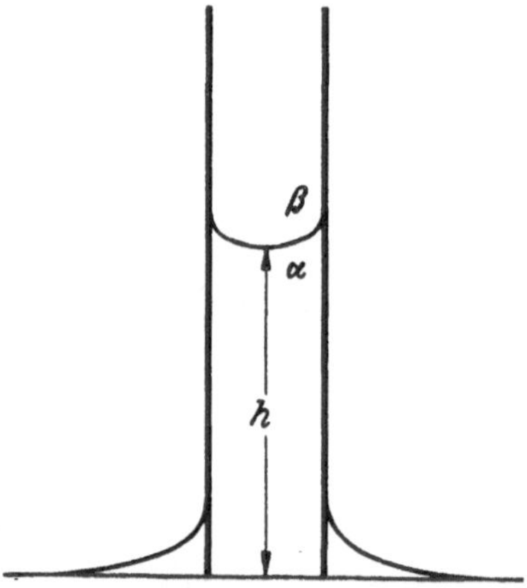

Fig. 2. Profile of liquid interface between flat parallel plates.

$$a^2 \frac{d \cos\psi}{dz} = -2z, \tag{8.3}$$

where

$$\frac{dz}{dx} = \tan\psi. \tag{8.4}$$

The first integral of Eq. (8.3) is obtained by inspection. When z_0 is the ordinate for which $\alpha = 0$, the result is

$$z = \sqrt{z_0^2 + 2a^2 \sin^2 \frac{\psi}{2}}. \tag{8.5}$$

The corresponding representation for x is obtained by the elimination of z between Eqs. (8.3) to (8.5). After rearrangement, it is found that

$$dx = -\frac{a k}{\sqrt{2}} \frac{(1 - 2\cos^2\varphi)\, d\varphi}{\sqrt{1 - k^2 \sin^2\varphi}}, \qquad k^2 = \frac{2a^2}{z_0^2 + 2a^2}; \qquad \varphi = \frac{\pi - \psi}{2}. \tag{8.6}$$

In case (a) $z_0 = 0$, so that $k = 1$. Eq. (8.5) immediately yields the maximum rise H at the plate, where $\psi = \frac{\pi}{2} - \Theta$,

$$H = a\sqrt{1 - \sin\Theta}. \tag{8.7}$$

The straightforward integration of (8.6) here leads to the profile

$$x + \text{const} = \sqrt{2a^2 - z^2} - \frac{a}{\sqrt{2}} \operatorname{Ar\,cosh} \frac{\sqrt{2}\,a}{z}, \qquad 0 \leqq z \leqq H \tag{8.8}$$

where the constant of integration is determined by the choice of origin.

In case (b) it is convenient to pass the line $x = 0$ through the lowest point of the meniscus, where $\psi = 0$ and $z_0 = h$. In the present example, the integration of Eq. (8.6) leads to elliptic integrals of the first and second kind. The parametric representation of the profile is given by

$$\left.\begin{aligned} z &= \sqrt{h^2 + 2a^2 \sin^2 \frac{\psi}{2}}\,; \\ x &= \frac{a\,k}{\sqrt{2}} \left[\left(\frac{2}{k^2} - 1 \right) \int\limits_{\frac{\pi-\psi}{2}}^{\frac{\pi}{2}} \frac{d\varphi}{\sqrt{1 - k^2 \sin^2 \varphi}} - \frac{2}{k^2} \int\limits_{\frac{\pi-\psi}{2}}^{\frac{\pi}{2}} \sqrt{1 - k^2 \sin^2 \varphi}\, d\varphi \right]. \end{aligned}\right\} \tag{8.9}$$

Although from the mathematical standpoint z is a periodic function of x which oscillates between h and $\sqrt{h^2 + 2a^2}$, the physically admissable solution is determined by the condition $0 \leqq |\psi| \leqq \frac{\pi}{2} - \Theta$. For very closely spaced plates the profile simplifies to a circular shape of radius $\frac{a^2}{2h}$.

Finally, we examine the meniscus which is obtained when a vertical cylindrical tube of internal radius R dips into a liquid. We assume that the angle of contact $0 \leqq \Theta < \frac{\pi}{2}$, so that capillary rise is encountered. The origin is located at the lowest point of the meniscus, which is at a height h above the horizontal plane. The Laplace equation for this surface of revolution $z = z(r)$ takes the form

$$\frac{a^2}{r} \frac{d(r \sin \psi)}{dr} = 2(h + z) \tag{8.10}$$

where

$$\frac{dz}{dr} = \tan \psi. \tag{8.11}$$

Since Eq. (8.10) can no longer be solved in closed form, we shall only consider the case of very small capillaries, which is of importance in surface tension determinations. In order to obtain iterative solutions, Eq. (8.10) is first integrated

$$r \sin \psi - \frac{h\,r^2}{a^2} = \frac{2}{a^2} \int\limits_0^r r\,z\,dr. \tag{8.12}$$

In the first approximation, the right-hand side of Eq. (8.12) is ignored, so that in conjunction with Eq. (8.11) it is seen that the profile is circular

$$\left(z - \frac{R}{\cos \Theta} \right)^2 + r^2 = \left(\frac{R}{\cos \Theta} \right)^2. \tag{8.13}$$

Use of Eq. (8.13) in Eq. (8.12) leads to the following approximation for h

$$h = \frac{a^2 \cos \Theta}{R} - \frac{R}{\cos \Theta} \left[1 - \frac{2}{3} \frac{(1 - \sin^3 \Theta)}{\cos^2 \Theta} \right]. \tag{8.14}$$

This procedure has been carried to higher orders by POISSON[1] and RAYLEIGH[2], and for liquids of zero contact angle in tubes with $R/a < 0.25$ leads to the follow-

[1] S. D. POISSON: Nouvelle théorie de l'action capillaire, p. 112. Paris 1831.
[2] Lord RAYLEIGH: Proc. Roy. Soc. Lond., Ser. A 92, 184 (1915).

ing formula for surface tension in terms of measurable quantities

$$a^2 = \frac{2\gamma}{(\varrho_\alpha - \varrho_\beta)\, g'} = R h \left\{ 1 + \frac{R}{3h} - 0.1288 \frac{R^2}{h^2} + 0.1312 \frac{R^3}{h^3} \right\}. \tag{8.15}$$

For higher values of R, the tables of BASHFORTH and ADAMS[1] must be employed, as well as the Rayleigh approximation for tubes of large radius.

9. Spherical droplets in one-component systems. As the size of droplets is diminished, the effect of the gravitational field becomes negligible and they will attain a spherical shape. The thermodynamic equilibrium of spherical drops in equilibrium with surrounding vapor was carefully examined by GIBBS [*4*]. An equivalent formulation expressed in terms of the dividing surface $\Gamma'' = 0$, is frequently advantageous[2] since its use directly relates the radius of the liquid drop R'' to its composition N_α.

The Gibbs-Kelvin equation may here be expressed in the alternative forms

$$\left.\begin{aligned} p_\alpha - p_\beta &= \frac{d\gamma'' s''}{dv''_\alpha}, \\ &= \frac{2\gamma'}{R'}. \end{aligned}\right\} \tag{9.1}$$

The pressure p_α within the drop is operationally defined in terms of the pressure p_β of the surrounding vapor by the condition of chemical equilibrium

$$\mu_\alpha(T, p_\alpha) = \mu_\beta(T, p_\beta). \tag{9.2}$$

With use of Eq. (9.2) and the power series expansion

$$\mu_\alpha(T, p_\alpha) = \mu_\alpha(T, p_\infty) + \bar{v}_\alpha(p_\alpha - p_\infty) - \frac{\varkappa \bar{v}_\alpha}{2}(p_\alpha - p_\infty)^2 + \cdots, \tag{9.3}$$

where $\bar{v}_\alpha$ and $\varkappa$ are the partial molar volume and isothermal compressibility coefficient of the liquid at the pressure p_∞ at which the two phases meet at a plane interface, the Gibbs-Thomson formula for the vapor pressure of the drop is immediately obtained,

$$\left.\begin{aligned} RT \log \frac{p}{p_\infty} &= \mu_\alpha(T, p) - \mu_\beta(T, p_\infty) \\ &= \bar{\bar{v}}_\alpha \left[(p_\alpha - p_\beta) - \frac{\varkappa}{2}(p_\alpha - p_\beta)^2 + \cdots \right] \end{aligned}\right\} \tag{9.4}$$

p/p_∞ is the supersaturation ratio. The usual assumption of incompressibility of the liquid phase corresponds to the deletion of the term ε^0, Eq. (5.3).

We shall now turn to the calculation of the isothermal work of formation of a droplet from a very large amount of vapor. Prior to formation of droplet, the free energy of the system is given by

$$E_{or} - T S_{or} = N \mu_\beta - p_\beta (v''_\alpha + v''_\beta), \tag{9.5}$$

while the free energy corresponding to the final state is given by

$$E_f - T S_f = N''_\alpha \mu_\alpha(T, p_\alpha) + (N - N''_\alpha)\, \mu_\beta(T, p_\beta) - p_\alpha v''_\alpha - p_\beta v''_\beta + \gamma'' s''. \tag{9.6}$$

Subtraction of Eq. (9.5) from Eq. (9.6) leads to the result

$$W^+ = N''_\alpha (\mu_\alpha - \mu_\beta) + \gamma'' s'' - (p_\alpha - p_\beta)\, v''_\alpha. \tag{9.7}$$

---

[1] R. BASHFORTH and J.C. ADAMS: An attempt to test the theories of capillary action. London: Cambridge Univ. Press 1883.

[2] F.P. BUFF: J. Chem. Phys. **19**, 1591 (1951).

Those droplets which are in *both* mechanical and chemical equilibrium with the surrounding vapor are characterized by $\mu_\alpha = \mu_\beta$, so that their work of formation is finally given by

$$W = \frac{\gamma' s'}{3} = \gamma'' s'' - v''_\alpha \frac{d\gamma'' s''}{d v''_\alpha}. \tag{9.8}$$

When the condition of chemical equilibrium is relaxed, it may be readily shown that

$$\left(\frac{dW^+}{dN''_\alpha}\right)_{\mu_\alpha=\mu_\beta} = 0; \qquad \left(\frac{d^2 W^+}{dN''^2_\alpha}\right)_{\mu_\alpha=\mu_\beta} < 0, \tag{9.9}$$

i.e., W^+ has a maximum for the droplet which is of such a size so as to be in equilibrium with the vapor phase, and where the formula for W^+ is rigorous. Eq. (9.9) shows that, for a one-component system, the equilibrium between droplet and vapor is metastable. For an incompressible liquid, the approximations

$$\mu_\beta = \mu_\alpha(T, p_\alpha^{\text{crit}}); \qquad p_\beta = p_\beta^{\text{crit}} \tag{9.11}$$

where p_α^{crit} and p_β^{crit} are the pressures in the respective bulk phases when the droplet is in complete equilibrium, lead to the following assessment for the free energy of formation of droplets in incomplete equilibrium:

$$W^+(R'') \approx \gamma'' s'' - (p_\alpha^{\text{crit}} - p_\beta^{\text{crit}})\, v''_\alpha. \tag{9.12}$$

These thermodynamic results which by the asymptotic significance of Ω, Eq. (5.2), are only rigorous for macroscopic systems, have been employed in theories of the kinetics of phase transition of supersaturated vapor to liquid [*13*]. Nucleation in a supersaturated vapor is initiated by statistical density fluctuations giving rise to molecular clusters, which decay unless they exceed a critical size. Cluster embryos exceeding the critical size serve as nuclei for the new phase through growth by accretion of molecules from the vapor. In the Becker-Döring theory[1], the clusters are idealized as minute spherical droplets possessing the properties of the liquid phase in bulk with a vapor pressure determined by the Gibbs-Thomson formula. The Becker-Döring kinetic considerations, which are based on a binary collision analysis, lead to Eq. (9.12) and finally yield a rate constant for the steady state rate of nucleation whose activation energy is given by GIBBS' expression for the work of formation of the critical embryo, Eq. (9.8).

10. The electrocapillary curve. The general thermodynamic treatment of electrocapillary phenomena was initiated by GIBBS and has been critically extended by KOENIG[2] and GRAHAME and WHITNEY[3]. The classical example of these phenomena is the variation of the interfacial tension of mercury-electrolytic solution systems with potential difference established across the interface. The thermodynamic treatment considers the electrode as ideally polarized, i.e., at equilibrium the concentration of every charged component is finite in one phase only. The desired variation of surface tension follows in a straightforward manner from the Gibbs adsorption equation, Eq. (4.24). In the present application to an ideally polarized electrode, it takes the form

$$\sum \Gamma_i^\alpha\, d\bar{\mu}_i + \sum \Gamma_j^\beta\, d\bar{\mu}_j + d\gamma + S_s\, dT = 0. \tag{10.1}$$

Since the system contains charged species, it is important to recognize: (1) that Eq. (10.1) must be subjected to the condition of electro-neutrality

$$F \sum z_i \Gamma_i^\alpha = - F \sum z_j \Gamma_j^\beta = q_\alpha, \tag{10.2}$$

[1] R. BECKER and W. DÖRING: Z. Physik **24**, 719 (1935).
[2] F. O. KOENIG: J. Phys. Chem. **38**, 111, 339 (1934).
[3] D. C. GRAHAME and R. B. WHITNEY: J. Amer. Chem. Soc. **64**, 1548 (1942).

and (2) the representative partial potential $\bar{\mu}_j$ of the charged species j of phase β, reduces to the electrochemical potential $\mu_j^\beta + z_j F\varphi^\beta$, where z_j is its valence, F is the Faraday constant, and ψ^β is the electrostatic potential of phase β. Upon transformation to chemical potentials and subsequent use of Eq. (10.2), the adsorption equation reduces to

$$\sum \Gamma_i^\alpha d\mu_i^\alpha + \sum \Gamma_j^\beta d\mu_j^\beta + q_\alpha d(\varphi^\alpha - \varphi^\beta) + d\gamma + S_s dT = 0. \tag{10.3}$$

Two pieces of the same metal are respectively connected (usually indirectly) to the metallic phase α and the electrolyte phase β. The potentiometer reading $\mathscr{E}$ is then related to the electrostatic potential difference between phases α and β, $\varphi_\alpha - \varphi_\beta$ by the relation

$$\mathscr{E} = \varphi^{\mathrm{II}} - \varphi^{\mathrm{I}} = (\varphi^{\mathrm{II}} - \varphi^\beta) - (\varphi^{\mathrm{I}} - \varphi^\alpha) - (\varphi^\alpha - \varphi^\beta). \tag{10.4}$$

Under conditions of constancy of temperature, pressure and composition, Eq. (10.4) implies that

$$d\mathscr{E} = -d(\varphi^\alpha - \varphi^\beta). \tag{10.5}$$

Under these thermodynamic conditions, Eq. (10.3) then reduces to the well-known Lippmann equation[1]

$$\left(\frac{\partial \gamma}{\partial \mathscr{E}}\right)_{T,\{\mu\}} = q_\alpha. \tag{10.6}$$

According to this relation, the slope of the electrocapillary curve is equal to the surface charge density q_α of the metallic surface. The effects of variations in composition can also be calculated from Eqs. (10.3) and (10.4). They are described in GRAHAME's comprehensive review [*14*].

General references

[*1*] BUFF, F.P.: J. Chem. Phys. **25**, 146 (1956).

[*2*] BUFF, F.P., and H. SALTSBURG: J. Chem. Phys. **26**, 23 (1957).

[*3*] BUFF, F.P., and H. SALTSBURG: J. Chem. Phys. **26**, 1526 (1957).

[*4*] GIBBS, J.W.: Collected Works, Vol. I, pp. 55—353. New Haven: Yale University Press 1948.

[*5*] GUGGENHEIM, E.A.: Thermodynamics. Amsterdam: North-Holland Publishing Company 1949.

[*6*] THOMSON, G.W.: In A. WEISSBERGER's Physical Methods of Organic Chemistry, Part I, pp. 413—425. New York: Interscience Publishers, Inc. 1949.

[*7*] KOENIG, F.O.: J. Chem. Phys. **18**, 449 (1950).

[*8*] ADAMS, N.K.: The Physics and Chemistry of Surfaces, 3rd edit. London: Oxford University Press 1941.

[*9*] BURDON, R.S.: Surface Tension and the Spreading of Liquids. Cambridge: Cambridge University Press 1949.

[*10*] HARKINS, W.D.: In A. WEISSBERGER's Physical Methods of Organic Chemistry, Part I, pp. 427—485. New York: Interscience Publishers, Inc. 1949.

[*11*] BAKKER, G.: In Handbuch der Experimentalphysik, Vol. VI. Leipzig: Akademische Verlagsgesellschaft 1928.

[*12*] HARKINS, W.D.: In A. WEISSBERGER's Physical Methods of Organic Chemistry, Part I, pp. 355—413. New York: Interscience Publishers, Inc. 1949.

[*13*] FRENKEL, J.: Kinetic Theory of Liquids. Oxford: Oxford University Press 1946.

[*14*] GRAHAME, D.C.: Chem. Rev. **41**, 441 (1947).

[1] G. LIPPMANN: Ann. Chim. Phys. **5**, 494 (1875).

Sachverzeichnis.

(Deutsch-Englisch.)

Bei gleicher Schreibweise in beiden Sprachen sind die Stichwörter nur einmal aufgeführt.

Adsorption 168, 169, 231, 235, 276.
— an einer ebenen Grenzschicht, *at a plane interface* 142.
—, relative, *relative* 144.
Adsorptionsaffinität, *affinity of adsorption* 168, 169.
Adsorptionsformel von GIBBS, *adsorption formula of Gibbs* 142, 170, 230, 267 ff.
Adsorptionsgleichung von GIBBS, *adsorption equation of Gibbs* 140, 150, 166, 205, 268.
Adsorptionsisotherme, *adsorption isotherm* 205.
Adsorptionspotential, *adsorption potential* 226.
Adsorptionswärme, *adsorption heat* 169.
Aktivität (Fugazität), *activity (fugacity)* 194.
—, thermodynamische, *thermodynamical* 30, 49, 69.
Anisotropie in der Oberflächenzone polarer Flüssigkeiten, *anisotropy in surface zone of polar liquids* 80.
—, optische, von Flüssigkeiten, *optical, of liquids* 22.
Arbeit der Tropfenbildung aus Dampf, *work of formation of a droplet from vapor* 302 bis 303.
Arbeitsleistung bei Oberflächenverschiebung, *work done by surface displacement* 286, 294.
Archimedisches Prinzip für eine Flüssigkeitslinse, *Archimedes' principles for a fluid lens* 298—299.
Argon, flüssiges, *liquid argon* 65, 67, 117.
—, Oberflächenspannung, *argon, surface tension* 217.
Assoziationsflüssigkeiten, *associated liquids* 4, 23, 25, 38, 80, 92.
Ausbreitung einer Flüssigkeitsschicht, *spreading of a liquid film* 299.

Bakker-Gleichung, *Bakker equation* 159.
Becker-Döringsche Theorie der Tropfenbildung, *Becker-Döring theory of droplet formation* 303.
Beltramischer (erster) Differentialparameter, *two-parametric gradient in a surface* 284.
B.E.T. (Brunauer-Emmet-Teller)-Flächen, *B.E.T. (Brunauer-Emmet-Teller) areas* 236.
Blasen, *bubbles* 12.
Blasen- und Tropfenbildung, *bubble and drop formation* 154, 157.
Bogoliubov-Green-Verfahren, *Bogoliubov-Green technique* 199.
Boltzmannsche Stoßgleichung, *Boltzmann equation* 104—107.
Born-Green-Yvon-Gleichung, *Born-Green-Yvon equation* 180, 185, 197, 210, 215 f.
Bornsche Näherung in der Streutheorie von Röntgenstrahlen oder Neutronen in Flüssigkeiten, *Born's approximation in scattering theory of x-rays or neutrons by liquids* 43.
Bose-Einstein-Statistik, *Bose-Einstein statistics* 122—123, 133.
Brechung in Flüssigkeiten, *diffraction in liquids* 15, 19, 43.
Brownsche Bewegung, *Brownian motion* 85, 112, 115.

Chemische Bindung, *chemical linkage* 23.
chemische Potentiale *chemical potentials* 82.
— —, laterale, *lateral chemical potentials* 170, 171, 272.
— —, vollständige, *complete chemical potentials* 169.
chemische Reaktionen in Flüssigkeiten, *chemical reactions in liquids* 40, 82, 92.
Cluster-Integral für Oberflächen, *superficial cluster integral* 202, 204 ff.
—, verallgemeinertes, *generalized cluster integral* 202.
Cluster-Modell der Flüssigkeiten, *cluster modell of liquids* 10—11.
Clusters (Molekülgruppen) in Flüssigkeiten, *clusters in liquids* 6, 30, 50.
Cybotaxis 7, 10.

Dampfdruck, *vapour pressure* 23, 133.
—, Krümmungsabhängigkeit, *curvature dependence* 154.
— über einem Tropfen, *of a drop* 302, 303.
Debye-Hückel-Theorie, *Debye-Hückel theory* 223.
Deformation von Flüssigkeiten durch Kraftfelder, *deformation of liquids by fields of forces* 67—69.
— der Molekülstruktur in Flüssigkeiten, *of molecular structure in liquids* 20, 67.
Dichte, molekulare, reduzierte, *reduced molecular density* 133.
Dichtematrix, *density matrix* 126, 237, 240.

Dichte-Schwankungen in Flüssigkeiten, *density fluctuations in liquids* 9, 11, 12, 15, 19, 30, 32, 43, 47.
Dichteübergangskurven, *density transition curves* 222.
dielektrisches Verhalten nicht-polarer Flüssigkeiten, *dielectric behaviour of non polar liquid* 71—74.
— — polarer Flüssigkeiten, *of polar liquids* 74—77.
Dielektrizitätskonstante von Flüssigkeiten, *dielectric constant of liquids* 22, 74, 81.
Diffusion, thermische, in Flüssigkeiten (Soret-Effekt), *thermal diffusion in liquids (Soret effect)* 2, 24, 82.
Diffusionskoeffizient, *diffusion coefficient* 84.
Dilatation 89, 118
Dipol-Kräfte in Flüssigkeiten, *dipolar forces in liquids* 23.
Dipolmoment, elektrisches, *electric dipole moment* 72.
Dipol-Wechselwirkung, *dipole interaction* 75.
Diskontinuitätsfläche, *surface of discontinuity* 216ff.
dissipative Prozesse, *dissipative processes* 112.
Dissoziation, *dissociation* 92.
Doppelschichtmodell, *two layer model* 269.
Drei-Phasen-Kontaktzone s. Kontaktzone, *three-phase confluent zone see confluent zone.*
Drei-Teilchen-Verteilungsfunktion, *triplet distribution function* 181, 197.
Druck, *pressure* 19, 30, 35, 49, 62, 67, 107, 119.
—, kinetischer, *kinetic* 132.
—, kritischer, *critical* 66, 133.
—, reduzierter, *reduced* 133.
—, thermodynamischer, *thermodynamic* 132.
Druckdiffusion, *pressure diffusion* 84.
Drucktensor, *pressure tensor* 33, 35, 82, 157, 159, 161, 164, 167, 208, 209, 210, 211, 212, 213.
—, Normalkomponente, *normal component* 158, 159, 164, 210, 218.
—, partieller, *partial* 41.
—, statistisch-mechanischer Ausdruck, *statistical-mechanical expression* 208f., 212, 215.
—, Tangentialkomponente, *tangential component* 158, 159, 164, 167, 210, 212, 214, 220, 222, 260, 261, 262.
Duhemscher Satz, *Duhem's theorem* 295.

ebene Grenzfläche, *planar interface* 142, 157, 165, 181, 184, 194, 210ff., 264f., 295—298.
Edelgase, flüssige, *liquid noble gases* 1, 2, 48, 53, 55, 65, 67, 80, 93, 117.
einatomige Flüssigkeiten, Berechnung der Oberflächenspannung, *monatomic liquids, surface tension calculation* 288.
einheitliche Flüssigkeiten, *uniform liquids* 47 bis 80.
Einschichtmodell, *monolayer model* 269, 270, 275.
Einteilchendichte, *singlet density* 282.
Ein-Teilchen-Verteilungsfunktion, *singlet distribution function* 179f., 184f., 194, 197, 200, 208, 210, 215f., 238, 240.
elastische Deformation von Flüssigkeiten, *elastic deformation of fluids* 91.
elastische Wellen, transversale und longitudinale, *transverse and longitudinal elastic waves* 104.
Elastizität und elastische Konstanten von Flüssigkeiten, *elasticity and elastic constants of liquids* 20, 69—71.
elektrokapillare Erscheinungen, *electrocapillary phenomena* 303—304.
Elektrolyte, Berechnung der Oberflächenspannung, *electrolytes, surface tension calculation* 288, 298.
elektromagnetische Kräfte, *electromagnetic forces* 67.
elliptische Integrale in der Theorie der Kapillarität, *elliptic integrals in capillarity theory* 301.
Energie s. auch freie und innere Energie, *energy see also free and internal energy.*
—, innere, an der Oberfläche, *superficial internal* 78, 79, 80.
Energiedichte, *energy density* 35.
Energiefluß, *energy flux* 35, 82, 96.
—, partieller, *partial* 41.
Energietransport, *energy transport* 87.
Entropie, *entropy* 67, 78, 81, 82, 120, 121.
—, gemeinschaftliche, *communal* 242f., 248, 250, 263, 273.
— der Oberfläche s. Oberflächenentropie und Oberflächendichte der Entropie, *entropy of surface see surface and superficial entropy.*
Entropieerzeugung, *entropy production* 83.
Entropiefluß, *entropy flux* 83.
Eötvössche Formel, *Eötvös formula* 174.
Erhaltung der Energie, *conservation of energy* 35, 120, 128.
— des Impulses, *of momentum* 33, 120, 128.

Fermi-Statistik, *Fermi statistics* 122.
Festkörperbegrenzung, *solid boundary* 69.
Filmbildung zwischen zwei flüssigen Phasen, *film formation between two fluid phases* 299.
Flächentheorie, grundlegende Sätze, *surface theory, basic theorems* 284.
flüssige Kristalle, *liquid crystals* 22, 81.
Flüssigkeit an ebenen, senkrechten Platten, *liquid at flat vertical plates* 300.
Flüssigkeiten mit ähnlicher molekularer Anordnung, *liquids with similar molecular constitution* 2.
—, angelagerte, *associated* 4, 23, 25, 38, 80.
—, normale, *normal* 120.
—, polare, *polar* 4, 21, 38, 80.
Flüssigkeitskugel, schwimmende, *floating globule of liquid* 299.
Flüssigkeitslinse, eingebettet zwischen zwei Flüssigkeitsphasen, *fluid lens embedded in two fluid phases* 282—283, 285, 298, 299.

Flüssigkeitsmodelle, Gitter-Zellen-Modell, *liquid models, lattice-cell model* 7—10, 24, 53.
—, kinetisches Modell, *kinetic model* 13—15.
—, Löcher-Modell, *hole model* 11—13.
—, Zellen-Cluster-Modell, *cell-cluster model* 10—11, 64—65.
Flüssigkeitszustand als ungeordneter Zustand, *liquid state as disordered state* 5.
freie Energie, *free energy* 62, 67, 78, 285, 286.
— — der Oberfläche, *superficial* 137, 149, 285, 288, 296.
freies Volumen, *free volume* 240, 242f., 247, 249f., 262, 265, 273.
— —, kugelförmiges, *sphericalized* 243, 250.
— —, verallgemeinertes, *generalized* 250.
— —, verschmiertes, *smeared* 243, 247f.
Fugazität (Aktivität), *fugacity (activity)* 194.
Fundamentalformen der Flächentheorie, *fundamental forms of surface theory* 284.

Gasadsorption, *gas adsorption* 205, 231ff.
— an fester Oberfläche, *on a solid surface* 231.
Gaußscher Satz in zwei Dimensionen, *divergence theorem in two dimensions* 284, 286.
Gaußsches Krümmungsmaß einer Fläche, *Gaussian curvature of a surface* 284.
Gemische von Flüssigkeiten, *mixtures of liquids* 2, 23—25, 40—43, 67.
Geschwindigkeits-Verteilungsfunktion, *velocity distribution function* 27—28, 39, 42, 127.
Gibbssche Adsorptionsformel, *Gibbs adsorption formula* 142, 170, 230, 267f., 269.
— — für eine reguläre Lösung, *for a regular solution* 268.
Gibbssche Adsorptionsgleichung, *Gibbs adsorption equation* 140, 150, 166, 205, 255, 268, 288, 292, 303.
— —, Verallgemeinerung, *generalization* 150, 168, 170.
Gibbssche freie Energie, *Gibbs free energy* 293.
Gibbssche Fundamentalgleichung, *Gibbs fundamental equation* 139, 170.
— — für die Grenzschicht, *for the interface layer* 140.
Gibbssche thermodynamische Formulierung der Kapillarität, *Gibbs' thermodynamical formulation of capillarity* 289—293.
Gibbssche Trennflächen, *Gibbs' dividing surfaces* 282, 286, 287, 289.
Gibbssches thermodynamisches Potential, *Gibbs' thermodynamic potential* 26, 107 bis 108.
Gibbs-Duhem-Gleichungen, *Gibbs-Duhem equations* 143, 150, 165, 166, 167, 219, 292, 297.
Gibbs-Helmholtz-Gleichung für eine ebene Grenzschicht, *Gibbs-Helmholtz equation for a plane interface* 142.
—, Oberflächenanalogon, *surface analogue* 296.
Gibbs-Thomsonsche Formel, *Gibbs-Kelvin formula* 288, 292.
— —, verallgemeinerte, *generalized* 155.
Gibbs-Tolman-Koenig-Buff-Gleichung, *Gibbs-Tolman-Koenig-Buff equation* 153.
Gitterebenen gleichen Potentials, *equipotential lattice planes* 252.
Gittermodell der Flüssigkeiten, *lattice model of liquids* 7—10, 24, 240, 242.
Gitterstruktur, *lattice structure* 123.
Gläser, *glasses* 4, 23, 69.
Gleichgewicht zwischen Tropfen und Dampf, *equilibrium between droplet and vapor* 303.
Gleichgewichtskriterium von Gibbs, *equilibrium criterion of Gibbs* 291.
Gravitationsfeld, Einfluß auf Flüssigkeiten, *gravitational field, influence on liquids* 15, 67.
Grenzfläche, *interface* 134, 136.
—, Adsorption an einer ebenen, *adsorption at a plane* 142.
—, ebene, *plane interface* 142, 157, 165, 181, 184, 194, 210ff., 264f., 295—298
—, —, Gibbs-Helmholtz-Gleichung, *Gibbs-Helmholtz equation* 142.
—, —, quasithermodynamische Beziehungen, *quasithermodynamic relations* 165.
Grenzfläche, kugelförmige, *spherical interface* 145, 146, 148, 149, 150, 159, 166, 213ff., 214, 215.
—, —, Oberflächenspannung, *surface tension* 146, 161, 187, 214.
—, —, quasithermodynamische Beziehungen, *quasithermodynamic relations* 166f.
—, —, thermodynamische Fundamentalgleichungen, *fundamental thermodynamic equations* 148.
Grenzflächennormale, Vorzeichenfestsetzung, *interface normal, sign convention* 283—284.
große Verteilungsfunktion, *grand partition function* 293.
großes Potential, *grand potential* 140, 192, 193.
Grundgleichung der Hydrostatik, *equation of hydrostatics* 283, 287.
— —, ein- und zweidimensionale Analoga, *one- and two-dimensional analogues* 285, 298.
Guggenheimsche Formel, *Guggenheim's formula* 155.
Guggenheimsches Mischungsgesetz, *Guggenheim's mixing law* 268.

Hauptkrümmungen einer Fläche, *principal curvatures of a surface* 284, 289—290.
Helium, flüssiges, *liquid helium* 119ff.
Helmholtzsche freie Energie, *Helmholtz free energy* 137, 149, 293, 296.
Hildebrandsche Gleichung, *Hildebrand's equation* 268, 269.
H-Theorem der Flüssigkeiten, *H-theorem of liquids* 107.
hydrostatische Grundgleichung, *hydrostatic fundamental equation* 283, 287.
— —, ein- und zweidimensionale Analoga, *one- and two-dimensional analogues* 285, 298.

hydrostatisches Modell der Kapillarität, *hydrostatic model of capillarity* 281.

Impulsdichte, *momentum density* 33.
Impulsfluß, *momentum flux* 33.
innere Eigenschaften, *internal modes* 95.
innere Energie, *internal energy* 20, 35—36, 68, 78, 82, 107, 119.
— — der Oberfläche, *superficial* 137, 141, 166, 167, 189, 190, 216, 244f., 257f.
Integrale über die Übergangszone hinweg, *integrals across transition zone* 285.
Ionenflüssigkeiten, *ionic liquids* 23.
irreversible Prozesse in Flüssigkeiten, *irreversible processes in liquids* 24, 25, 38, 81 bis 89, 92.
Isothermen in Flüssigkeiten, nach der Überlagerungsnäherung, *isotherms in liquids, superposition approximation* 53.
Isotropie des Spannungstensors, *isotropy of stress tensor* 283.

kanonische Gesamtheit, *canonical ensemble* 177, 178, 179, 237.
— —, große, *grand* 191, 192, 194.
Kapillarität, Differentialgleichung von Laplace, *capillarity, differential equation of Laplace* 286, 288, 294.
—, historische Bemerkungen zu ihrer Theorie, *capillarity theory, historical remarks* 281.
—, Randbedingungen, *boundary conditions* 286, 288, 289.
—, thermodynamische Formulierung von Gibbs, *thermodynamical formulation of Gibbs* 289—293.
Kapillare, Meniskusbildung, *capillary, meniscus formation* 301.
Kapillarkonstante, *capillary constant* 300.
Katayama-Guggenheim-Formel, *Katayama-Guggenheim formula* 174, 175, 176, 297.
Katayama-Guggenheim-Formeln, modifizierte, *modified Katayama-Guggenheim formulae* 177.
Keimbildung, *nucleation* 157, 294, 303.
Kelvin-Beziehung, *Kelvin relation* 134, 147, 154, 160, 161, 162.
kinetische Energie von Molekülen, *kinetic energy of molecules* 27.
kinetische Theorie der Flüssigkeiten, Überlagerungsnäherung, *kinetic theory of liquids, superposition approximation* 52, 53, 59, 66, 74, 111.
kinetisches Modell der Flüssigkeiten, *kinetic model of liquids* 13—15, 52—61, 65—67.
Kirkwood-Müller-Gleichung, *Kirkwood-Müller equation* 234.
Koexistenz zweier Phasen, *coexistence of two phases* 252.
Kompressibilität von Flüssigkeiten, *compressibility of liquids* 19, 20, 66, 69, 119, 126.
Kondensation, *condensation* 6, 33.
Konfigurationsintegral, *configuration integral* 179, 182, 188, 241f., 249ff., 253f., 263f.
konservative Kräfte, *conservative system of forces* 26, 107.
Kontaktzone von drei Flüssigkeitsphasen, *confluent zone of three fluid phases* 282, 283, 285.
Koordinationsschalen, *coordination shells* 17, 89, 124, 125.
Korrelation, binäre, *binary correlation* 65.
Korrelations-Funktion, *correlation function* 26, 47, 62, 79, 181, 216.
korrespondierende Zustände, Prinzip der, *principle of corresponding states* 172, 174, 175, 177, 186, 187.
— —, quantenmechanisches Prinzip der, *quantum mechanical principle of* 176.
Kräfte, zwischenmolekulare, *intermolecular forces* 2, 3.
Kraftfelder in Flüssigkeiten, Deformation *fields of forces in liquids, deformation* 67 bis 69.
kristalliner Zustand, *crystalline state* 6, 7, 17, 22.
kritischer Punkt, *critical point* 19, 66, 172, 248, 297.
Krümmungsfunktion von Gibbs, *curvature term of Gibbs* 290.
Krümmungsmaß einer Fläche, *curvature of a surface* 284
kugelförmige Grenzfläche (s. auch Tropfen), *spherical interface (see also droplet)* 145, 148, 150, 159, 166, 213ff., 292.

Längenparameter in der Theorie der Kapillarität, *length parameter in capillarity theory* 285, 286, 291—293.
λ-Phänomen, *λ-phenomenon* 94, 119, 122.
λ-Punkt des Heliums, *λ-point of helium* 119, 122, 133.
Laplacesche Gleichung der Kapillarität, *Laplace equation of capillarity* 286, 288, 294.
— — —, dimensionslose Form, *dimensionless form* 300.
— — —, ebener Fall, *planar case* 295.
— — —, einfache Lösungen, *simple solutions* 299—302.
Laplacesche Gleichungen zur Bestimmung der Gestalt einer Flüssigkeitslinse, *Laplace equations determining shape of fluid lens* 298.
Leitfähigkeit, elektrische, *electric conductivity* 81.
Lennard-Jones-Devonshire-Modell, *Lennard-Jones and Devonshire model* 242.
Lennard-Jones-Potential, *Lennard-Jones potential* 173, 217, 244, 246.
—, modifiziertes, *modified* 181.
—, Rushbrookesche Modifikation, *Rushbrooke's modification* 217.
Linse einer Flüssigkeit zwischen zwei flüssigen Phasen, *lens of liquid embedded in two fluid phases* 282—283, 285, 298—299.
Liouvillesche Gleichung, *Liouville's equation* 97.
Lippmannsche Gleichung, *Lippmann equation* 304.

Löchermodell der Flüssigkeiten, *hole model of liquids* 11—13, 242, 249f., 258, 261.
Lösungen, athermische polymere, *athermal polymer solutions* 272f.
— von Nichtelektrolyten, *solutions of nonelectrolytes* 262f.
—, Oberflächenspannung, *surface tension* 298.
—, polymere, *polymer* 272f.
—, reguläre, *regular* 262, 264f.
—, —, mit ebenen Grenzflächen, *with plane interface* 264.
London-Konstante, *London constant* 234.
London-Kräfte, *London forces* 233.

Massendichte, *mass density* 33.
Massenverhältnis, *mass fraction* 41, 82.
McLeodsche Gleichung, *McLeod's equation* 174, 175.
Mechanik, molekulare, *molecular mechanics* 33—47.
mehratomige Flüssigkeiten, *polyatomic liquids* 65.
Membranverhalten einer Trennfläche, *membrane behavior of dividing surface* 286.
Meniskus, *meniscus* 301.
mittlere Dicke der Ionenatmosphäre, *mean thickness of the ionic atmosphere* 224.
mittleres Krümmungsmaß einer Fläche, *mean curvature of a surface* 284, 288, 300.
Modelle von Flüssigkeiten s. Flüssigkeitsmodelle, *models of liquids see liquid models.*
Moleküle als harte Kugeln, *hard sphere molecules* 221.
—, starr kugelförmig, *rigid spherical molecules* 48, 59.
Molekülgruppen (clusters) in Flüssigkeiten, *clusters of molecules* 6, 30—33, 50.
molekulare Mechanik von Flüssigkeiten, *molecular mechanics of liquids* 33—47.
molekulare Polarisierbarkeit, *molecular polarizability* 72.
molekulare Statistik von Flüssigkeiten, *molecular statistics of liquids* 28—33.
molekulare Verteilungsfunktion, *molecular distribution function* 25, 27, 39, 47, 179, 182, 194, 195.
— —, nichtlineare Integralgleichung, *nonlinear integral equation* 79.

Natrium, flüssiges, *liquid sodium* 56.
Neumannsches Dreieck, *Neumann triangle* 286, 288, 292, 294, 298—299.
Neutronen-Brechung in Flüssigkeiten, *neutron diffraction in liquids* 43—47, 124.
Newtonsche Flüssigkeiten, *Newtonian liquids* 15.
nicht-einheitliche Flüssigkeiten, *non-uniform liquids* 81ff.
— —, nicht-stationäre Zustände, *unsteady states* 101—104.
— —, stationäre Zustände, *steady states* 99.
Normale zur Grenzfläche, Vorzeichenfestsetzung, *normal to interface, sign convention* 283—284.

Oberflächendichte der Entropie, *surface entropy* 141, 171, 248, 249, 259, 260.
— der Helmholtzschen freien Energie, *superficial density of the Helmholtz free energy* 139.
— der inneren Energie, *surface energy* 141, 185, 190, 216, 244, 245, 246, 247, 248, 257f., 259.
— der Moleküle, *of molecules* 267, 269.
— der potentiellen Energie, *of potential energy* 245, 246.
— der Teilchen, *superficial number density* 141, 142, 145, 166, 167, 171, 184, 185, 202, 227.
Oberflächenenergie von Argon, *surface energy of argon* 217.
—, freie, *superficial free energy* 137, 149, 285, 288, 296.
—, innere, *superficial internal energy* 137, 166, 167, 185.
Oberflächenentropie, *superficial entropy* 78, 137, 166, 167, 177.
Oberflächengrößen, *superficial quantities* 137, 139, 149, 166.
Oberflächenkoeffizient der Aktivität, *superficial activity coefficient* 205, 208.
Oberflächenkraft, *surface force* 286.
Oberflächenladung, *surface charge* 226.
Oberflächenspannung, *surface tension* 23, 77, 81, 138, 139, 148, 149, 158, 161f., 166, 167, 171, 172f., 177, 179, 182, 186, 190, 192, 193, 198, 200, 201, 205, 211, 213, 215, 216, 218, 219, 237—239, 240, 246, 248, 257f., 259, 260, 266, 268, 270, 271, 275, 276, 277.
—, Anstieg, *increase* 229.
— von Argon, *of argon* 217.
— athermischer polymerer Lösungen, *of athermal polymer solutions* 273.
—, Definition, *definition* 288, 290.
—, Differenz, *difference* 197, 202, 227f., 230f.
—, dynamische, *dynamic* 271f.
— einer ebenen Grenzfläche, *of a plane interface* 181, 184.
—, empirische Formel, *empirical formula* 172, 174.
—, Korrekturen für hohe Temperatur, *high temperature corrections* 288.
—, Krümmungsabhängigkeit, *curvature dependence* 152, 153, 215.
— der kugelförmigen Grenzschicht, *of the spherical interface* 146, 159, 161, 187, 214.
— von Lösungen, *of solutions* 201.
—, mechanische Definition, *mechanical definition* 157, 158, 161, 166, 167, 261.
— polymerer Lösungen, *of polymer solutions* 272.
—, Quanteneffekte, *quantum effects* 237.
—, reduzierte, *reduced* 172, 173, 175, 176, 187.
— starker Elektrolyte, *of strong electrolytes* 223.
—, statische, *static* 271f.
—, Temperaturabhängigkeit, *temperature dependence* 151, 259, 297.

Oberflächenspannung verdünnter Lösungen, *surface tension of dilute solutions* 298.
— einer vollkommenen Lösung, *of a perfect solution* 268.
Oberflächenzone in polaren und angelagerten Flüssigkeiten, *surface zone in polar and associated liquids* 80.
offenes System, *open system* 139.
Onsager-Beziehung, *Onsager relation* 110.
Opaleszenz in Flüssigkeiten, *opalescence in liquids* 47.
optische Anisotropie von Flüssigkeiten, *optical anisotropy of liquids* 22.
Ordnung-Unordnung-Übergang = Schmelzerscheinung, *order-disorder transition = phenomenon of melting* 5, 7.
Orientierung, kooperative, *cooperative orientation* 80.
Orientierungsverschmelzung, *orientation fusion* 81.
osmotische Bedingungen, *osmotic conditions* 196f., 200.
osmotische Lösung, *osmotic solution* 198, 201.
osmotischer Druck, *osmotic pressure* 196f., 201, 228.
— —, normaler, *normal* 228.
— —, tangentialer, *tangential* 228.

Parachor 174.
phänomenologische Kapillaritätsgleichungen, *phenomenological capillarity equations* 284 bis 286.
Phononen, *phonons* 123.
piezoelektrischer Effekt in Flüssigkeiten, *piezoelectric effect in liquids* 70.
Poisson-Boltzmann-Gleichung, *Poisson-Boltzmann equation* 223.
polare Flüssigkeiten, *polar liquids* 4, 21, 38, 80.
Polarisierbarkeit, molekulare, *molecular polarizability* 72.
Potential, großes, *grand potential* 140, 192, 193.
— der mittleren Kräfte, *potential of average forces* 196, 200f., 223, 226.
—, thermodynamischer, *thermodynamic* 26, 107—108.

Quanteneffekte in Flüssigkeiten, *quantum effects in liquids* 20, 28, 93, 119—133.
quantenmechanischer Parameter, *quantum mechanical parameter* 176.
Quantenstatistik, *quantum statistics* 122.
quasi-chemische Näherung, *quasi-chemical approximation* 8.
quasithermodynamische Theorie, *quasithermodynamic theory* 163, 172, 218, 219ff., 260f.
— —, Ausdruck für das Gleichgewicht, *expression for equilibrium* 165, 166, 260, 261.
— —, Beziehungen für die ebene Grenzfläche, *relations for plane interface* 165.
quasithermodynamische Theorie, Beziehungen für die kugelförmige Grenzfläche, *quasithermodynamic theory, relations for spherical interface* 166, 167.
— —, Grundpostulat, *fundamental postulate* 163, 165, 166.

Radiale Verteilungsfunktion, *radial distribution function* 15—17, 19, 26, 47—61, 65, 125.
Randbedingung für Kontakt von zwei flüssigen mit einer festen Phase, *boundary condition at intersection of two fluid phases with a solid phase* 289, 295.
— für Kontaktstelle dreier Flüssigkeitsphasen, *at confluence of three fluid phases* 286, 288, 294.
reduzierte Variable, *reduced variables* 172, 173, 182, 186, 187, 188.
Reibungskoeffizient, *friction constant* 85, 112, 118.
Reibungstensor, *friction tensor* 114.
Relaxationsfrequenz, *relaxation frequency* 103, 114.
Relaxationsprozesse, *relaxation processes* 82, 95.
Relaxationszeit, *relaxation time* 71.
Reversibilität, mikroskopische, *microscopic reversibility* 83.
reversible Änderung einer Grenzfläche, *reversible change of an interface* 289—290.
Riesen-Zellen-Modell, *giant-cell model* 11.
Röntgenstrahlbrechung in Flüssigkeiten, *x-ray diffraction in liquids* 7, 16, 19, 43 bis 47, 55, 56, 81, 124.
Rotonen, *rotons* 123.

Schall, Ausbreitung und Absorption in Flüssigkeiten, *sound, propagation and absorption in liquids* 91, 92, 95.
—, Geschwindigkeit in Flüssigkeiten ohne Absorption, *velocity in liquids without absorption* 95.
Scherungsmodul der Elastizität, *shear modulus of elasticity* 92.
— von Flüssigkeiten, *of liquids* 20, 71.
Scherungsspannung, *shearing stress* 91.
Scherungsviskosität, *shearing viscosity* 83, 89, 91, 92.
—, Koeffizient, *coefficient* 91.
Schmelzerscheinung = Übergang von Ordnung zu Unordnung, *melting phenomenon = order-disorder transition* 5, 7.
Schmelzpunkt, *melting point* 66.
Schrödinger-Gleichung für Quantenflüssigkeiten, *Schrödinger wave equation for quantum liquids* 127.
Schwankungen der Massendichte, *fluctuations of mass density* 282.
Schwerefeld, *gravitational field* 283, 287, 295.
second sound in Helium II, *second sound in He II* 94.
Semiinvariante, *semi-invariant* 202.

senkrechte Platten, Kapillarerscheinungen einer Flüssigkeit, *vertical plates, capillarity effect of liquid* 300.
Siedeprozeß in Flüssigkeiten, *boiling process in liquids* 18.
Soret-Effekt in Flüssigkeiten, *Soret effect in liquids* 2, 24, 82.
Spannungsfläche, *surface of tension* 147, 148, 155, 160, 161, 163, 212, 213, 216, 217, 218, 287, 290.
Spannungstensor, *stress tensor* 283, 286.
spezifische Wärme, *specific heat* 123.
Starrheit von Flüssigkeiten, *rigidity of liquids* 5, 69—71, 81.
stationäre Zustände einer nicht-einheitlichen Flüssigkeit, *steady states of a non-uniform liquid* 99—101.
Statistik, molekulare, *molecular statistics* 28 bis 33.
Stoßwellen, *shock waves* 35.
Streuung von Licht in Flüssigkeiten, *scattering of light in liquids* 30, 32, 47.
Strömungsgeschwindigkeit, *velocity of flow* 33, 41.
—, laminare, *fluid velocity potential* 41.
Struktur von Flüssigkeiten, quantitative Beschreibung, *structure of liquids, quantitative description* 25—47, 47—81, 81 bis 113.
Superpositionsnäherung der kinetischen Theorie, *superposition approximation of kinetic theory* 52, 53, 59, 66, 74, 111, 181, 197, 199, 201.
Suprafluidität von Helium, *superfluidity of helium* 119, 122.

Teilchendichte, *number density* 26, 67.
—, Ortsabhängigkeit in der Übergangszone, *local variation in transition zone* 282.
Temperatur, kritische, *critical temperature* 66, 133, 297.
—, reduzierte, *reduced temperature* 133.
Temperaturabhängigkeit der Oberflächenspannung, *temperature dependence of surface tension* 151, 259, 288, 297.
thermische Diffusion in Flüssigkeiten (Soret-Effekt), *thermal diffusion in liquids (Soret effect)* 2, 24, 82.
Thermodynamik irreversibler Prozesse, *thermodynamics of irreversible processes* 24, 25, 81, 168, 169, 271.
—, lokale Formulierung, *local formulation* 163.
thermodynamische Fundamentalparameter der Kapillarität, *thermodynamic basic parameter of capillarity* 287.
thermodynamische Hauptkrümmungsfunktionen, *thermodynamic principal curvature terms* 290.
thermodynamische Temperatur einer ruhenden Flüssigkeit, *thermodynamic temperature for a liquid at rest* 28.
thermodynamisches Potential, *thermodynamic potential* 26, 107—108.
thermodynamisches Verhalten von Flüssigkeiten, *thermodynamic behaviour of liquids* 61—67.
Tolman-Buff-Gleichung, *Tolman-Buff equation* 162, 163.
Tonkssche Gleichung, *Tonks' equation* 221.
Traubesche Regel, *Traube's rule* 276f.
Trennfläche, *dividing surface* 135, 136, 139, 142—144, 146, 147, 152, 158, 160—162, 166, 167, 171, 210, 213, 214, 216, 225, 232, 244, 248, 259, 264, 267, 274.
—, äquimolekulare, *equimolecular* 137, 139, 142, 152, 186, 213, 216, 217, 218, 244, 257.
Trennfläche von Gibbs, *dividing surface of Gibbs* 282, 286, 287, 289.
Tropfen, kugelförmiger, *spherical droplet* 293, 302—303.
Tropfenbildung, *droplet formation* 302—303.
Tropfen- und Blasenbildung, *drop and bubble formation* 154, 157.

Übergangszone, *transition zone* 282, 284, 286.
—, Form des Spannungstensors, *form of stress tensor* 283, 286.
—, Integrale darüber hinweg, *intrgrals across it* 285.
Überschuß der freien Energie an einer Grenzfläche, *excess free energy at an interface* 286, 296.
Überschußentropie an einer Oberfläche, *excess entropy at a surface* 296.
Überschußfunktionen, thermodynamische, an einer Grenzfläche, *excess thermodynamic functions at an interface* 290.
Überschußvolumen, *excess volume* 232, 233.
Ultraschallabsorption in Flüssigkeiten, *ultrasonic absorption in liquids* 35, 91.
Umfangsspannung, *peripheral tension* 286.
Unterkühlung von Flüssigkeiten, *supercooled state of liquids* 9.

van der Waals-Gleichung, *van der Waals equation* 172, 219, 220, 221.
van der Waals-Schleife, *van der Waals loop* 172, 244, 252.
van der Waalssche Kräfte, *van der Waals' forces* 23, 119, 223.
Variationsprinzip der Kapillarität, *variational principle of capillarity* 293—295.
Verformungstensor, *strain tensor* 83, 90.
Verteilungsfunktion, *distribution function* 25ff.
— s. auch Ein-, Zwei- und Drei-Teilchen-Verteilungsfunktion, *see also singlet, pair, and triplet distribution function.*
— s. auch Geschwindigkeits-, molekulare, radiale Verteilungsfunktion, *see also velocity, molecular, radial distribution function.*
— der molekularen Verschiebungen, *displacement distribution function* 47—50.
Vielkristalle, *polycrystals* 10.
Virialkoeffizient, zweiter, *second virial coefficient* 66.
Virial-Koeffizienten, *virial coefficients* 52, 66.

Viskosität von Flüssigkeiten, *viscosity of liquids* 14, 20, 81, 82, 89—91, 110, 111, 115, 119.
Volumen, kritisches, *critical volume* 66.
Volumenphase, *bulk phase* 282, 283.
—, Eliminierung, *elimination* 290.
Volumenviskosität, *volume viscosity* 83, 89, 91, 92, 118.
—, Koeffizient, *coefficient* 91.

Wärmefluß-Vektor, *thermal flux vector* 36, 42, 82.
Wärmeleitung in Flüssigkeiten, *thermal conduction of liquids* 14, 20, 38, 82, 86—89, 111, 115.
Wahrscheinlichkeitsdichte, *probability density function* 178, 191, 237.
Wechselwirkungsenergie in Flüssigkeiten, *interaction energy in liquids* 3, 39.
Winkel in einer Flüssigkeitslinse, *angle in a fluid lens* 299.
Winkelgeschwindigkeit, Einfluß auf Hydrostatik, *angular velocity, effect on hydrostatics* 283.

Youngsche Gleichung, *Young equation* 289, 295.

Zähigkeit s. Viskosität.
Zellenmodell der Flüssigkeiten, *cell model of liquids* 7—10, 47—51, 61—65, 242.
Zentrifugalkraftfeld, *centrifugal field* 283.
Zerreißfestigkeit, *tensile strength* 22.
Zustandsfunktion, *partition function* 12, 79, 131, 177, 178, 237, 238, 240, 241, 242.
—, große, *grand* 190, 191f., 194f., 206.
—, Zellenmethode für die Berechnung, *cell method for calculating* 240.
Zustandsgleichung für Flüssigkeiten, *equation of state for liquids* 13, 19, 52, 55.
zweiatomige Flüssigkeiten, *diatomic liquids* 66.
Zwei-Phasen-Übergangszone s. Übergangszone, *two-phase transition zone see transition zone.*
Zweiteilchen-Oberflächendichte, *superficial pair density* 184.
Zwei-Teilchen-Verteilungsfunktion, *pair distribution function* 179ff., 184, 186f., 194, 196, 200, 204, 208, 215, 238.
zylindrisches Rohr, Meniskus einer Flüssigkeitsoberfläche, *cylindrical tube, meniscus of a liquid surface* 301—302.

Subject Index.

(English-German).

Where English and German spelling of a word is identical the German version is omitted.

Activity (fugacity), *Aktivität (Fugazität)* 194.
—, thermodynamical, *thermodynamische Aktivität* 30, 49, 69.
Adsorption 168, 169, 231, 235, 276.
Adsorption equation of GIBBS, *Adsorptionsgleichung von Gibbs* 140, 150, 166, 205, 268.
Adsorption formula of GIBBS, *Adsorptionsformel von Gibbs* 142, 170, 230, 267 seq.
Adsorption heat, *Adsorptionswärme* 169.
Adsorption isotherm, *Adsorptionsisotherme* 205.
Adsorption at plane interface, *Adsorption an einer ebenen Grenzschicht* 142.
Adsorption potential, *Adsorptionspotential* 226.
Adsorption, relative, *relative Adsorption* 144.
Affinity of adsorption, *Adsorptionsaffinität* 168, 169.
Angle in a fluid lens, *Winkel in einer Flüssigkeitslinse* 299.
Angular velocity, effect on hydrostatics, *Winkelgeschwindigkeit, Einfluß auf Hydrostatik* 283.
Anisotropy, optical, of liquids, *optische Anisotropie von Flüssigkeiten* 22.
— in surface zone of polar liquids, *Anisotropie in der Oberflächenzone polarer Flüssigkeiten* 80.
ARCHIMEDES' principle for a fluid lens, *Archimedisches Prinzip für eine Flüssigkeitslinse* 298—299.
Argon, liquid, *flüssiges Argon* 65, 67, 117.
—, surface tension, *Argon, Oberflächenspannung* 217.
Associated liquids, *Assoziationsflüssigkeiten* 4, 23, 25, 38, 80, 92.

Bakker equation, *Bakker-Gleichung* 159.
B.E.T. (Brunauer-Emmet-Teller) areas, *B.E.T. (Brunauer-Emmet-Teller)-Flächen* 236.
Becker-Döring theory of droplet formation, *Becker-Döringsche Theorie der Tropfenbildung* 303.
Bogoliubov-Green technique, *Bogoliubov-Green-Verfahren* 199.
Boiling process in liquids, *Siedeprozeß in Flüssigkeiten* 18.
Boltzmann equation, *Boltzmannsche Stoßgleichung* 104—107.
Born-Green-Yvon equation, *Born-Green-Yvon-Gleichung* 180, 185, 197, 210, 215 seq.
BORN's approximation in scattering theory of x-rays or neutrons by liquids, *Bornsche Näherung in der Streutheorie von Röntgenstrahlen oder Neutronen in Flüssigkeiten* 43.
Bose-Einstein statistics, *Bose-Einstein-Statistik* 122—123, 133.
Boundary condition at confluence of three fluid phases, *Randbedingung für Kontaktstelle dreier Flüssigkeitsphasen* 286, 288, 294.
— — at intersection of two fluid phases with a solid phase, *für Kontakt von zwei flüssigen mit einer festen Phase* 289, 295.
Brownian motion, *Brownsche Bewegung* 85, 112, 115.
Bubbles, *Blasen* 12.
Bubble and drop formation, *Blasen- und Tropfenbildung* 154, 157.
Bulk phase, *Volumenphase* 282, 283.
— —, elimination, *Eliminierung* 290.

Canonical ensemble, *kanonische Gesamtheit* 177, 178, 179, 237.
— —, grand, *große* 191, 192, 194.
Capillarity, boundary conditions, *Kapillarität Randbedingungen* 286, 288, 289.
—, differential equation of LAPLACE, *Differentialgleichung von Laplace* 286, 288, 294.
—, meniscus formation, *Meniskusbildung* 301.
—, thermodynamical formulation of GIBBS, *thermodynamische Formulierung von Gibbs* 289—293.
Capillarity theory, historical remarks, *Kapilrarität, historische Bemerkungen zu ihre-Theorie* 281.
Capillary constant, *Kapillarkonstante* 300.
Cell model of liquids, *Zellenmodell der Flüssigkeiten* 7—10, 47—51, 61—65, 242.
Centrifugal field, *Zentrifugalkraftfeld* 283.
Chemical linkage, *chemische Bindung* 23.
Chemical potential, *chemisches Potential* 82.
Chemical potentials, complete, *vollständige chemische Potentiale* 169.
— —, lateral, *laterale chemische Potentiale* 170, 171, 272.
Chemical reactions in liquids, *chemische Reaktionen in Flüssigkeiten* 40, 82, 92.
Cluster integral, generalized, *verallgemeinertes Cluster-Integral* 202.
— —, superficial, *Cluster-Integral für Oberflächen* 202, 204 seq.

Clusters in liquids, *Clusters (Molekülgruppen) in Flüssigkeiten* 6, 30, 50.
Cluster model of liquids, *Cluster-Modell der Flüssigkeiten* 10—11.
Coexistence of two phases, *Koexistenz zweier Phasen* 252.
Compressibility of liquids, *Kompressibilität von Flüssigkeiten* 19, 20, 66, 69, 119, 126.
Condensation, *Kondensation* 6, 33.
Conductivity, electric, *elektrische Leitfähigkeit* 81.
Configuration integral, *Konfigurationsintegral* 179, 182, 188, 241seq., 249seq., 253seq., 263seq.
Confluent zone of three fluid phases, *Kontaktzone von drei Flüssigkeitsphasen* 282, 283, 285.
Conservation of energy, *Erhaltung der Energie* 35, 120, 128.
— of momentum, *des Impulses* 33, 120, 128.
Conservative system of forces, *konservative Kräfte* 26, 107.
Coöperative orientation, *kooperative Orientierung* 80.
Coordination shells, *Koordinationsschalen* 17, 89, 124, 125.
Correlation, binary, *binäre Korrelation* 65.
Correlation function, *Korrelations-Funktion* 26, 47, 62, 79, 181, 216.
Corresponding states, principle of *Prinzip der korrespondierenden Zustände* 172, 174, 175, 177, 186, 187.
— —, quantum mechanical principle of, *quantenmechanisches Prinzip der korrespondierenden Zustände* 176.
Critical point, *kritischer Punkt* 19, 66, 172, 248, 297.
Critical volume, *kritisches Volumen* 66.
Crystalline state, *kristalliner Zustand* 6, 7, 17, 22.
Curvature of a surface, *Krümmungsmaß einer Fläche* 284.
Curvature term of Gibbs, *Krümmungsfunktion von Gibbs* 290.
Cybotaxis 7, 10.
Cylindrical tube, meniscus of a liquid surface, *zylindrisches Rohr, Meniskus einer Flüssigkeitsoberfläche* 301—302.

Debye-Hückel theory, *Debye-Hückel-Theorie* 223.
Deformation of liquids by fields of forces, *Deformation von Flüssigkeiten durch Kraftfelder* 67—69.
— of molecular structure in liquids, *der Molekülstruktur in Flüssigkeiten* 20, 67.
Density fluctuations in liquids, *Dichte-Schwankungen in Flüssigkeiten* 9, 11, 12, 15, 19, 30, 32, 43, 47.
Density matrix, *Dichtematrix* 126, 237, 240.
Density, molecular, reduced, *reduzierte molekulare Dichte* 133.
Density transition curves, *Dichteübergangskurven* 222.
Diatomic liquids, *zweiatomige Flüssigkeiten* 66.
Dielectric behaviour of liquids, non polar liquid, *dielektrisches Verhalten von Flüssigkeiten, nicht-polare Flüssigkeiten* 71 to 74.
— — —, polar liquids, *polare Flüssigkeiten* 74—77.
Dielectric constant of liquids, *Dielektrizitätskonstante von Flüssigkeiten* 22, 74, 81.
Diffraction in liquids, *Brechung in Flüssigkeiten* 15, 19, 43.
Diffusion coefficient, *Diffusionskoeffizient* 84.
Diffusion, thermal, in liquids (Soret effect), *thermische Diffusion in Flüssigkeiten (Soret-Effekt)* 2, 24, 82.
Dilatation 89, 118.
Dipolar forces in liquids, *Dipol-Kräfte in Flüssigkeiten* 23.
Dipole interaction, *Dipol-Wechselwirkung* 75.
Dipole moment, electric, *elektrisches Dipolmoment* 72.
Displacement distribution function, *Verteilungsfunktion der molekularen Verschiebungen* 47—50.
Dissipative processes, *dissipative Prozesse* 112.
Dissociation, *Dissoziation* 92.
Distribution function, *Verteilungsfunktion* 25ff.
— — see also singlet, pair, and triplet distribution function, *Verteilungsfunktion s. auch Ein-, Zwei- und Drei-Teilchen-Verteilungsfunktion.*
— — see also velocity, displacement, molecular, radial distribution function, *s. auch unter Geschwindigkeits-, molekulare und radiale Verteilungsfunktion.*
Divergence theorem in two dimensions, *Gaußscher Satz in zwei Dimensionen* 284, 286.
Dividing surface, *Trennfläche* 135, 136, 139, 142—144, 146, 147, 152, 158, 160—162, 166, 167, 171, 210, 213, 214, 216, 225, 232, 244, 248, 259, 264, 267, 274.
— —, equimolecular, *äquimolekulare* 137, 139, 142, 152, 186, 213, 216, 217, 218, 244, 257.
— — of Gibbs, *Trennfläche von Gibbs* 282, 286, 287, 289.
Drop and bubble formation, *Tropfen- und Blasenbildung* 154, 157.
Droplet formation, *Tropfenbildung* 302—303.
Droplet, spherical, *kugelförmiger Tropfen* 293, 302—303.
Duhem's theorem, *Duhemscher Satz* 295.

Elastic deformation of fluids, *elastische Deformation von Flüssigkeiten* 91.
Elastic waves, transverse and longitudinal, *transversale und longitudinale elastische Wellen* 104.
Elasticity and elastic constants of liquids, *Elastizität und elastische Konstanten von Flüssigkeiten* 20, 69—71.
Electrocapillary phenomena, *elektrokapillare Erscheinungen* 303—304.

Electrolytes, surface tension calculation, *Elektrolyte, Berechnung der Oberflächenspannung* 288, 298.
Electromagnetic forces, *elektromagnetische Kräfte* 67.
Elliptic integrals in capillarity theory, *elliptische Integrale in der Theorie der Kapillarität* 301.
Energy see also free and internal energy, *Energie s. auch freie und innere Energie.*
Energy density, *Energiedichte* 35.
Energy flux, *Energiefluß* 35, 82, 96.
— —, partial, *partieller* 41.
Energy, superficial internal, *innere Energie an der Oberfläche* 78, 79, 80.
Energy transport, *Energietransport* 87.
Entropy, *Entropie* 67, 78, 81, 82, 120, 121.
—, communal, *gemeinschaftliche* 242seq., 248, 250, 263, 273.
Entropy flux, *Entropiefluß* 83.
Entropy production, *Entropieerzeugung* 83.
Entropy of surface see surface and superficial entropy, *Entropie der Oberfläche s. Oberflächenentropie und Oberflächendichte der Entropie.*
Eötvös formula, *Eötvössche Formel* 174.
Equation of hydrostatics, *Grundgleichung der Hydrostatik* 283, 287.
— —, one- and two-dimensional analogues, *ein- und zweidimensionale Analoga* 285, 298.
Equation of state for liquids, *Zustandsgleichung für Flüssigkeiten* 13, 19, 52, 55.
Equilibrium criterion of GIBBS, *Gleichgewichtskriterium von Gibbs* 291.
Equilibrium between droplet and vapor, *Gleichgewicht zwischen Tropfen und Dampf* 303.
Excess entropy at a surface, *Überschußentropie an einer Oberfläche* 296.
Excess free energy at an interface, *Überschuß der freien Energie an einer Grenzfläche* 286, 296.
Excess thermodynamic functions at an interface, *thermodynamische Überschußfunktionen an einer Grenzfläche* 290.
Excess volume, *Überschußvolumen* 232, 233.

Fermi statistics, *Fermi-Statistik* 122.
Fields of forces in liquids, deformation, *Kraftfelder in Flüssigkeiten, Deformation* 67 to 69.
Film formation between two fluid phases, *Filmbildung zwischen zwei flüssigen Phasen* 299.
Fluctuations of mass density, *Schwankungen der Massendichte* 282.
Fluid lens embedded in two fluid phases, *Flüssigkeitslinse, eingebettet zwischen zwei Flüssigkeitsphasen* 282—283, 285, 298, 299.
Fluid, normal, *normale Flüssigkeit* 120.
Fluid velocity potential, *laminare Strömungsgeschwindigkeit* 41.
Free energy, *freie Energie* 62, 67, 78, 285, 286.
— —, superficial, *der Oberfläche* 137, 149, 285, 288, 296.
Free volume, *freies Volumen* 240, 242seq., 247, 249seq., 262, 265, 273.
— —, generalized, *verallgemeinertes* 250.
— —, smeared, *verschmiertes* 243, 247seq.
— —, sphericalized, *kugelförmiges* 243, 250.
Friction constant, *Reibungskoeffizient* 85, 112, 118.
Friction tensor, *Reibungstensor* 114.
Fugacity (activity), *Fugazität (Aktivität)* 194.
Fundamental forms of surface theory, *Fundamentalformen der Flächentheorie* 284.

Gas adsorption, *Gasadsorption* 205, 231seq.
— — on a solid surface, *an fester Oberfläche* 231.
Gaussian curvature of a surface, *Gaußsches Krümmungsmaß einer Fläche* 284.
Giant-cell model, *Riesen-Zellen-Modell* 11.
Gibbs adsorption equation, *Gibbssche Adsorptionsgleichung* 140, 150, 166, 205, 255, 268, 288, 292, 303.
— — —, generalization, *Verallgemeinerung* 150, 168, 170.
Gibbs adsorption formula, *Gibbssche Adsorptionsformel* 142, 170, 230, 267seq., 269.
— — — for a regular solution, *für eine reguläre Lösung* 268.
GIBBS' dividing surfaces, *Gibbssche Trennflächen* 282, 286, 287, 289.
Gibbs-Duhem equations, *Gibbs-Duhem-Gleichungen* 143, 150, 165, 166, 167, 219, 292, 297.
Gibbs free energy, *Gibbssche freie Energie* 293.
Gibbs fundamental equation, *Gibbssche Fundamentgleichung* 139, 170.
— — — for the interface layer, *für die Grenzschicht* 140.
Gibbs-Helmholtz equation for a plane interface, *Gibbs-Helmholtz-Gleichung für eine ebene Grenzschicht* 142.
— —, surface analogue, *Oberflächenanalogon* 296.
Gibbs-Kelvin formula, *Gibbs-Thomsonsche Formel* 288, 292.
GIBBS' thermodynamic potential, *Gibbssches thermodynamisches Potential* 26, 107—108.
GIBBS' thermodynamical formulation of capillarity, *Gibbssche thermodynamische Formulierung der Kapillarität* 289—293.
Gibbs-Thomson formula, generalized, *verallgemeinerte Gibbs-Thomson-Formel* 155.
Gibbs-Tolman-Koenig-Buff equation, *Gibbs-Tolman-Koenig-Buff-Gleichung* 153.
Glasses, *Gläser* 5, 23, 69.
Globule of liquid, floating, *schwimmende Flüssigkeitskugel* 299.
Gradient, two-parametric, in a surface, *Beltramischer (erster) Differentialparameter* 284.
Grand partition function, *große Verteilungsfunktion* 293.

Grand potential, *großes Potential* 140, 192, 193.
Gravitational field, *Schwerefeld* 283, 287, 295.
— —, influence on liquids, *Einfluß auf Flüssigkeiten* 15, 67.
GUGGENHEIM's formula, *Guggenheimsche Formel* 155.
GUGGENHEIM's mixing law, *Guggenheimsches Mischungsgesetz* 268.

H-theorem of liquids, *H-Theorem der Flüssigkeiten* 107.
Hard sphere molecules, *Moleküle als harte Kugeln* 221.
Helium, liquid, *flüssiges Helium* 119seq.
Helmholtz free energy, *Helmholtzsche freie Energie* 137, 149, 293, 296.
HILDEBRAND's equation, *Hildebrandsche Gleichung* 268, 269.
Hole model of liquids, *Löchermodell der Flüssigkeiten* 11—13, 242, 249seq., 258, 261.
Hydrostatic fundamental equation, *hydrostatische Grundgleichung* 283, 287.
— — —, one- and two-dimensional analogues, *ein- und zweidimensionale Analoga* 285, 298.
Hydrostatic model of capillarity, *hydrostatisches Modell der Kapillarität* 281.

Integrals across transition zone, *Integrale über die Übergangszone hinweg* 285.
Interaction energy in liquids, *Wechselwirkungsenergie in Flüssigkeiten* 3, 39.
Interface, *Grenzfläche* 134, 136.
—, adsorption at a plane, *Adsorption an einer ebenen* 142.
Interface normal, sign convention, *Grenzflächennormale, Vorzeichenfestsetzung* 283 to 284.
Interface, plane, *ebene Grenzfläche* 142, 157, 165, 181, 184, 194, 210seq., 264seq.
—, —, Gibbs-Helmholtz equation, *Gibbs-Helmholtz-Gleichung* 142.
—, —, quasithermodynamic relations, *quasithermodynamische Beziehungen* 165.
Interface, spherical, *kugelförmige Grenzfläche* 145, 146, 148, 149, 150, 159, 166, 213seq., 214, 215.
—, —, fundamental thermodynamic equations, *thermodynamische Fundamentalgleichungen* 148.
—, —, quasithermodynamic relations, *quasithermodynamische Beziehungen* 166seq.
—, —, surface tension, *Oberflächenspannung* 146, 161, 187, 214.
Intermolecular forces, *zwischenmolekulare Kräfte* 2, 3.
Internal energy, *innere Energie* 20, 35—36, 68, 78, 82, 107, 119.
— —, superficial, *der Oberfläche* 137, 141, 166, 167, 189, 190, 216, 244seq., 257seq.
Internal modes, *innere Eigenschaften* 95.
Ionic liquids, *Ionenflüssigkeiten* 23.
Irreversible processes in liquids, *irreversible Prozesse in Flüssigkeiten* 24, 25, 38, 81 to 89, 92.
Isotherms in liquids, superposition approximation, *Isothermen in Flüssigkeiten, nach der Überlagerungsnäherung* 53.
Isotropy of stress tensor, *Isotropie des Spannungstensors* 283.

Katayama-Guggenheim formula, *Katayama-Guggenheim-Formel* 174, 175, 176, 297.
Katayama-Guggenheim formulae, modified, *modifizierte Katayama-Guggenheim-Formeln* 177.
Kelvin relation, *Kelvin-Beziehung* 134, 147, 154, 160, 161, 162.
Kinetic energy of molecules, *kinetische Energie von Molekülen* 27.
Kinetic model of liquids, *kinetisches Modell der Flüssigkeiten* 13—15, 52—61, 65—67.
Kinetic theory of liquids, superposition approximation, *kinetische Theorie der Flüssigkeiten, Überlagerungsnäherung* 52, 53, 59, 66, 74 seq, 111.
Kirkwood-Müller equation, *Kirkwood-Müller-Gleichung* 234.

λ-phenomenon, *λ-Phänomen* 94, 119, 122.
λ-point of helium, *λ-Punkt des Heliums* 119, 122, 133.
Laplace equation of capillarity, *Laplacesche Gleichung der Kapillarität* 286, 288, 294.
— — —, dimensionless form, *dimensionslose Form* 300.
— — —, planar case, *ebener Fall* 295.
— — —, simple solutions, *einfache Lösungen* 299—302.
Laplace equations determining shape of fluid lens, *Laplacesche Gleichungen zur Bestimmung der Gestalt einer Flüssigkeitslinse* 298.
Lattice model of liquids, *Gittermodell der Flüssigkeiten* 7—10, 24, 240, 242.
Lattice planes, equipotential, *Gitterebenen gleichen Potentials* 252.
Lattice structure, *Gitterstruktur* 123.
Length parameter in capillarity theory, *Längenparameter in der Theorie der Kapillarität* 285, 286, 291—293.
Lennard-Jones and Devonshire model, *Lennard-Jones-Devonshire-Modell* 242.
Lennard-Jones potential, *Lennard-Jones-Potential* 173, 217, 244, 246.
— —, modified, *modifiziertes* 181.
— —, RUSHBROOKE's modification, *Rushbrookesche Modifikation* 217.
Lens of liquid embedded in two fluid phases, *Linse einer Flüssigkeit zwischen zwei flüssigen Phasen* 282—283, 285, 298—299.
LIOUVILLE's equation, *Liouvillesche Gleichung* 97.
Lippmann equation, *Lippmannsche Gleichung* 304.
Liquid crystals, *flüssige Kristalle* 22, 81.
Liquid at flat vertical plates, *Flüssigkeit an ebenen, senkrechten Platten* 300.
Liquid models, cell-cluster model, *Flüssigkeitsmodelle, Zellen-Cluster-Modell* 10—11, 64—65.

Liquid models, hole model, *Flüssigkeitsmodelle, Löcher-Modell* 11—13.
— —, kinetic model, *kinetisches Modell* 13 to 15.
— —, lattice-cell model, *Gitter-Zellen-Modell* 7—10, 24, 53.
Liquid state as disordered state, *Flüssigkeitszustand als ungeordneter Zustand* 5.
Liquids, associated, *angelagerte Flüssigkeiten* 4, 23, 25, 38, 80.
—, ionic, *Ionenflüssigkeiten* 23.
—, polar, *polare Flüssigkeiten* 21, 38, 80.
— with similar molecular constitution, *Flüssigkeiten mit ähnlicher molekularer Anordnung* 2.
London constant, *London-Konstante* 234.
London forces, *London-Kräfte* 233.

Mass density, *Massendichte* 33.
Mass fraction, *Massenverhältnis* 41, 82.
Mean curvature of a surface, *mittleres Krümmungsmaß einer Fläche* 284, 288, 300.
Mean thickness of the ionic atmosphere, *mittlere Dicke der Ionenatmosphäre* 224.
McLEOD's equation, *McLeodsche Gleichung* 174, 175.
Mechanics, molecular, *molekulare Mechanik* 33—47.
Melting phenomenon = order-disorder transition, *Schmelzerscheinung = Übergang von Ordnung zu Unordnung* 5, 7.
Melting point, *Schmelzpunkt* 66.
Membrane behavior of dividing surface, *Membranverhalten einer Trennfläche* 286.
Meniscus, *Meniskus* 301.
Mixtures of liquids, *Gemische von Flüssigkeiten* 2, 23—25, 40—43, 67.
Models of liquids see liquid models, *Modelle von Flüssigkeiten, s. Flüssigkeitsmodelle.*
Molecular distribution function, *molekulare Verteilungsfunktion* 25, 27, 39, 47, 179, 182, 194, 195.
— — —, nonlinear integral equation, *nichtlineare Integralgleichung* 79.
Molecular mechanics of liquids, *molekulare Mechanik von Flüssigkeiten* 33—47.
Molecular polarizability, *molekulare Polarisierbarkeit* 72.
Molecular statistics of liquids, *molekulare Statistik von Flüssigkeiten* 28—33.
Molecules, rigid spherical, *starre kugelförmige Moleküle* 48, 59.
Momentum density, *Impulsdichte* 33.
Momentum flux, *Impulsfluß* 33.
Monatomic liquids, surface tension calculation, *einatomige Flüssigkeiten, Berechnung der Oberflächenspannung* 288.
Monolayer model, *Einschichtmodell* 269, 270, 275.

Neumann triangle, *Neumannsches Dreieck* 286, 288, 292, 294, 298—299.
Neutron diffraction in liquids, *Neutronen-Brechung in Flüssigkeiten* 43—47, 124.
Newtonian liquids, *Newtonsche Flüssigkeiten* 15.
Noble gases, liquid, *flüssige Edelgase* 1, 2, 48, 53, 55, 65, 67, 80, 93, 117.
Non-uniform liquids, *nicht-einheitliche Flüssigkeiten* 81 seq.
— —, steady states, *stationäre Zustände* 99.
— —, unsteady states, *nicht-stationäre Zustände* 101.
Normal to interface, sign convention, *Normale zur Grenzfläche, Vorzeichenfestsetzung* 283—284.
Nucleation, *Keimbildung* 157, 294, 303.
Number density, *Teilchendichte* 26, 67.
— —, local variation in transition zone, *Ortsabhängigkeit in der Übergangszone* 282.
— —, superficial, *an der Oberfläche* 141, 142, 145, 166, 167, 171, 184, 185, 202, 227.

Onsager relation, *Onsager-Beziehung* 110.
Opalescence in liquids, *Opaleszenz in Flüssigkeiten* 47.
Open system, *offenes System* 139.
Optical anisotropy of liquids, *optische Anisotropie von Flüssigkeiten* 22.
Order-disorder transition = phenomenon of melting, *Ordnung-Unordnung-Übergang = Schmelzerscheinung* 5, 7.
Orientation, coöperative, *kooperative Orientierung* 80.
Orientation fusion, *Orientierungsverschmelzung* 81.
Osmotic conditions, *osmotische Bedingungen* 196 seq., 200.
Osmotic pressure, *osmotischer Druck* 196 seq., 201, 228.
— —, normal, *normaler* 228.
— —, tangential, *tangentialer* 228.
Osmotic solution, *osmotische Lösung* 198, 201.

Pair distribution function, *Zwei-Teilchen-Verteilungsfunktion* 179 seq., 184, 186 seq., 194, 196, 200, 204, 208, 215, 238.
Parachor 174.
Partition function, *Zustandsfunktion* 12, 79, 131, 177, 178, 237, 238, 240, 241, 242.
— —, cell method for calculating, *Zellenmethode für die Berechnung* 240.
— —, grand, *große* 190, 191 seq., 194 seq., 206.
Peripheral tension, *Umfangsspannung* 286.
Phenomenological capillarity equations, *phänomenologische Kapillaritätsgleichungen* 284—286.
Phonons, *Phononen* 123.
Piezoelectric effect in liquids, *piezoelektrischer Effekt in Flüssigkeiten* 70.
Planar interface, *ebene Grenzfläche* 142, 157, 165, 181, 184, 194, 210 seq., 264 seq., 295—298.
Poisson-Boltzmann equation, *Poisson-Boltzmann-Gleichung* 223.
Polar liquids, *polare Flüssigkeiten* 4, 21, 38, 80.

Polarizability, molecular, *molekulare Polarisierbarkeit* 72.
Polyatomic liquids, *mehratomige Flüssigkeiten* 65.
Polycrystals, *Vielkristalle* 10.
Potential of average forces, *Potential der mittleren Kräfte* 196, 200seq., 223, 226.
—, grand, *großes* 140, 192, 193.
—, thermodynamic, *thermodynamisches* 26, 107—108.
Pressure, *Druck* 19, 30, 35, 49, 62, 67, 107, 119.
—, critical, *kritischer* 66, 133.
—, kinetic, *kinetischer* 132.
—, reduced, *reduzierter* 133.
—, thermodynamic, *thermodynamischer* 132.
Pressure diffusion, *Druckdiffusion* 84.
Pressure tensor, *Drucktensor* 33, 35, 82, 157, 159, 161, 164, 167, 208, 209, 210, 211, 212, 213.
— —, normal component, *Normalkomponente* 158, 159, 164, 210, 218.
— —, partial, *partieller* 41.
— —, statistical-mechanical expression, *statistisch-mechanischer Ausdruck* 208seq., 212, 215.
— —, tangential component, *Tangentialkomponente* 158, 159, 164, 167, 210, 212, 214, 220, 222, 260, 261, 262.
Principal curvatures of a surface, *Hauptkrümmungen einer Fläche* 284, 289—290.
Probability density function, *Wahrscheinlichkeitsdichte* 178, 191, 237.

Quantum effects in liquids, *Quanteneffekte in Flüssigkeiten* 20, 28, 93, 119—133.
Quantum mechanical parameter, *quantenmechanischer Parameter* 176.
Quantum statistics, *Quantenstatistik* 122.
Quasi-chemical approximation, *quasi-chemische Näherung* 8.
Quasithermodynamic theory, *quasithermodynamische Theorie* 163, 172, 218, 219seq., 260seq.
— —, expression for equilibrium, *Ausdruck für das Gleichgewicht* 165, 166, 260, 261.
— —, fundamental postulate, *Grundpostulat* 163, 165, 166.
— —, relations for plane interface, *Beziehungen für die ebene Grenzfläche* 165.
— —, relations for spherical interface, *Beziehungen für die kugelförmige Grenzfläche* 166, 167.

Radial distribution function, *radiale Verteilungsfunktion* 15—17, 19, 26, 47—61, 65, 125.
Relaxation frequency, *Relaxationsfrequenz* 103, 114.
Relaxation processes, *Relaxationsprozesse* 82, 95.
Relaxation time, *Relaxationszeit* 71.
Reduced variables, *reduzierte Variable* 172, 173, 182, 186, 187, 188.
Reversibility, microscopic, *mikroskopische Reversibilität* 83.
Reversible change of an interface, *reversible Änderung einer Grenzfläche* 289—290.
Rigidity of liquids, *Starrheit von Flüssigkeiten* 5, 69—71, 81.
Rotons, *Rotonen* 123.

Scattering of light in liquids, *Streuung von Licht in Flüssigkeiten* 30, 32, 47.
Schrödinger wave equation for quantum liquids, *Schrödinger-Gleichung für Quantenflüssigkeiten* 127.
Second sound in He II, *second sound in Helium II* 94.
Semi-invariant, *Semiinvariante* 202.
Shear modulus of elasticity, *Scherungsmodul der Elastizität* 92.
— — of liquids, *von Flüssigkeiten* 20, 71.
Shearing stress, *Scherungsspannung* 91.
Shearing viscosity, *Scherungsviskosität* 83, 89, 91, 92.
— —, coefficient, *Koeffizient* 91.
Shock waves, *Stoßwellen* 35.
Singlet density, *Einteilchendichte* 282.
Singlet distribution function, *Ein-Teilchen-Verteilungsfunktion* 179seq., 184seq., 194, 197, 200, 208, 210, 215seq., 238, 240.
Sodium, liquid, *flüssiges Natrium* 56.
Solid boundary, *Festkörperbegrenzung* 69.
Solutions, athermal polymer, *athermische polymere Lösungen* 272seq.
Solutions of non-electrolytes, *Lösungen von Nichtelektrolyten* 262seq.
Solutions, regular, *reguläre Lösungen* 262, 264seq.
—, —, with plane interface, *mit ebener Grenzfläche* 264.
—, polymer, *polymere* 272seq.
—, surface tension, *Oberflächenspannung* 298.
Soret effect in liquids, *Soret-Effekt in Flüssigkeiten* 2, 24, 82.
Sound, propagation and absorption in liquids, *Schall, Ausbreitung und Absorption in Flüssigkeiten* 91, 92, 95.
—, velocity in liquids without absorption, *Geschwindigkeit in Flüssigkeiten ohne Absorption* 95.
Specific heat, *spezifische Wärme* 123.
Spherical interface (see also droplet), *kugelförmige Grenzfläche (s. auch Tropfen)* 145, 148, 150, 159, 166, 213seq., 292.
Spreading of a liquid film, *Ausbreitung einer Flüssigkeitsschicht* 299.
Statistics, molecular, *molekulare Statistik* 28 to 33.
Steady states of a non-uniform liquid, *stationäre Zustände einer nicht-einheitlichen Flüssigkeit* 99—101.
Strain tensor, *Verformungstensor* 83, 90.
Stress tensor, *Spannungstensor* 283, 286.
Structure of liquids, quantitative description, *Struktur von Flüssigkeiten, quantitative Beschreibung* 25—47, 47—81, 81—113.
Supercooled state of liquids, *Unterkühlung von Flüssigkeiten* 9.

Superficial activity coefficient, *Oberflächenkoeffizient der Aktivität* 205, 208.
Superficial density of the Helmholtz free energy, *Oberflächendichte der Helmholtzschen freien Energie* 139.
— — of internal energy, *der inneren Energie* 245.
— — of molecules, *der Moleküle* 267, 269.
— — of potential energy, *der potentiellen Energie* 245, 246.
Superficial entropy, *Oberflächenentropie* 78, 137, 166, 167, 177.
Superficial free energy, *freie Oberflächenenergie* 137, 149, 285, 288, 296.
Superficial internal energy, *innere Oberflächenenergie* 137, 166, 167, 185.
Superficial number density, *Oberflächendichte der Teilchen* 141, 142, 145, 166, 167, 171, 184, 185, 202, 227.
Superficial pair density, *Zweiteilchen-Oberflächendichte* 184.
Superficial quantities, *Oberflächengrößen* 137, 139, 149, 166.
Superfluidity of helium, *Suprafluidität von Helium* 119, 122.
Superposition approximation of kinetic theory, *Superpositionsnäherung der kinetischen Theorie* 52, 53, 59, 66, 74, 111, 181, 197, 199, 201.
Surface charge, *Oberflächenladung* 226.
Surface of discontinuity, *Diskontinuitätsfläche* 216seq.
Surface energy, *Oberflächendichte der inneren Energie* 141, 185, 190, 216, 244, 245, 246, 247, 248, 257seq., 259.
— — of argon, *von Argon* 217.
Surface entropy, *Oberflächendichte der Entropie* 141, 171, 248, 249, 259, 260.
Surface force, *Oberflächenkraft* 286.
Surface tension, *Oberflächenspannung* 23, 77, 81, 138, 139, 148, 149, 158, 161seq., 166, 167, 171, 172seq., 177, 179, 182, 186, 190, 192, 193, 198, 200, 201, 205, 211, 213, 215, 216, 218, 219, 237—239, 240, 246, 248, 257seq., 259, 260, 266, 268, 270, 271, 275, 276, 277.
— — of argon, *von Argon* 217.
— — of athermal polymer solutions, *athermischer polymerer Lösungen* 273.
— —, curvature dependence, *Krümmungsabhängigkeit* 152, 153, 215.
— —, definition, *Definition* 288, 290.
— —, difference, *Differenz* 197, 202, 227seq., 230seq.
— —, dilute solution, *verdünnte Lösung* 298.
— —, dynamic, *dynamische* 271seq.
— —, empirical formula, *empirische Formel* 172, 174.
— —, high temperature corrections, *Korrekturen für hohe Temperatur* 288.
— —, increase, *Anstieg* 229.
— —, mechanical definition, *mechanische Definition* 158, 161, 166, 167, 261.
— — of a perfect solution, *einer vollkommenen Lösung* 268.
Surface tension of a plane interface, *Oberflächenspannung einer ebenen Grenzfläche* 181, 184.
— — — — mechanical definition, *mechanische Definition* 157.
— — of polymer solutions, *polymerer Lösungen* 272.
— —, quantum effects, *Quanteneffekte* 237.
— —, reduced, *reduzierte* 172, 173, 175, 176, 187.
— — of solutions, *von Lösungen* 201.
— — of the spherical interface, *der kugelförmigen Grenzschicht* 146, 159, 161, 187, 214.
— — — —, mechanical definition, *mechanische Definition* 159, 160.
— —, static, *statische Oberflächenspannung* 271seq.
— — of strong electrolytes, *starker Elektrolyte* 223.
— —, temperature dependence, *Temperaturabhängigkeit* 151, 259, 297.
Surface theory, basic theorems, *Flächentheorie, grundlegende Sätze* 284.
Surface zone in polar and associated liquids, *Oberflächenzone in polaren und angelagerten Flüssigkeiten* 80.

Temperature, critical, *kritische Temperatur* 66, 133, 297.
Temperature dependence of surface tension, *Temperaturabhängigkeit der Oberflächenspannung* 151, 259, 288, 297.
Temperature, reduced, *reduzierte Temperatur* 133.
Tensile strength, *Zerreißfestigkeit* 22.
Tension, surface of, *Spannungsfläche* 147, 148, 155, 160, 161, 163, 212, 213, 216, 217, 218, 287, 290.
Thermal conduction of liquids, *Wärmeleitung in Flüssigkeiten* 14, 20, 38, 82, 86—89, 111, 115.
Thermal diffusion in liquids (Soret effect), *thermische Diffusion in Flüssigkeiten (Soret-Effekt)* 2, 24, 82.
Thermal flux vector, *Wärmefluß-Vektor* 36, 42, 82.
Thermodynamic basic parameters of capillarity, *thermodynamische Fundamentalparameter der Kapillarität* 287.
Thermodynamic behaviour of liquids, *thermodynamisches Verhalten von Flüssigkeiten* 61—67.
Thermodynamic potential, *thermodynamisches Potential* 26, 107—108.
Thermodynamic principal curvature terms, *thermodynamische Hauptkrümmungsfunktionen* 290.
Thermodynamic temperature for a liquid at rest, *thermodynamische Temperatur einer ruhenden Flüssigkeit* 28.
Thermodynamics, local formulation, *Thermodynamik, lokale Formulierung* 163.
Thermodynamics of irreversible processes, *Thermodynamik irreversibler Prozesse* 24, 25, 81, 168, 169, 271.

Three-phase confluent zone see confluent zone, *Drei-Phasen-Kontaktzone s. Kontaktzone.*
Transition zone, *Übergangszone* 282, 284, 286.
— —, form of stress tensor, *Form des Spannungstensors* 283, 286.
— —, integrals across it, *Integrale darüber hinweg* 285.
Tolman-Buff equation, *Tolman-Buff-Gleichung* 162, 163.
TONKS' equation, *Tonkssche Gleichung* 221.
TRAUBE's rule, *Traubesche Regel* 276seq.
Triplet distribution function, *Drei-Teilchen-Verteilungsfunktion* 181, 197.
Two layer model, *Doppelschichtmodell* 269.
Two-phase transition zone see transition zone, *Zwei-Phasen-Übergangszone s. Übergangszone.*

Ultrasonic absorption in liquids, *Ultraschallabsorption in Flüssigkeiten* 35, 91.
Uniform liquids, *einheitliche Flüssigkeiten* 47 to 80.
Unsteady states of a non-uniform liquid, *nicht-stationäre Zustände einer nicht-einheitlichen Flüssigkeit* 101—104.

Van der Waals equation, *van der Waals-Gleichung* 172, 219, 220, 221.
VAN DER WAALS' forces, *van der Waalssche Kräfte* 23, 119, 223.
Van der Waals loop, *van der Waals-Schleife* 172, 244, 252.
Vapour pressure, *Dampfdruck* 23, 133.
— —, curvature dependence, *Krümmungsabhängigkeit* 154.
Vapour pressure of a drop, *Dampfdruck über einem Tropfen* 302, 303.
Variational principle of capillarity, *Variationsprinzip der Kapillarität* 293—295.
Velocity distribution function, *Geschwindigkeits-Verteilungsfunktion* 27—28, 39, 42, 127.
Velocity of flow, *Strömungsgeschwindigkeit* 33, 41.
Vertical plates, capillarity effect of liquid, *senkrechte Platten, Kapillarerscheinungen einer Flüssigkeit* 300.
Virial coefficients, *Virial-Koeffizienten* 52, 66.
Virial coefficient, second, *zweiter Virialkoeffizient* 66.
Viscosity of liquids, *Viskosität von Flüssigkeiten* 14, 20, 81, 82, 89—91, 110, 111, 115, 119.
Volume, critical, *kritisches Volumen* 66.
Volume viscosity, *Volumenviskosität* 83, 89, 91, 92, 118.
— —, coefficient, *Koefficient* 91.

Work done by surface displacement, *Arbeitsleistung bei Oberflächenverschiebung* 286, 294.
Work of formation of a droplet from vapor, *Arbeit der Tropfenbildung aus Dampf* 302 to 303.

X-ray diffraction in liquids, *Röntgenstrahlbrechung in Flüssigkeiten* 7, 16, 19, 43 to 47, 55, 56, 81, 124.

Young equation, *Youngsche Gleichung* 289, 295.